Unconformities and Porosity in Carbonate Strata

Edited by

David A. Budd
Arthur H. Saller
and
Paul M. Harris

AAPG Memoir 63

Published by
The American Association of Petroleum Geologists
Tulsa, Oklahoma, U.S.A.
Printed in the U.S.A.

ISBN: 0-89181-342-X

Association Editor: Kevin T. Biddle
Science Director: Richard Steinmetz
Publications Manager: Kenneth M. Wolgemuth
Special Projects Editor: Anne H. Thomas
Production: Custom Editorial Productions, Inc., Cincinnati, Ohio

This and other AAPG publications are available from:
The AAPG Bookstore
P.O. Box 979
Tulsa, OK 74101-0979
Telephone: (918) 584-2555; (800) 364-AAPG (USA—book orders only)
FAX: (918) 584-0469; (800) 898-2274 (USA—book orders only)

About the Editors

David A. Budd is an Associate Professor of Geological Sciences at the University of Colorado, Boulder. He received B.A., M.S., and Ph.D. degrees in geology from The College of Wooster, Duke University, and The University of Texas at Austin, respectively. Between 1983 and 1986 he was employed by ARCO Exploration and Production Technology Company where his primary duties involved reservoir characterization studies. Since 1987 he has been a professor in the Department of Geological Science at the University of Colorado. His research interests include the origin and diagenesis of carbonates, with special emphasis on the geochemistry of limestones, the relations between carbonate alteration and diagenetic pore fluids, and the application of diagenesis to the understanding of pore-system evolution and porosity heterogeneity in carbonate reservoirs and aquifers.

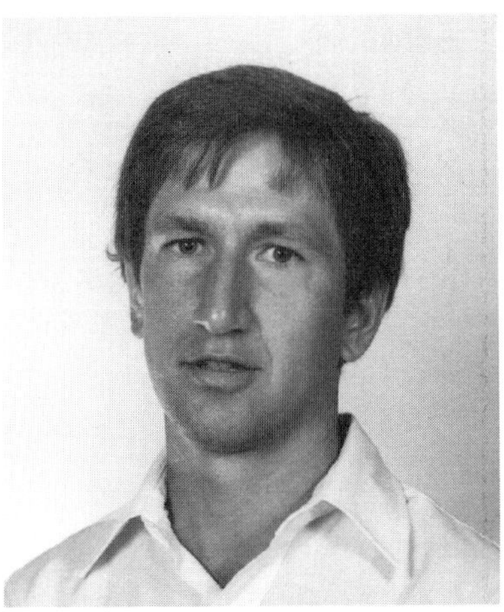

Arthur H. Saller currently works as a carbonate sedimentologist for UNOCAL Energy Resources in Brea, California. He did undergraduate studies at the University of Kansas (1974–1978), received a Master's degree from Stanford University in 1980, and a Ph.D. in geology from Louisiana State University in 1984. From 1984 to 1986, he worked as a Research Geologist with Cities Service Oil and Gas in Tulsa, Oklahoma, and he joined UNOCAL in 1986. At UNOCAL, Art teaches courses, performs technical service work, and conducts research related to exploration and development in carbonate rocks.

Paul M. (Mitch) Harris, a Senior Research Associate with Chevron Petroleum Technology Company in La Habra, California, does carbonate technical support projects, research, consulting, and training for the various operating units of Chevron. His work centers on facies-related, stratigraphic, and diagenetic problems that pertain to carbonate reservoirs and exploration plays. Mitch received his B.S. and M.S. degrees from West Virginia University and his Ph.D. from the University of Miami, Florida. He has worked in the oil industry since 1977 doing projects in most carbonate basins worldwide. He is active in AAPG and SEPM, having published numerous papers and edited several volumes on carbonates.

AAPG
Wishes to thank the following
for their generous contribution
to

Unconformities and Porosity in Carbonate Strata

❖

AMOCO Production Company

❖

Marathon Oil Company

❖

Shell Research

❖

Contributions are applied against the production costs of the publication, thus directly reducing the book's purchase price and making the volume available to a greater audience.

Table of Contents

Foreword

Advances in carbonate sedimentology, cyclostratigraphy, and seismic/sequence stratigraphy have made carbonate depositional facies more predictable in the subsurface. However, predicting porosity in subsurface carbonates in frontier basins remains difficult because current diagenetic models are largely qualitative, rather than quantitative. Dissolution associated with subaerial exposure is thought to be responsible for much of the secondary porosity in many large oil and gas fields around the world including Arun field, Indonesia (Jordan and Abdullah, 1988), Yates field, west Texas (Craig, 1988), Horseshoe atoll fields, west Texas (Vest, 1970; Schatzinger, 1983), Golden Lane fields, Mexico (Coogan et al., 1972), numerous Lower Cretaceous fields of the Middle East (Wilson, 1975; Harris et al., 1984), and Casablanca field, offshore Spain (Esteban, 1991; Lomando et al., 1993). Unfortunately, subaerial exposure is not always present as predicted, and subsurface porosity is not always associated with subaerial exposure.

An AAPG Hedberg Research Conference was held in July 1993 in Vail, Colorado, to discuss detection of unconformities and porosity associated with unconformities in carbonate strata. AAPG Memoir 63 contains papers derived from presentations at that conference. Four major topics are addressed in this memoir: (1) detection of unconformities and subaerial exposure, (2) modification of porosity and permeability during subaerial exposure, (3) preservation of exposure-related porosity during burial, and (4) influence of unconformities on subsequent depositional and diagenetic patterns.

DETECTION OF SUBAERIAL UNCONFORMITIES

Techniques for detecting subaerial exposure and unconformities discussed in this memoir include seismic stratigraphy, petrologic features observed in cores and/or outcrops, cycle stacking patterns (abrupt facies offsets), and stable isotope geochemistry. Sarg (1988) and Loucks and Sarg (1993) describe examples of subaerial exposure associated with seismic onlap and erosional truncation. However, similar seismic geometries can occur without subaerial exposure (Erlich et al., 1990; Schlager, 1991; Saller et al. 1993). In this volume, Moldovanyi et al. show how several seismic reflection geometries are indicators of unconformities and subaerial exposure including chaotic reflection intervals, concave-up "sink-hole" reflectors, back-stepped margins, and truncated surfaces. In contrast, Hurley et al. (this volume) show an example where the main subaerial exposure surface associated with reservoir porosity was not identified in studies which relied solely on seismic data.

A number of petrologic features can be used to identify subaerial exposure surfaces in core and/or outcrop. Palmer and Mylroie and Carew (this volume) review the processes that lead to the formation of various forms of karst. Irregular karst surfaces and solution vugs are described below subaerial exposure surfaces by Beach, Lucia, Moldovanyi et al., and Wendte and Muir (this volume). Caliches, paleosols, and soil residues are discussed by Beach, Dickson and Saller, and Mylroie and Carew (this volume) as criteria to identify subaerial exposure surfaces. Dissolution of carbonate by fresh water is commonly observed below subaerial exposure surfaces. In this volume, selective dissolution of depositional grains is reported below exposure surfaces by Beach, Mutti, Hurley et al., and Dickson and Saller, and selective dissolution of evaporites is shown by Vahrenkamp. However, fabric selective dissolution can also occur in near-surface hypersaline environments (Sun, 1992), deep marine environments (Saller, 1986; Dix and Mullins, 1988, 1992; Budd, 1989; Saller and Koepnick, 1990), and burial environments (Moore and Druckman, 1981; Jameson, 1994; Mazzullo and Harris, 1992). Cavernous pore networks are also an important product of subaerial exposure as reported in this volume by Tinker et al. and Lucia. However, vugs, caves, and breccias can form by dissolution in basinal fluids independent of subaerial exposure (Hill, Palmer, this volume; Mazzullo and Harris, 1992; Dravis and Muir, 1993).

Cycle-stacking patterns are commonly used to identify major subaerial unconformities. One type of pattern involves abrupt landward and/or basinward shifts in depositional facies, especially shelf margin facies. Basinward shifts of depositional facies are used to predict "sequence boundaries" and/or infer major subaerial exposure surfaces (Van Wagoner et al., 1988; Sarg, 1988; Mutti, this volume). However, Wendte and Muir (this volume) show an example where the major subaerial exposure surface occurs in an interval in which depositional facies have "stepped back" landward.

Stable carbon and oxygen isotope profiles have been used to identify subaerial exposure surfaces (Allan and Matthews, 1982). In other areas, stable carbon and oxygen isotope profiles failed to detect major exposure events (Moshier, 1989; Vahrenkamp, 1994). Several studies in this volume (Wagner et al., Moldovanyi et al., Dickson and Saller) show examples where deflections in stable carbon isotope profiles correspond with subaerial exposure surfaces. Dickson and Saller (this volume) also show intervals where stable carbon isotope profiles show little or no affect of subaerial exposure, and they attempt to explain why characteristic isotope profiles occur in some limestones, but not others.

Several other methods and techniques can be used to detect subaerial unconformities, but are not discussed in detail in this memoir: eustatic sea-level curves, interpretation of wireline logs, biostratigraphy and other methods of dating strata, and computer modeling of tectonics, sea-level and basin evolution (Saller et al., 1994). An integrated approach using all available data is best for recognizing and predicting subaerial exposure because all methods have some pitfalls.

EFFECT OF SUBAERIAL EXPOSURE ON POROSITY

Most of the papers in this volume illustrate how freshwater and mixing-zone diagenesis during subaerial exposure rearranged pore networks, thereby changing porosity and permeability. Effects of subaerial exposure depend on many interrelated factors including: (1) climate, (2) reactive potential of groundwaters, (3) mineralogy, (4) duration of-exposure, (5) existing pore networks, (6) depositional facies and stratigraphy, (7) hydrologic system, (8) size and topography of the exposed area, (9) base-level changes, and (10) tectonic setting (Saller et al. 1994). Many of these factors are discussed in papers in this memoir.

1. Climate, especially amount of rainfall, largely controls the intensity of dissolution in meteoric systems. Dissolution increases markedly with annual precipitation (Mylroie and Carew, Wagner et al., Palmer, this volume). Wagner et al. propose that at moderate to low rainfall levels, porosity will decrease during subaerial exposure, but in climates with high rainfall, porosity below the soil zone will increase.

2. The reactive potential of groundwaters is considered in several papers in this memoir and is the focus of the geochemical models discussed by Palmer. Mixing of fresh water and seawater can make groundwaters more corrosive as can addition of dissolved CO_2 (Matsuda et al., Mylroie and Carew, Wagner et al., Palmer, this volume). Remarkably little diagenetic alteration occurs in some confined aquifers because the waters have a low reactive potential (Budd et al., 1993; Cander, Palmer, this volume).

3. Mineralogy greatly influences the style and ultimate impact of freshwater diagenesis during subaerial exposure (Palmer, Mylroie and Carew, Wagner et al., this volume). During initial subaerial exposure, aragonitic grains commonly dissolve producing molds, which, in grainstones, are generally surrounded by intergranular cements (Dickson and Saller, this volume). In contrast, depositional sediments dominated by calcite may retain depositional pore geometries (Wendte and Muir, this volume). Where mixtures of calcite and dolomite or dolomite and evaporites are present, calcites or evaporites may be preferentially dissolved during subaerial exposure creating intercrystalline porosity (Hurley et al., this volume), vuggy porosity (Vahrenkamp, this volume), or cavernous porosity (Lucia, this volume).

4. Duration of exposure is important as pore systems evolve during subaerial exposure (Mylroie and Carew, this volume). Brief periods of subaerial exposure (10,000–400,000 yr) may be better for development of matrix porosity as shown in studies of Mutti and Dickson and Saller (this volume). Prolonged subaerial exposure (1–40 m.y.) may reduce matrix porosity, but increase fissure and cavernous porosity (Lucia, Tinker, this volume). Prolonged subaerial exposure may change permeability less than porosity because high-permeability karst-related conduits can form quickly and persist for millions of years.

5. As discussed by Palmer (this volume), existing pore networks determine where fresh water flows and hence the location of dissolution and cementation. Beach (this volume) shows how cementation at subaerial exposure surfaces caused perched meteoric phreatic zones and associated intense diagenetic alteration in overlying, distinctly younger limestones. In aquifers with conduit flow (fracture, fissure, and/or cavernous porosity), Cander (this volume) and Palmer (this volume) indicate that diagenesis is localized in the rock immediately adjacent to the conduit, but that the rest of the rock, even where very porous, is not affected by meteoric alteration.

6. After subaerial exposure, matrix porosity is still commonly correlated to depositional facies and stratigraphy, with grainstones commonly having the greatest porosity (Dickson and Saller, Wagner et al., Hurley et al., Lucia, Wendte and Muir, Mutti, this volume).

7. Nature, size, and configuration of the hydrologic system often determine how and where pore systems are modified (Beach, Mylroie and Carew, this volume). Matsuda et al. and Wagner et al. (this volume) show that systematic variations in amounts of dissolution and cementation cause porosity to decrease in the upper meteoric phreatic zone and porosity to increase in the vadose and mixing zones. Diagenetic alteration in confined aquifers can be quite minor (Cander, this volume). Variations in the location of the meteoric phreatic and mixing zones greatly affected the location of cavernous porosity in Paleozoic carbonates (Lucia, Tinker, this volume).

8. Size and topography of the exposed area influence the type of hydrologic system present and amount of rock affected by subaerial exposure (Mylroie and Carew, Palmer, this volume). In larger

systems, freshwater flux increases and groundwater flow becomes dominated by conduits like fractures, fissures, and caves.

9. Base-level (commonly sea level) changes will determine when and where subaerial exposure will occur, and the level of associated water tables. High-amplitude sea level fluctuations can cause repeated episodes of subaerial exposure and meteoric diagenesis (Beach, Mylroie and Carew, this volume). In this volume, papers by Tinker and by Lucia note preferential occurrence of caves at different levels and relate those levels to different positions of sea level.

10. Tectonic setting is commonly the ultimate control on many of the factors mentioned above including climate, duration of exposure, size and topography of exposed area, and base-level changes. In tectonically active areas, subaerial exposure, erosion, and deposition can also create unconformity-related reservoirs in structurally low areas (Vahrenkamp, this volume).

In summary, subaerial exposure commonly does not increase total subsurface porosity; however, it does rearrange pores and hence modifies permeability at a variety of scales (Saller et al. 1994). Diagenesis associated with subaerial exposure makes porosity and permeability more heterogeneous. As Matsuda et al. and Wagner et al. (this volume) show, some areas and intervals lose porosity and/or permeability, while other zones gain porosity and/or permeability. In a few cases, subaerial exposure has very little effect on pore networks (Wendte and Muir, this volume). Climate and duration of exposure are very important in determining the ultimate effect of subaerial exposure. High rainfall will cause dissolution to dominate over cementation, and overall porosity may increase. In areas with moderate to low rainfall, cementation will exceed dissolution below the soil zone, and overall porosity will decrease (Wagner et al., this volume).

PRESERVATION OF EXPOSURE-RELATED POROSITY DURING BURIAL

Development of pore systems below unconformities does not guarantee that early, unconformity-related porosity will be preserved in the subsurface after deep burial. Pressure solution and cementation will greatly reduce matrix porosity with burial (Bathurst, 1984; Scholle and Halley, 1985), and compaction during deeper burial can cause collapse and reduction of unconformity-related cavernous and matrix porosity. Lithification (cementation) during subaerial exposure may be critical to retarding compaction and preserving matrix porosity in the moderately deep subsurface (Dickson and Saller, this volume). Caverns are very rare in carbonates more than 2000–3000 m deep, apparently due to cavern collapse during burial. Cavernous porosity in Permian dolomites at Yates field (west Texas) remains open at relatively shallow depths (500 m) and contributes greatly to production in the field (Tinker, this volume). Many caverns developed in the Ordovician El Paso and Ellenburger groups during subaerial exposure (southwestern United States), but most caverns collapsed during moderate burial (Lucia, this volume). Former cavernous areas are commonly tight zones composed of fine cave fill and collapse breccias (Kerans, 1988; Canter et al., 1993; Lucia, this volume).

INFLUENCE OF UNCONFORMITIES ON SUBSEQUENT DEPOSITIONAL AND DIAGENETIC PATTERNS

Some unconformities have no exposure-derived porosity associated with them, yet are significant because they influenced subsequent depositional and diagenetic patterns which were critical to later porosity development. For example, an intra-Devonian unconformity in the Swan Hills Formation (Alberta) has little directly associated porosity, but greatly influenced depositional patterns in overlying strata which have porosity (Wendte and Muir, this volume). Similarly, much reservoir porosity in the Clear Fork Formation (Permian, west Texas) is related to depositional patterns above third-order sequence boundaries (Ruppel, 1992). Lithologic changes at unconformities can influence the flow of subsurface fluids during deeper burial, resulting in deep burial diagenesis and sometimes dissolution localized along unconformities. In this volume, Lucia describes how fluids moved along karst-related conduits during deep burial and dolomitized adjacent strata at elevated temperatures. Several oil fields have reservoir porosity formed by deep burial fluids that moved along subaerial unconformities (Jameson, 1994; Kirkby and Simo, 1994).

NONEXPOSURE RELATED CAVERNOUS FEATURES

Features similar to subaerial karst can form by other processes, but be misidentified. Palmer (this volume) reviews this phenomena of "hypogenetic caves" and discusses the various geochemical processes that are unrelated to aggressive meteoric infiltration, but can lead to the formation of vugs and caves. Hill (this volume) raises the possibility that cavernous porosity in some of the world's giant oil fields may be the result of such a process, in particular the oxidation of upward-moving H_2S-rich fluids.

IMPORTANT TOPICS NOT ADDRESSED IN THIS MEMOIR

Several important topics discussed at the 1993 Hedberg Conference are not discussed in this memoir. These include: (1) karst-like subsurface breccias formed in association with hot burial fluids

(Packard et al., 1990; Dravis and Muir, 1993; Saller and Yaremko, 1994); (2) stratigraphic traps created by subaerial exposure, erosion, and karsting, good examples of which are given by Christensen et al. (1994); and (3) the positive attributes (Goldhammer et al., 1990; Montañez and Osleger, 1993) and potential problems (Drummond and Wilkinson, 1993; Gianniny and Simo, 1993; Harris et al., 1993; Kirkby and Simo, 1993) of identifying major subaerial exposure surfaces using variations in depositional cycle thicknesses as indicated in cycle stacking patterns.

CONCLUSIONS

Predicting and detecting subaerial unconformities and associated porosity are not straightforward. All methods for detecting subaerial unconformities have shortcomings, and individually can result in either the misidentification of an exposure surface, or failure to detect a surface. Hopefully the papers of this memoir will provide insight into methods that geologists can use for predicting or identifying subaerial exposure surfaces.

Subaerial exposure alone is not a reliable mechanism to produce porosity that will be preserved in the moderately deep subsurface. As many of the papers in this memoir demonstrate, diagenesis below subaerial exposure surfaces is highly variable. Subaerial exposure alters and redistributes porosity more than it increases porosity. Contributions in this memoir demonstrate the many factors controlling the effect of subaerial exposure on pore networks. Important factors include amount of rainfall, mineralogy, duration of exposure, existing pore networks, and depositional facies and stratigraphy. Furthermore, porosity generally decreases during burial. Preservation of porosity during deeper burial requires a rigid, mineralogically stable framework that resists physical and chemical compaction.

We do not have all of the answers relative to prediction of subaerial unconformities and associated porosity. In the future, existing methods for predicting subaerial exposure need to be further tested, and new methods developed. Porosity prediction in carbonates will remain difficult. More quantitative studies of diagenetic processes occurring during subaerial exposure are needed, especially with regard to net flux of calcium carbonate in and out of various meteoric and mixing-zone environments. Processes affecting porosity during burial also need to be understood more quantitatively, and hopefully future research will move in that direction.

REFERENCES

Allan, J.R., and R.K. Matthews, 1982, Isotopic signatures associated with early meteoric diagenesis: Sedimentology, v. 29, p. 797–817.

Bathurst, R.G.C., 1984, The integration of pressure-solution and mechanical compaction and cementation, *in* ADREF, eds., Stylolites and associated phenomena—relevance to hydrocarbon reservoirs: Abu Dhabi National Reservoir Research Foundation Special Publication, p. 41–56.

Budd, D.A., 1989, Diagenesis of aragonitic and high Mg calcite sands with burial in seawater: Geological Society of America, Abstracts with Program, v. 21, p. 76.

Budd, D.A., U. Hammes, and H.L. Vacher, 1993, Calcite cementation in the upper Floridan aquifer: a modern example for confined-aquifer cementation models?: Geology, v. 21, p. 33–36.

Canter, K.L., D.B. Stearns, R.C. Geesaman, and J.L. Wilson, 1993, Paleostructural and related paleokarst controls on reservoir development in the Lower Ordovician Ellenburger Group, Val Verde basin, Texas, *in* R.D. Fritz, J.L. Wilson, and D.A. Yurewicz, eds., Paleokarst Related Hydrocarbon Reservoirs: SEPM Core Workshop No. 18, p. 61–100.

Christensen, R.J., M.L. Hendricks, and J.D. Eisel, 1994, Mississippian buried hills reservoirs along the northeastern flank of the Williston basin, Canada and United States, *in* J.C. Dolson, ed., Unconformity-related Hydrocarbons in Sedimentary Sequences: Denver, Rocky Mountain Association of Geologists, p. 245–258.

Coogan, A.H., D.G. Bebout, and C. Maggio, 1972, Depositional environments and geological history of Golden Lane and Poza Rica trends, Mexico, an alternative view: AAPG Bulletin, v. 56, p. 1419–1447.

Craig, D.H., 1988, Caves and other features of the Permian karst in San Andres dolomite, Yates field reservoir, west Texas, *in* N.P. James, and P.W. Choquette, eds., Paleokarst: New York, Springer-Verlag, p. 342–363.

Dix, G.R., and H.T. Mullins, 1988, Rapid burial diagenesis of deep-water carbonates: Exuma Sound, Bahamas: Geology, v. 16, p. 680–683.

Dix, G.R., and H.T. Mullins, 1992, Shallow-burial diagenesis of deep-water carbonates, northern Bahamas: results from deep-ocean drilling transects: Geological Society of America Bulletin, v. 104, p. 303–315.

Dravis, J., and I. Muir, 1993, Deep brecciation in the Devonian Upper Elk Point Group, Rainbow basin, Alberta, western Canada, *in* R.D. Fritz, J.L. Wilson, and D.A. Yurewicz, eds., Paleokarst Related Hydrocarbon Reservoirs: SEPM Core Workshop 18, p. 119–166.

Drummond, C.N., and B.H. Wilkinson, 1993, On the use of cycle thickness diagrams as records of long-term sealevel change during accumulation of carbonate sequences: Journal of Geology, v. 101, p. 687–702.

Erlich, R.N., S.F. Barrett, and B.J. Guo, 1990, Seismic and geological characteristics of drowning events on carbonate platforms: AAPG Bulletin, v. 74, p. 1523–1537.

Esteban, M., 1991, Chapter 4: Palaeokarst: Case histories, *in* V.P. Wright, M. Esteban, and P.L. Smart, eds., Palaeokarsts and Palaeokarstic

Reservoirs: Postgraduate Research Institute for Sedimentology, University of Reading, p. 120–146.

Gianniny, G.L., and J.A. Simo, 1993, Kilometer-scale facies variability on a low angle carbonate/siliciclastic ramp, lower Desmoinesian of the Paradox basin, SE Utah: AAPG 1993 Annual Convention Program, p. 108.

Goldhammer, R.K., P.A. Dunn, and L.A. Hardie, 1990, High-frequency glacio-eustatic sea-level oscillations with Milankovitch characteristics recorded in Middle Triassic platform carbonates in northern Italy: American Journal of Science, v. 287, p. 853–892.

Harris, P.M., S.H. Frost, G.A. Seglie, and N. Schneidermann, 1984, Regional unconformities and depositional cycles, Cretaceous of the Arabian peninsula, *in* J.S. Schlee, ed., Interregional Unconformities and Hydrocarbon Accumulation: AAPG Memoir 36, p. 67–80.

Harris, P.M., C. Kerans, D.G. Bebout, 1993, Ancient outcrop and modern examples of platform carbonate cycles—implications for subsurface correlation and understanding reservoir heterogeneity, *in* R.G. Loucks and J.F. Sarg, eds., Carbonate Sequence Stratigraphy: Recent Developments and Applications: AAPG Memoir 57, p. 475–492

Jameson, J., 1994, Models of porosity formation and their impact on reservoir description of Lisburne field, Prudoe Bay, Alaska: AAPG Bulletin, v. 78, p. 1651–1658.

Jordan, C.F., and M. Abdullah,, 1988, Lithofacies analysis of the Arun reservoir, north Sumatra, Indonesia, *in* A.J. Lomando and P.M. Harris, eds., Giant Oil and Gas Fields: A Core Workshop: Society of Economic Paleontologists and Mineralogists Core Workshop 12, p. 89–118.

Kerans, C., 1988, Karst-controlled reservoir heterogeneity in Ellenburger Group carbonates of west Texas: AAPG Bulletin, v. 72, p. 1160–1183.

Kirkby, K.C., and J.A. Simo, 1993, Differences in geometry and stacking patterns along a carbonate ramp margin: Lower Carboniferous Pekisko Formation, west-central Alberta: AAPG 1993 Annual Convention Program, p. 108.

Kirkby, K.C., and J.A. Simo, 1994, Disparate roles of unconformity surfaces in porosity generation—an example from the Pekisko Formation, west Canadian sedimentary basin: AAPG 1994 Annual Convention Program, p. 188.

Lomando, A.J., P.M. Harris, and D.E. Orlopp, 1993, Casablanca field, Tarragon Basin, offshore Spain, a karsted carbonate reservoir, *in* R.D. Fritz, J.L. Wilson, and D.A. Yurewicz, eds., Paleokarst Related Hydrocarbon Reservoirs: SEPM Core Workshop 18, p. 201–225.

Loucks, R.G., and J.F. Sarg, eds., 1993, Carbonate sequence stratigraphy: recent developments and applications: AAPG Memoir 57, 545p.

Mazzullo, S.J., and P.M. Harris, 1992, Mesogenetic dissolution: its role in porosity development in carbonate reservoirs: AAPG Bulletin, v. 76, p. 607–620.

Montañez, I.P., and D.A. Osleger, 1993, Parasequence stacking patterns, third-order accommodation events, and sequence stratigraphy of middle to upper Cambrian platform carbonates, Bonanza King Formation, southern Great Basin, *in* R.G. Loucks and J.F. Sarg, eds., Carbonate sequence stratigraphy: recent developments and advancements: AAPG Memoir 57, p. 305–326.

Moore, C.H., and Y. Druckman, 1981, Burial diagenesis and porosity evolution, Upper Jurassic Smackover, Arkansas and Louisiana: AAPG Bulletin, v. 65, p. 597–628.

Moshier, S.O., 1989, Development of microporosity in a micritic limestone reservoir, Lower Cretaceous, Middle East: Sedimentary Geology, v. 63, p. 217–240.

Packard, J.J., G.J. Pellegrin, I.S. Al-Aasm, I. Samson, and J. Gagnon, 1990, Diagenesis and dolomitization associated with hydrothermal karst in Famennian upper Wabamun ramp sediments, north-central Alberta, *in* G.R. Bloy, and M.G. Hadley, eds., The Development of Porosity in Carbonate Reservoirs: Canadian Society of Petroleum Geologists Continuing Education Short Course, Section 9.

Ruppel, S.C., 1992, Expression of high frequency sea level cyclicity on shallow carbonate platforms: the Leonardian of west Texas, *in* C. Kerans and S.C. Ruppel, Course Notes: High Frequency Sequence and Cycle Stratigraphy for Description of Clearfork, San Andres and Grayburg Reservoirs: Midland Texas, Permian Basin Graduate Center, p. 6-1– 6-25.

Saller, A.H., 1986, Radiaxial calcite in lower Miocene strata, subsurface Enewetak atoll: Journal of Sedimentary Petrology, v. 56, p. 743–762.

Saller, A.H., and R.B. Koepnick, 1990, Eocene to early Miocene growth of Enewetak Atoll: Insight from strontium isotope data: Geological Society of America Bulletin, v. 102, p. 381–390.

Saller, A.H., and K. Yaremko, 1994, Dolomitization and porosity development in the middle and upper Wabamun Group, southeast Peace River arch, Alberta, Canada: AAPG Bulletin, v. 78, p. 1406–1430.

Saller, A.H., R.A. Armin, L.O. Ichram, C. Glenn-Sullivan, 1993, Sequence stratigraphy of aggrading and backstepping carbonate shelves, Oligocene, Central Kalimantan, Indonesia, *in* R.G. Loucks and J.F. Sarg, eds., Carbonate Sequence Stratigraphy: Recent Developments and Applications: AAPG Memoir 57, p. 267–290.

Saller, A.H., D.A. Budd, and P.M. Harris, 1994, Unconformities and porosity development in carbonate strata: ideas from a Hedberg conference: AAPG Bulletin, v. 78, p. 857–872.

Sarg, J.F., 1988, Carbonate sequence stratigraphy, *in* C.K. Wilgus, B.S. Hastings, C.G.St.C. Kendall, H.W. Posamentier, C.A. Ross, and J.C. Van Wagoner, eds., Sea-Level Changes: An Integrated Approach: Society of Economic Paleontologists and Mineralogists Special Publication 42, p. 156–181.

Schatzinger, R.A., 1983, Phylloid algal and sponge-bryozoa mound-to-basin transition: a late Paleozoic facies tract from the Kelly-Snyder field, west Texas, *in* P.M. Harris, ed., Carbonate Buildups—A Core Workshop: Society of Economic Paleontologists and Mineralogists Core Workshop 4, p. 244–303.

Schlager, W., 1991, Depositional bias and environmental change—important factors in sequence stratigraphy: Sedimentary Geology, v. 70, p. 109–130.

Scholle, P.A., and R.B. Halley, 1985, Burial diagenesis: out of sight, out of mind, *in* N. Scheidermann and P.M. Harris, Carbonate Cements: Society of Economic Paleontologists and Mineralogists Special Publication 36, p. 309–335.

Sun, S.Q., 1992, Skeletal aragonite dissolution from hypersaline seawater: a hypothesis: Sedimentary Geology, v. 77, p. 249–257.

Vahrenkamp, V., 1994, A major unconformity and not much to show for it: the early Aptian Shuiaba Formation of Al Huwaisah field, Oman: AAPG 1993 Annual Convention Program, v. 3, p. 274.

Van Wagoner, J.C., H.W. Posamentier, R.M. Mitchum, P.R. Vail, J.F. Sarg, T.S. Loutit, and J. Hardenbol, 1988, An overview of the fundamentals of sequence stratigraphy and key definitions, *in* C.K. Wilgus, B.S. Hastings, C.G.St.C. Kendall, H.W. Posamentier, C.A. Ross, and J.C. Van Wagoner, eds., Sea-Level Changes: An Integrated Approach: Society of Economic Paleontologists and Mineralogists Special Publication 42, p. 39–45.

Vest, E.L., 1970, Oil fields of Pennsylvanian–Permian, Horseshoe atoll, west Texas, *in*, Halbouty, M.T., ed., Geology of giant petroleum fields: AAPG Memoir 14, p. 185–203.

Wilson, J.L., 1975, Carbonate Facies in Geologic History: New York, Springer-Verlag, 471p.

Chapter 1

Controls and Effects of Subaerial Exposure on Cementation and Development of Secondary Porosity in the Subsurface of Great Bahama Bank

David K. Beach
Marathon Petroleum Ireland, Ltd.
Mahon Industrial Estate
Blackrock, Cork, Ireland

ABSTRACT

Cementation trends and porosity profiles across multiple subaerial unconformities demonstrate how induration created during initial subaerial exposure played an important role in controlling fluid flow in shallow subsurface Pliocene–Pleistocene carbonate rocks on Great Bahama Bank (GBB). This control over fluid flow helped govern loci of dissolution and cementation during shallow burial of these metastable carbonates. Its role varied between meteoric vadose and phreatic, and mixing-zone diagenetic environments. Early induration also resulted in preferential preservation of subaerial unconformities in the subsurface.

This study of cementation and porosity trends also revealed gradual changes in diagenetic maturity of the rocks and progressive evolution of the pore systems with increasing depth of burial. Subsurface cementation and secondary porosity development occurred primarily during emergence and subaerial exposure of the bank top. Three diagenetic stages were recognized, and were related to changing diagenetic environments regulated by changing Pliocene–Pleistocene sea level and slow bank subsidence. Stage I, dominated by vadose diagenesis, commenced with initial subaerial exposure of metastable sediments, and ended with development of an indurated surface breached locally by vertical solution pipes. In Stage II, with shallow burial (surface to variably 12 to 20 m) and under ephemeral freshwater phreatic conditions, metastable carbonate sediments completed alteration to low-Mg calcite, and porosity inverted from primary interparticle and intraparticle to moldic. Relatively uniform cementation by equant calcite also occurred. In Stage III (depths to 150 to 200 m), subjection of deeper subsurface rocks to prolonged episodes of corrosive bank-wide freshwater phreatic and mixing-zone conditions during bank emergence resulted in extensive dissolution. Because GBB is comprised of carbonate rock and lacks siliciclastic aquitards, freshwater lenses and underlying mixing zones fluctuated freely with changing sea level. This allowed shallow-meteoric and mixing-zone processes to modify rocks and porosity at considerable depths within the subsurface during sea level lowstands.

INTRODUCTION

Numerous workers have described shallow core holes from various areas of south Florida and the Bahamas (Field and Hess, 1933; Supko, 1970; Perkins, 1977; Beach and Ginsburg, 1980; Beach, 1982; Pierson, 1982; Kaldi and Gidman, 1982; Williams, 1985; McNeill et al., 1988; Vahrenkamp, 1988; Melim et al., 1994). The emphases of these studies varied; however, most stressed especially stratigraphic and depositional aspects (Field and Hess, 1933; Perkins, 1977; Beach and Ginsburg, 1980; Beach, 1982; McNeill et al., 1988) and/or dolomitization (Supko, 1970, 1977; Kaldi and Gidman, 1982; Williams, 1985; Vahrenkamp, 1988; Vahrenkamp and Swart, 1991).

Besides dolomitization, the other diagenetic process often described in some detail was alteration associated with subaerial unconformities (Perkins, 1977; Beach, 1982; Pierson, 1982; Williams, 1985; McNeill et al., 1988). These workers recognized that subaerial unconformities both provide useful chronostratigraphic horizons and are important to understanding the early diagenesis of these rocks. Although observations and descriptions were generally included, these reports did not stress development of porosity and calcite cements.

In this paper, trends of cementation and porosity development in shallow subsurface Pliocene–Pleistocene carbonate rocks of GBB are described and related both to initial subaerial exposure and to the subsequent history of the bank. Factors having the greatest influence on subsequent subsurface diagenesis and porosity changes were induration created by case hardening during initial subaerial exposure, creation of bank-wide freshwater lenses during falling and lowstand sea levels, migration of these lenses with changing glacial-eustatic sea level, and ongoing bank subsidence. Three diagenetic stages are recognized, with subaerial unconformities preferentially preserved through each.

LOCATION AND METHODS

This study described sixteen core holes on GBB (Figure 1). Nine are located on northwestern Great Bahama Bank (NWGBB) between Morgan's Bluff on Andros Island and Orange Cay, five are on Cat Island, and two are on Long Island. Table 1 shows depth drilled and distance from the nearest bank edge for each location. Coring at locations U-1, U-2, and U-3 used a 10 cm diameter, 3.0 m long core barrel; at ABM, OJ-1, and OJ-3 a 5 cm, 1.5 m core barrel was used; all other coring utilized a 10 cm, 1.5 m core barrel. Figure 2 shows core recovery. Recovery was generally best in the upper portions of all core holes

Table 1. Distances of core locations from nearest platform edge, elevation of surface locations, and depths penetrated.

Core Hole	Distance to Nearest Platform Edge* (km)	Elevation at Surface (m above sea level)	Total Depth Penetrated (m from surface)
OJ-3	2.5L	1.8	50.3
OJ-1	12L	–5.8	40.5
ABM	24.5L	–7.6	31.4
U-3	52L, 58W	0.0	75.3
AN-66	36.5W	0.8	44.7
U-2	32W	0.9	74.2
AN-46	27.5W	1.5	34.4
AN-5	12.5W	1.6	30.5
U-1	2.5W	1.5	71.2
C-71	5.5W	4.4	30.5
C-70	5W	1.4	30.5
C-72	6.5W	1.3	30.5
C-73	5.5W	3.2	30.5
C-74	4.5W	1.3	30.2
LO-39	4.5W	2.1	30.5
LO-12	2W	2.0	50.6

* W=Windward, L=Leeward.

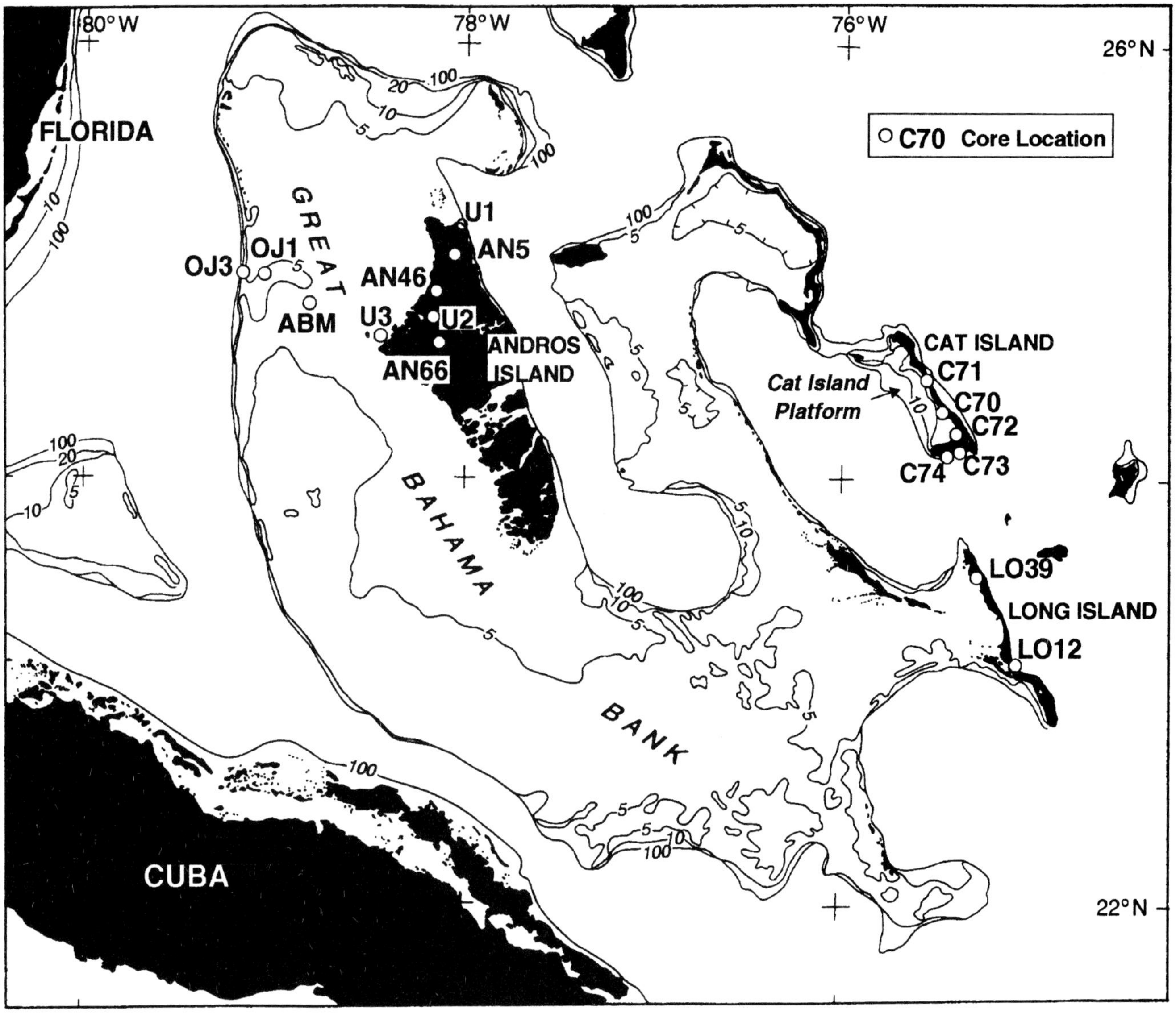

Figure 1. Bathymetric map of Great Bahama Bank showing the location of core holes described in this paper. Bathymetric contours in meters below sea level.

(above 12 to 20 m depth), being particularly good in the Cat Island locations. Recovery was also excellent in the dolomitized portion of U-1 (the basal 17.7 m). Recovery was generally better where the 10 cm diameter, 1.5 m core barrel was used. Where recovery was poor, lithology and drill time helped position core sections.

Cores were slabbed, photographed, and described using a hand lens and binocular microscope. Thin sections of 860 samples were prepared and described from core intervals of 0.3 to 1.0 m. Impregnation of rock samples with blue plastic before sectioning aided determination of original porosity. Staining techniques included Feigle's solution for aragonite, Alizarine Red S for dolomite (Friedman, 1959; Warne, 1962), and Clayton Yellow for high-Mg calcite (Choquette and Trusell, 1978). Observed and noted from thin-section analysis were texture, composition, pore types, cement types, mineralogy (from staining), and estimated percentage porosity. GRAPE logs (Gamma Ray Attenuation Porosity Evaluation, see Evans, 1965; Harms and Choquette, 1965) from eight core holes (U-1, U-2, U-3, AN-5, AN-46, and AN-66 on Andros Island, and LO-12 and LO-39 on Long Island) provided quantitative whole core porosity measurements. Analysis of selected "perm plugs" from U-1 on Andros Island, and C-70, C-71; C-72, C-73, and C-74 on Cat Island provided additional porosity and permeability data.

STRATIGRAPHY, SEDIMENTATION, AND DEPOSITIONAL UNITS

Lithology

Except for U-1, all rocks cored are limestone. Massive dolomite occurs in U-1 at depths below 51 m (Beach, 1982, 1993).

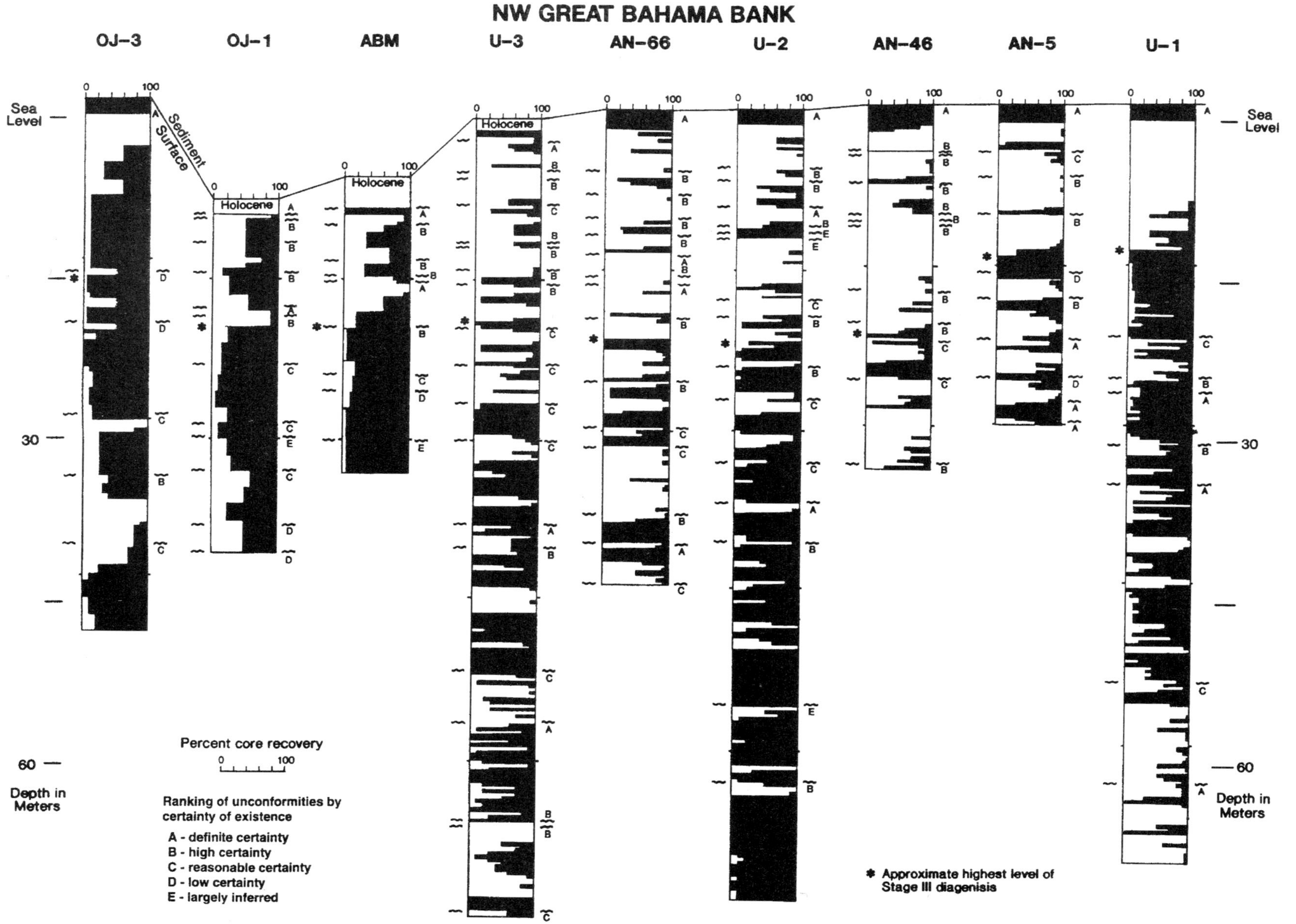

Figure 2. Cross sections of cores showing core recovery (white) and positions of recognized and inferred subaerial unconformities. Unconformities are ranked (Table 2) based on certainty of presence after Beach (1982). Rankings do not necessarily equate to length of subaerial exposure.

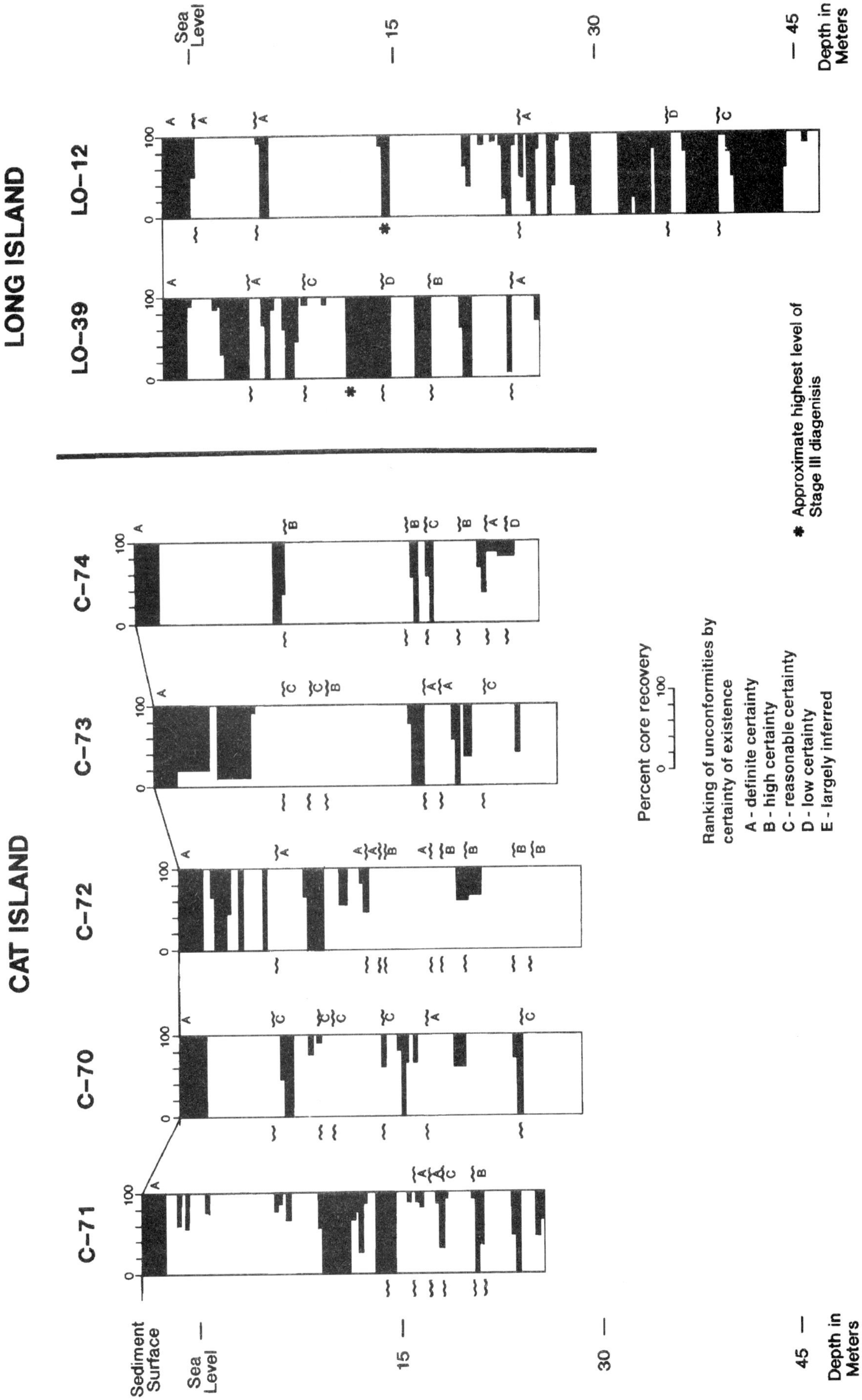

Figure 2 (*continued*).

Age

Coring at locations U-3, ABM, and OJ-1 began beneath Holocene sediments. Coring in all other locations began about 2 m below the Pleistocene limestone surface and extended into at least lower Pleistocene and upper Pliocene rocks. At most locations, surface rocks are correlatable to Sangamonian (120,000–132,000 yr b.p.) deposits (Neuman and Moore, 1975; Chen et al., 1991). During the Sangamonian, sea level reached at least 5.6 m above present sea level (Neuman and Moore, 1975; Garrett and Gould, 1984; Chen et al., 1991; Williams et al., 1993; Sherman et al., 1993). Only the cored portion of U-1 and OJ-3 begin in older and younger Pleistocene aeolian sediments, respectively (Beach, 1982). The deeper portions of four core holes, U-1, U-2 and U-3 on Andros Island, and LO-12 on Long Island penetrate into lower Pliocene sediments based on the common presence of *Stylophora* spp., *Amphistegina angulata* and Bowden bed equivalent age molluscs (Beach and Ginsburg, 1980; Beach, 1982; Williams et al., 1983; Williams, 1985; McNeill et al., 1988; Vahrenkamp and Swart, 1991).

Bank Subsidence

GBB subsided throughout the Pliocene–Pleistocene. Rates of subsidence have been estimated at between 10 and 20 Bubnoffs (1 Bubnoff is equivalent to 1 micron per year) (Paulus, 1972; Pierson and Beach, 1980; Beach, 1982; Carew and Mylroie, 1985). Subsidence was essentially uniform across the bank, but was comparatively greater than that for other Bahamian platforms (Pierson and Beach, 1980; Beach, 1982; Pierson, 1982; Williams, 1985). As a result of subsidence, accommodation space slowly developed over the top of the bank. Shallow-water sediments accumulated in this space during sea level highstands, were subaerially exposed during the ensuing lowstand, and eventually buried by later highstand deposition.

Stratigraphy

The upper Pliocene and Pleistocene rocks across the interior of GBB constitute the Lucayan Limestone of Beach and Ginsburg (1980). As defined, this formation is predominately non-skeletal, tan, and mottled limestone. Lucayan sediments grade laterally into reefal facies along the bank margins (Cant, 1977; Beach, 1982). A basal subaerial unconformity marks the sharp contact with underlying lower Pliocene sub-Lucayan deposits. These rocks are commonly poorly stratified, skeletal limestone in the bank interior, grading to reefal limestone and dolomite at the margins (Beach, 1982, 1993). All cored intervals are in the upper portion of Eberli and Ginsburg's (1987, 1989) flat-lying "A" megasequence.

Depositional Facies

As illustrated in cross section (Figure 3), the generalized depositional facies pattern of the upper Pliocene and Pleistocene (Lucayan and stratigraphic equivalent) is similar to the Holocene (Enos, 1974; Beach and Ginsburg, 1980; Beach, 1982). Coral-coralline algal framestone and bafflestone predominate along the outer edges of the bank, grading to ooidal and peloidal grainstone and packstone along the inner bank margin. Grainstone accumulations are generally well sorted, with variously low- to high-angle unidirectional and herringbone cross-bedding. The interior of the bank is generally burrow-mottled peloidal and skeletal packstone and wackestone, with mud-rich sediments common over the leeward half of the bank, and skeletal grains more abundant below about 10 m depth.

Based on the deeper penetration of cores U-1, U-2, U-3, and LO-12, lower Pliocene deposits are coral-coralline algal framestone, bafflestone, and rudstone along windward bank margins, and poorly stratified, marine-cemented skeletal-rich packstone and grainstone across the bank interior (Beach, 1982, 1993). Cores along the leeward margin did not penetrate into lower Pliocene rocks.

Shallow subtidal depositional environments characterized the interior of the bank throughout the Pliocene–Pleistocene (Figure 3). During late Pliocene and Pleistocene, cross-bank circulation was partially restricted, water depths were usually less than 10 m, and sedimentation rates were moderate to rapid (Beach and Ginsburg, 1980; Beach, 1982, 1993). Depositional environments along the bank margins were more variable, with sedimentation on open-water subtidal reefs and grainstone shoals, in protected subtidal lagoons, and as littoral and aeolian deposits. Except for deeper shelf-edge reefs, sedimentation was rapid. In contrast, during the pre-late Pliocene, the interior of the bank was more open, water depths usually exceeded 10 m, and rates of sedimentation were slower. Windward margins were predominately reefal, but generally did not form effective barriers to cross-bank circulation.

Subaerial Unconformities

Zones of heavily altered sediments punctuate all cores. The features observed in these zones are similar to those described from modern subaerial exposure surfaces in south Florida and the Bahamas and suggest similar origins (calichification—Kornicker, 1958; Multer and Hoffmeister, 1968; Kahle, 1977; Robbin and Stipp, 1979; Beier, 1987; Bain and Foos, 1993; karstification—Benjamin, 1970; Dill, 1977; Little et al., 1977; Gascoyne et al., 1979; Smart and Whitaker, 1988; Whitaker and Smart, in press; residual soil development—Ahmad and Jones, 1969; Little et al., 1977; Carew and Mylroie, 1991; Rossinsky and Wanless, 1992; Bain and Foos, 1993; and erosion—Illing, 1954; Doran, 1955; Newell and Rigby, 1957; Little et al., 1977; Rossinsky and Wanless, 1992). Figures 2 and 3 show the distribution of postulated subaerial unconformities in the 16 cores studied, whereas Table 2 (after Beach, 1982) lists the more important attributes recognized. Based on the observed features listed in Table 2, each unconformity is ranked from A to E according to certainty of its existence (after Beach,

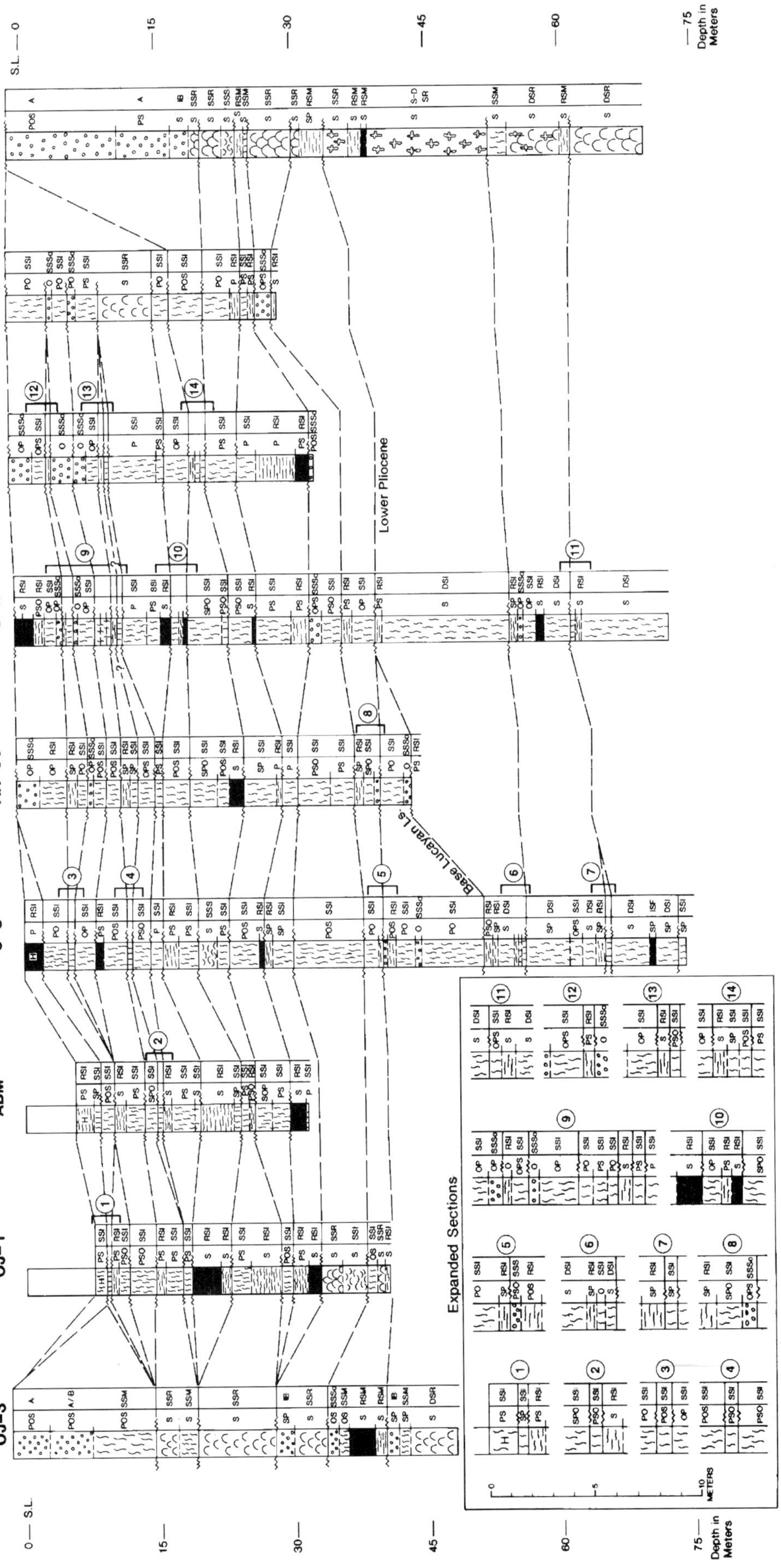

Figure 3. Cross sections of cores showing depositional texture, composition, interpreted depositional environments, and subaerial unconformities. Correlation of sedimentary units on NWGBB are after Beach (1982). The lower Pliocene is based mostly on the presence of abundant *Stylophora* spp. The base of the Lucayan Limestone is defined on the facies change from predominantly nonskeletal above to skeletal below.

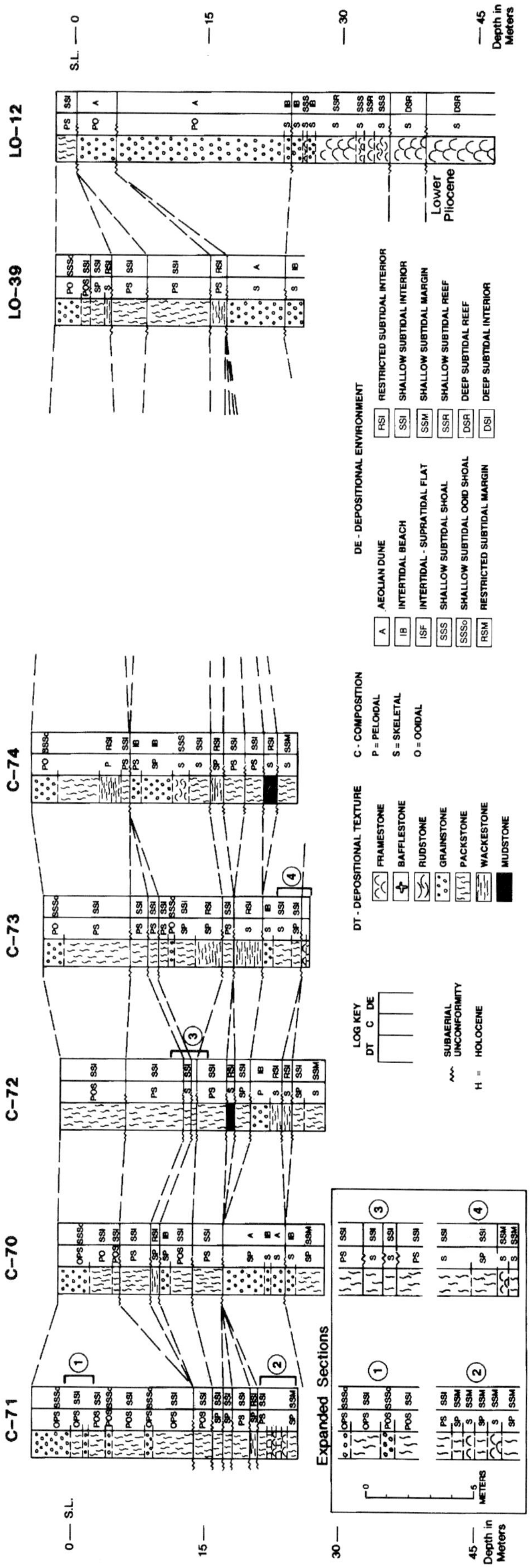

Figure 3 *(continued)*.

Table 2. Features associated with subaerial unconformities recognized in cores, and ranking of unconformities based on certainty of existence.

Features Observed	
red, gray, or black staining	needle-fiber cement
secondary unlaminated micrite	rhizomorphs
secondary laminated micrite	microcodium
secondary micritic pisolites	boring, macro and/or micro
extensive leaching above unconformity	marked increase in induration
localized extensive secondary alteration of primary sediments	solution pipes
paleosol	sediment fill from overlying unit
lithoclasts, including darkened clasts	facies change
Ranking by Certainty of Existence	
A) Definite certainty	D) Low certainty
B) High certainty	E) Largely inferred
C) Reasonable certainty	

1982; Beach and Ginsburg, 1994). Figure 2 includes the ranking for each unconformity.

The typical variety of features preserved along unconformities is exhibited in Figure 4. It shows a section of core from AN-46 containing three closely spaced subaerial unconformities. The upper unconformity (10.0 m, 33 ft) is denoted by a sharp change from white ooidal and peloidal packstone with occasional darkened lithoclasts, leached bivalve shells, and *Porites porites* above the unconformity, to red and dark gray, altered, foraminiferal-rich skeletal wackestone, with common *Porites porites* below it. Secondary micritization occurs along the unconformity. Alteration gradually decreases beneath the unconformity and color lightens to tan. Black lithoclasts occur near the base of this unit within a 2 cm thick yellow-brown to reddish-brown zone of paleosol. A thin, laminated crust occurs beneath the middle unconformity (10.6 m, 34.8 ft) with associated microborings, rhizomorphs, and needle-fiber cement. Underlying sediments are peloidal, ooidal, and skeletal packstone. These are in sharp contact with the third unconformity (11.0 m, 36.2 ft). This unconformity is marked by extensive alteration and erosion along a thin, brown to yellow-brown laminated crust. Solution pipes extend downward from the unconformity and are filled by sediments from the overlying unit. A small coral polyp grew from the side of the solution pipe (see arrow in Figure 4). Rhizomorphs, microborings, and needle-fiber cement are abundant immediately beneath this unconformity. Sediment is predominately a peloidal packstone.

Buried subaerial unconformities serve as boundaries dividing distinct depositional events into lithostratigraphic units (Beach, 1982, 1993). The average

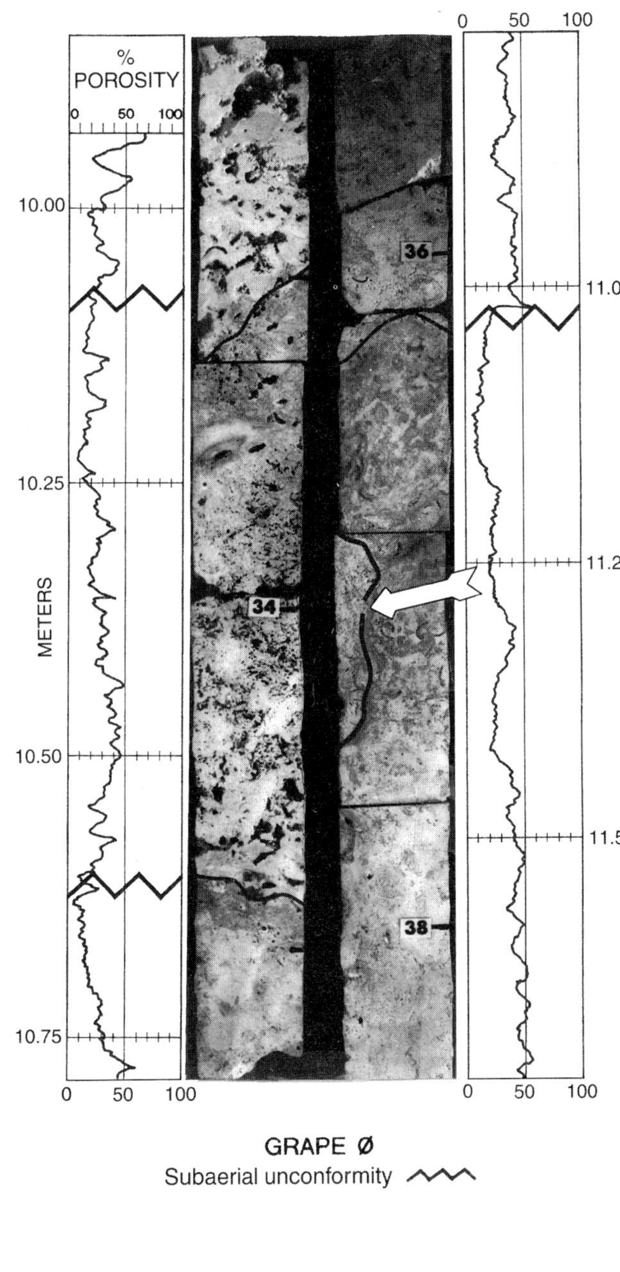

Figure 4. Section of core AN-46 between 9.8 and 11.7 m depth below ground surface containing three closely spaced subaerial unconformities and the corresponding GRAPE log response. Core is scaled in feet below surface.

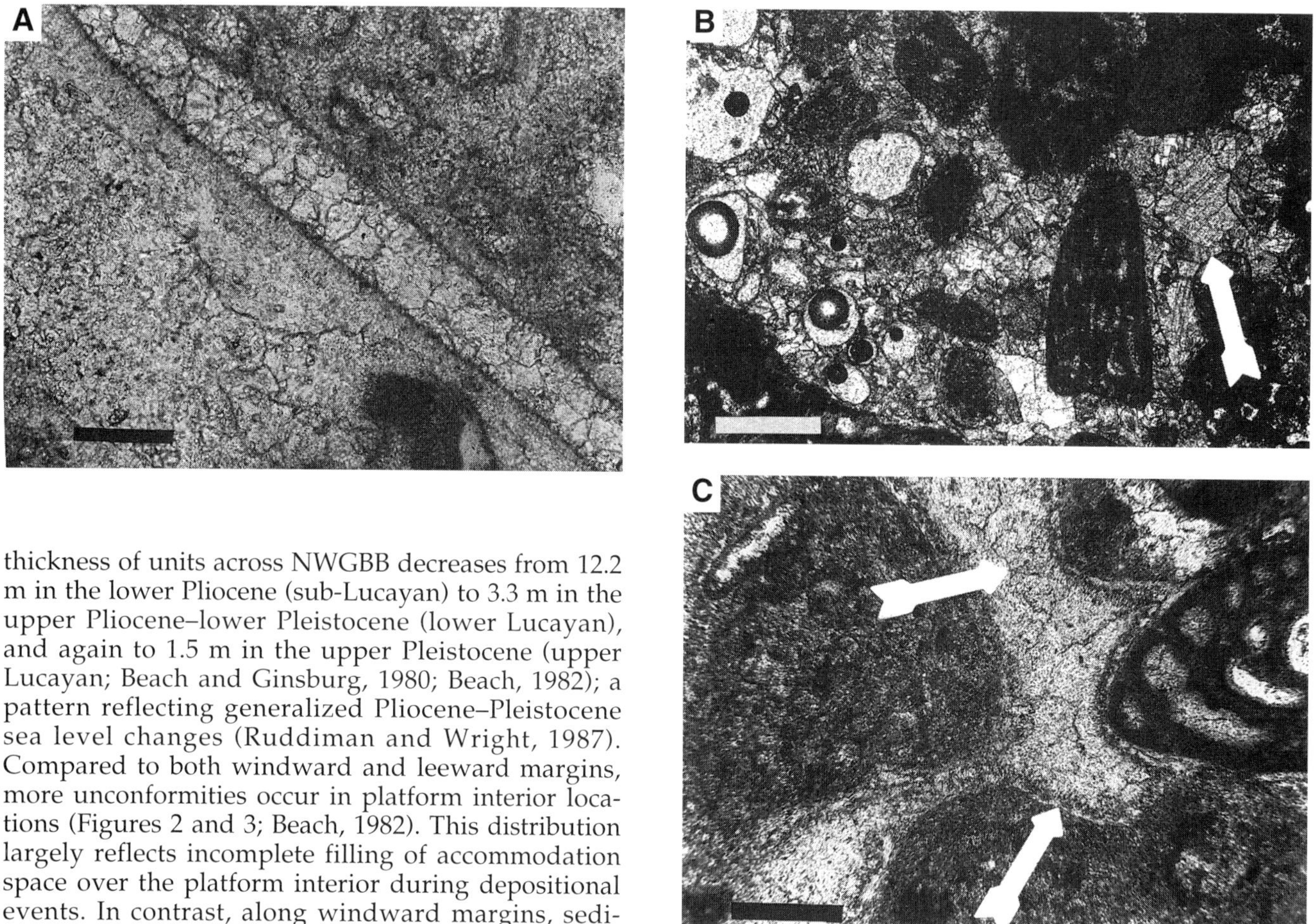

Figure 5. Photomicrographs in plane light of various cement textures. (A) Equant spar from 72.8 m depth in U-3 filling leached bivalve mold. Outside of mold is lined by isopachous cement; interparticle pore is filled by irregular spar cement. Bar scale is 0.5 mm. (B) Coarse equant spar (see arrow) from AN-66, 5.6 m below ground level and immediately above a subaerial unconformity. Several moldic pores occur to the left. Bar scale is 0.5 mm. (C) Inclusion-rich isopachous and irregular spar cements from U-2, 64 m below ground level. Inclusions outline relict fibrous texture (lower arrow). Upper arrow points to irregular contact of crystals of irregular spar. Intraparticle porosity is retained near the center of the soritid foraminifer in the far right. Bar scale is 0.5 mm.

thickness of units across NWGBB decreases from 12.2 m in the lower Pliocene (sub-Lucayan) to 3.3 m in the upper Pliocene–lower Pleistocene (lower Lucayan), and again to 1.5 m in the upper Pleistocene (upper Lucayan; Beach and Ginsburg, 1980; Beach, 1982); a pattern reflecting generalized Pliocene–Pleistocene sea level changes (Ruddiman and Wright, 1987). Compared to both windward and leeward margins, more unconformities occur in platform interior locations (Figures 2 and 3; Beach, 1982). This distribution largely reflects incomplete filling of accommodation space over the platform interior during depositional events. In contrast, along windward margins, sedimentary deposits generally accumulated up to or above sea level. Little or no unfilled accommodation space remained, and additional sedimentation could only occur above sea level, during a subsequent higher sea level, or following an extended period of bank subsidence. Leeward margins reveal a less consistent pattern, as periods of little or no accumulation below sea level interchanged with episodes of deposition building up to or above sea level.

DIAGENESIS

Alteration of Metastable Sediments

Metastable sediments (aragonite and high-Mg calcite) persist only in the uppermost depositional units (approximately the top 10 m). Residual unaltered sediments are mostly aragonite, though locally miliolid foraminifers and fragments of coralline algae and echinoids retain some high-Mg calcite. There is progressive loss of aragonite in successively deeper units. Below about 10 m, cores are almost entirely low-Mg calcite, and, in U-1, dolomite. Even where present in near-surface units, aragonite is uncommon within heavily altered zones immediately underlying unconformities.

Cementation

Cement Types

The more abundant cement fabrics observed include equant spar, isopachous cement, irregular spar, microspar, and secondary micrite. Less common are meniscus cement, needle-fiber cement, and coarse-bladed spar. Typical crystal size, occurrence, and inferred environment of precipitation for each of these cement fabrics is summarized in Table 3. Examples of each are included in Figures 5–8.

Trends of Cementation

The limestone in these cores shows varying degrees and patterns of cementation. There are, however, three notable trends in these variations: (1) In the uppermost sections of cores, there is a gradual

Table 3. Cements described, their occurrence, and interpreted environment of precipitation.

Cement Fabric	Crystal Size	Occurrence	Environment of Precipitation
Equant Spar (Figure 5A and B)	16–250μm	Ubiquitous, commonly filling primary interparticle, intraparticle and sheltered pores, and, increasingly below ≈15 meters, secondary pores. Coarse equant spar is common immediately overlying unconformities and as syntaxial overgrowths.	Mostly in meteoric phreatic zone (Land, 1970; Halley and Harris, 1979)
Isopachous Cement (Figure 5A and C)	16–250μm	Fringe lining primary interparticle pores, almost exclusively in Pre-Lucayan rocks.	Neomorphosed or recrystallized from marine cements (Beach, 1982, 1993)
Irregular Spar (Figure 5A and C)	100–300μm	Fills or partially fills primary pores, almost exclusively in Pre-Lucayan rocks.	Neomorphosed or recrystallized from marine cements (Beach, 1982, 1993)
Microspar (Figure 6A)	<20μm	As a light brown clotted texture, associated with silt-size fossil fragments in rocks of all depositional textures except grainstone.	Neomorphic alteration of lime mud (Beach, 1982)
Secondary Micrite (Figure 6B and C)	<20μm	Usually below unconformities as aggrading crusts, stringers, or in patches usually <2 centimeters in diameter, locally laminated. Little or no internal structure, but may retain ghosts or windows of former sediment or rock, also filamentous microborings.	Vadose zone, below and near subaerial exposure surfaces (Esteban and Klappa, 1983), locally deeper, especially along rhizomorphs (Rossinsky et al., 1992)
Meniscus Cement (Figure 7A)	50–100μm	Partial filling of primary interparticle pores, restricted mostly to the uppermost depositional units.	Vadose zone (Dunham, 1971)
Needle-Fiber Cement (Figure 7B)	<5 × 200μm	Clusters or individual randomly oriented crystals in open, mostly primary pores, usually below subaerial unconformities	Vadose zone near subaerial unconformities (James, 1972; Ward, 1975)
Coarse-Bladed Spar (Figure 8A, B, and C)	.25mm–>1.0cm	Frequently an irregular lining of large vugs and channels, especially along either floors of cavities (where crystals have flat or blunt upper surfaces), or roofs (where terminations are euhedral). Bands of silt inclusions often present.	Crystals of irregular occurrence, having geopetal growth, blunt terminations, and/or bands of silt inclusions suggest a vadose origin. Uniform development and euhedral crystal terminations suggest a meteoric phreatic origin. (Beach, 1982)

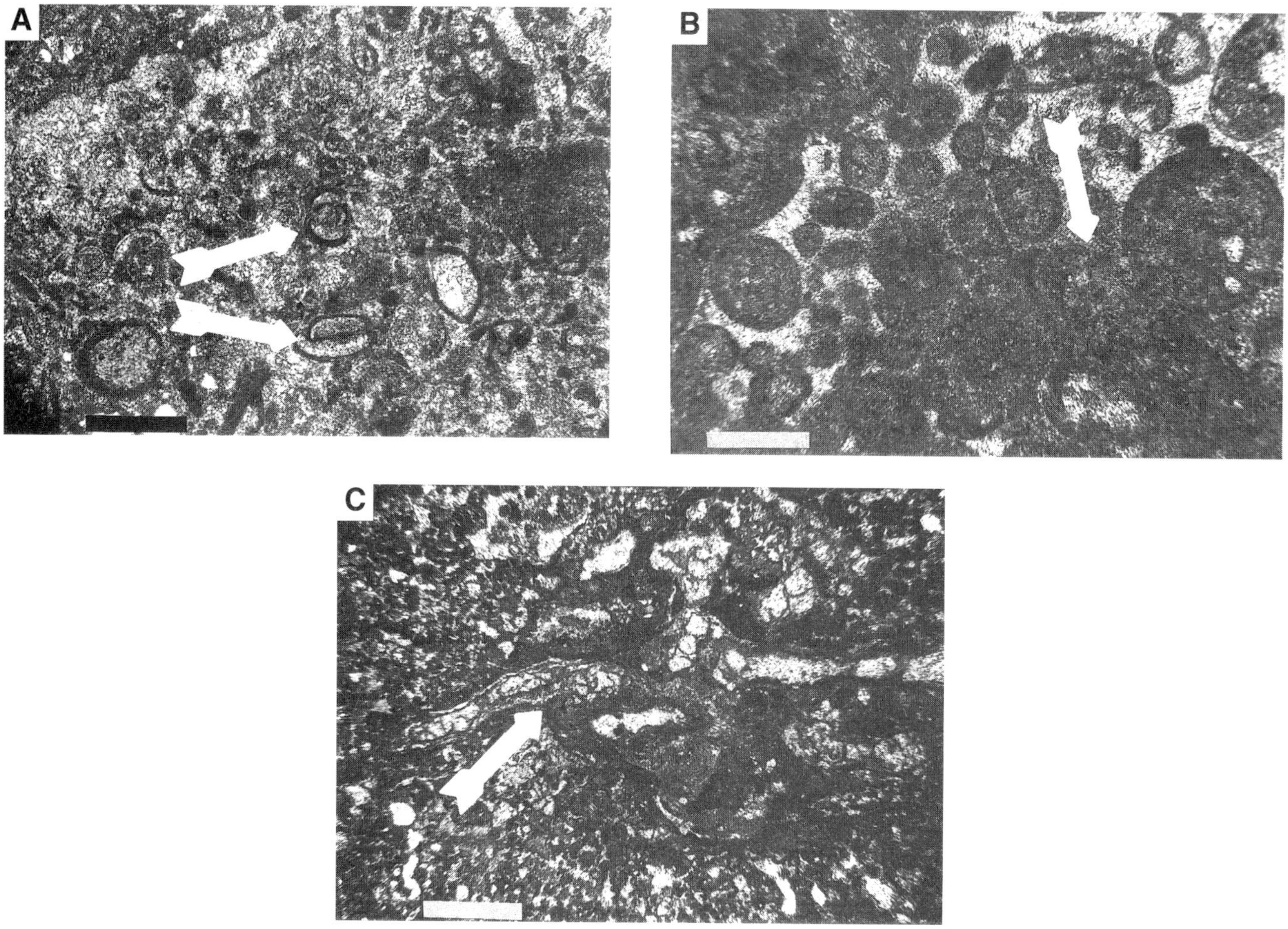

Figure 6. Photomicrographs in plane light of various cement textures. (A) Skeletal wackestone from AN-66, 5.9 m below ground level, containing abundant small miliolid foraminifers (arrows) and matrix of microspar. Bar scale is 0.5 mm. (B) Aggrading secondary micritization (arrow), associated with subaerial unconformity, altering lithified ooidal grainstone (micritized ooid occurs to right of arrow), from AN-66, 41.3 m below ground level. Bar scale is 0.5 mm. (C) Rhizomorphs (root casts) from AN-46, 41.3 m below ground level. Arrow indicates juncture of original root bifurcation. Shown is clear equant spar-filled center bounded by vaguely laminated secondary micrite. Bar scale is 0.5 mm.

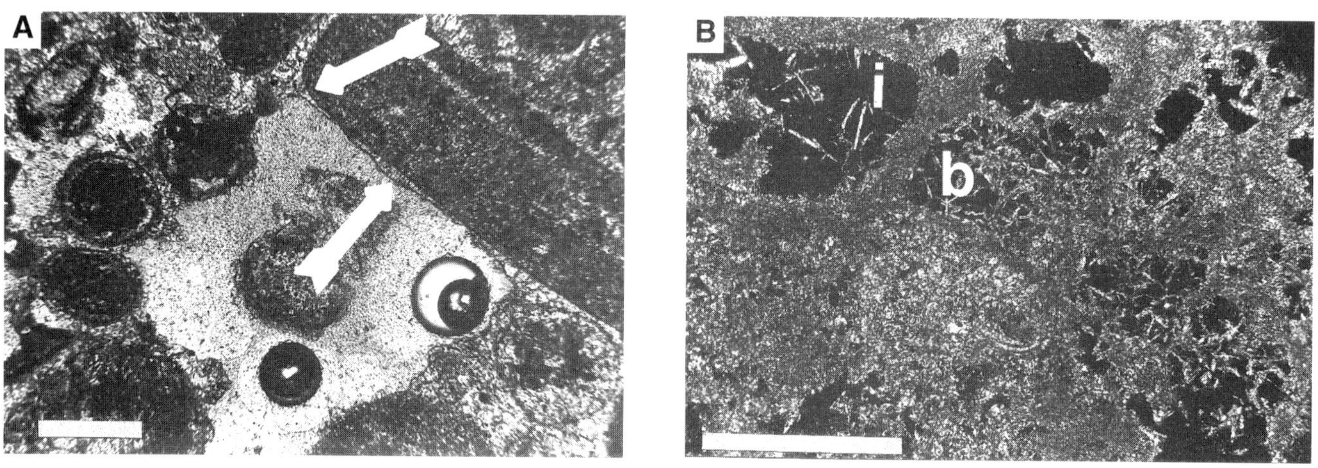

Figure 7. Photomicrographs of meniscus and needle-fiber cements. (A) Meniscus cement partially filling interparticle pore in peloidal and superficial ooid grainstone from U-2, 5.3 m below ground level. Equant spar fills most of the interparticle porosity in this slide. Bar scale is 0.5 mm. (B) Cross-polarized light view of needle-fiber cement occurring as both individual needles (i) and in clusters of needles (b) from U-2, 4.1 m below ground level. Bar scale is 0.25 mm.

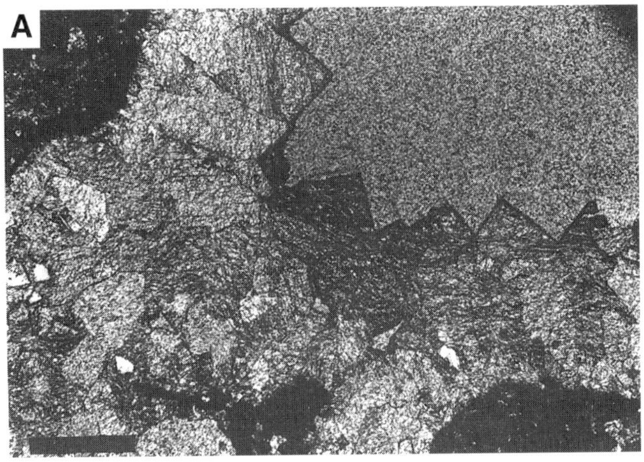

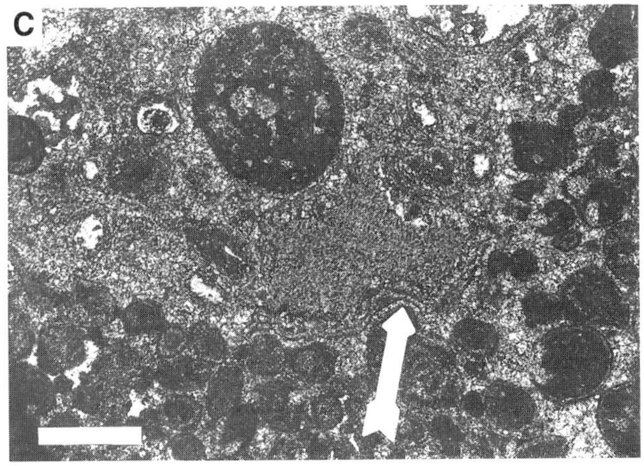

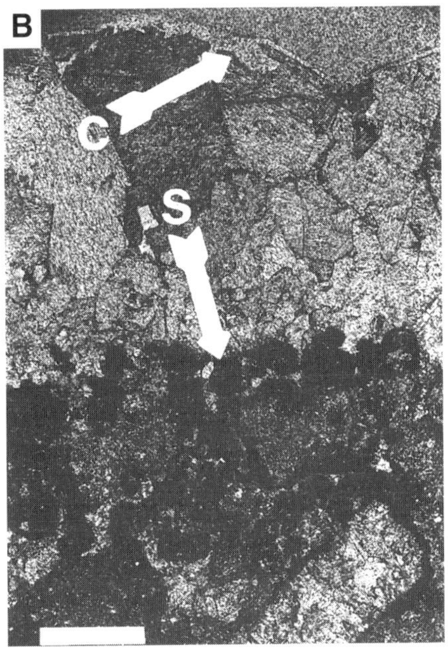

Figure 8. Photomicrographs of various cement textures. (A) Plane light view of coarse-bladed spar with euhedral crystal terminations lining vug in AN-46, 19.8 m below ground level. Bar scale is 0.5 mm. (B) Photomicrograph in plane light, from AN-46, 20.7 m below ground level, showing blunt terminations of coarse calcite crystals (c) in contact with open vug (arrow above) and vaguely clotted to peloidal textureless sediment (paleosol, s—lower arrow). Bar scale is 2 mm. (C) Plane light view of multiple generations of cement (arrow) separated by inclusion-rich zones (dark bands) surrounding large interparticle pore in ooidal grainstone in AN-66, 14.6 m below ground level. Large crustacean pellet at upper left also shows at least two generations of evenly developed cement. Silt results either as residuum from leached cements or from deposition of impurities over crystals by downward-flowing ground water. Bar scale is 0.5 mm.

increase downward in degree of cementation. (2) Within individual depositional units, there is a gradual decrease downward in cementation. (3) Mud-rich sediments are preferentially indurated. The first two trends are shown schematically in Figure 9. Figure 10A shows an example of the third trend.

Above about 10 m, the gradual increase in cementation downward through the uppermost sections or depositional units in each core corresponds both to increasingly complete mineralogical stabilization of metastable sediments, and to inversion of porosity from primary interparticle and intraparticle to secondary moldic. Grain-supported sediments are typically friable and poorly lithified in these uppermost units. Locally, meniscus cements are observed. Below about 10 m, the rocks are generally moderately well cemented by equant spar. The degree of induration remains fairly constant, but large vugs, channels, and cavernous voids become increasingly widespread. Only in U-1 and LO-12, where older, well-lithified aeolian deposits occur near the surface, is this trend not observed.

Within individual units the gradual decrease in cementation or induration downward from unconformities is independent of the mineralogical state of sediments within units (Figure 9). In profile, well-indurated rock occurs just under the capping subaerial unconformity. Induration gradually decreases downward to where rocks are variously slightly to moderately well cemented. This trend is particularly important in the uppermost depositional unit, where in LO-39 and C-73, the basal portion of this unit is very slightly cemented, and in OJ-3 it is uncemented. The well-indurated nature of the upper portion of each unit results from near-total occlusion of both primary and secondary porosity by cement, especially equant spar and secondary micrite.

The third trend is preferential lithification of mud-rich sediments. In the shallower portions of cores, evidence for this includes pervasive cementation and alteration to low-Mg calcite microspar of mud-supported sediments, whereas grain-supported deposits remain poorly cemented and, in part, retain aragonite. This trend is also reflected in apparent

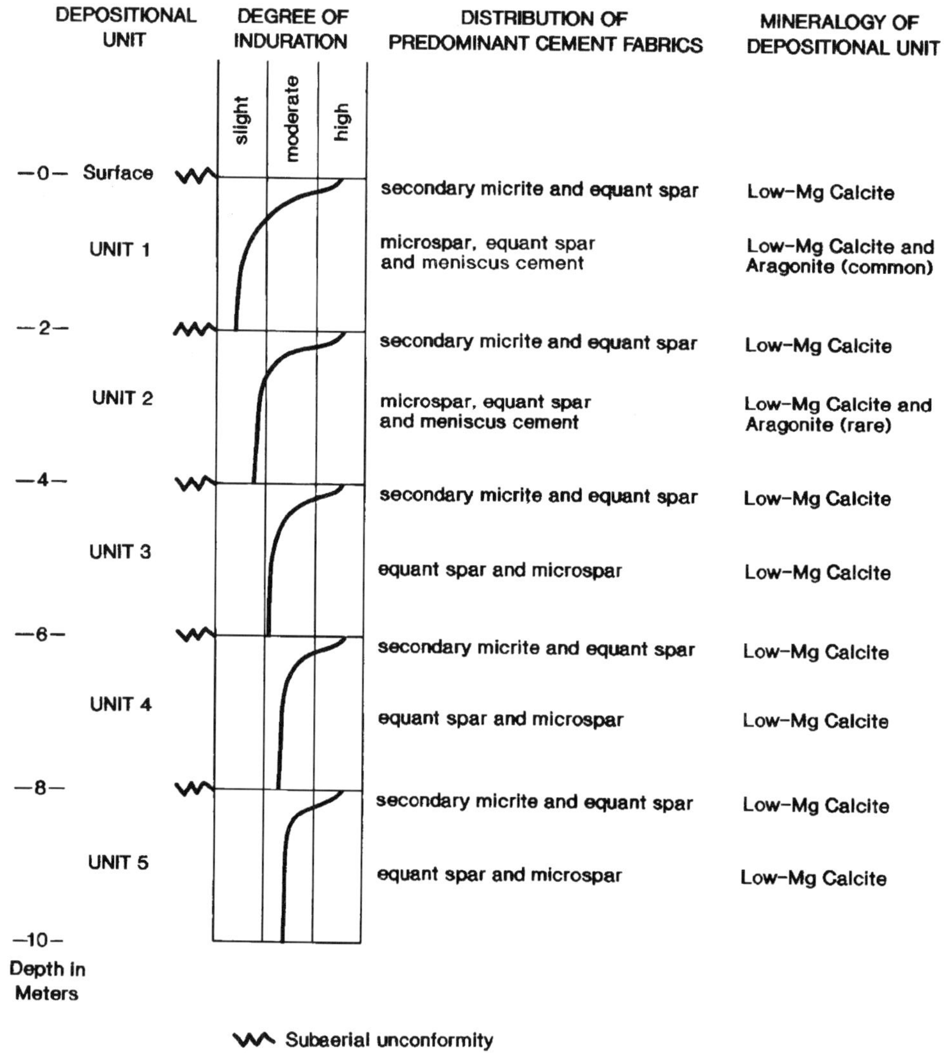

Figure 9. Schematic illustration of trends of cementation observed in the upper depositional units.

greater resistance to dissolution at all depths. For example, there is a tendency for preservation of muddier portions of burrows, whereas more grain-rich portions were leached (Figure 10A).

Porosity Development

Pore Types

Pore types and sizes (after Choquette and Pray, 1970) are highly variable both at macroscopic (greater than 0.5 mm) and microscopic (less than 0.5 mm) levels. Whereas macroscopic pores are largely secondary, including vugs, channels, and inferred caverns, preserved primary pores include especially sheltered and (commonly solution enhanced) burrow porosity. Proving connectivity of channel systems in cores is difficult. Therefore, vugs are identified by apparent isolation; channels are identified by apparent continuity (Figure 10). Cavernous pores are simply large channels (at least 0.5 m, after Choquette and Pray, 1970), determined by bit drops and zones of no recovery. Microscopic porosity includes primary interparticle (Figure 8C) and intraparticle (Figure 5A), and secondary moldic pores (Figure 5B). Intercrystalline and fracture pores are important in the dolomitized portions of U-1.

In the upper portions of most cores, particularly grain-rich cores from the bank interior, there is a gradual change in pore types from primary interparticle and intraparticle to secondary moldic. This trend is evident in Figure 11, where below depths of 10 to 12 m pores are predominately secondary, exclusive of the reefal portions of U-1. This inversion largely coincides with the progressive loss of aragonite (Halley and Beach, 1979). Sediments originally of calcitic composition (sub-Lucayan in U-2 and U-3; Beach, 1982, 1993) retain more primary porosity than those that were aragonite-rich.

Large secondary pores formed preferentially along either vertical or horizontal orientations. Vertically

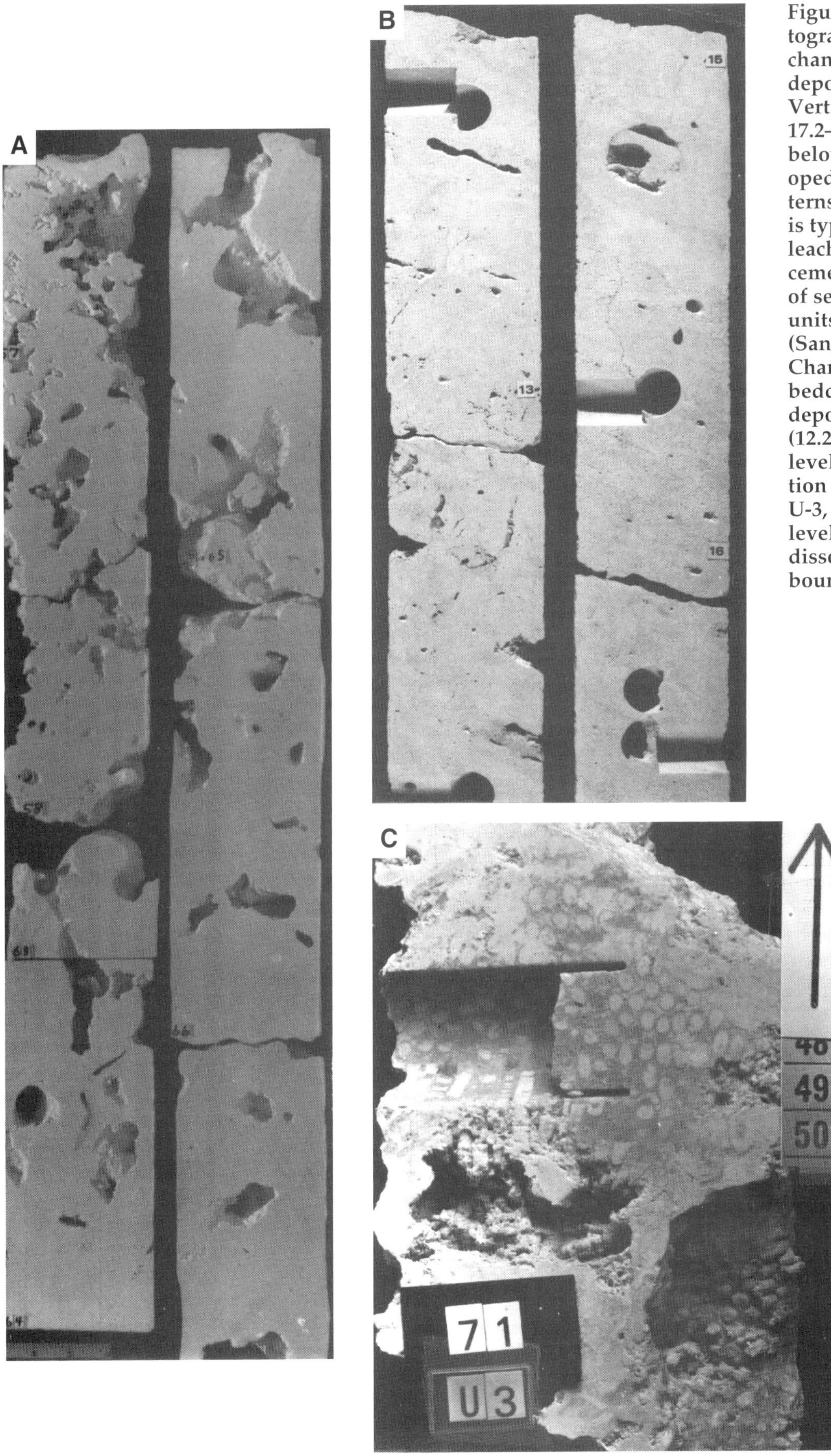

Figure 10. Slabbed core photographs showing control of channel development by depositional fabric. (A) Vertical channels in C-71, 17.2–20.3 m (56.5–66.5 ft) below ground level, developed along burrowing patterns. Coarser grained fabric is typically selectively leached, muddier fabric cemented. Selective leaching of sediments occurs in all units including the youngest (Sangamonian) deposits. (B) Channels developed along bedding planes in aeolian deposits in U-1, 3.7–5.0 m (12.2–16.5 ft) below ground level. (C) Irregular dissolution of massive corrals from U-3, 71 m below ground level. Irregular patterns of dissolution are common in boundstone sections.

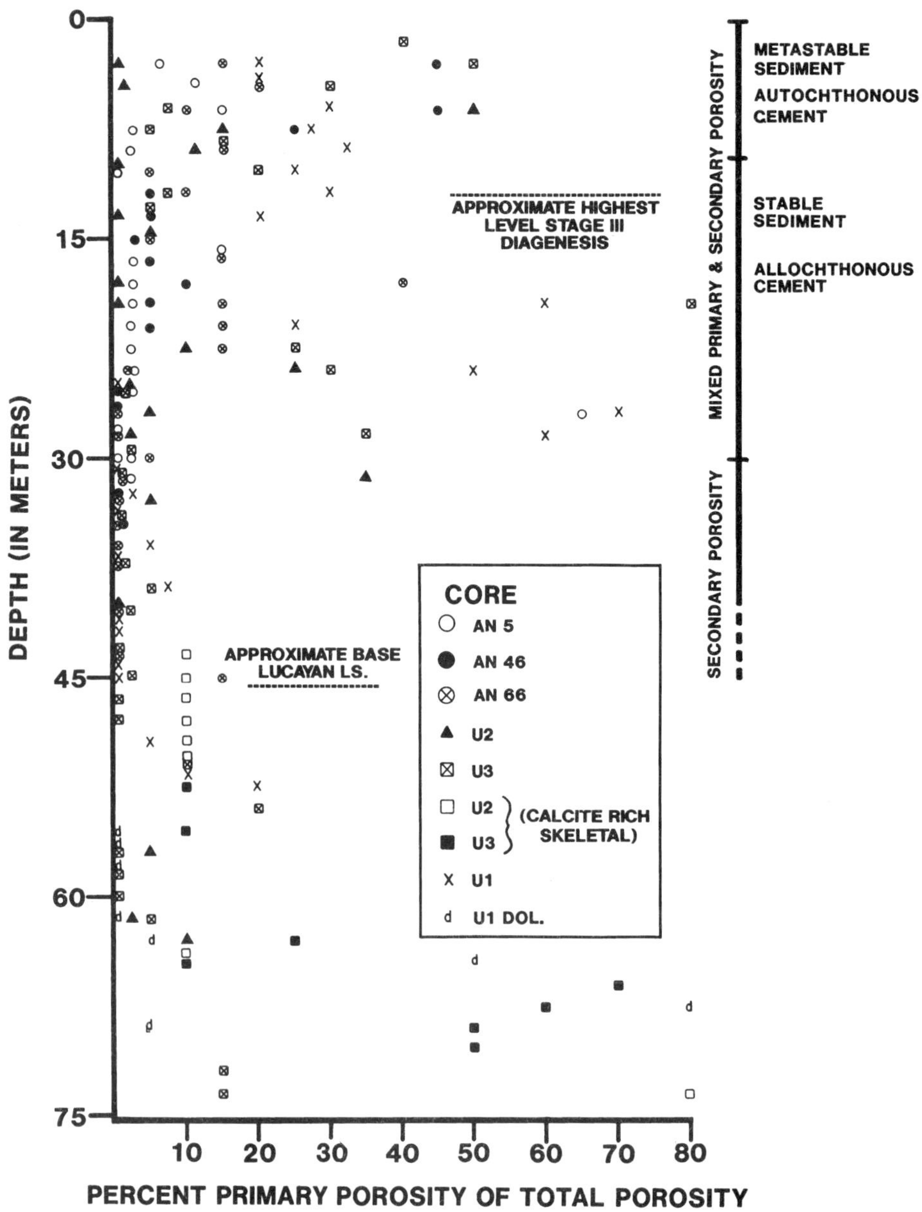

Figure 11. Plot of primary versus secondary porosity, derived from thin sections, per 1.5 m interval of core.

oriented solution channels are present in even the youngest units. They commonly developed along vertical burrowing patterns in bank interior locations (Figure 10A), followed bedding surfaces in high-angle cross-bedded units (Figure 10B), or more random patterns in reefal deposits (Figure 10C). Solution channels commonly initiated at and cut downward from subaerial unconformities (Figure 12A).

Extensively leached intervals, up to several tens of centimeters thick and characterized by vugs, channels, and small caverns, commonly developed immediately above subaerial unconformities (intervals of no recovery above unconformities in Figure 2). In deeper intervals, the orientation of most solution-enhanced porosity is inferred to be horizontal also. Solution-enhanced pores, notably large vugs (Figure 12B, C), channels, and caverns (again reflected by zones of poor core recovery in Figure 2), are evident below depths of about 12 m near the margins of NWGBB and Long Island, and 20 m across the interior of NWGBB. Pore size is extremely variable in these sections. This trend is not evident in the Cat Island cores.

Percentage Porosity

GRAPE logs provide a continuous measure of whole core porosity at eight locations (Figure 13). Figure 14 shows that average core porosity per 1.5 m interval typically ranges from 25 to 50%. There is a slight decrease in porosity at depths below about 10 m. This shift approximates the change from metastable to stable mineralogy, but may also reflect a compositional change to slightly more skeletal-rich sediments (Beach, 1982). Mean porosity values of rocks altered to low-Mg calcite are about 35%,

Figure 12. Slabbed core sections showing effects of extensive dissolution. (A) Vertical solution channel developed beneath subaerial unconformity at 36 m (118 ft) extending downward to at least 41.5 m (136 ft) through well-indurated rock. (B and C) "Swiss cheese" solution pattern from cores AN-46, 33.2–33.6 m (109–110.2 ft) below ground level, and AN-66, 37.3–37.8 m (122.3–124 ft). In AN-46, this section is skeletal and peloidal mudstone and overlies a 20 cm thick interval of no core recovery above a subaerial unconformity at 33.8 m (110.9 ft). In AN-66, this interval occurs in peloidal and skeletal packstone, with the base 0.5 m above a subaerial unconformity at 38.3 m (125.7 ft).

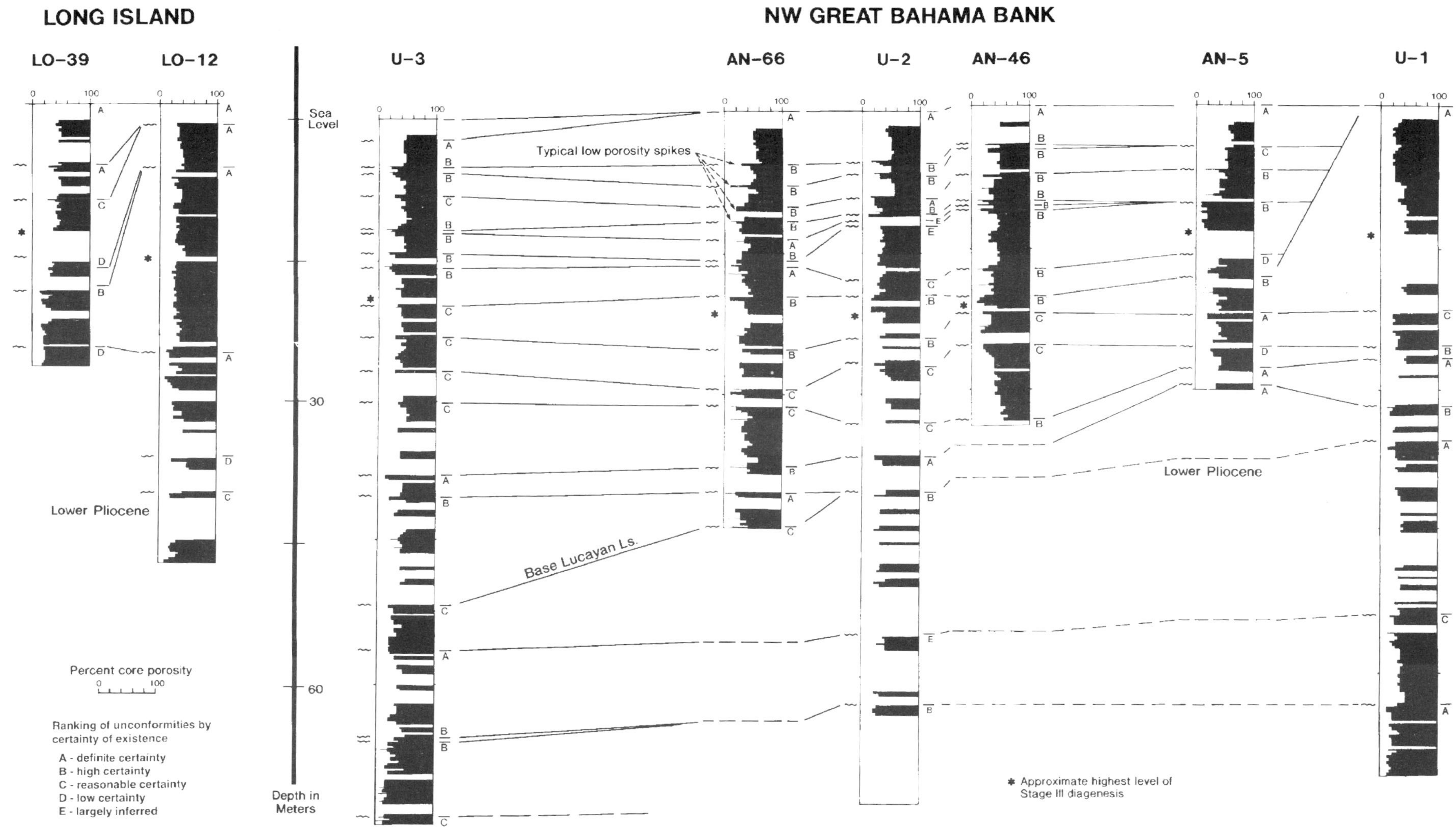

Figure 13. GRAPE whole core porosity (white) measured from eight locations and averaged over 0.3 m intervals. Plots are in 0.15 m intervals in sections of markedly decreased porosity. Also shown are positions of recognized and inferred subaerial unconformities and comparative ranking of each based on certainty of existence (Table 2).

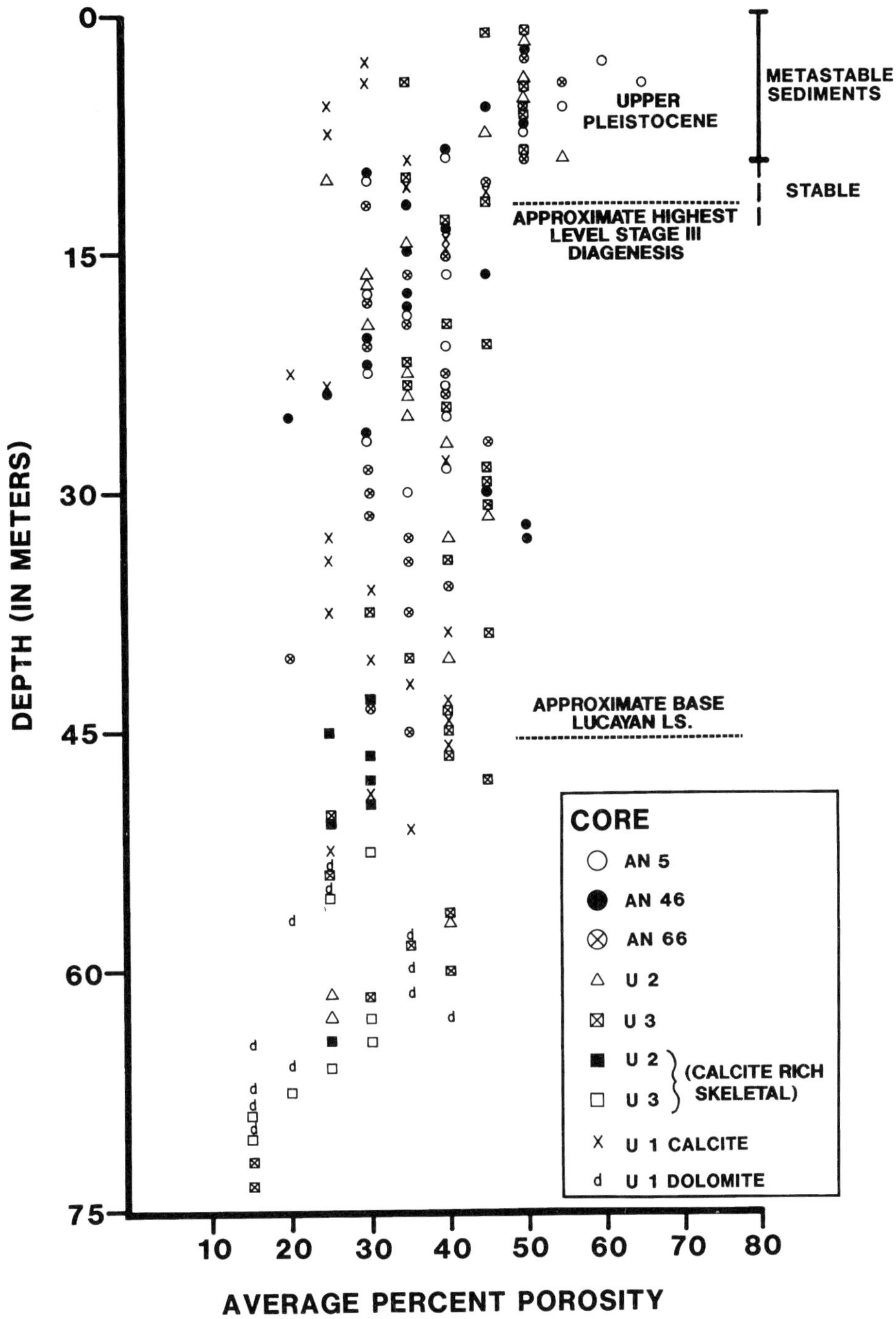

Figure 14. Plot of average porosity per 1.5 m core interval from GRAPE logs. Values have been rounded to the nearest 5%.

compared to 45% for shallower rocks containing appreciable metastable components.

Low-Porosity Spikes

Spikes of decreased porosity punctuate all cores and are independent of the change from metastable to stable mineralogy or from calcite to dolomite. Figure 13 quantitatively depicts these spikes at the eight locations having porosity logs. This figure displays plots of average porosity over 0.3 m log intervals, but also includes plots over 0.15 m intervals in sections of markedly decreased porosity. Low-porosity spikes are most apparent in subtidal bank interior sediments of the Lucayan Limestone, where intervals of low porosity frequently coincide with cemented (well-indurated) and altered sediments lying immediately below subaerial unconformities (e.g., AN-46, U-2, AN-66, and U-3). Typically, porosity of 30 to 50% abruptly decreases to 10 to 15% below an unconformity and only gradually returns to higher values with depth. Well-indurated intervals usually vary from 15 to 75 cm in thickness. Figure 4 shows three excellent examples of this relationship. This porosity trend is logically opposite to that previously noted for cementation.

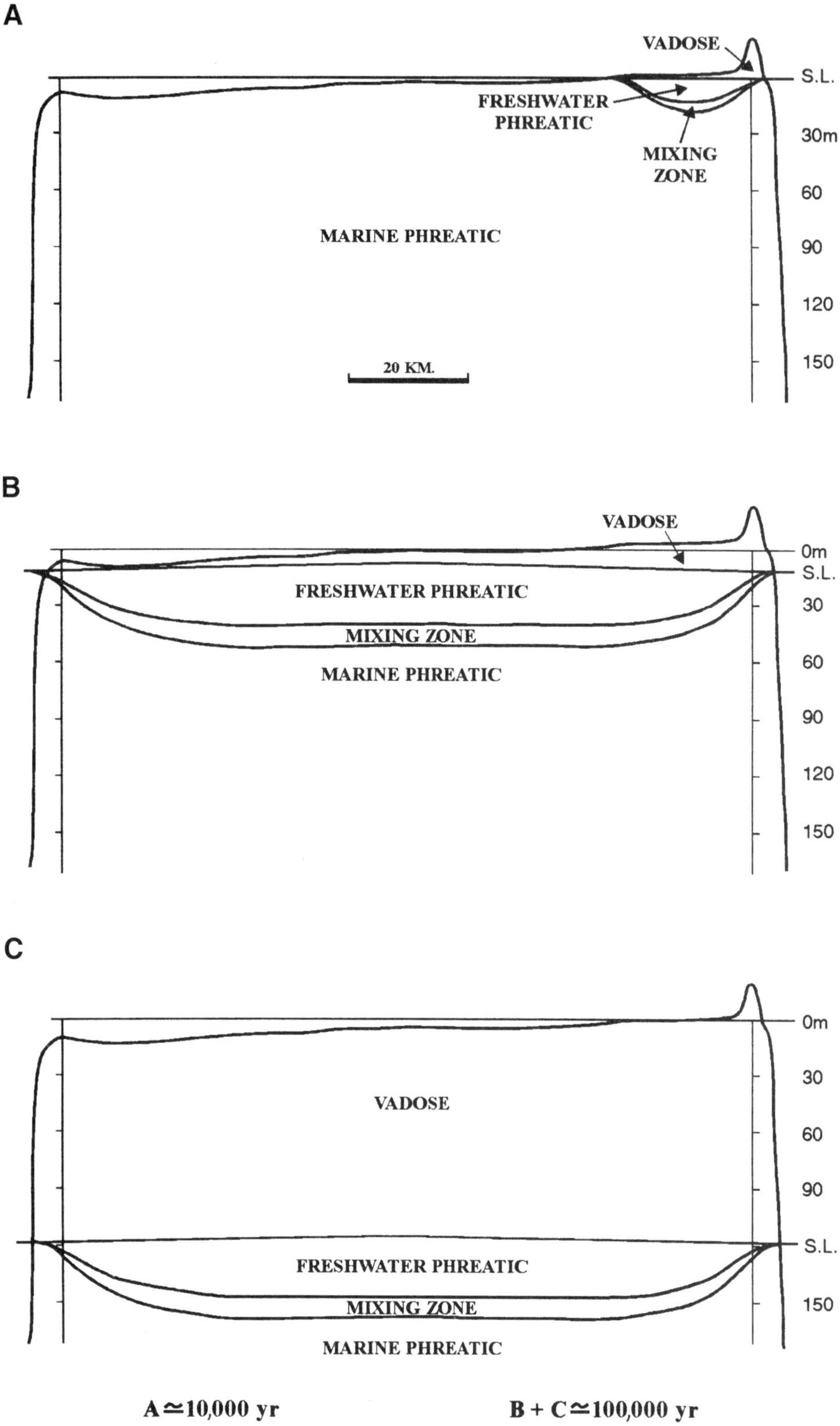

Figure 15. Schematic diagram showing relationship between various Pliocene–Pleistocene sea levels and diagenetic environments on Great Bahama Bank. (A) Interglacial highstand, similar to the present. (B) Intermediate (falling and rising) sea levels. (C) Glacial maxima lowstands.

Influences and Controls on Cementation and Porosity Development

Changing Sea Levels—Changing Diagenetic Environments

In the shallow subsurface of GBB, Pliocene–Pleistocene glacio-eustatic sea level changes resulted in significant concomitant changes in diagenetic environments (Figure 15). Increased resolution of the upper Pliocene and Pleistocene record of climatic fluctuations has been discussed by Imbrie et al. (1984), Ruddiman and Wright (1987), Raymo et al. (1989), and Farrell and Prell (1991), among others (Figure 16). This record indicates few periods of sea level stability

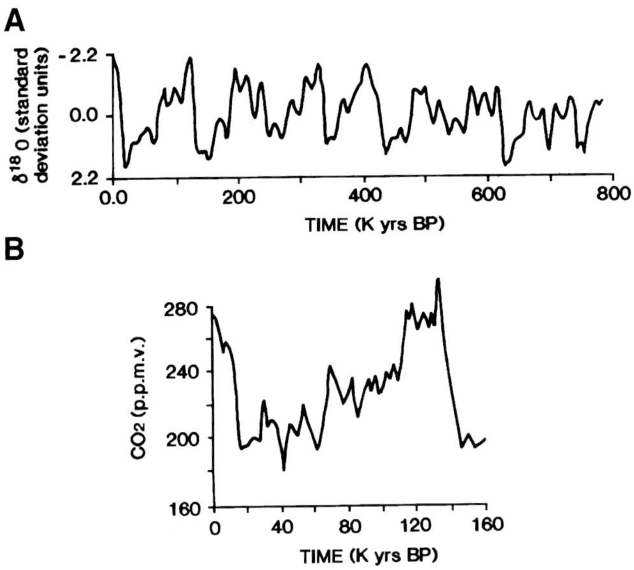

Figure 16. Records of climatic fluctuations. (A) After Imbrie et al. (1984). (B) After Barnola et al. (1987).

during the mid- to late Pleistocene (Figure 16A, Imbrie et al., 1984; Farrell and Prell, 1989). Periods of relative stability (or more accurately, comparatively slower rates of sea level change) occurred during interglacial sea level highstands, such as the last 5000 years (Lighty et al., 1982), interstadial highstands (Chappell and Veeh, 1978), and glacial lowstands (Imbrie et al., 1984). Sea level fell in stages following interglacial highstands, with minor stillstands and interstadial highstand reversals (Chappell and Veeh, 1978; Imbrie et al., 1984; Barnola et al., 1987; Humphrey and Kimbell, 1990; Richards et al., 1994), but rose precipitously following glacial lowstands (Figure 16B, Imbrie, 1984; Barnola et al., 1987).

The bathymetric contours included in Figure 1 show GBB to be very steep sided, flat topped, and, at present, mostly covered by shallow marine water. Because of the flat-topped nature of the bank, broad areas of the bank surface became abruptly exposed during sea level falls, but resubmerged only near the end of ensuing sea level rises. During high sea levels, the bank, like today, was mostly submerged with only localized islands and shoals subaerially exposed (Figures 1 and 15A). During glacial maxima, sea level stood several tens of meters to greater than a 100 m below the top of the bank (Curray and Shepard, 1972; Gascoyne et al., 1979; James and Ginsburg, 1979; Mylroie and Carew, 1988; Humphrey and Kimbell, 1990), and GBB became a vast subaerially exposed limestone plateau (Figures 15C and 17). Today, a 10 m fall in sea level would subaerially expose over 95% of the bank top (Figures 1 and 15B). With a drop of 24 m, the Cat Island platform becomes a peninsular extension of the bank. During the late Pleistocene, periods of bank submergence were on the order of 10,000 years; bank emergence, 100,000 years (Figure 15, Imbrie et al., 1984; Barnola et al., 1987; Chen et al., 1991).

Also depicted in Figure 15 is the extent of various diagenetic environments across the main body of GBB at different sea levels. Lacking regional subsurface aquitards, and assuming reasonable meteoric recharge, groundwater lenses should have migrated vertically in step with sea level. During interglacial highstands, with most of the bank submerged, diagenetic environments were similar to those found today. In submerged areas, new carbonate sediments accumulated, and all sediments, both new and preexisting, were bathed in marine phreatic waters. Only on islands, generally located along windward margins and of restricted size, were sediments subaerially exposed and subjected to meteoric diagenesis. Like today, Ghyben-Hertzberg groundwater lenses were generally thin and of limited areal extent (Little et al., 1977; Cant and Weech, 1986; Vacher, 1988; Wallis and Vacher, 1990, in Vacher and Mylroie, 1991; Budd and Vacher, 1991). Current lens thicknesses are generally between 5 and 15 m, the thickest being 34.1 m on North Andros, with underlying mixing zones of 1 to 10 m (Cant and Weech, 1986; Whitaker and Smart, in press). Most sediments situated above sea level resided in a vadose environment. In submergent regions, marine diagenesis occurred mostly at or near the sediment surface.

As the surface of the bank became subaerially exposed in response to falling sea level, dramatically different hydrologic conditions prevailed. A bank-wide freshwater lens probably developed exceeding in thickness the 34 m currently found on Andros Island. The vertical thickness of the freshwater lens would have varied through time based upon rainfall, vertical permeability of the subsurface rocks, and position on the bank (Cant and Weech, 1986; Budd and Vacher, 1991; Mylroie and Balcerzak, 1992). The diagenetic impact of this lens and underlying mixing zone would have affected given stratigraphic intervals or locations differently based upon rates of sea level change, position relative to upper and lower boundaries of the lens and mixing zone, location relative to bank margins, depth, and preceding diagenetic history (Choquette and James, 1988). Lens development was probably much more extensive on the main body of GBB than on the more isolated Cat Island platform.

Stages of Cementation and Porosity Development

Based upon the observed trends of cementation and porosity development, three diagenetic stages are proposed for Pliocene–Pleistocene depositional units on GBB. Schematically shown in Figure 18, these include: Stage I—initial exposure of metastable marine sediments to chiefly vadose meteoric water (restricted to uppermost depositional unit); Stage II—shallow burial with intermittent subjection to meteoric phreatic conditions in vertically migrating freshwater lenses (surface to variably 12 to 20 m); and Stage III—deeper burial with periodic exposure to caustic phreatic and mixing-zone conditions (to depths of 150 to 200 m). With the exception of subaerially exposed Holocene and some uppermost Pleistocene sediments (uppermost depositional unit in OJ-1), all subsurface intervals represented by cores in this study currently reside in either Stage II or III.

Figure 17. Sea cliff along the north coast of Mona Island located in the Mona Passage between Puerto Rico and Hispaniola. Cliff rises from about 20 m below sea level to between 30 and 45 m above sea level. During Pleistocene glacial periods, with resultant lowered sea level, the Bahama Banks undoubtedly had a similar appearance. The dark ribbon at the base of the cliff is a notch caused by the combined action of waves, boring animals, and possibly a freshwater-marine mixing zone (Mylroie and Carew, 1990). Undermining has produced large slides along this notch.

During Stage I, sea level dropped exposing metastable, predominately aragonitic sediments to meteoric diagenesis for the first time (Figures 15B and 18A). Typically, the drop was of sufficient rate and duration to rapidly place all sediments in the newly exposed unit within a vadose environment (Chen et al., 1991). The impact of any short-lived groundwater lens appears minimal. Ideally, rainwater initially percolated evenly downward through newly exposed sediment until encountering an impediment. This ideal situation would exist in vadose zone ooidal grainstone like the Holocene Joulters Cays in the Bahamas (Halley and Harris, 1979). However, most sediments are not homogeneous, and contain natural heterogeneities, such as bedding surfaces, burrows, or root traces, which result in preferential channeling of vadose waters. The efficiency of these channels to drain rainfall should have improved through time as connectivity or permeability improved (Palmer, 1991). The most effective channels preferentially enlarged at the expense of less effective conduits. Also, through time, due to alternate wetting and drying episodes, a calcrete hard pan or case hardening eventually developed. Its depth, thickness, and degree of induration were related to rates and seasonal patterns of rainfall, temperature, and evaporation (Harrison, 1977; Esteban and Klappa, 1983; James and Choquette, 1984). This hard pan acted to impede diffuse infiltration, concentrated meteoric recharge, and further enhanced preferential channeling. After development of this hard pan and an effective vertical-passage system for meteoric waters through the vadose zone, vadose cementation should effectively have ceased away from those vertical conduits open to the surface. At this point, the first diagenetic stage ended. Resulting rocks were cemented mostly by vadose cement, and the degree of lithification varied considerably both laterally and vertically. Because a hard

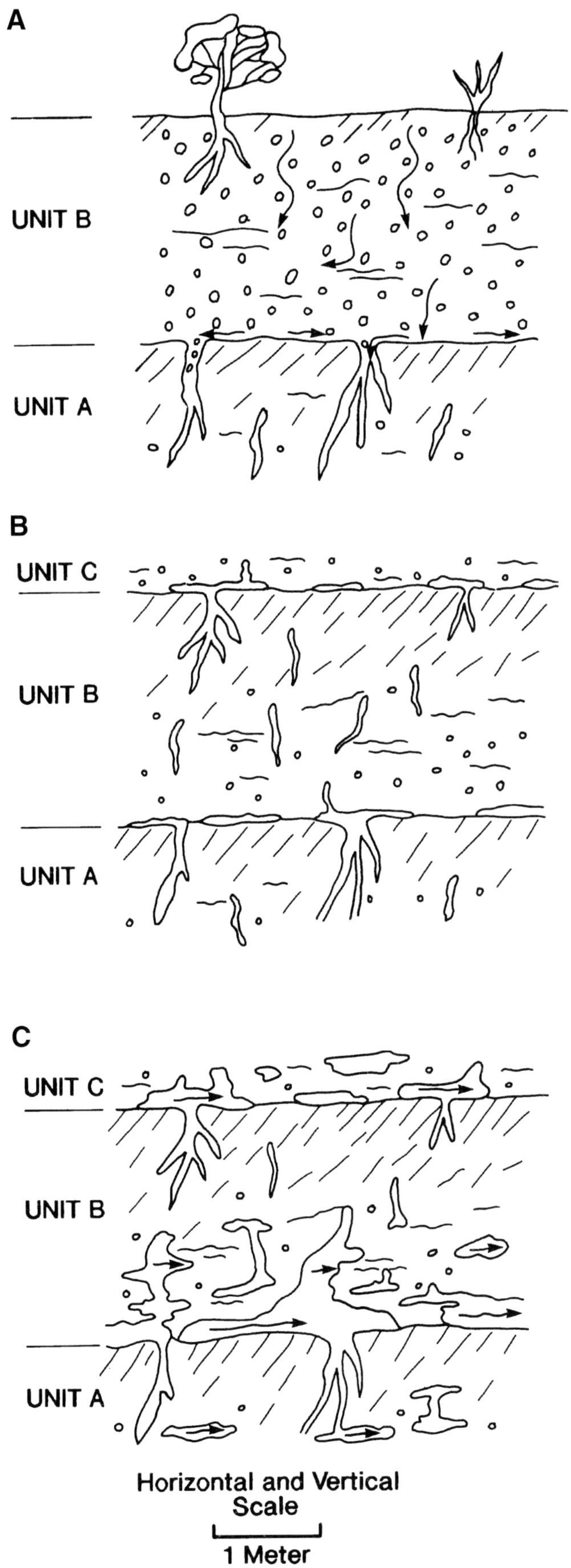

Figure 18. Three stages of cementation and porosity development recognized for depositional units in the shallow subsurface of GBB. These are related to changes in sea level, concomitant changes in diagenetic environments, and continuous bank subsidence. (A) Stage I—Dominated by vadose conditions, occurs with initial subaerial exposure of metastable marine sediments. It ends with development of surface case hardening and localized vertical solution pipes. (B) Stage II—Occurs with shallow burial (surface to 12–20 m) and is dominated by meteoric phreatic diagenesis (alteration of aragonite to low-Mg calcite, cementation, and inversion of porosity). (C) Stage III—With deeper burial (depths to 200 m) rocks are subjected to large-scale dissolution by corrosive waters near the top of the freshwater lens and within the meteoric and marine mixing zone during bank exposure. Arrows indicate direction of groundwater flow.

pan could develop rapidly, the duration of this stage may have been less than a few thousand years (Robbin and Stipp, 1979).

Most deposits of Sangamonian age seen in the Bahamas today have passed through this stage of diagenesis. In these rocks (represented by the uppermost depositional unit in Figures 2, 3, and 13), a well-indurated case hardening of equant spar and secondary micrite exists at the surface, whereas the deeper portion of the unit remains slightly cemented mostly by microspar, equant spar, and meniscus cements. Porosity remains high and retains a significant proportion of primary pore types (Figures 7A, 11, and 13). Poor recovery of this unit reflects its variable lithification (Figure 2). Only the very thick top unit in OJ-3 resides in Stage I. Here, the deepest portion of this unit remains uncemented with no evidence of vertical solution pipes which would have restricted diffuse flow during lower sea level. This unit is interpreted to be post-Sangamonian, having been deposited during a Wisconsin interstadial highstand (Beach, 1982).

A corollary development during Stage I was leaching of sediment at the base of freshly exposed units (Figures 2, 13, and 18A), with local precipitation of coarse sparry calcite within these leached zones (Figures 5B and 8B). Downward-infiltrating water eventually encountered the hard pan developed previously on the underlying depositional unit. Except for chance coincidence with preexisting solution channels, this hard pan formed an impermeable barrier or partial aquitard. Consequently, water ponded up and flowed laterally along this surface as a thin (millimeters to a few centimeters) perched water lens where leaching and precipitation occurred. Finally, this water entered older, preexisting vertical channels through which it continued its downward journey to the water table.

Stage II encompassed the time from the end of Stage I through shallow burial (approximately 12 m near margins to 20 m over the bank interior, Figure 18B). During Stage II, sediments completed mineralogical stabilization and inversion of porosity (Figure 11), and underwent a slight overall increase in cementation (Figure 14) in response to transitory subjection to marine phreatic, and meteoric vadose and phreatic diagenesis occurring through repeated fluctuations of sea level. Little alteration probably occurred during exposure to vadose and marine phreatic conditions. The freshwater phreatic environment, when in a position similar to that of Figure 15B, apparently had the most important lasting effect on these relatively young rocks: namely, precipitation of equant cements, and continuing development of secondary (especially moldic) porosity (Land, 1970; Steinen and Matthews, 1973; Steinen, 1974; Budd, 1988). Extensive large-scale dissolution, notable in its absence, was unimportant in these shallow phreatic zones. Coincidentally, due to short residence times, mixing zones apparently also had little effect during this stage. Thick bank-wide phreatic zones quickly disintegrated as rapidly rising sea level initially topped and flooded large areas of the bank during interglacial sea level highstands. This can be envisioned between Figures 15B and 15C where a rise of only a few meters drastically changes the hydrology of the bank. As a result, dissolution by large corrosive mixing zones operative at lower sea levels could not significantly affect rocks lying at less than about 12 m depth near bank margins, or 20 m over the interior (Figure 2).

Stage III diagenesis (Figure 18C) began where more deeply buried rocks (depths below 12 to 20 m subsea) were subjected to large-scale horizontally oriented dissolution, which led to the development of channels and caves. In spite of this large-scale dissolution, whole-core cementation and porosity generally remained constant (Figure 14). During falling sea level, the establishment of areally extensive groundwater lenses was accompanied by equally laterally extensive mixing zones of marine and fresh water (Figure 15B and C). The most intensive local dissolution probably occurred during stillstands (glacial lowstands and interstadial highstands) near the top of the water table (James and Choquette, 1984; Mylroie and Carew, 1988) or, arguably more importantly, within the mixing zone (Hanshaw and Back, 1980; James and Choquette, 1984; Back et al., 1986; Smart et al., 1988a; Bottrell et al., 1991; Canter and Humphrey, 1994). The effects of dissolution were cumulative through time and repeated sea level fluctuations. Therefore, greater dissolution should be expected with greater depth (Whitaker and Smart, in press). Because of the range of glacial-related lowstand sea levels (10 to at least 120 m), thick stratigraphic sections could be affected by Stage III dissolution. A 120 m lowstand with a 30 to 70 m thick freshwater lens and underlying mixing zone could impact rocks 150 to 200 m beneath the bank surface. Within the thick overlying vadose interval, diagenetic alteration was restricted mostly to enlargement of vertical channels and/or creation of sinks and sink holes through dissolution and collapse. Stage III diagenesis effectively ended once strata had subsided to a depth beneath which they could not be immersed within lowstand mixing zones.

DISCUSSION

Large-Scale Dissolution in the Subsurface

Large-scale dissolution of shallow subsurface carbonate rocks occurred primarily during Stage I, under vadose conditions, and Stage III, under phreatic/mixing-zone conditions (Figure 19). Dissolution during Stage I was mostly vertical and controlled largely by burrowing and bedding trends inherited from sediment deposition, and to a lesser extent from root molds (Smart and Whitaker, 1988). Evidence for its early development includes its common observation in the youngest depositional units (Figure 10A).

The most significant large-scale dissolution occurred in the deeper subsurface and is attributed to freshwater phreatic and mixing-zone conditions developed during sea level lowstands. The key to determining the importance of this effect lies in establishing that zones of poor core recovery result primarily from dissolution, and not from poor lithification or loss of rock during coring operations. Core recovery can vary as a result of differences in drilling techniques (different drilling rigs or drillers, variable weight on the drill bit, different length core barrels, etc.), lithology (especially rock fabric and induration), and subsurface dissolution. Poor recovery in the uppermost cores of C-71, C-72, C-73, and LO-39 reflects limited induration of Sangamonian age deposits (Figure 2). Likewise, the top unit of OJ-3 (post-Sangamonian) was also poorly lithified between 10 and 17 m. In contrast, at most locations core recovery between the base of the uppermost unit and depths of 12 to 20 m subsea was generally good with rocks moderately well cemented. In this intermediate level, as noted previously, intervals of poor recovery mostly overlie and are inferred to result from localized dissolution above case-hardened unconformities.

Beneath these depths, recovery was often poor (Figure 2). This poor recovery, based both on the nature of recovered core adjacent to zones of no recovery and on the pattern of recovery, also resulted largely from extensive dissolution. Throughout this interval recovered core is typically moderately well cemented by equant spar, and is generally similar to core from above 20 m. Core-derived porosity remains on the average 35% or less (Figure 14). Commonly, however, intervals of recovered core also evidence extensive dissolution similar to that described by Back et al. (1986) and Smart et al. (1988a). Features include extensive leaching of massive corals (Figure 10C), "Swiss cheese" texture especially in burrowed wackestone and mudstone (Figure 12B, C), and sponge-like vuggy texture. Coring technique does influence recovery. Recovery of fragile, brittle rock containing extensive dissolution features is better from larger diameter (10 cm rather than 5 cm) but shorter (1.5 m rather than 3 m) core barrels.

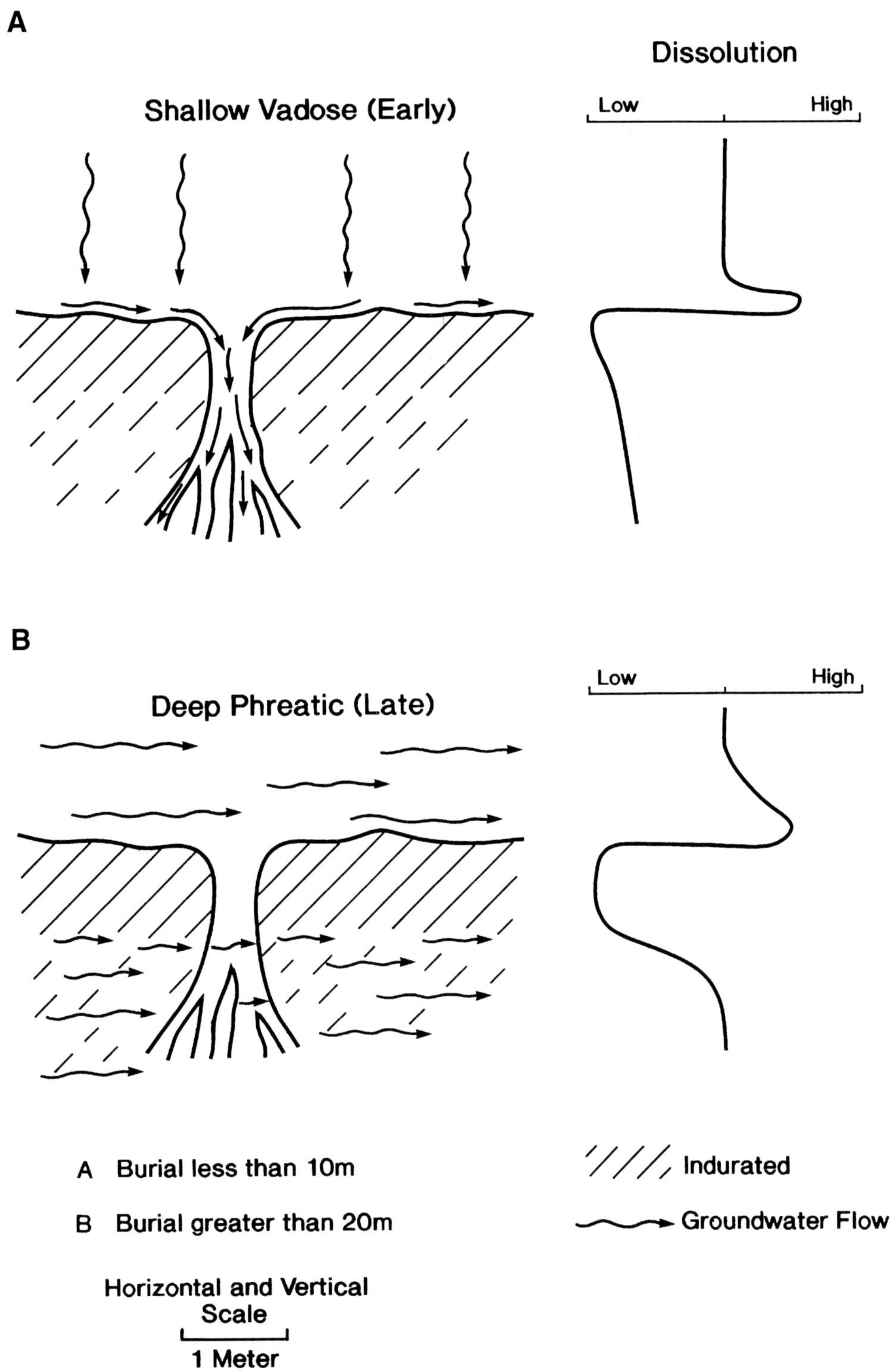

Figure 19. Selective preservation of subaerial unconformities in the subsurface of GBB resulting from exposure-related induration and subsequent control of groundwater flow. Diagram A shows effect under vadose conditions; diagram B shows effect within corrosive phreatic/mixing-zone environment. Schematic profile to the right depicts relative dissolution occurring across unconformities within the two settings.

The distribution of dissolution features and zones of poor recovery reveals that extensive large-scale dissolution did not occur equally in all locations or intervals of cores. This is demonstrated by the three graphs in Figure 20. Figure 20A compares average limestone core recovery from NWGBB and Long Island to both Cat Island and the dolomitized portion of U-1. Figure 20B compares average core recovery from Cat Island to similar depths from cores U-1, AN-5, LO-12, and LO-39, locations all within 13 km of windward bank margins. Figure 20C compares average recovery from Cat Island with similar depths of the five AN and LO series cores. These cores were all drilled using basically the same coring techniques. Cores most extensively affected by large-scale dissolution are from NWGBB and Long Island, locations on the main body of GBB. All large corrosive groundwater lenses developed across the broad expanse of

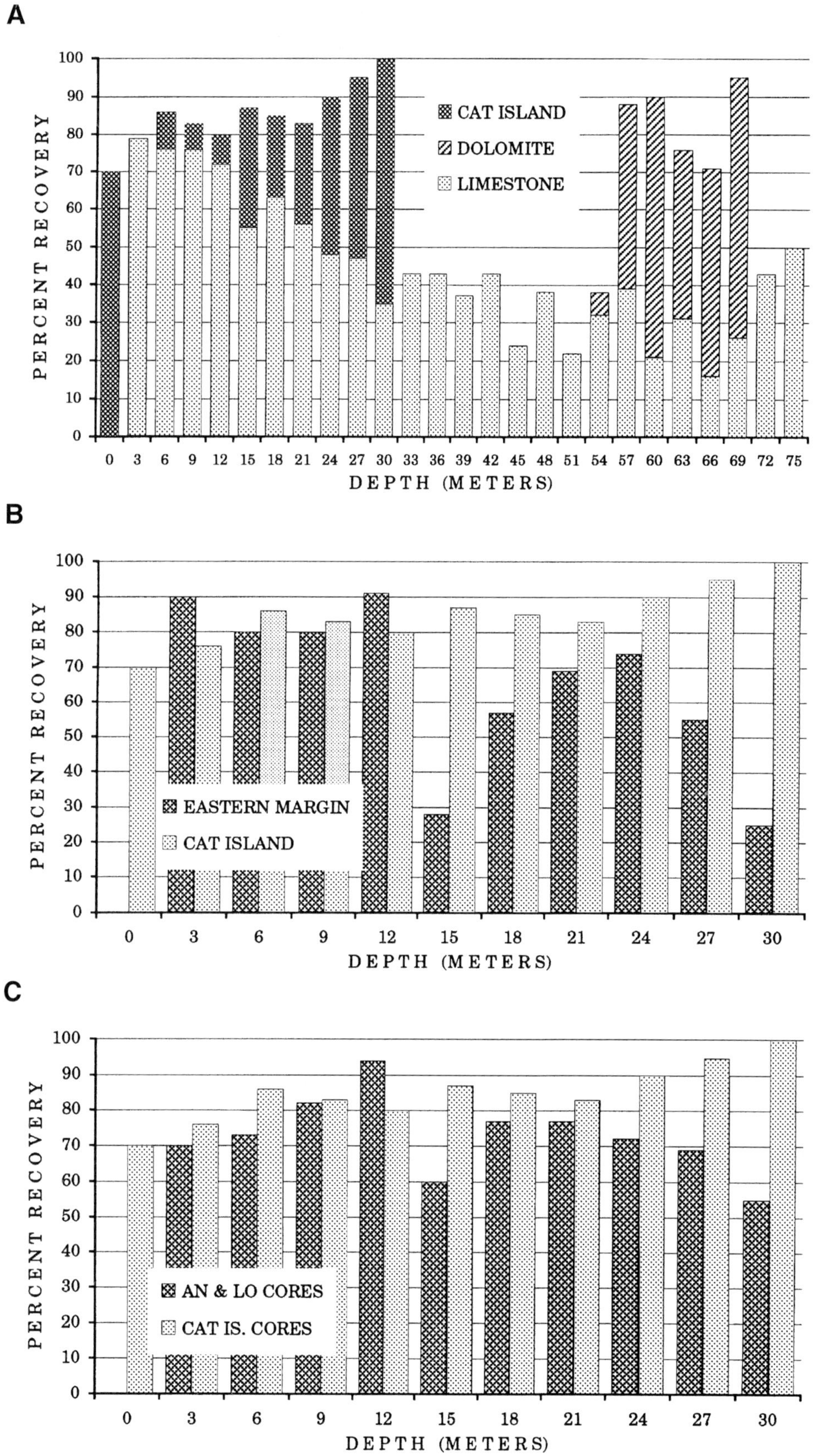

Figure 20. Graphs comparing average percent core recovery over 3 m intervals to depth for different locations and lithologies. (A) Compares all limestone cores from the main body of GBB (Long Island and NWGBB) with the Cat Island cores and the dolomite portion of U-1. (B) Compares the Cat Island cores to those from Long Island and NWGBB within 13 km of the windward bank margin. (C) Compares the Cat Island cores with all AN and LO series cores. These cores were acquired using the same coring techniques.

the bank during lowered sea levels affected these locations. In contrast, cores from Cat Island show less significant dissolution (Figures 2 and 20B). During periods of lowered sea level, the Cat Island platform remained isolated as a long, narrow, peninsular extension of the bank (Figure 1). As elsewhere on the bank, sufficient effective permeability existed in deeper, more lithified rocks (Little et al., 1977; Cant and Weech, 1986), so that because of its narrow width, ready exchange between fresh and ocean water occurred within short distances on three sides. As a result, it appears that comparatively thin freshwater lenses and mixing zones developed within this portion of GBB.

Core U-1 provides further evidence of significant though variable dissolution under the main mass of GBB. In this core, poor recovery occurs in more soluble limestone intervals between 12 and 51 m subsea (Figure 2). Evidence of extensive leaching is abundant in recovered core from this interval. In contrast, in deeper, less soluble dolomite intervals, evidence of leaching is much less apparent and recovery is markedly improved (Figure 20A). This same observation was reported from the cored hole on New Providence in 1932 by Field and Hess (1933), "Alternating beds of slightly cemented calcareous sand and very porous cavernous limestone were found until dolomite was reached at 160± ft [48.8 m]." Thus, zones of poor recovery, with or without significant bit drops, generally correlate with intervals of extensive dissolution.

The potential importance of Stage III–style dissolution in the Bahamas is underscored in papers by Agassiz (1894), Doran (1955), Newell and Rigby (1957), Benjamin (1970), Dill (1977), Gascoyne et al. (1979), Smart et al. (1988b), and Whitaker and Smart (1993) describing the thickness and extent of cavernous systems across the various platforms of the Bahamas. Figures 2 and 21 and the discussion by Newell and Rigby (1957, their p. 28 and pl. 17) show that extensive dissolution and cavernous systems were certainly not simply restricted to the edges of the bank, but extended across the platform. The overall depth of extensive dissolution on GBB is considerable. Spencer (1967) noted that during drilling of the Bahamas Oil Company, Ltd., Andros No. 1 (Superior Oil Co.) deep test, located on North Andros Island approximately 12 km from the bank margin, lost circulation caused by an extremely porous section was a problem above 1280 m depth. Drillers' logs from other wells in the Bahamas indicate bit drops to depths of 3000 m (Whitaker and Smart, 1990).

Gauged from reported core recovery, dissolution on other platforms in the Bahamas varies considerably. As in the difference between Cat Island and the rest of GBB, development of groundwater lenses and mixing zones was influenced by platform size, climate, and lithology (Cant and Weech, 1986). Core recovery from limestone sections of banks in the southeastern Bahamas is generally good (Supko, 1970; Pierson, 1982). These platforms are considerably smaller, and the present-day climate is drier than GBB. Core recovery on Little Bahama Bank shows a pattern more similar to that of GBB, with generally good recovery above 10 to 12 m along the margin, and above 18 m in the interior (Williams, 1985). Dolomite occurs across the bank at shallow depths, and core recovery is significantly improved in this deeper section.

Stage III dissolution may also have been important in other relatively pure carbonate platforms in the geologic past. Similar extensive dissolution has been reported at depths over 1000 m across the slowly subsiding South Florida platform (Kohout, 1967). Thick intervals of extensive leaching were reported by Lomando et al. (1993) in the Tarragona Basin offshore Spain. Craig (1988) describes thick intervals of extensive large-scale dissolution in Permian San Andres carbonates on the Central Basin platform of west Texas. Similar thick, large-scale dissolution is reported from the Knox Group of the eastern United States by Mussman et al. (1988) and Wilson and Medlock

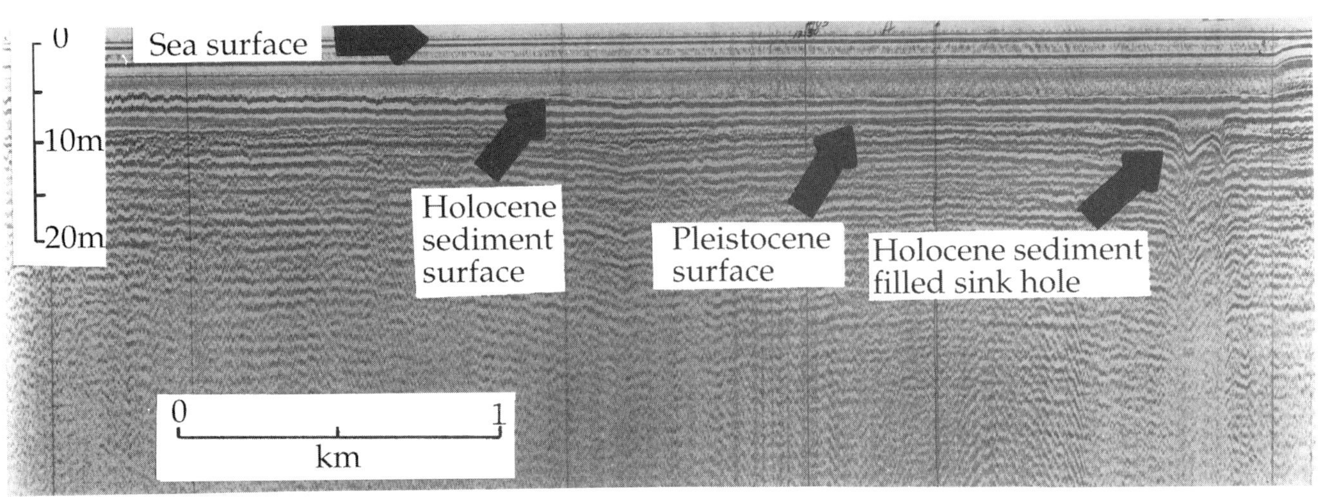

Figure 21. Boomer seismic display of Holocene sediment-filled sink from the middle of NWGBB near location ABM. Karst is common over the entire surface of GBB.

(1993). Mussman et al. suggest that at least a portion of the dissolution they observed developed in a meteoric-marine mixing zone. In these and similar settings, development of areally extensive freshwater phreatic lenses and mixing zones during periods of relative sea level fall and subaerial exposure could have led to similar large-scale Stage III dissolution. Within tectonically active platforms (either by slow uplift, or subsidence with alternating deposition and intermittent subaerial exposure), this dissolution could through time modify thick stratigraphic sections. Because calcite and aragonite are susceptible to dissolution not only along the water table but also within the mixing zone (Back et al., 1986; Smart et al., 1988a), and because the rocks affected on GBB had already been altered to low-Mg calcite, Stage III dissolution could have occurred throughout the geologic past.

Preservation of Unconformities

One attribute commonly associated with modern subaerial exposure surfaces is a high degree of induration or case hardening relative to underlying sediments (James, 1972; Yaalon and Singer, 1974; Harrison, 1977). Case hardening is common across subaerially exposed Pleistocene limestone in the Bahamas today (Illing, 1954). It was produced by rapid and pervasive cementation and concomitant loss of porosity in the uppermost portions of freshly exposed sediment. Because of similar differential induration, rocks directly underlying unconformities were preferentially preserved in the rock record of GBB (Figures 2, 13, and 19). Core recovery from the deeper intervals subjected to Stage III dissolution particularly reflects this selective preservation. Well-indurated rocks underlying unconformities are preserved along with other sediments in the shallower intervals not affected by Stage III diagenesis. Except for leaching of less cemented rocks overlying them (Figure 19A), and enlargement of local vertical solution pipes extending through them, indurated intervals underlying subaerial unconformities are not greatly impacted by dissolution occurring during Stage I or II. In contrast, corrosive waters in bank-wide phreatic and mixing zones of Stage III do constitute a threat to preservation of subaerial exposure horizons, as to all carbonates, during sea level lowstands. This threat is mitigated, however, because of their greater induration relative to adjacent, less cemented intervals (Figure 19B). Corrosive horizontal fluid flow opportunistically occurs through overlying and underlying rocks having higher initial permeability (Wright, 1991). With time, the flow capacity through these adjacent intervals increases as leaching occurs and creates more effective horizontal conduits (Palmer, 1991). The result is preferential preservation of indurated, subaerial, unconformity zones in the rock record. Because induration beneath unconformities observed on GBB results mostly from intense localized alteration of aragonite-rich sediments, this degree of selective preservation may pertain mostly to "aragonite sea" intervals (Sandberg, 1983).

Preservation with Deeper Burial

Given significantly deeper burial, what could be expected of the Pliocene–Pleistocene record described for GBB? Because no ancient analogs are known where deeply buried rocks preserve the degree of open porous and permeable rock observed in the shallow subsurface record of GBB (Saller et al., 1994), it is doubtful that this record would survive deeper burial intact. It is likely that most of the open vug, channel, and cavernous system will compact or collapse, but leave a carbonate platform showing evidence of widespread subaerial exposure and a preexisting bank-wide cavernous system. The Ordovician Ellenburger Group of west Texas and New Mexico may represent an ancient example of deeper burial and collapse of a similar carbonate platform (Kerans, 1988; Loucks and Anderson, 1988; Loucks and Handford, 1992; Kerans, 1993).

CONCLUSIONS

1. Subsurface Pliocene–Pleistocene carbonate rocks on GBB reflect a complex diagenetic history occurring during slow bank subsidence and frequent high amplitude changes of glacio-eustatic sea level. During sea level highstands, aragonite-rich sediments accumulated on the bank. Subaerial unconformities formed with ensuing sea level falls. Freshwater vadose and phreatic, mixed freshwater and marine, and marine phreatic diagenetic environments migrated upward and downward with sea level, affecting sediments and rocks differently through time depending upon their mineralogical stability, fabric, location, and preceding diagenetic history.

2. Three progressive stages or periods of diagenesis are recognized for the Pliocene–Pleistocene rocks of GBB. Stage I was dominated mostly by vadose diagenesis occurring from the initial subaerial exposure of a sedimentary unit through development of an indurated surface pierced locally by vertical solution pipes. In Stage II, during shallow burial to depths of 12 to 20 m, sediments completed both alteration to low-Mg calcite and inversion from primary interparticle and intraparticle to secondary moldic porosity under ephemeral freshwater phreatic conditions. Relatively uniform cementation by equant calcite occurred. Stage III, marked by extensive dissolution, took place deeper in the subsurface (to depths of 200 m) where rocks were exposed to prolonged episodes of corrosive bank-wide freshwater phreatic and mixing-zone conditions during bank emergence.

3. Porosity profiles reflect the marked induration occurring across unconformities. Average core porosity within sedimentary units is 30 to 40%. Immediately above unconformities a thin, centimeters thick, heavily leached zone is common, giving way to a sharp decrease in porosity (10 to 15%) at the unconformity. At depths of 15 to 75 cm below unconformities, porosity gradually returns to between 30 and 40%. Similar profiles may have formed across buried

unconformities developed on aragonitic sediments in other regions and at other times.

4. Upon burial, well-indurated, case-hardened intervals developed immediately beneath subaerial unconformities and formed effective controls over subsequent movement of groundwater. In the vadose zone, they formed partial aquitards above which groundwater would locally pond and move laterally. Localized, thin, perched water tables resulted with accompanying leaching and cementation. In phreatic zones, water preferentially coursed laterally through more permeable beds above or below these indurated intervals. Subaerial unconformities may have exerted similar controls within the subsurface in other areas, particularly during "aragonite sea" geologic periods.

5. The Pliocene–Pleistocene record from GBB chronicles regional large-scale dissolution of moderately cemented low-Mg calcite limestone. Corrosion is most pervasive over the main body of GBB below depths ranging from about 12 m near bank margins to 20 m over the interior. More isolated locations and dolomitized intervals were less significantly affected.

6. The depth and pattern of corrosion on GBB testifies to the development of bank-wide phreatic lenses during glacial sea level lowstands. Large lenses developed only across the main body of GBB. Lenses were thinner and less extensive on isolated banks and peninsulas, such as the Cat Island platform.

7. Because of their well-indurated characteristics, subaerial unconformities are preferentially preserved in the Pliocene–Pleistocene rock record of GBB.

ACKNOWLEDGMENTS

This paper represents a portion of my Ph.D. dissertation from the University of Miami, Rosenstiel School of Marine and Atmospheric Science. My committee included Robert N. Ginsburg (chairman), Wolfgang Schlager, Harold R. Wanless, Philip W. Choquette, and Robert B. Halley. The research was supported by the National Science Foundation (Grant OCE 76-21932) to Robert N. Ginsburg and by members of the Cooperative Research in Comparative Sedimentology (Amoco Production Co., Chevron Oil Field Research Co., Cities Service Oil Co., Continental Oil Co., Exxon Production Research Co., Getty Oil Co., Gulf Energy and Minerals, Marathon Oil Co., Pennzoil Co., Phillips Petroleum Co., Shell Development Co., and Union Oil Co. of California). Ten cores described for this paper were provided by Richard Cant and the Ministry of Works and Utilities of the Bahamas; and three by Union Oil Co. of California, acquired under the direction of Perry O. Roehl. Three cores were also obtained with the help of George Sauter and the crew of the Humble CT-1, and Robert N. Ginsburg. Slabbing and photographing of cores and GRAPE logs were provided by Jim Miller and the Union Oil Research Center. Thin sections were prepared by Union Oil Co. of California and Marathon Oil Co. research centers. I extend special thanks to Philip W. Choquette for help in core descriptions, and to him and Robert N. Ginsburg for stimulating discussions and helpful suggestions. David A. Budd, Harris Cander, Karen L. Canter, and Fiona F. Whitaker provided critical reviews of this paper. Thanks also to Angela M. Miller and Brendan Golden for drafting of figures used in this text.

REFERENCES CITED

Agassiz, A., 1894, A reconnaissance of the Bahamas and of the elevated reefs of Cuba: Museum of Comparative Zoology Bulletin 26, p. 1–203.

Ahmad, H., and R.L. Jones, 1969, Occurrence of luminous lateritic soils (bauxites) in Bahamas and Cayman Islands: Economic Geology, v. 64, p. 804–808.

Back, W., B.B. Hanshaw, J.S. Herman, and J.N. Van Driel, 1986, Differential dissolution of a Pleistocene reef in the ground-water mixing zone of coastal Yucatan, Mexico: Geology, v. 14, p. 137–140.

Bain, R.J., and A.M. Foos, 1993, Carbonate microfabrics related to subaerial exposure and paleosol formation, *in* R. Rezak and D.L. Lavoie, eds., Carbonate Microfabrics: New York, Springer-Verlag, p. 19–27.

Barnola, J.M., D. Raynaud, Y.S. Korotkevich, and C. Lorius, 1987, Vostok ice core provides 160,000-year record of atmospheric CO_2: Nature, v. 329, p. 408–414.

Beach, D.K., 1982, Depositional and Diagenetic History of Pliocene–Pleistocene Carbonates of Northwestern Great Bahama Bank; Evolution of a Carbonate Platform: Ph.D. dissertation, University of Miami, Coral Gables, Florida, v. 1, 447 p.

Beach, D. K., 1993, Submarine cementation of subsurface Pliocene carbonates from the interior of Great Bahama Bank: Journal of Sedimentary Petrology, v. 63, p. 1059–1069.

Beach, D.K., and R.N. Ginsburg, 1980, Facies succession of Pliocene–Pleistocene carbonates, northwestern Great Bahama Bank: AAPG Bulletin, v. 64, p. 1634–1642.

Beach, D.K., and R.N. Ginsburg, 1994, Recognition of subaerial unconformities in cores: NW Great Bahama Bank (abs.): AAPG 1994 Annual Convention, p. 101.

Beier, J.A., 1987, Petrographic and geochemical analysis of caliche profiles in a Bahamian Pleistocene dune: Sedimentology, v. 34, p. 991–998.

Benjamin, G. J., 1970, Blue Holes of the Bahamas: National Geographic Magazine, v. 138, p. 346–363.

Bottrell, S.H., P.L. Smart, F. Whitaker, and R. Raiswell, 1991, Geochemistry and isotope systematics of sulphur in the mixing zone of Bahamian blue holes: Applied Geochemistry, v. 6, p. 97–103.

Budd, D.A., 1988, Aragonite-to-calcite transformation during fresh-water diagenesis of carbonates: insights from pore-water chemistry: Geological Society of America Bulletin, v. 100, p. 1260–1270.

Budd, D.A., and H.L. Vacher, 1991, Predicting the thickness of fresh-water lenses in carbonate paleo-islands: Journal of Sedimentary Petrology, v. 61, p. 43–53.

Cant, R. V., 1977, Role of coral deposits in building the margins of the Bahama Banks, *in* D.L. Taylor, ed., Proceedings Third International Coral Reef Symposium, v. 2, Geology, Miami Florida, p. 9–13.

Cant, R.V., and P.S. Weech, 1986, A review of the factors affecting the development of Ghyben-Hertzberg lenses in the Bahamas: Journal of Hydrology, v. 84, p. 333–343.

Canter, K.L., and J.D. Humphrey, 1994, Carbonate dissolution within the meteoric and mixing zone diagenetic environments: porosity development within late Pleistocene reef and reef-associated lithologies, southeastern Barbados (abs.): AAPG 1994 Annual Convention, p. 115.

Carew, J.L., and J.E. Mylroie, 1985, The Pleistocene and Holocene stratigraphy of San Salvador Island, Bahamas, with reference to marine and terrestrial lithofacies at French Bay, *in* H.A. Curran, ed., Pleistocene and Holocene Environments on San Salvador Island, Bahamas: Geological Society of America Field Guidebook, p. 11–61.

Carew, J.L., and J.E. Mylroie, 1991, Some pitfalls in paleosol interpretation in carbonate sequences: Carbonates and Evaporites, v. 6, p. 69–74.

Chappell, J., and H.H. Veeh, 1978, Late Quaternary tectonic movements and sea-level changes at Timor and Atauro Island: Geological Society of America Bulletin, v. 89, p. 356–368.

Chen, J.H., H.A. Curran, B. White, and G.J. Wasserburg, 1991, Precise chronology of the last interglacial period: ^{234}U–^{230}Th data from fossil coral reefs in the Bahamas: Geological Society of America Bulletin, v. 103, p. 82–97.

Choquette, P.W., N.P. James, 1988, Introduction, *in* N.P. James and P.W. Choquette, eds., Paleokarst: New York, Springer Verlag, p. 1–57.

Choquette, P.W., and L.C. Pray, 1970, Geologic nomenclature and classification of porosity in sedimentary carbonates: AAPG Bulletin, v. 54, p. 207–250.

Choquette, P. W., and F.C. Trusell, 1978, A procedure for making the Titan Yellow stain for Mg-calcite permanent: Journal of Sedimentary Petrology, v. 48, p. 639–641.

Craig, D.H., 1988, Caves and other features of Permian karst in San Andres dolomite, Yates field reservoir west Texas, *in* N.P. James and P.W. Choquette, eds., Paleokarst: New York, Springer Verlag, p. 342–363.

Curray, J.R., and F.P. Shepard, 1972, Some major problems of Holocene sea levels (abs.): American Quaternary Association Second National Conference, p. 16–18.

Dill, R., 1977, The blue holes—geologically significant submerged sink holes and caves off British Honduras and Andros, Bahama Islands, *in* D.L. Taylor, ed., Proceedings of the Third International Coral Reef Symposium, v. 2, Geology, Miami, Florida, p. 237–242.

Doran, E., 1955, Landforms of the southeast Bahamas: University of Texas Publ. 5509, Austin, Texas, 38 p.

Dunham, R.J., 1971, Meniscus cement, *in* O.P. Bricker, ed., Carbonate Cements: Johns Hopkins Univ. Studies in Geology 19, p. 297–300.

Eberli, G.P, and R.N. Ginsburg, 1987, Segmentation and coalescence of Cenozoic carbonate platforms, northwestern Great Bahama Bank: Geology, v. 15, p. 75–79.

Eberli, G.P, and R.N. Ginsburg, 1989, Cenozoic progradation of northwestern Great Bahama Bank, a record of lateral platform growth and sea-level fluctuations, *in* P.D. Crevello, J.L. Wilson, J.F. Sarg, and J.F. Read, eds., Controls on Carbonate Platform and Basin Development: SEPM Special Publication 44, p. 339–351.

Enos, P., 1974, Surface sediment facies of the Florida-Bahama plateau: Geological Society of America Map 5, 2 p.

Esteban, M., and C.F. Klappa, 1983, Subaerial exposure environment, *in* P.A. Scholle, D.G. Bebout, and C.H. Moore, eds., Carbonate Depositional Environments: Tulsa, Oklahoma, AAPG, p. 1–54.

Evans, H. B., 1965, GRAPE: A device for continuous determination of material density and porosity: Transactions of Society of Professional Well Log Analysts, 6th Logging Symposium, Dallas, v. 2, p. b1–b25.

Farrell, J.W., and W.L. Prell, 1989, Climatic change and $CaCO_3$ preservation: an 800,000 year bathymetric reconstruction from the central equatorial Pacific Ocean: Paleoceanography, v. 4, p. 447–466.

Farrell, J.W., and W.L. Prell, 1991, Pacific $CaCO_3$ preservation and $\delta^{18}O$ since 4 Ma: Paleoceanic and paleoclimatic implications: Paleoceanography, v. 6, p. 485–498.

Field, R.M., and H.H. Hess, 1933, A bore hole in the Bahamas: American Geophysical Union, Transactions Annual Meeting, p. 234–235.

Friedman, G. M., 1959, Identification of carbonate minerals by staining methods: Journal of Sedimentary Petrology, v. 29, p. 87–97.

Garrett, P., and S.J. Gould, 1984, Geology of New Providence Island, Bahamas: Geological Society of America Bulletin, v. 95, p. 209–220.

Gascoyne, M., G.J. Benjamin, H.P. Schwarcz, and D.C. Ford, 1979, Sea-level lowering during the Illinoian glaciation: Evidence from a Bahama "blue hole": Science, v. 205, p. 806–808.

Halley, R. B., and D.K. Beach, 1979, Porosity preservation and early freshwater diagenesis of marine carbonate sands (abs.): AAPG Bulletin, v. 63, p. 460.

Halley, R.B., and P.M. Harris, 1979, Freshwater cementation of a 1000 year-old oolite: Journal of Sedimentary Petrology, v. 49, p. 969–988.

Hanshaw, B.B., and W. Back, 1980, Chemical mass-wasting of the northern Yucatan Peninsula by groundwater dissolution: Geology, v. 8, p. 222–224.

Harms, J. C., and P.W. Choquette, 1965, Geologic evaluation of a gamma-ray porosity device: Transactions of Society of Professional Well Log Analysts, 6th Logging Symposium, v. 2, p. b1–b37.

Harrison, R. S., 1977, Caliche profiles: indicators of near-surface subaerial diagenesis Barbados, West Indies: Bulletin of Canadian Petroleum Geology, v. 25, p. 123–173.

Humphrey, J.D., and T.N. Kimbell, 1990, Sedimentology and sequence stratigraphy of upper Pleistocene carbonates of southeastern Barbados, West Indies: AAPG Bulletin, v. 74, p. 1671–1684.

Illing, L. M., 1954, Bahaman calcareous sands: AAPG Bulletin, v. 38, p. 1–95.

Imbrie, J., J.D. Hays, D.G. Martinson, A. McIntyre, A.C. Mix, J.J. Morley, N.G. Pisias, W.L. Prell, and N.J. Shackleton, 1984, The orbital theory of Pleistocene climate: support from a revised chronology of the marine $\delta^{18}O$ Record, *in* A.L. Berger, et al., eds., Milankovitch and Climate, Part 1, p. 269–305.

James, N.P., 1972, Holocene and Pleistocene calcareous crust (caliche) profiles: criteria for subaerial exposure: Journal of Sedimentary Petrology, v. 42, p. 817–836.

James, N.P., and P.W. Choquette, 1984, Diagenesis 9: limestones—the meteoric diagenetic environment: Geoscience Canada, v. 11, p. 161–194.

James, N.P., and R.N. Ginsburg, 1979, The seaward margin of Belize barrier and atoll reefs: International Association of Sedimentologists Special Publication 3, p. 191.

Kahle, C. F., 1977, Origin of subaerial Holocene calcareous crusts: role of algae, fungi and sparmicritisation: Sedimentology, v. 24, p. 413–435.

Kaldi, J., and J. Gidman, 1982, Early diagenetic dolomite cements: examples from the Permian Lower Magnesian Limestone of England and the Pleistocene carbonates of the Bahamas: Journal of Sedimentary Petrology, v. 52, p. 1073–1085.

Kerans, C., 1988, Karst-controlled reservoir heterogeneity in Ellenburger Group carbonates of west Texas: AAPG Bulletin, v. 72, p. 1160–1183.

Kerans, C., 1993, Description and interpretation of karst-related breccia fabrics, Ellenburger Group, west Texas, *in* R.D. Fritz, J.L. Wilson and D.A. Yurewicz, eds., Paleokarst Related Hydrocarbon Reservoirs: SEPM Core Workshop 18, Society for Sedimentary Geology, New Orleans, p. 181–200.

Kohout, F.A., 1967, Ground water flow and the geothermal regime of the Florida plateau: Transactions Gulf Coast Association of Geological Societies, v. 17, p. 339–354.

Kornicker, L.S., 1958, Bahamian limestone crusts: Transactions Gulf Coast Association of Geological Societies, v. 8, p. 167–170.

Land, L. S., 1970, Phreatic versus vadose meteoric diagenesis of limestones: evidence from a fossil water table: Sedimentology, v. 14, p. 175–185.

Lighty, R.G., I.G. Macintyre, and R. Stuckenrath, 1982, *Acropora palmata* reef framework: a reliable indicator of sea level in the western Atlantic for the past 10,000 years: Coral Reefs, v. 1, p. 125–130.

Little, B.G., D.K. Buckley, R. Cant, P.W.T. Henry, A. Jefferiss, J.D. Mather, J. Stark, and R.N. Young, 1977, Land resources of the Bahamas: a summary: Land Resource Study 27, Land Resources Division, Surrey, England, Ministry of Overseas Development, p. 133.

Lomando, A.J., P.M. Harris, and D.E. Orlopp, 1993, Casablanca field, Tarragona Basin, offshore, Spain, a karsted carbonate reservoir, *in* R.D. Fritz, J.L. Wilson and D.A. Yurewicz, eds., Paleokarst Related Hydrocarbon Reservoirs: SEPM Core Workshop 18, Society of Sedimentary Geology, New Orleans, p. 201–225.

Loucks, R.G., and J.H. Anderson, 1988, Depositional facies, diagenetic terranes, and porosity development in Lower Ordovician Ellenburger Dolomite, Puckett field, west Texas, *in* P.O. Roehl, and P.W. Choquette, eds., Carbonate Petroleum Reservoirs: New York, Springer-Verlag, p. 1–19.

Loucks, R.G., and C.R. Handford, 1992, Origin and recognition of fractures, breccias, and sediment fills in paleocave reservoir networks, *in* M.P. Candelaria, and C.L. Reed, eds., Paleokarst, Paleokarst Related Diagenesis and Reservoir Development: Examples from Ordovician–Devonian Age Strata of West Texas and the Mid-Continent: West Texas Geological Society Publication 92–33, p. 31–44.

McNeill, D.F., R.N. Ginsburg, S.-B.R. Chang, and J.L. Kirschvink, 1988, Magnetostratigraphic dating of shallow-water carbonates from San Salvador, Bahamas: Geology, v. 16, p. 8–12.

Melim, L.A., G.P. Eberli, and Swart, P.K., 1994, The correlation between sequence stratigraphy and diagenesis in Quaternary to Neogene carbonate sediments, subsurface Great Bahama Bank (abs.): AAPG 1994 Annual Convention, p. 214.

Multer, H.G., and J.E. Hoffmeister, 1968, Subaerial laminated crusts of the Florida Keys: Geological Society of America Bulletin, v. 79, p. 183–192.

Mussman, W.J., I.P. Montanez, F.J. Read, 1988, Ordovician Knox paleokarst unconformity, Appalachians, *in* N.P. James and P.W. Choquette, eds., Paleokarst: New York, Springer Verlag, p. 211–228.

Mylroie, J.E., and J.L. Carew, 1988, Solution conduits as indicators of late Quaternary sea level position: Quaternary Science Reviews, v. 7, p. 55–64.

Mylroie, J.E., and W.J. Balcerzak, 1992, Interaction of microbiology and karst processes in Quaternary carbonate island aquifers, *in* J.A. Stanford and J.J. Simons, eds., Proceedings of the First International Conference on Ground Water Ecology, U.S. Environmental Protection Agency, American Water Resources Association, and Ecological Society of America, p. 37–46.

Mylroie, J.E., and J.L. Carew, 1990, The flank margin model for dissolution cave development in carbonate platforms: Earth Surface Processes and Landforms, v. 15, p. 413–424.

Neuman, A.C., and W.S. Moore, 1975, Sea level events and Pleistocene coral ages in the northern Bahamas: Quaternary Research, v. 5, p. 215–224.

Newell, N.D., and K. Rigby, 1957, Geological studies on the Great Bahama Bank, *in* R.J. LeBlanc, and

J.G. Breeding, eds., Regional Aspects of Carbonate Deposition: SEPM Special Publication 5, p. 15–72.

Palmer, A.N., 1991, Origin and morphology of limestone caves: Geological Society of America Bulletin, v. 103, p. 1–21.

Paulus, F.J., 1972, The geology of site 98 and the Bahama Platform, *in* C.D. Hollister, J.I. Ewing, and others, Initial reports of the Deep Sea Drilling Project, v. 11, U.S. Govt. Printing Office, Washington, D. C., p. 877–897.

Perkins, R.D., 1977, Depositional framework of Pleistocene rocks in south Florida, Part 2, *in* P. Enos and R.D. Perkins, Quaternary Sedimentation in South Florida: Geological Society of America Memoir 147, p. 131–198.

Pierson, B.J., 1982, Cyclic sedimentation, limestone diagenesis and dolomitization in Upper Cenozoic carbonates of southeastern Bahamas: Ph.D. dissertation, University of Miami, 295 p.

Pierson, B., and D. Beach, 1980, Late Cenozoic stratigraphy and structure of the Bahama archipelago: Abstracts of the 26th International Geological Congress, Paris, p. 530.

Raymo, M.E., W.F. Ruddiman, J. Backman, B.M. Clement, and D.G. Martinson, 1989, Late Pliocene variation in Northern Hemisphere ice sheets and North Atlantic deep water circulation: Paleoceanography, v. 4, p. 413–446.

Richards, D.A., P.L. Smart, and R.L. Edwards, 1994, Maximum sea levels for the last glacial period from U-series ages of submerged speleothems: Nature, v. 367, p. 357–360.

Robbin, D.M., and Stipp, J.J., 1979, Depositional rate of laminated soilstone crusts, Florida Keys: Journal of Sedimentary Petrology, v. 49, p. 175–180.

Rossinsky, V., Jr., and H.R. Wanless, 1992, Topographic and vegetative controls on calcrete formation, Turks and Caicos Islands, British West Indies: Journal of Sedimentary Petrology, v. 62, p. 84–98.

Rossinsky, V., Jr., H.R. Wanless, and P.K. Swart, 1992, Penetrative calcretes and their stratigraphic implications: Geology, v. 20, p. 331–334.

Ruddiman, W.F., and H.E. Wright Jr., 1987, Introduction, *in* W.F. Ruddiman, and H.E. Wright, Jr., eds., North America and adjacent oceans during the last deglaciation: Geological Society of America, The Geology of North America, v. K-3, p. 1–12.

Saller, A.H., D.A. Budd, and P.M. Harris, 1994, Unconformities and porosity development in carbonate strata: ideas from a Hedberg Research Conference: AAPG Bulletin, v. 78, p. 857–871.

Sandberg, P.A., 1983, An oscillating trend in Phanerozoic non-skeletal carbonate mineralogy: Nature, v. 305, p. 19–22.

Sherman, C.E., C.R. Glenn, A.T. Jones, W.C. Burnett, and H.P. Schwarcz, 1993, New evidence for two highstands of the sea during the last interglacial, oxygen isotope substage 5e: Geology, v. 21, p. 1079–1082.

Smart, P.L., and Whitaker, F.F., 1988, Controls on the rate and distribution of carbonate bedrock dissolution in the Bahamas, *in* J.E. Mylroie, ed., Proceedings of the 4th Symposium of the Geology of the Bahamas, Bahamian Field Station, San Salvador, p. 313–322.

Smart, P.L., J.M. Dawans, and F. Whitaker, 1988a, Carbonate dissolution in a modern mixing zone: Nature, v. 335, p. 811–813.

Smart, P.L., R.J. Palmer, F. Whitaker, and V.P. Wright, 1988b, Neptunian dikes and fissure fills: an overview and account of some modern examples, *in* N.P. James and P.W. Choquette, eds., Paleokarst: New York, Springer Verlag, p. 149–163.

Spencer, M., 1967, Bahama deep test: AAPG Bulletin, v. 51, p. 263–268.

Steinen, R.P., 1974, Phreatic and vadose diagenetic modification of Pleistocene limestone: petrographic observations from subsurface Barbados, West Indies: AAPG Bulletin, v. 58, p. 1008–1024.

Steinen, R.P., and R.K. Matthews, 1973, Phreatic vs. vadose diagenesis: stratigraphy and mineralogy of a cored borehole on Barbados, W. I.: Journal of Sedimentary Petrology, v. 43, p. 1012–1020.

Supko, P.R., 1970, Depositional and Diagenetic Patterns in Subsurface Bahamian Rocks: Ph.D. dissertation, University of Miami, Coral Gables, Florida, 80 p.

Supko, P.R., 1977, Subsurface dolomites, San Salvador, Bahamas: Journal of Sedimentary Petrology, v. 47, p. 1063–1077.

Vacher, H.L., 1988, Dupuit-Ghyben-Herzberg analysis of strip-island lenses: Geological Society of America Bulletin, v. 100, p. 580–591.

Vacher, H.L., and J.E. Mylroie, 1991, Geomorphic evolution of topographic lows in Bermudian and Bahamian islands: effect of climate, *in* R.J. Bain, ed., Proceedings of the Fifth Symposium on the Geology of the Bahamas, Bahamian Field Station, San Salvador, Bahamas, p. 221–234.

Vahrenkamp, V.C., 1988, Constraints on the formation of platform dolomites: a geochemical study of Late Tertiary dolomite from Little Bahama Bank, Bahamas: Ph.D. dissertation, University of Miami, Coral Gables, Florida, 434 p.

Vahrenkamp, V.C., and P.K. Swart, 1991, Episodic dolomitization of late Cenozoic carbonates in the Bahamas: evidence from strontium isotopes: Journal of Sedimentary Petrology, v. 61, p. 1002–1014.

Wallis, T.N., and H.L. Vacher, 1990, Shape of freshwater lens in small carbonate islands: Exuma vs. Bermuda and the effect of climate, *in* J.H. Krishna, V. Quinones-Aponte, F. Gomez-Gomez, and G. Morris, eds., Tropical Hydrology and Caribbean Water Resources: Proceedings of the International Symposium on Tropical Hydrology and 4th Caribbean Islands Water Resources Congress, San Juan, Baltimore, Maryland, American Water Resources Association, p. 317–326.

Ward, W.C., 1975, Petrology and diagenesis of carbonate aeolianites of northeastern Yucatan Peninsula, Mexico, *in* K.F. Wantland, and W.C. Pusey, eds., Belize Shelf-Carbonate Sediments, Clastic Sediments, Ecology: AAPG Studies in Geology 2, p. 500–571.

Warne, S.St. J., 1962, A quick laboratory staining scheme for the differentiation of the major carbonate minerals: Journal of Sedimentary Petrology, v. 32, p. 29–38.

Whitaker, F.F., and P.L. Smart, 1990, Active circulation of saline ground waters in carbonate platforms: evidence from the Great Bahama Bank: Geology, v. 18, p. 200–203.

Whitaker, F.F., and Smart, P.L., 1993, Circulation of saline groundwaters through carbonate build-ups; an overview and case study from the Bahamas, *in* A.D. Horbury and A.G. Robinson, eds., Diagenesis and Basin Development, AAPG Studies in Geology 36, p. 113–132.

Whitaker, F.F., and Smart, P.L., in press, Hydrogeology and hydrology of the Bahamian archipelago, *in* L.H. Vacher and T.M. Quinn, eds., Hydrology and Hydrogeology of Carbonate Islands: New York, Elsevier.

Williams, S.C., 1985, Stratigraphy, facies evolution, and diagenesis of Late Cenozoic limestones and dolomites, Little Bahama Bank, Bahamas: Ph.D. dissertation, University of Miami, Coral Gables, Florida, 462 p.

Williams, S.C., D. Choi, and D. Wilson, 1983, Neogene molluscan biostratigraphy of little Bahama Bank (abs.): Geological Society of America, Southeastern Section Abstracts with Programs, v. 15, p. 99.

Williams, M.A.J., D.L. Dunkerley, P. De Deckker, A.P. Kershaw, and T. Stokes, 1993, Quaternary Environments: London, Edward Arnold, 256 p.

Wilson, J.L., and P. Medlock, 1993, Paleokarst within the Knox Group of Alabama, east side of Black Warrior basin, *in* R.D. Fritz, J.L. Wilson and D.A. Yurewicz, eds., Paleokarst Related Hydrocarbon Reservoirs, SEPM Core Workshop No. 18, Society for Sedimentary Geology, New Orleans, p. 245–275.

Wright, V.P., 1991, Paleokarst: types, recognition, controls and associations, *in* V.P. Wright, M. Esteban, and P.L. Smart, eds., Palaeokarsts and Palaeokarstic Reservoirs: Postgraduate Research Institute for Sedimentology, University of Reading, p. 120–146.

Yaalon, D.H., and S. Singer, 1974, Vertical variation in strength and porosity of calcrete (Nari) on chalk, shefela, Israel and interpretation of its origin: Journal of Sedimentary Petrology, v. 44, p. 1016–1023.

Chapter 2

Early Diagenesis of Pleistocene Carbonates from a Hydrogeochemical Point of View, Irabu Island, Ryukyu Islands: Porosity Changes Related to Early Carbonate Diagenesis

Hiroki Matsuda
Yoshihiro Tsuji
Japan National Oil Corporation
Chiba, Japan

Nobuyuki Honda
United Petroleum Development Co. Ltd.
Tokyo, Japan

Jun-ichi Saotome
Japan Oil Development Co. Ltd.
Tokyo, Japan

ABSTRACT

To elucidate the early diagenesis of carbonates, a hydrogeochemical study was carried out on groundwater in shallow-marine Pleistocene limestones on Irabu Island, southwestern Japan. Carbonate diagenetic processes and porosity changes within vadose, freshwater phreatic, and mixing zones were determined on the basis of hydrogeochemical analyses. Net dissolution of limestone occurs mainly in the vadose and upper mixing zones where dissolution results in an increase in porosity of 1.98% per 10 k.y. and 2.10% per 10 k.y., respectively. Calcite precipitation occurs in the upper 1.5 m of the freshwater phreatic zone, and causes a decrease in porosity of 5.36% per 10 k.y. Significant dissolution and precipitation apparently do not occur in the middle to lower part of the freshwater phreatic zone, even though that zone has active groundwater flow. The main factor controlling diagenetic reactions on Irabu Island is considered to be CO_2 fluxes into the groundwater system. Differences in solubility of carbonate minerals are not significant for the diagenesis on Irabu Island because the island is composed almost entirely of low-Mg calcite.

INTRODUCTION

Carbonate rock properties, such as porosity and permeability, are influenced by primary carbonate sedimentary facies reflecting their sedimentary environment and components. These properties are subsequently modified by various diagenetic processes, especially early diagenetic processes which change the rock properties drastically by dissolution, precipitation, and stabilization of carbonate minerals. It is, therefore, very important in petroleum exploration and development to understand the early diagenesis of carbonates.

Studies on carbonate diagenesis have been carried out by many workers (e.g., Moore, 1989), and, consequently, the diagenetic processes and products in various diagenetic settings are known. However, quantitative aspects of diagenetic processes, such as the precipitation-dissolution and transformation rates of carbonate minerals within each diagenetic environment, are not known well because the chemical reactions which govern early diagenesis are very slow. Therefore, it is difficult to measure diagenetic changes proceeding in present-day diagenetic environments by a direct petrological approach. It is also difficult to examine diagenetic processes under normal temperature and pressure conditions experimentally. Quantitative evaluation of early carbonate diagenesis from hydrogeochemical analyses has been actively carried out in recent years (e.g., Plummer et al., 1976; Budd, 1988a; Anthony et al., 1989; Sanford and Konikow, 1989a, b; Stoessell et al., 1989; McClain et al., 1992). The hydrogeochemical approach provides a relatively easy way to quantitatively assess diagenetic processes proceeding presently within vadose, freshwater phreatic, mixing, and marine phreatic zones.

The Technology Research Center of Japan National Oil Corporation has been studying Quaternary and modern carbonates in the Ryukyu Islands, southwestern Japan, as an analogue to ancient rocks, with a focus on processes which affect carbonate reservoir development (Obata and Tsuji, 1992: Tsuji, 1993; Honda et al., 1994). In this project, we have carried out a hydrogeochemical study of groundwater in limestones of the Pleistocene Ryukyu Group to elucidate porosity development and destruction during early diagenesis. Hydrogeochemical data from this research provide basic information on diagenetic environments and processes in a carbonate province. In this paper, we describe hydrogeochemical features in the Pleistocene limestones, discuss early diagenesis of the carbonates, and estimate rates of porosity change within each diagenetic zone on the basis of the hydrogeochemical data.

HYDROGEOLOGIC FRAMEWORK

Geology

This study was carried out mainly on Irabu Island, southern Ryukyu Islands, with supplementary surveying of spring waters on Kikai Island, northern Ryukyu Islands (Figure 1). The Ryukyu Island Arc, which extends approximately 1200 km from Kyushu southwestward to Taiwan, rims the northwestern Pacific Ocean, and is bounded by the Okinawa Trough to the northwest and the Ryukyu Trench to the southeast. Irabu Island is situated in a subtropical region, and has an average temperature of 23°C and an average annual rainfall of 2200 mm (Maritime Safety Agency of Japan, 1986). The island is underlain by the uplifted Pleistocene Ryukyu Group, which is about 100 m in thickness and consists of shallow shelf facies (coral framestone/rudstone/floatstone and bioclastic grainstone/packstone/wackestone) and deep shelf facies (rhodolith rudstone/floatstone, larger foraminiferal rudstone/floatstone, and bioclastic grainstone/packstone/wackestone) (Figure 2; Yuki et al., 1988; Honda et al., 1994). The age of the Ryukyu Group in Irabu Island is 1.36–0.39 Ma as indicated by nannofossil biostratigraphy (Sado et al., 1992; Honda et al., 1994). Limestones of the Ryukyu Group on the island are composed almost entirely of low-Mg calcite as a result of diagenesis and include a variety of diagenetic products, such as equant mosaic cements, neomorphosed isopachous fringing fibrous to bladed cements, meniscus cements, neomorphic calcites replacing bioclastic grains, moldic and vuggy porosity, pressure solution and so on (Tsuji et al., 1990). The Ryukyu Group lies unconformably on the Pliocene Shimajiri Group, composed mainly of sand-silt alternations. The Shimajiri Group forms an aquiclude on Irabu Island.

Kikai Island is in the northern part of the Ryukyu Island Arc, and is also underlain by the uplifted Pleistocene Ryukyu Group with a Pliocene siliciclastic basement. The Ryukyu Group is subdivided into two units in age. The older unit is middle Pleistocene (0.20 Ma to more than 0.25 Ma, Omura et al., 1985; Omura, 1988), is widely distributed across the whole island, and covers the Pliocene basement unconformably. The unit consists mainly of coral boundstone/rudstone/floatstone, rhodolith rudstone/floatstone, and bioclastic grainstone/packstone/wackestone, and its thickness is about 50 m (Konishi et al., 1970). The younger, locally exposed unit is upper Pleistocene (41,000 to 129,000 yr, Konishi et al., 1974; Omura et al., 1985; Omura, 1988), and was deposited as small-scale fringing reefs on the older, middle Pleistocene limestones. Limestones of the older unit are composed mainly of low-Mg calcite with subordinate high-Mg calcite, aragonite, and dolomite. The younger limestone unit is composed of low-Mg calcite, high-Mg calcite, and aragonite.

Hydrology

Distribution of groundwater in Irabu Island was monitored in November, 1989, utilizing a series of research wells (Figure 1). The island is underlain by a freshwater lens (less than 1% seawater), which has a maximum thickness of 12 m in the central part of the island and is absent in the coastal area (Figure 3). The freshwater lens overlies a freshwater-seawater mixing zone (1–90% seawater) ranging in thickness from 7 m

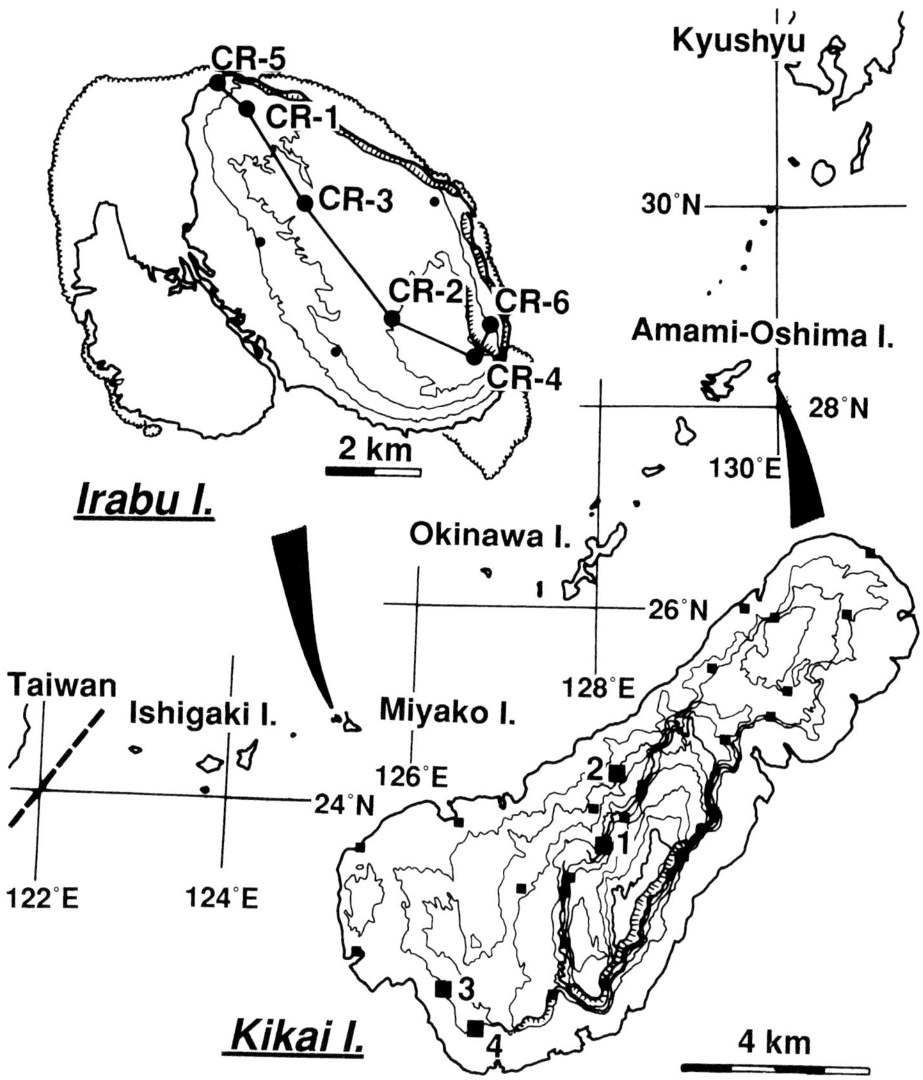

Figure 1. Index map showing Irabu and Kikai Islands, Ryukyus. On Irabu Island, large dots are analyzed research wells, and small dots are other research wells. On Kikai Island, large squares are analyzed springs (1–Takikawa, 2–Ooasato, 3–Tekutsuku, 4–Kami-katetsu) and small squares are other springs.

to more than 30 m, with the thickness increasing toward the coast. In the coastal area, the thickness of the mixing zone varies with tidal changes. The marine phreatic zone is between the mixing zone and the Pliocene basement.

Contrary to Irabu Island, some springs exist above sea level in Kikai Island (Noma, 1978; Figure 1). The springs are located around the central part of the island that forms a topographic high, and the altitudes of the springs range from 15 to 110 m. A steady-state aquifer is distributed in the coastal area and in the western part of the island.

METHODS

Hydrogeochemical analyses on Irabu Island were carried out by using three wells (CR-1, CR-3, and CR-5). These wells were established by drilling 116 mm holes and then inserting open-ended PVC pipes with numerous small holes through the wall. This work was carried out in 1986 and 1987. Prior to water sampling, electrical conductivity surveys using cable tools were performed to establish hydrologic zonation within the aquifer in August and October, 1989. On the basis of these surveys, the water sampling points were chosen at 2.5 m intervals in freshwater phreatic and mixing zones, and at hydrologic subzones in the marine phreatic zone, as defined by differences of electrical conductivity, water temperature, pH, and oxidation-reduction potentials. In boreholes in which tidal fluctuations were observed, sampling was carried out at both high and low tides. It was impossible to collect vadose water samples from Irabu Island because of lack of springs above the water table; thus the vadose zone waters were collected from springs on Kikai Island. Water samples in a freshwater phreatic zone in Kikai Island were also taken for the comparison with Irabu Island.

Water samples were collected by using a downhole water sampler in November, 1989. In order to avoid disturbing the water column in the well, the water sampler was moved at a rate of 2 m per min, and the upper samples in a well were collected first. Sampling of spring water in Kikai Island was carried out with a hand sampler in June, 1990. The pH, total alkalinity, and concentration of dissolved oxygen were measured

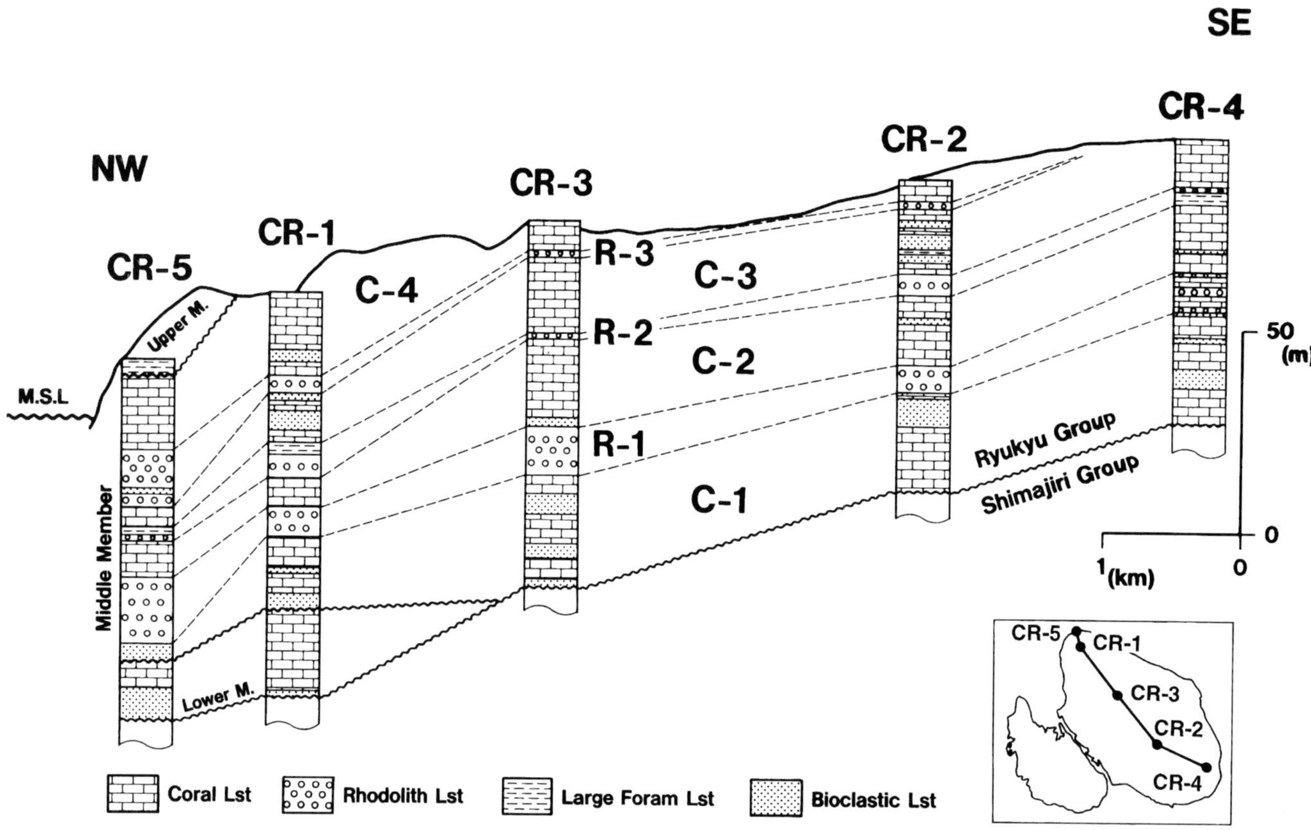

Figure 2. Geological cross section of Irabu Island, southwestern Ryukyu Islands. Units denoted by "C" are coral limestone units; units denoted by "R" are rhodolith limestone units.

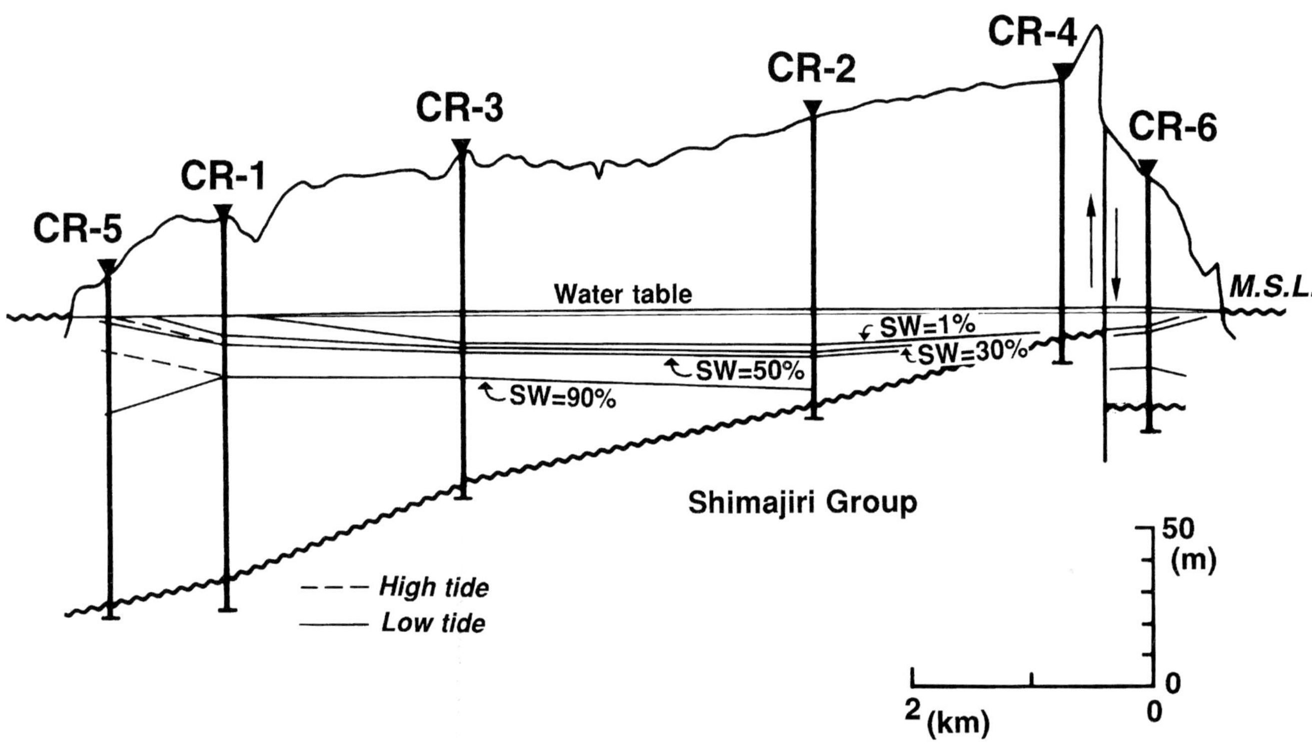

Figure 3. Hydrologic profile of Irabu Island. Lines denoted by "SW" represent the proportion of admixed seawater.

in the field by a portable pH meter (HM-5S type, TOA Co. Ltd.), the hydrochloric acid titration technique, and the Winkler method, respectively. Concentrations of major cations and anions were determined in the laboratory, and the analytical methods for each species are listed in Table 1; reproducibility is 3% for Na^+ and K^+, 5% for Sr^{2+}, and 2% for other cations and anions.

The distribution and activities of all aqueous species, the partial pressure of CO_2, ion-activity products, and saturation states of carbonate minerals were calculated using the aqueous equilibrium model PHREEQE (Parkhurst et al., 1980, revised 1985). Major cation and anion concentrations, calculated P_{CO_2} values, and the log of calcite indices are shown in Table 2.

Table 1. Analytical methods for chemical components.

	Analytical Methods
Ec*	Digital Ec meter
pH	pH meter
DO**	Winkler method
Na^+	Ion chromatography
K^+	Ion chromatography
Ca^{2+}	Atomic absorption spectrophotometry
Mg^{2+}	Atomic absorption spectrophotometry
Sr^{2+}	Atomic absorption spectrophotometry
Cl^-	Ion chromatography
HCO_3^-	Hydrochloric acid titrimetry
SO_4^{2-}	Ion chromatography
PO_4^{3-}	Molybdenum-blue method

* Ec—Electric conductivity, ** DO—Dissolved oxygen.

CONTRIBUTION OF WATER-ROCK INTERACTION TO HYDROCHEMISTRY OF GROUNDWATER

Results of the Hydrogeochemical Analyses

Hydrogeochemical analyses of Irabu and Kikai waters are shown in Table 2 and Figures 4 and 5. The mixing ratio of seawater is calculated from the ratio of Cl^- ion concentration in each sample to that in seawater around the islands. As shown in Figure 5, sodium and sulfate are conserved with respect to chloride. The concentration of potassium also shows positive correlation with that of chloride, although the data is scattered compared to those of sodium and sulfate. These results indicate that the mixing of rainwater and seawater controls the concentrations of these species in the aquifer.

Regarding the concentrations of calcium, strontium, and magnesium, there is a positive correlation between the concentrations of these species and chloride. However, enrichment in calcium and strontium with respect to chloride is observed in low-salinity groundwater samples, and high-salinity groundwater samples are depleted in magnesium with respect to chloride. Further, the concentration of bicarbonate does not show a positive correlation with that of chloride, and low salinity groundwater samples are enriched in bicarbonate with respect to chloride. The concentrations of these species can not be explained by only the mixing of rainwater and seawater; their concentrations in the groundwater samples must be influenced by interactions between the groundwater and the host limestones of the Ryukyu Group.

Calculation of Ion Concentration Supplied by Water-Rock Interaction

In the case of a small carbonate-dominated island like Irabu Island, the concentration of each ion in the groundwater is defined by the amount of ions derived from seawater and rainwater, and the amount of ions supplied by dissolution of host carbonates (or removed by precipitation of carbonate minerals). Therefore, the concentration of each ion is represented as follows:

$$I\text{-}MEASURE = I\text{-}RW + I\text{-}SW + I\text{-}WRI \quad (1)$$

where *I-MEASURE* is the measured concentration of ion I in groundwater, *I-RW* is the concentration of ion I derived from rainwater, *I-SW* is the concentration of ion I derived from seawater, and *I-WRI* is the concentration of ion *I* supplied by water-rock interaction. The concentration of ion I due to water-rock interactions can be estimated by subtracting the concentration of ion I originating in rainwater and seawater from the measured concentration in the groundwater. Positive *I-WRI* values indicate the supply of ion *I* into groundwaters due to limestone dissolution, and negative values indicate the removal of ion I from groundwater by carbonate precipitation. The concentrations of ion *I* derived from seawater and rainwater were calculated by the following equations:

$$I\text{-}SW = I\text{-}ASW \times (Cl\text{-}MEASURE/Cl\text{-}ASW) \quad (2)$$

$$I\text{-}RW = I\text{-}ARW \times (1 - Cl\text{-}MEASURE/Cl\text{-}ASW) \quad (3)$$

where *I-ASW* is the average concentration of ion I in seawater around Irabu Island, *I-ARW* is the average concentration of ion I in rainwater on Irabu Island, *Cl-MEASURE* is the measured concentration of Cl in sample, and *Cl-ASW* is the average concentration of Cl in seawater around Irabu Island.

Distribution of *Ca-WRI* and *Mg-WRI*

Ca-WRI and *Mg-WRI* values calculated are listed in Table 2, and their distribution within the Irabu aquifer is shown in Figure 6. The *Ca-WRI* value is greater than zero in the low-salinity portion of the groundwater. In the freshwater phreatic zone, the *Ca-WRI* value ranges from 1.91 to 2.12 mM/L, and the value increases with salinity such that it is 2.49 to 2.80 mM/L in the upper mixing zone. However, the *Ca-WRI* value decreases with greater salinity within the lower mixing and

Table 2. Hydrogeochemical data on groundwater, seawater, and rain water in and near Irabu and Kikai islands, Ryukyu Islands.

	Depth (m)	Altitude (m)	Mixing (%)	pH	DO (mg/L)	Na^+ (mM/L)	K^+ (mM/L)	Ca^{2+} (mM/L)	Mg^{2+} (mM/L)	Sr^{2+} (μM/L)	Cl^- (mM/L)	HCO_3^- (mM/L)	SO_4^{2-} (mM/L)	PO_4^{3-} (μM/L)	Ca-WRI (mM/L)	Mg-WRI (mM/L)	SI-CAL*	$\log P_{CO_2}$
IRABU GROUNDWATER																		
CR-1-1	32.0	−1.2	1.12	7.4	6.9	5.70	0.10	2.69	0.996	5.0	6.08	5.6	0.49	0.48	2.58	0.422	0.415	−1.90
CR-1-2	32.5	−1.7	1.67	7.3	6.6	7.04	0.15	2.97	1.23	7.0	9.07	5.7	0.66	0.43	2.80	0.374	0.341	−1.80
CR-1-3	35.0	−4.2	3.32	7.2	6.5	14.1	0.28	2.94	2.03	6.0	18.0	5.8	1.12	0.45	2.62	0.333	0.190	−1.70
CR-1-4	37.5	−6.7	25.5	7.3	4.6	109	1.85	4.99	14.4	23	138	5.4	6.93	0.84	2.49	1.392	0.192	−1.92
CR-1-5	40.0	−9.2	60.8	7.4	1.9	260	6.52	7.01	28.6	41	330	3.4	16.4	0.93	1.06	−2.46	0.094	−2.27
CR-1-6A	42.5	−11.7	75.9	7.5	1.7	321	7.16	8.38	33.5	58	411	2.7	20.7	0.83	0.952	−5.32	0.134	−2.49
CR-1-6B	42.5	−11.7	82.2	7.5	1.0	359	6.96	8.75	37.6	59	445	2.5	22.3	0.73	0.715	−4.39	0.105	−2.54
CR-1-7A	45.0	−14.2	76.4	7.4	1.5	327	8.01	8.65	35.5	59	414	2.7	20.4	0.83	1.18	−3.56	0.049	−2.40
CR-1-7B	45.0	−14.2	79.0	7.5	1.5	340	7.80	8.75	36.5	55	428	2.5	21.4	0.78	1.02	−3.95	0.111	−2.53
CR-1-8	47.5	−16.7	83.7	7.3	1.0	370	8.31	9.10	40.9	63	454	2.3	22.8	0.62	0.912	−1.85	−0.114	−2.37
CR-1-9	50.0	−19.2	91.0	7.5	0.7	385	6.80	9.25	38.2	63	493	2.2	24.7	0.72	0.349	−8.83	0.063	−2.60
CR-1-10	55.0	−24.2	98.3	7.3	0.2	402	9.31	9.55	49.4	74	532	2.2	27.1	0.82	−0.0638	−0.862	−0.134	−2.41
CR-1-11	61.0	−30.2	96.7	7.4	0.3	406	8.85	9.38	40.8	68	524	2.2	26.2	0.60	−0.0858	−8.62	−0.035	−2.50
CR-1-12	70.0	−39.2	97.2	7.4	0.4	406	7.24	9.48	37.8	64	527	2.1	26.4	0.73	−0.0369	−11.9	−0.049	−2.52
CR-1-13	85.0	−54.2	97.2	7.4	0.7	425	10.0	8.78	37.8	74	527	2.2	26.1	0.79	−0.735	−11.9	−0.062	−2.50
CR-1-14	97.5	−66.7	96.7	7.5	1.1	410	8.29	9.70	39.8	75	524	2.3	26.0	0.91	0.238	−9.65	0.095	−2.58
CR-1-15	108.0	−77.2	97.8	7.5	0.2	404	8.59	10.0	44.4	80	530	2.3	26.4	0.76	0.436	−5.53	0.103	−2.59
CR-1-16	115.0	−84.2	90.0	7.5	0.6	396	8.16	9.85	36.5	72	487	2.2	24.0	0.62	1.05	−9.49	0.092	−2.60
CR-3-1	49.0	−0.5	0.169	7.5	7.5	0.856	0.01	1.93	0.262	3.4	0.915	4.3	0.11	0.37	1.91	0.176	0.343	−2.10
CR-3-2	50.0	−1.5	0.261	7.3	7.1	1.25	0.02	2.16	0.317	3.4	1.41	4.4	0.14	0.28	2.14	0.184	0.193	−1.89
CR-3-3	52.5	−4.0	0.359	7.4	6.9	1.30	0.03	1.95	0.344	4.6	1.95	4.3	0.16	0.30	1.91	0.160	0.238	−2.00
CR-3-4	55.0	−6.5	0.291	7.4	7.1	1.36	0.03	2.16	0.308	3.4	1.58	4.4	0.14	0.35	2.13	0.159	0.290	−1.99
CR-3-5	57.5	−9.0	0.473	7.3	6.6	2.11	0.04	2.08	0.395	3.4	2.56	4.5	0.19	0.39	2.04	0.154	0.174	−1.88
CR-3-6	60.0	−11.5	24.5	7.4	4.8	105	2.23	4.94	10.7	23	133	3.8	6.62	0.75	2.54	−1.89	0.151	−2.16
CR-3-7	62.5	−14.0	60.3	7.4	2.9	255	5.73	7.01	27.4	48	327	2.9	16.1	0.88	1.11	−3.39	0.029	−2.34
CR-3-8	65.0	−16.5	82.7	7.5	1.4	347	8.06	8.75	34.4	63	448	2.4	22.2	0.95	0.664	−7.82	0.092	−2.55
CR-3-9	67.5	−19.0	87.4	7.4	0.9	357	8.08	9.35	35.8	61	473	2.3	23.4	0.78	0.805	−8.82	−0.001	−2.50
CR-3-10	70.0	−21.5	91.0	7.4	1.0	386	8.39	9.50	36.1	65	493	2.2	24.9	0.80	0.599	−10.4	−0.020	−2.49
CR-3-11	75.0	−26.5	91.0	7.5	0.4	403	8.34	9.73	34.2	66	493	2.2	24.5	0.76	0.823	−1.3	0.087	−2.60
CR-3-12	83.0	−34.5	95.7	7.5	0.4	406	7.70	9.55	37.9	66	518	2.2	25.5	0.65	0.191	−11.0	0.073	−2.60
CR-3-13	90.0	−41.5	96.7	7.5	0.5	439	10.0	10.4	39.8	63	524	2.3	26.2	0.77	0.962	−9.65	0.122	−2.59
CR-3-14	95.0	−46.5	96.7	7.5	0.7	416	9.59	9.73	39.0	66	524	2.2	26.3	0.63	0.263	−10.4	0.077	−2.60
CR-3-15	105.0	−56.5	95.7	7.6	0.3	403	8.21	9.73	37.8	70	518	2.2	25.8	0.70	0.365	−11.1	0.175	−2.70

Table 2 (*continued*).

	Depth (m)	Altitude (m)	Mixing (%)	pH	DO (mg/L)	Na^+ (mM/L)	K^+ (mM/L)	Ca^{2+} (mM/L)	Mg^{2+} (mM/L)	Sr^{2+} (μM/L)	Cl^- (mM/L)	HCO_3^- (mM/L)	SO_4^{2-} (mM/L)	PO_4^{3-} (μM/L)	Ca-WRI (mM/L)	Mg-WRI (mM/L)	SI-CAL*	$logP_{CO_2}$
IRABU GROUNDWATER																		
CR-5-N1**	12.0	–0.5	45.4	7.6	7.6	221	4.58	5.74	23.6	—	246	3.1	14.2	0.34	1.29	0.372	0.195	–2.51
CR-5-N2	15.0	–3.5	53.0	7.3	6.3	258	5.14	6.73	27.1	48	287	3.4	16.6	0.18	1.54	0.003	–0.014	–2.17
CR-5-N3	17.5	–6.0	64.5	7.5	5.9	282	6.98	7.16	29.9	51	349	3.3	17.7	0.46	0.849	–3.05	0.176	–2.40
CR-5-N4	20.0	–8.5	80.6	7.5	5.0	349	6.52	8.58	34.0	55	437	2.7	22.0	0.44	0.693	–7.21	0.136	–2.50
CR-5-N5	22.5	–11.0	78.5	7.5	5.4	353	8.08	7.91	34.4	58	425	2.6	20.7	0.29	0.223	–5.74	0.087	–2.52
CR-5-N6	25.0	–13.5	82.2	7.5	5.0	377	6.70	8.35	35.9	71	445	2.4	21.4	0.29	0.316	–6.08	0.069	–2.56
CR-5-N7	30.0	–18.5	80.6	7.6	5.2	353	5.24	8.13	35.3	59	437	2.5	21.3	0.38	0.244	–5.86	0.175	–2.64
CR-5-N8	40.0	–28.5	88.4	7.7	4.9	375	8.49	8.35	35.7	71	479	2.4	23.4	0.25	–0.294	–9.47	0.254	–2.77
CR-5-N9	47.5	–36.0	94.6	7.7	4.4	403	7.42	9.08	36.8	68	513	2.2	25.6	0.21	–0.182	–11.6	0.242	–2.81
CR-5-N10	57.5	–46.0	99.3	7.9	4.8	412	7.26	9.65	39.1	66	538	2.2	26.7	0.24	–0.0658	–11.7	0.444	–3.03
CR-5-P1	12.0	–0.5	53.0	7.5	7.3	222	4.71	6.16	—	42	287	3.2	14.7	0.38	0.971	—	0.124	–2.40
CR-5-P2	15.0	–3.5	60.8	7.5	6.0	266	6.57	7.06	27.2	46	330	3.4	16.6	0.38	1.11	–3.86	0.194	–2.38
CR-5-P3	17.5	–6.0	69.7	7.5	5.6	294	5.58	7.06	28.5	53	377	3.2	18.7	0.35	0.240	–7.10	0.152	–2.41
CR-5-P4	22.5	–11.0	97.8	7.7	4.5	414	8.95	9.40	40.5	67	530	2.1	26.3	0.24	–0.163	–9.48	0.229	–2.84
CR-5-P5	30.0	–18.5	96.2	7.6	4.6	425	9.97	9.33	41.6	74	521	2.2	25.2	0.31	–0.0848	–7.62	0.152	–2.71
CR-5-P6	45.0	–33.5	96.7	7.8	4.9	430	10.3	9.55	40.9	72	524	2.1	25.4	0.24	0.0888	–8.54	0.328	–2.95
CR-5-P7	70.0	–58.5	101	7.7	4.6	430	10.5	9.43	35.8	70	546	2.0	27.3	0.32	–0.443	–15.8	0.211	–2.85
CR-5-P8	90.0	–78.5	102	7.7	4.4	430	9.92	9.73	36.5	70	552	2.0	27.5	0.19	–0.245	–15.6	0.224	–2.85
CR-5-P9	97.5	–86.0	101	7.6	4.7	421	9.44	9.63	37.0	63	546	1.9	27.4	0.22	–0.243	–14.5	0.103	–2.77
IRABU RAIN WATER																		
IRABU RW-2				6.0	8.6	0.100	0.0036	0.0501	0.0078	0.1>	0.13	0.1>	0.019	0.55			–4.155	–2.15
IRABU SEAWATER																		
AVERAGE (no. = 12)				8.2	7.1	449	9.61	9.85	51.2	75	542	2.2	26.4	0.10			0.688	–3.39
KIKAI SPRING WATER																		
TAKIKAWA		110.0		7.8	8.5	0.7	0.0133	1.30	0.186	3.4	0.656	2.9	0.065	0.47	1.28	0.124	0.333	–2.56
OOASATO		35.0		7.5	8.4	0.787	0.0164	1.97	0.206	4.6	0.637	3.9	0.110	0.46	1.96	0.146	0.316	–2.14
TEKUTSUKU		17.5		7.3	7.4	0.674	0.0110	2.67	0.226	13	0.532	4.8	0.350	0.39	2.66	0.175	0.308	–1.85
KAMIKATETSU		15.0		7.1	7.6	1.07	0.0159	3.17	0.392	11	0.600	5.9	0.500	0.44	3.16	0.335	0.246	–1.57
KIKAI GROUNDWATER																		
90-34	4.5	–1.7	0.681	7.4	7.6	3.730	0.0959	2.09	0.737	16	3.69	4.6	0.360	0.21	2.02	0.388	0.259	–1.98

* log of calcite saturation index.

** N- and P-series samples of CR-5 were collected at low and high tide, respectively.

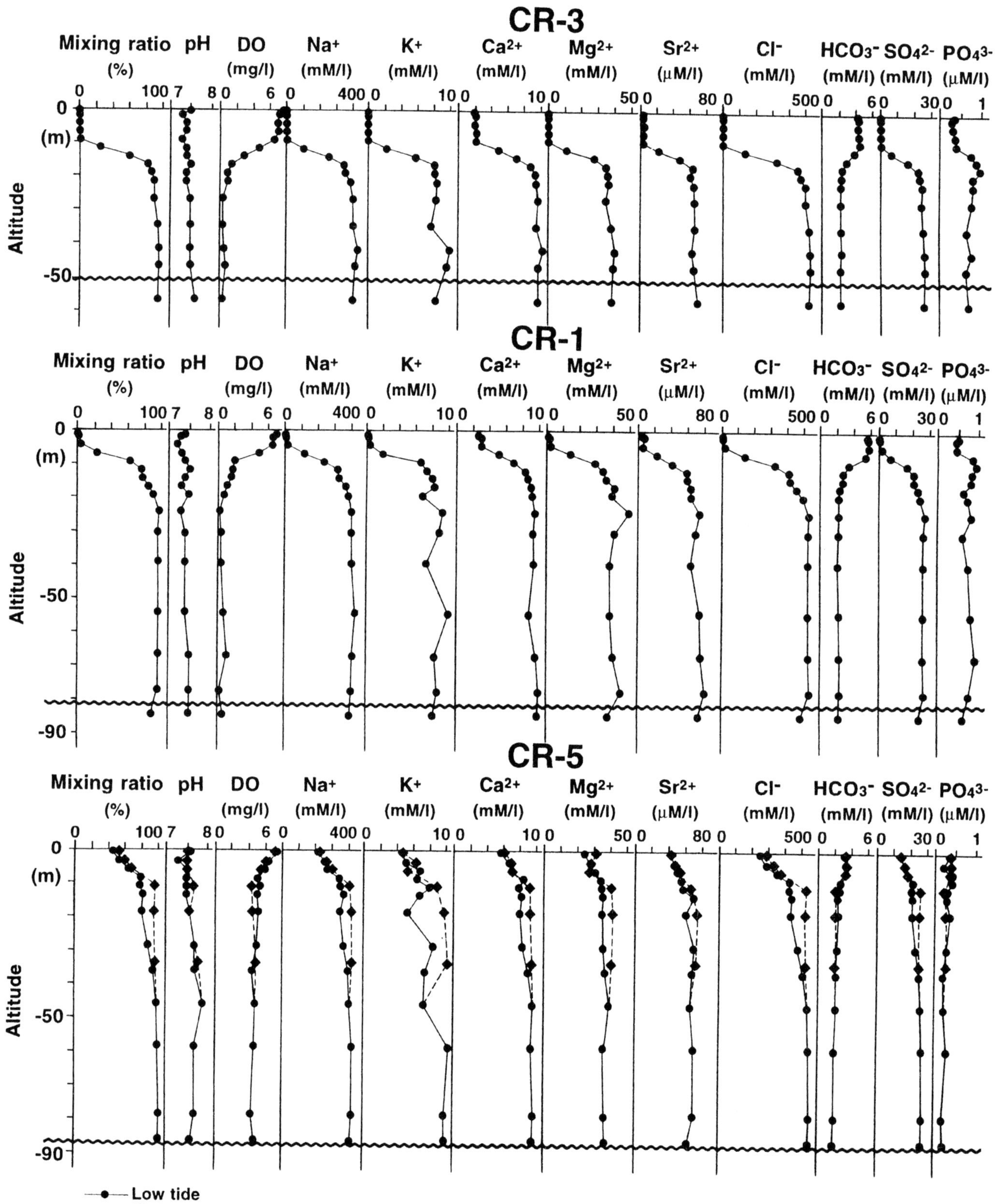

Figure 4. Depth profile of each major component in the groundwater from wells CR-3, CR-1, and CR-5. Wavy lines represent unconformity between the Ryukyu and Shimajiri groups, and 0 m lines in altitude indicate mean sea level.

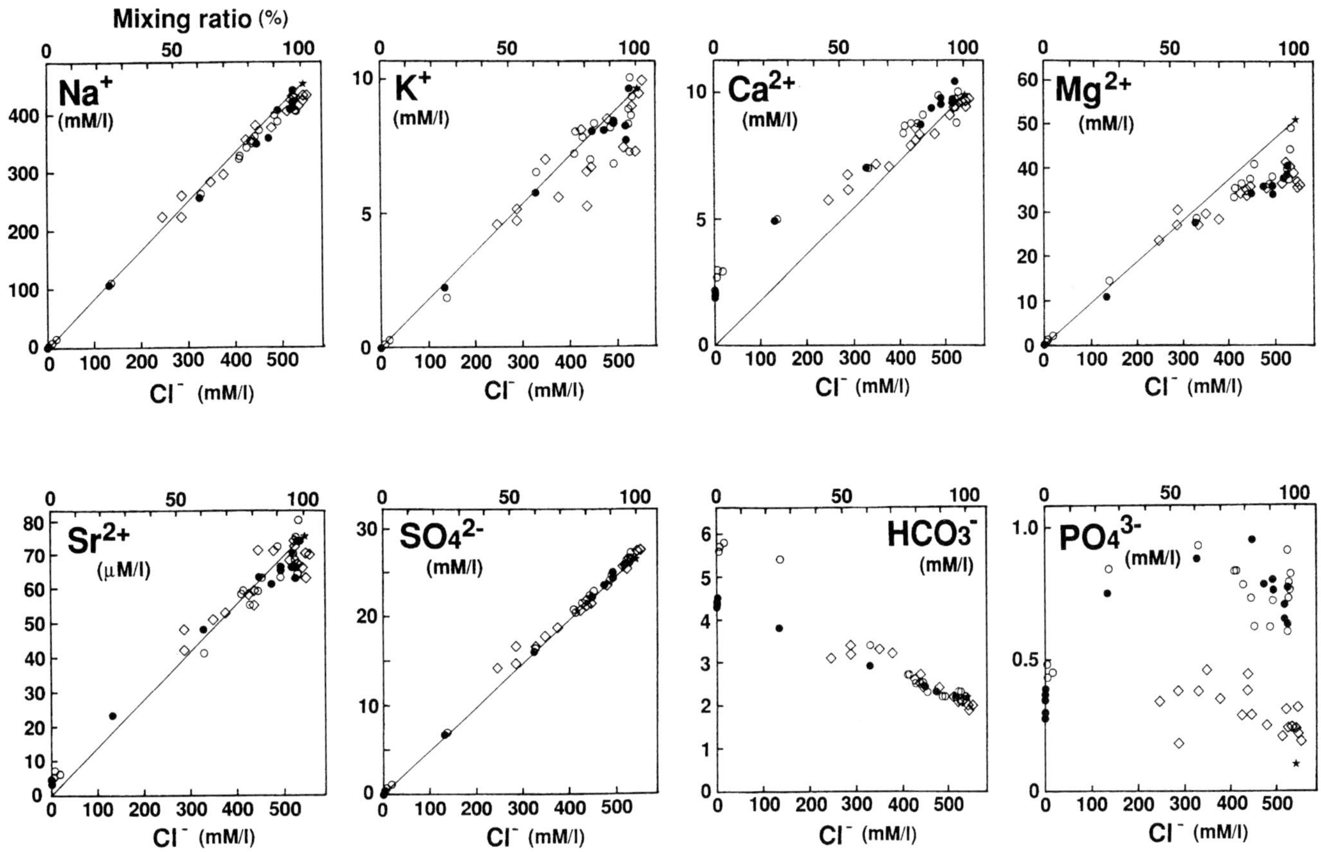

Figure 5. Cross plots of major cations and anions versus Cl^- concentration in Irabu groundwater (open circle = CR-1, solid circle = CR-3, open diamond = CR-5, star = average Irabu seawater).

marine phreatic zones, with values near zero in the marine phreatic zone.

The *Mg-WRI* value distribution is different from that of the *Ca-WRI* value. The *Mg-WRI* value is almost zero in the freshwater and upper mixing zones. However, the *Mg-WRI* value decreases abruptly with salinity such that it is –9.65 to –12.3 mM/L in the marine phreatic zone. The groundwater in the marine phreatic zone is much more deficient in Mg^{2+} ion than predicted from the mixing ratio of seawater. At –58.5 m mean sea level (MSL) in CR-5 located near the coast, the groundwater has minimum value of –15.8 mM/L .

DIAGENESIS WITHIN EACH DIAGENETIC ZONE

The Irabu groundwater is characterized by enrichment in Ca^{2+} ion in the low-salinity portion, and depletion of Mg^{2+} ion with high-salinity portion, as shown in Figure 6. The depletion of Mg^{2+} ions reaches about ten times the enrichment in Ca^{2+} ion. Therefore, a mechanism for the depletion of Mg^{2+} ion in the groundwater is discussed first.

Mechanism for the Depletion of Mg2+ Ion

The *Mg-WRI* values are strongly negative in the lower mixing and marine phreatic zones, and the Mg^{2+} ion has much lower concentrations than those calculated from the seawater mixing ratio. Where and how are the Mg^{2+} ions depleted in the groundwater? The only viable mechanism for the selective removal of Mg^{2+} ion from groundwater in this situation is dolomitization. Dolomites are not detected in bulk samples from research wells on Irabu Island. However, a few insoluble residue samples contain very small amounts of dolomites (Tsuji et al., 1990). Matsuda and Iijima (1986) also reported a minor amount of dolomite in cores from research wells for water resource developments on Irabu Island. Unfortunately, their occurrence has not been confirmed petrographically because the dolomites are extremely tiny. Therefore, it has not been decided whether or not dolomite formation is related to present-day diagenetic environments. Matsuda et al. (1994) reported modern dolomites in Holocene carbonate sands and reefal sediments from research wells on an insular shelf, off Irabu Island (CR-7, water depth = 49 m; CR-8, water depth = 15 m; CR-9, water depth = 46 m). The euhedral dolomites, a few microns across, grew on isopachous rims of high-Mg calcite cement in intergranular pores of the grainstone matrix of coral rudstones to framestones, which have corals dated as 7000 to 9000 years before present. Although the dolomites are sporadically distributed and are not observed in all intergranular pore spaces, the occurrence of dolomite supports dolomitization of

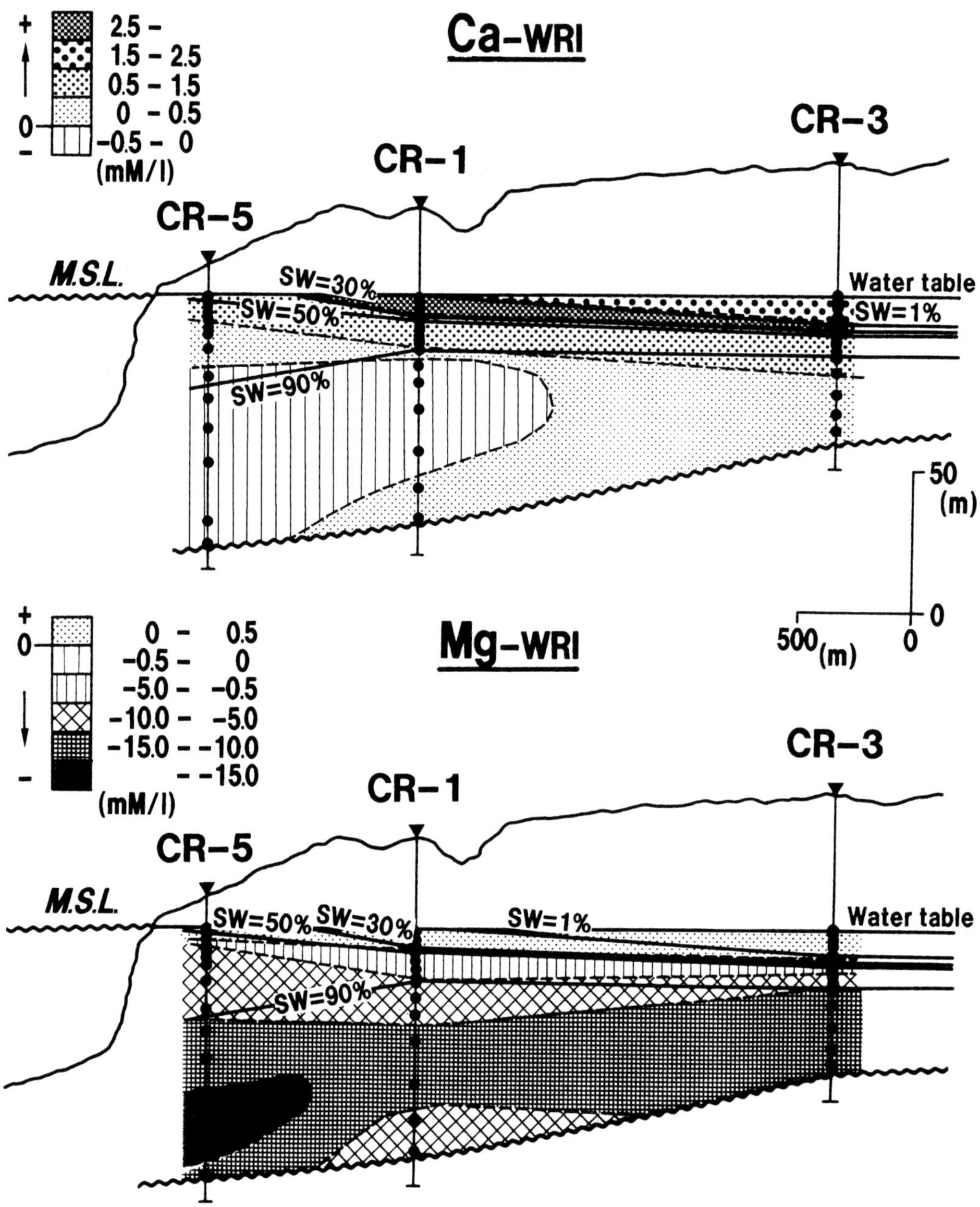

Figure 6. Profiles of *Ca-WRI* and *Mg-WRI* distribution in Irabu Island groundwater. *Ca-WRI* values are positive in low-salinity groundwater with a maximum value of 2.80 mM/L in the upper mixing zone (mixing ratio = 1.67%) of CR-1. On the other hand, *Mg-WRI* values are negative in high-salinity groundwater with a minimum value of –15.8 mM/L in the marine phreatic zone of CR-5 in the coastal area.

Holocene sediments on the outer shelf in modern times.

The groundwater is most depleted in the Mg^{2+} ion in coastal areas (Figure 6). In the coastal groundwater, other cations are conserved with respect to chloride. At high tide, when seawater infiltrates below the island, the *Mg-WRI* value is more negative than at low tide (Figure 7). Seawater flows into the aquifer below Irabu Island across the sea floor on an insular shelf. The huge depletion in Mg^{2+} ions in the high-salinity groundwater is interpreted to result from the selective consumption of Mg^{2+} ions in seawater during dolomitization, and, hence, seawater was deficient in Mg^{2+} before infiltrating below Irabu Island. Furthermore, the existence of dolomites in the core samples on Irabu Island supports the possibility that the dolomites formed in the present-day lower mixing and marine phreatic zones below Irabu Island.

Evaluation of Diagenetic Processes by the *Ca-WRI* Values

Ca-WRI values show the amount of calcium ions added to (or removed from) groundwater by water-rock interaction, and the values directly represent diagenetic processes proceeding within the present

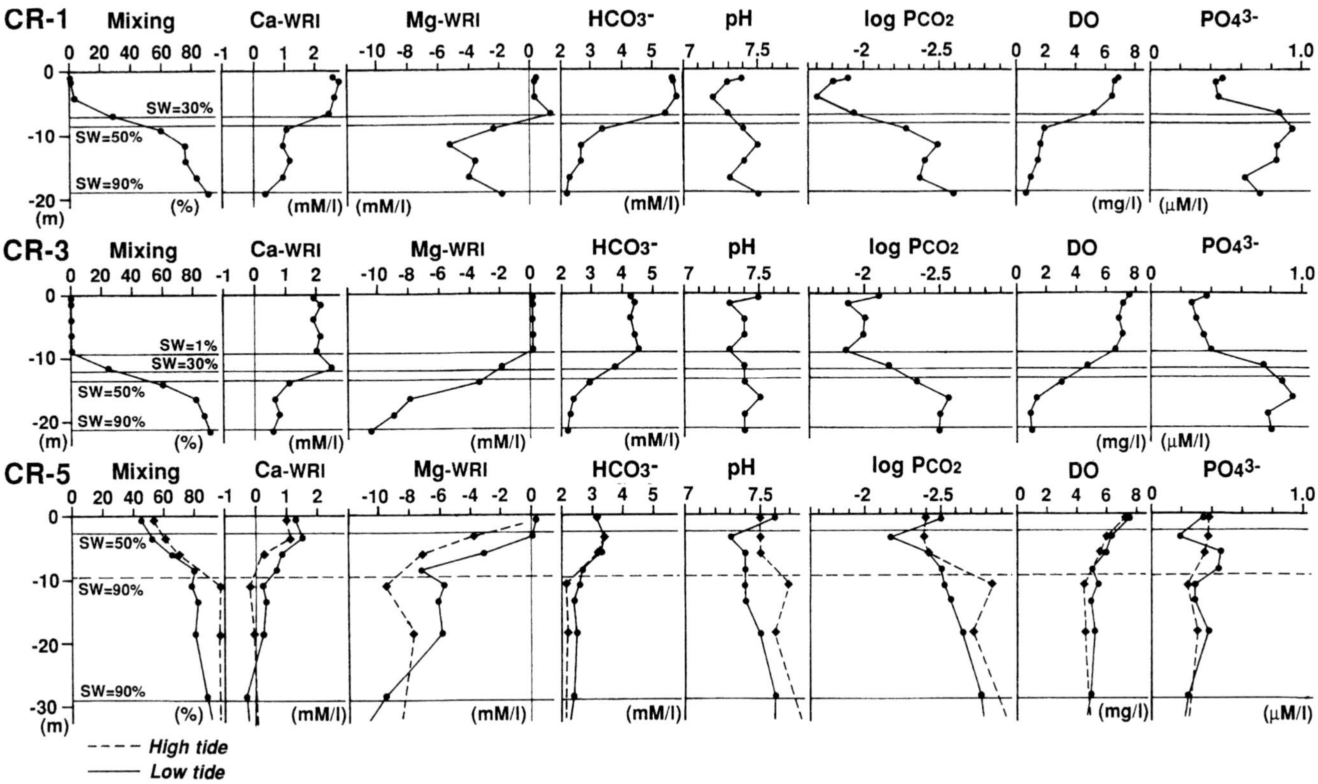

Figure 7. Depth profiles of each major component of Irabu groundwater within the freshwater phreatic and mixing zones in CR-1, CR-3, and CR-5. Lines across profiles represent the proportion of admixed seawater, and 0 m lines indicate mean sea level.

groundwater. As shown in Figure 6, their distribution within the aquifer is closely related to the distribution of the freshwater phreatic, mixing, and marine phreatic zones. The precipitation or dissolution of calcium carbonate is controlled primarily by the following reaction:

$$CaCO_3 + CO_2 + H_2O = Ca^{2+} + 2HCO_3^- \qquad (4)$$

The limestones in Irabu Island are composed almost entirely of low-Mg calcite, so the reaction could be estimated by the above equation in vadose, freshwater phreatic, and upper mixing zones in which the Mg^{2+} value shows no significant change. Diagenetic processes proceeding within each diagenetic zone with low salinity are discussed in detail on the basis of these *Ca-WRI* values.

Diagenetic Process in the Vadose Zone

In the Irabu groundwater system, Ca^{2+} ion is in excess of that inferred from the mixing of freshwater and seawater in the freshwater phreatic zone. The excess Ca^{2+} ion is derived from the host limestones when groundwater flows down through the vadose zone and into the freshwater phreatic zone. The character of groundwater flowing through the vadose zone into a freshwater lens can be understood from hydrogeochemical analyses of the spring water in Kikai Island.

In the Kikai groundwater system, Cl^- ion concentrations in spring water samples are almost 0.6 mM/L, and are constant regardless of altitude of the springs (Figure 8). On the other hand, Ca^{2+} ion concentrations change with altitude from 1.30 mM/L at 110 m to 3.17 mM/L at 15 m. *Ca-WRI* values are positive in all spring water samples, with values in lower altitude samples greater than higher altitude samples. Thus, the concentration of Ca^{2+} ion increases gradually by dissolution of the host limestones while the groundwater flows down through the vadose zone.

The above scenario is supported by variations in the saturation index (*SI*), which represents the degree of saturation of minerals in solution. Rainwater is strongly undersaturated with respect to calcite, and its *SI* value is –4.155 to –2.075. The groundwater samples in the vadose zone are, however, supersaturated with respect to calcite, and the *SI* values range from 0.246 to 0.333. These values correspond to those of the freshwater phreatic zone in Irabu Island (0.174 to 0.415). These data also suggest that the dissolution of limestone proceeds in the vadose zone.

Solubility of calcite is in part dependent on the amount of dissolved CO_2 in solution, which is controlled by the partial pressure of CO_2 in the atmosphere in contact with the groundwater. In a carbonate province, the partial pressure of CO_2 is generally highest in a soil zone because of respiration of plant roots and decay of organic matter. Therefore, the groundwater gains CO_2 when it flows down through the soil

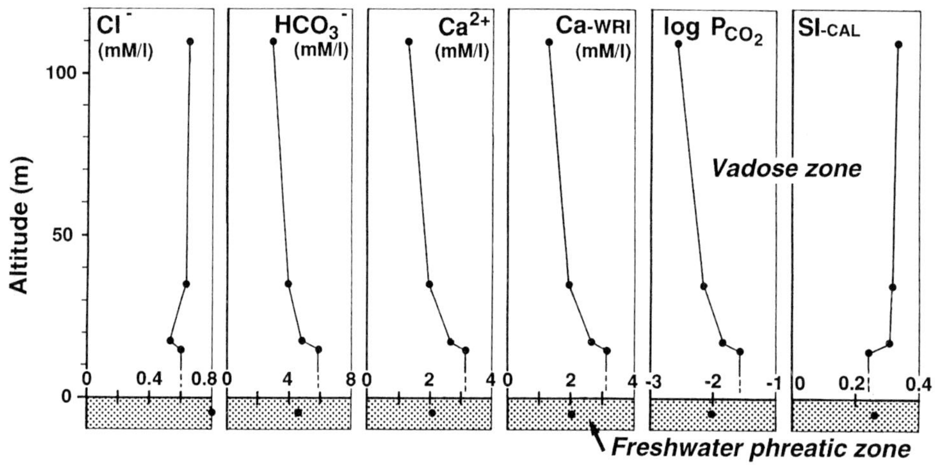

Figure 8. Concentration of major components of Kikai spring waters versus altitude of analyzed spring site. The spring water of the lowest altitude is most enriched in Ca^{2+} and HCO_3^- and shows the highest *Ca-WRI* value.

zone, and, consequently, the dissolution of limestones is increased by the groundwater having a higher P_{CO_2} level than that of rainwater. As shown in Figure 8, the P_{CO_2} level in the spring water samples tends to increase from $10^{-2.56}$ atm at 110 m to $10^{-1.57}$ atm at 15 m. Because all spring water samples are near surface, lower altitude samples would have more chance to be in contact with the high partial pressure of CO_2 of the adjacent soil zone for a longer period of time. Thus, CO_2 is supplied to the groundwater from the soil zone during the movement of groundwater from surface to the aquifer, and the elevated P_{CO_2} promotes the dissolution of the limestones.

Diagenetic Process in the Freshwater Phreatic Zone

The *Ca-WRI* values in the Irabu freshwater phreatic zone range from 1.91 to 2.13 mM/L, with an average of 2.03 mM/L. As mentioned previously, the groundwater passing through the vadose zone reaches the aquifer with an excess of Ca^{2+} ions derived from the host limestones. According to the Kikai spring waters, the groundwater just above the water table has a *Ca-WRI* value of 3.16 mM/L, which is greater than the average in the Irabu freshwater phreatic zone by 1.13 mM/L. The *Ca-WRI* value in the Kikai freshwater phreatic zone is 2.02 mM/L and is almost equal to that in the Irabu freshwater phreatic zone. Because both islands are in a similar geological setting, the groundwater in Irabu Island just above the water table would be inferred to have a similar value to that on Kikai.

Compared with the groundwater in the vadose zone, the groundwater in the Irabu freshwater phreatic zone has low P_{CO_2}, HCO_3^-, and *Ca-WRI* values, and has a high pH (Figure 7). The *Ca-WRI* value is fairly uniform through the freshwater phreatic zone. The P_{CO_2} is lowest just below the water table. To the contrary, the pH value is highest just below the water table. Lowering the P_{CO_2} value moves the reaction in equation (4) to the left, promoting the precipitation of calcium carbonate. The decrease in *Ca-WRI* across the water table thus indicates that calcium carbonate is precipitated at the top of the freshwater phreatic zone. On the other hand, no significant change of the *Ca-WRI* value through the deeper portion of the freshwater phreatic zone means that little precipitation or dissolution occurs in the deeper part of the zone. The calcium carbonate precipitated in the zone is low-Mg calcite, as shown by limestones in Irabu Island consisting almost entirely of low-Mg calcite.

Calcite cementation in freshwater phreatic zones has been studied by many workers (e.g., Halley and Harris, 1979; Budd, 1988a, b; McClain et al., 1992). From results of studies at Ocean Bight, Great Exuma, Bahamas, McClain et al. (1992) concluded that cementation and dissolution reactions were controlled by CO_2 effects in a vadose environment and the upper 1.5 m of the phreatic zone, and that diagenetic processes controlled by different solubilities of metastable carbonates do not play important roles in the deeper part of the freshwater zone, even where metastable carbonates are abundant.

Calcite cementation on Irabu Island is restricted to the top of the freshwater phreatic zone, just below the water table, similar to Ocean Bight. As indicated by the P_{CO_2} data (Figures 7 and 8), the major factor controlling the precipitation of calcite at the top of the freshwater phreatic zone is the lowering of P_{CO_2}. The partial pressure of CO_2 in the groundwater is controlled by the P_{CO_2} value of the atmosphere in contact with the groundwater, by addition of CO_2 from recharging water, and by decay of organic material carried in the recharging water. Excess CO_2 in the groundwater is degassed at the water table (Hanor, 1978), lowering P_{CO_2} just below the water table.

Carbonate cementation and dissolution are very minor in the deeper part of the freshwater phreatic zone. Partial pressure of CO_2 is almost constant throughout a 10 m thick freshwater zone, indicating no addition and removal of CO_2. In contrast to Holocene carbonate sediments examined in previous studies (e.g., Halley and Harris, 1979; Budd, 1988a, b;

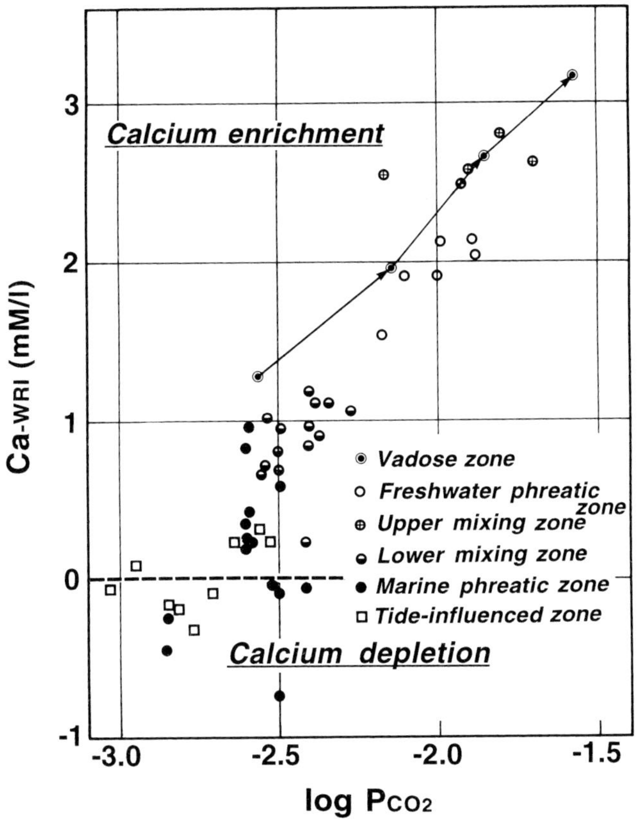

Figure 9. Cross plot of *Ca-WRI* value versus log P_{CO_2}. Groundwater having higher P_{CO_2} level has higher *Ca-WRI* values. Arrows indicate decrease in altitude of the spring sites.

McClain et al., 1992), the host limestones in the Irabu freshwater phreatic zone contain no metastable carbonate minerals, such as aragonite and high-Mg calcite. Therefore, diagenesis driven by the differences in solubility of carbonate minerals would not be expected in the Irabu freshwater phreatic zone. CO_2-driven diagenesis operated extensively in the uppermost part of the freshwater zone where CO_2 fluxes are active.

Diagenetic Process in the Upper Mixing Zone

In the Irabu groundwater system, *Ca-WRI* values in the upper mixing zone (1–30% seawater) range from 2.49 to 2.80 mM/L (average = 2.61 mM/L) and are greater than values in the freshwater phreatic zone. The increase in the *Ca-WRI* value suggests that the host limestones are dissolving in the upper mixing zone. Dissolution of limestone in a mixing zone has been documented by many researchers (e.g., Badiozamani, 1973). Thermodynamic calculations show that a mixture of freshwater and seawater, both supersaturated with respect to calcite, becomes undersaturated over a range of mixing ratios (Plummer, 1975; Wigley and Plummer, 1976). The mixtures undersaturated with respect to calcite vary due to hydrochemistry of each solution, including saturation state, P_{CO_2}, pH, temperature, and ionic strength. Dissolution in mixing zones has been reported from many carbonate areas (Yucatan Peninsula—Back et al., 1986; Andros Island—Smart et al., 1988; Schooner Cays, Bahamas—Budd, 1988a; Majuro Atoll—Anthony et al., 1989; Nauru Island—Jankowsky and Jacobson, 1991).

In the upper mixing zone on Irabu Island, the *Ca-WRI* value increases by 0.58 mM/L compared with that in the freshwater phreatic zone. A drop in pH and an increase in the partial pressure of CO_2 are also observed (Figure 7). In equation (4), the increase in P_{CO_2} moves the reaction to the right and promotes the dissolution of calcite. Furthermore, the drop in pH increases the solubility of calcite. Figure 9 shows the relation between the *Ca-WRI* value and the partial pressure of CO_2 in each diagenetic zone. A diagenetic zone at a higher P_{CO_2} level tends to have a higher *Ca-WRI* value, and therefore, the increase in P_{CO_2} results in the dissolution of calcite.

Smart et al. (1988) argued that the partial pressure of CO_2 in a mixing zone increases by decay of organic matter. The content of dissolved oxygen in Irabu mixing-zone waters decreases abruptly with increasing salinity below the 1% seawater isohaline (Figure 4). This decrease in dissolved oxygen suggests that oxygen is consumed by the decay of organic matter in the upper mixing zone, and as a result, the evolved CO_2 increases P_{CO_2} values in the upper mixing zone. The above consideration is also supported by high concentrations of PO_4^{3-} in this zone (Figure 7).

Diagenetic Process in the Lower Mixing Zone

The *Ca-WRI* value decreases markedly with increasing salinity in the middle to lower mixing zone (30–80% seawater). This decrease could be attributed to: (1) dilution by mixing with groundwater from the marine phreatic zone that has low *Ca-WRI* values, and/or (2) precipitation of carbonate within the middle to lower mixing zone. Figure 10 shows Ca^{2+} concentrations and *Ca-WRI* values in the middle to lower mixing zone. If there is no diagenesis in the middle and lower mixing zones, then the concentration of Ca^{2+} and *Ca-WRI* values in the middle to lower mixing zone should lie on a mixing line connecting the values of the upper mixing and upper marine phreatic zones. However, if any diagenetic reactions occur in the middle to lower mixing zone, then these data points would be apart from a mixing line. As shown in Figure 10, data points of Ca^{2+} concentrations and *Ca-WRI* values are scattered in the middle to lower mixing zone, and it difficult to judge whether these data are on or apart from a mixing line.

As mentioned before, the Irabu aquifer below the 50% seawater isohaline is characterized by significant depletion in Mg^{2+} concentrations and strongly negative *Mg-WRI* values (Figure 11). The depletion in Mg^{2+} ion is probably due to selective removal of the ion by dolomite formation in seawater on the insular shelf adjacent to or below Irabu Island. Thus, seawater invading an aquifer below Irabu Island has a different composition from that of normal seawater, and cannot be specified at this stage. Therefore, this study could not determine whether the decrease in the *Ca-WRI*

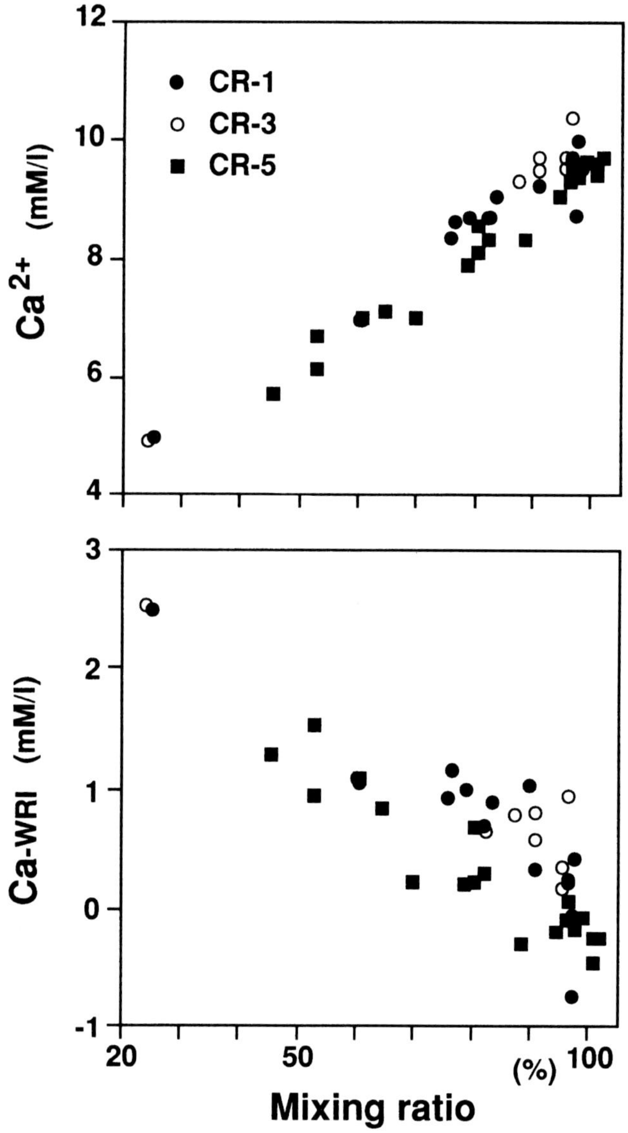

Figure 10. Cross plots of Ca^{2+} ion concentration and *Ca-WRI* value versus the proportion of admixed seawater within the middle to lower mixing and marine phreatic zones.

values in the middle to lower mixing zone was due to dilution by mixing or carbonate precipitation.

Relationship Between Saturation Index and Precipitation-Dissolution Reactions

All Irabu groundwater samples show supersaturation or near saturation with respect to calcite (Figure 12). To clarify the relationship between observed *SI* values and diagenetic processes inferred from changes in *Ca-WRI* values between diagenetic zones, theoretical *SI* values with respect to calcite were calculated as a function of freshwater-seawater mixing. The aqueous equilibrium model PHREEQE (Parkhurst et al., 1980) was applied to simulate the mixing and calculate calcite saturation states. The average chemical composition of freshwater samples in the freshwater

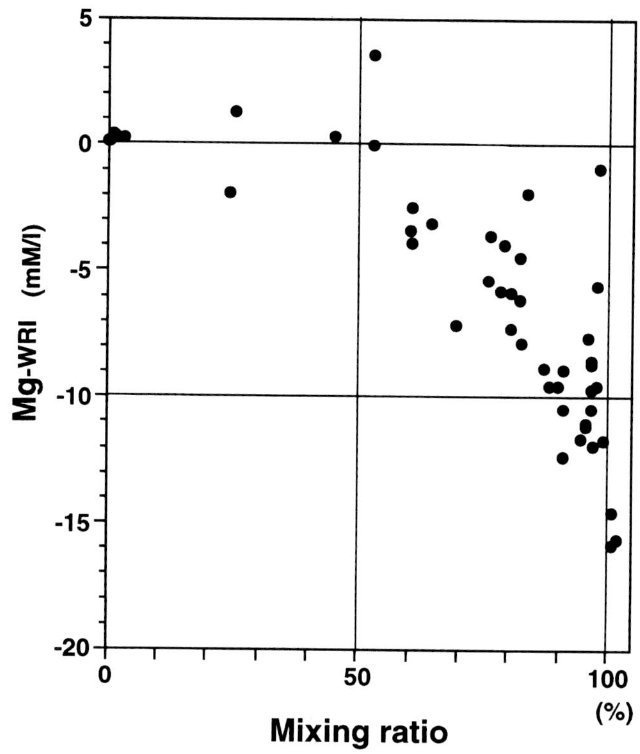

Figure 11. Cross plots of *Mg-WRI* value versus the proportion of admixed seawater. The *Mg-WRI* values show strongly negative in high-salinity groundwater more than 50% seawater.

phreatic zone in Irabu Island and the average composition of seawater around Irabu Island were used as the two end members. Because the Irabu groundwater system is apparently not closed to CO_2 (Figure 12), the observed partial pressure of CO_2 through the mixing zone (Figure 13) was used for the calculations, and the modeled results (Figure 12) do not pass through the seawater end member.

According to the PHREEQE calculations, the mixture of freshwater with seawater should result in calcite undersaturation in the upper mixing zone. The groundwater is supersaturated with respect to calcite in the freshwater phreatic zone. The calcite is undersaturated in mixtures containing 1.6 to 60% seawater (Figure 12). The results indicate that the theoretical values are approximately coincident with the observed values in mixtures containing greater than about 70% seawater, but that there is a significant difference between the theoretical and observed values in mixtures containing less than about 70% seawater. Furthermore, the observed values are lower than the theoretical values in the top of the freshwater phreatic zone where calcite precipitation is inferred from the changes in *Ca-WRI* values, and the relation is opposite in the upper mixing zone where calcite dissolution occurs.

Theoretical and observed saturation indices, and the *Ca-WRI* values in the low-salinity mixtures, show the chemical condition of the groundwater at different stages of the diagenesis. Groundwater passing

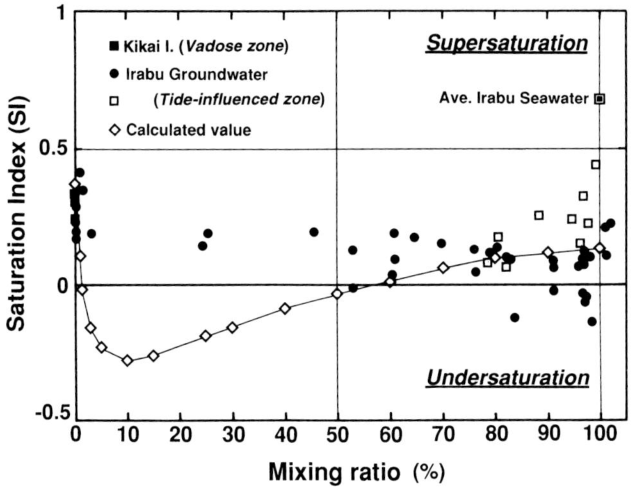

Figure 12. Relationship between calcite saturation index and the proportion of admixed seawater. Solid line shows theoretical mixing curve. Observed *SI* values at high salinity coincide with theoretical values, but observed *SI* values at low salinity do not match theoretical values. Theoretical mixing curve is obtained from modeling using PHREEQE (Parkhurst et al., 1980) and a system open to CO_2.

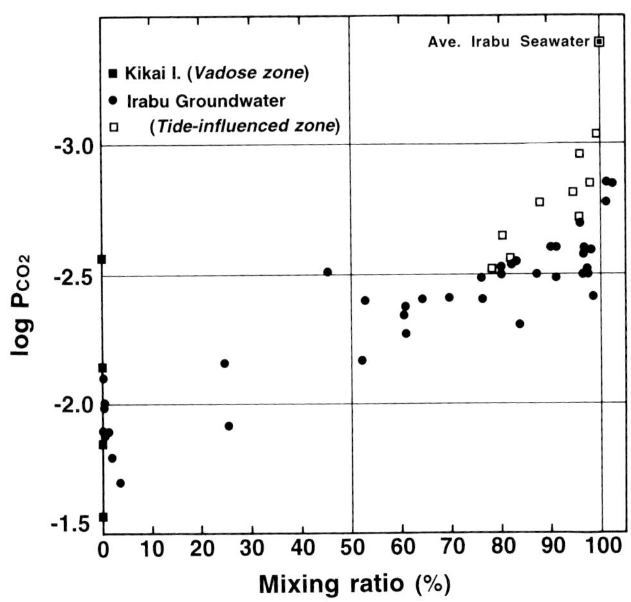

Figure 13. Relationship between log P_{CO_2} and the proportion of admixed seawater. The P_{CO_2} level is highest in the upper part of the mixing zone (mixing ratio = 1–5% seawater) and P_{CO_2} levels become lower with increases in the proportion of admixed seawater.

through the vadose zone has high partial pressure of CO_2. However, the lowering of P_{CO_2} at the water table raises the theoretical *SI* values and causes calcite precipitation, shown by *Ca-WRI* values. As a result, the observed *SI* values are lowered. In the upper mixing zone, the theoretical *SI* values would have been lowered, becoming undersaturated with respect to calcite, by addition of CO_2 from the decay of organic matter and by a mixing of freshwater with seawater. Consequently, calcite dissolution occurs, and it increases the *Ca-WRI* value and raises the observed *SI* values to supersaturation. Therefore, the theoretical and observed *SI* values, and the *Ca-WRI* values, express different stages of diagenetic reactions. The theoretical *SI* values represent chemical conditions before the diagenetic reaction, the *Ca-WRI* values represent the diagenetic reaction itself, and the observed *SI* values show the chemical condition after the reaction.

As pointed out before, the diagenesis on Irabu Island is controlled by CO_2 fluxes. The calculated calcite saturation state is related to the partial pressure of CO_2. A positive correlation between the saturation state and the partial pressure of CO_2 is observed in parts of the aquifer which are not influenced by tidal exchange. However, a negative correlation is observed in the groundwater samples strongly influenced by tidal exchange (Figure 14). This would suggest that the saturation state in the system is controlled by input/output of CO_2, which is associated with: (1) supply from a soil zone, (2) formation of CO_2 by decay of organic matter by bacteria in water, (3) degassing from water table, and (4) exchange of CO_2 between the groundwater and seawater.

POROSITY CHANGE WITH EARLY DIAGENESIS

Porosity changes drastically by dissolution and precipitation of carbonate minerals during diagenesis. From the petroleum geological point of view, diagenetic changes in porosity influence ultimate reservoir properties of the carbonate rock. Therefore, it is very important for petroleum exploration to clarify the change of porosity and porosity distribution by diagenesis. Amounts of dissolution/precipitation within each diagenetic zone were determined by changes in the *Ca-WRI* values during groundwater flow, and the change of porosity was estimated from the dissolution/precipitation rate of carbonate minerals. Estimates of porosity change in the vadose, freshwater phreatic, and upper mixing zones are discussed in this section.

Porosity Change in the Vadose Zone

The vadose zone is considered to be a site of overall dissolution. The excess Ca^{2+} in the freshwater phreatic zone apparently results from dissolution of limestone within the vadose zone during the movement of groundwater from the subaerial surface to the water table. The dissolution rate in the vadose zone in Irabu Island could not be determined directly. However, because the *Ca-WRI* values in the freshwater phreatic zone are almost the same on Irabu and Kikai islands and both islands are in the same geological setting, it is assumed here that the dissolution rate in the vadose

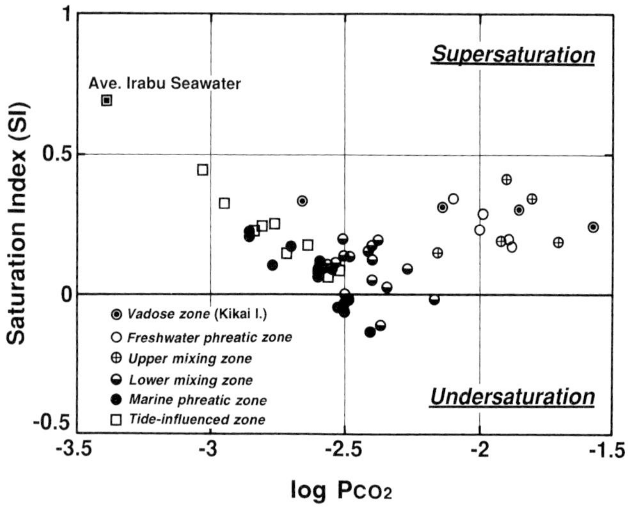

Figure 14. Relationship between calcite saturation index and the log P_{CO_2} of Irabu groundwater. The groundwater within higher P_{CO_2} realm tends to be supersaturated with respect to calcite.

zone on Irabu Island is the same as that on Kikai Island. Although the Kikai groundwater just above the water table has the *Ca-WRI* value of 3.16 mM/L, all groundwaters do not necessarily reach the maximum value just above the water table because each groundwater passes through different flow paths requiring different amounts of time to move through the vadose zone. Therefore, the average *Ca-WRI* value of four springs on Kikai Island (2.27 mM/L) is used for calculations.

The annual recharge into the vadose zone can be calculated from an annual rainfall of 2200 mm in the adjacent Miyako Island (Maritime Safety Agency of Japan, 1986), a recharge rate of 40% (Okinawa General Bureau, 1983) and 2.95×10^7 m^2 for the gross area of Irabu Island. The calculation shows that the annual recharge into the vadose zone on Irabu Island is 2.60×10^7 m^3. Consequently, the amount of dissolved calcite in a year is 5.90×10^7 moles (M). This molar volume is equivalent to 2.18×10^3 m^3 of dissolved limestones per year. The volume of Irabu Island above mean sea level is 1.10×10^9 m^3, thus the rate of volume reduction is 1.98×10^{-4}% in a year. This value suggests that the rate of porosity increase in the vadose zone is 1.98% in 10,000 yr. This is a maximum rate because the above calculation does not separate dissolution in the subsurface vadose zone from dissolution at the subaerial exposure surface. Dissolution on the surface does not increase subsurface porosity.

Anthony et al. (1989) performed a similar calculation for Laura Island of Majuro Atoll, Marshall Islands. They found a rate of porosity increase of 1.1×10^{-2}% per year (=110% in 10,000 yr), which is 55 times as much as that on Irabu Island. Laura Island is underlain by Holocene calcareous sediments composed of aragonite and high-Mg calcite with minor amounts of low-Mg calcite. Annual rainfall on Laura Island is 3404 mm, which is about 1.5 times as much as that on Irabu Island. The difference in the rate of porosity increase between the two islands is probably caused by the difference in the annual rainfall and the mineralogy of the sediments.

Porosity Change in the Freshwater Phreatic Zone

Many of the Ca^{2+} ions from the vadose zone precipitate at the water table as low-Mg calcite due to P_{CO_2} changes in the freshwater phreatic zone, and, consequently, *Ca-WRI* values decrease. The rate of calcite precipitation in the freshwater phreatic zone is calculated as follows. Groundwater passing through the vadose zone has an excess of Ca^{2+} ions. The *Ca-WRI* value averages 2.27 mM/L just above the water table in Kikai Island, as assumed before, and the average *Ca-WRI* value in the freshwater phreatic zone in Irabu Island is 2.03 mM/L. The difference between these (Δ*Ca-WRI* = –0.24 mM/L) may represent the amount removed by calcite precipitation. As mentioned above, calcite precipitation does not occur throughout the freshwater phreatic zone, but is probably limited to just below the water table. Therefore, we assume that calcite precipitation occurs in only the uppermost 1.5 m of the freshwater phreatic zone.

Total flow into the freshwater phreatic zone can be estimated from the total annual recharge into the area of the freshwater lens. According to this study and Okinawa General Bureau (1983), the area of the freshwater phreatic zone is 2.04×10^7 m^2. The annual specific recharge is, as described before, 880 mm/cm^2. Therefore, the total annual recharge into the freshwater phreatic zone is 1.80×10^7 m^3. From the total annual recharge and Δ*Ca-WRI* value, the amount of calcite precipitation in a year is calculated to be 1.64×10^2 m^3. The volume of the upper 1.5 m of the freshwater zone is 3.06×10^7 m^3, thus calcite precipitation increases that volume by 5.36×10^{-4}% per year. That is, a porosity decrease of 5.36% per 10 k.y.

Some workers have attempted to estimate the precipitation rate of calcites in the freshwater phreatic zone. Halley and Harris (1979) pointed out from a study of Holocene oolites on Joulters Cays, Bahamas, that calcite cement in the freshwater phreatic zone precipitated at an average rate of 27–55% for 10 k.y. Budd (1988a) studied the transformation rate of aragonite to low-Mg calcite in the Bahamas, and derived rates of 4.8 to 119.6 cm^3/m^3/yr (4.8–119% per 10 k.y.). Furthermore, McClain et al. (1992) indicated from the data of Holocene biogenic carbonates on Ocean Bight that cumulative magnitudes of low-Mg calcite precipitation range between 6.03 mM/L and 24.04 mM/L, with an average of 15.37 mM/L. The calculated rates from a series of studies on Holocene carbonates are much larger than that on Irabu Island. Low-Mg calcite precipitation in Holocene carbonates is associated with dissolution of metastable carbonate minerals with different solubilities. Therefore, the supply of Ca^{2+} ions by the dissolution of metastable carbonate minerals would make the groundwater more supersaturated with respect to low-Mg calcite, and would promote the precipitation of low-Mg calcite. Thus, the precipitation rate is lower on Irabu Island because limestones are almost entirely low-Mg calcite.

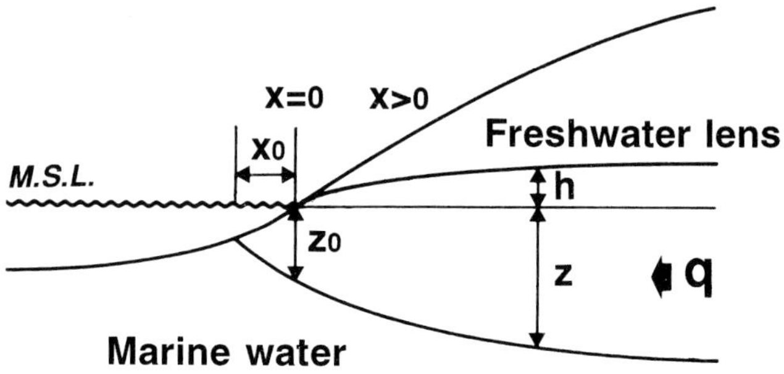

Figure 15. Schematic profile of groundwater distribution and movement in the coastal area. See text for explanation of parameters.

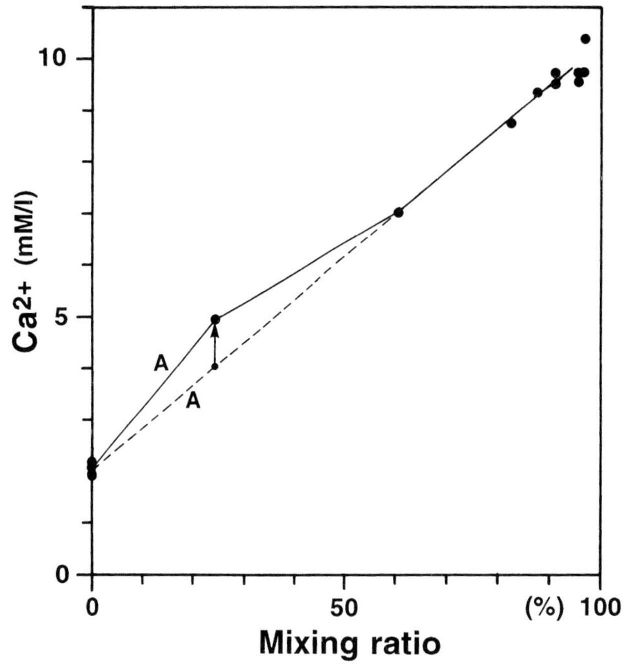

Figure 16. Cross plot of Ca^{2+} ion versus the proportion of admixed seawater in CR-3 in the inland area. Dashed lines reflect ideal concentrations inferred from simple diffusive mixing, and solid lines are actual concentrations. Arrows indicate shifts of actual concentration from ideal concentration, which correspond to the amounts of dissolution of calcite. See text for explanation of A symbols.
Figure 17. Schematic profile of early diagenetic

Porosity Change in the Upper Mixing Zone

As described before, calcite dissolution occurs in the upper mixing zone. The porosity change within the upper mixing zone in coastal and inland areas will be discussed separately.

Porosity Change in the Coastal Mixing Zone

Porosity change associated with calcite dissolution in the upper mixing zone can be estimated from the amount of groundwater passing through the upper mixing zone. In Irabu Island, the freshwater phreatic zone is floating on the marine phreatic zone as a "freshwater lens," and the rim of the lens in the coastal area shows a wedge-like shape (Figure 15). In the wedge-shaped coastal mixing zone, vertical flow velocity of groundwater can be neglected relative to horizontal flow velocity, and the flow of groundwater can be modeled as horizontal, quasi-uniform flow (Tamai, 1989). When the position of an interface between the freshwater and marine phreatic zones is stable, the shape of the interface is defined by Dupuit-Ghyben-Herzberg equation as follows (Vacher, 1988):

$$z^2 = 2qx/\gamma K + q^2/\gamma^2 K^2 \quad (5)$$

where q is specific discharge, K is hydraulic conductivity, γ is $\rho s - \rho f/\rho s$ (ρs = density of seawater, ρf = density of freshwater), x is distance from the shoreline, and z is depth of the interface. Assuming that the interface is the 50% seawater isohaline, the specific discharge, q, is calculated to be 9.81×10 m/yr (= 0.269 m/day) at well CR-1 (x = 1100 m, z = 8.5 m, and average K = 0.354 cm/sec; Okinawa General Bureau, 1983).

The *Ca-WRI* value at the base of the upper mixing zone increases by 0.58 mM/L relative to the freshwater phreatic zone. On the basis of this Δ*Ca-WRI* value, the specific discharge, and a horizontal flow distance of 1000 m, the annual rate of volume reduction in the upper mixing zone is estimated to be 2.10×10^{-4}%, which indicates a porosity increase of 2.10% per 10 k.y. This value is concordant with a 2.30% porosity increase estimated by Sanford and Konikow (1989a, b) in mixing zones by using a flow simulation model combined with an aqueous equilibrium model and a mass transport model.

Porosity Change Rate in the Inland Mixing Zone

In an inland area, the interface between freshwater and marine water is horizontal. As groundwater flows horizontally in the inland area, the flow direction is normal to the gradient of concentration. Therefore, mass transfer is not controlled by the movement of the groundwater, but by diffusion. The rate of porosity change should also be estimated on the basis of diffusive phenomena.

Mass transfer by diffusion is generally represented by Fick's Law as follows (Berner, 1980):

$$\text{flux} = -\phi Db\ (\delta C/\delta z) \quad (7)$$

where ϕ is effective porosity, Db is bulk diffusion coefficient, and $\delta C/\delta z$ is the concentration gradient. If the Ca^{2+} ion concentration in a mixing zone is due to the simple diffusive mixing of freshwater and seawater, the concentration gradient, $\delta C/\delta z$, would be represented by dashed line A in Figure 16, or 0.799 mM/m. The flux (= *F-IDEAL*) would then be 2.21×10^{-7} M/cm^2/yr (where 0.25 and 3.5×10^{-6} cm^2/sec were used as effective porosity and bulk diffusion coefficient, respectively). However, the actual concentration gradient in the upper mixing zone (1–30% seawater) is 1.15 mM/m (Figure 16; solid line A), and the actual flux (= *F-TRUE*) is thus 3.19×10^{-7} M/cm^2/yr. Therefore, the difference between the *F-IDEAL* and *F-TRUE*

Table 3. Porosity data of limestones from the Ryukyu Group.

	CR-1			CR-3			CR-5			Total Average
	Max.	Min.	Avg.	Max.	Min.	Avg.	Max.	Min.	Avg.	
Vadose zone	39.3	10.7	25.2 (no. = 27)	43.2	9.0	23.7 (no. = 36)	42.2	14.8	31.6 (no. = 9)	25.3 (no. = 72)
Freshwater phreatic zone	—	—	—	30.6	18.0	23.4 (no. = 9)	—	—	—	23.4 (no. = 9)
Upper mixing zone	38.6	10.7	29.8 (no. = 6)	24.4	20.4	22.4 (no. = 2)	—	—	—	28.0 (no. = 8)
Lower mixing zone	39.3	13.2	24.7 (no. = 9)	31.2	17.5	24.7 (no. = 10)	28.2	16.6	22.4 (no. = 2)	24.5 (no. = 21)
Tide-influenced zone	—	—	—	—	—	—	36.5	12.3	26.5 (no. = 35)	26.5 (no. = 35)
Marine phreatic zone	46.5	7.8	23.8 (no. = 61)	42.0	7.0	23.4 (no. = 30)	40.2	10.3	27.3 (no. = 50)	25.0 (no. = 141)

(0.98×10^{-7} M/cm^2/yr) indicates an increase in the flux of Ca^{2+} ions by dissolution of limestones. From this value, the rate of porosity increase can be calculated to be 1.45×10^{-6}% per year, which corresponds to a porosity increase of 0.0145% per 10 k.y.

Geological Evaluation on the Calculated Porosity Change Rates

The above estimates indicate that porosity changes in different diagenetic environments vary from an increase in porosity of 2.10% to a decrease of 5.36% in 10,000 years. These rates may seem too large for geological time scale because, for example, the limestones in a vadose zone would reach 100% porosity within about 500,000 years. However, the hydrologic framework is controlled by many factors such as rainfall, evaporation, flow path, and sea level; consequently, it is never steady. The present hydrologic framework on Irabu Island was probably established a few thousand years ago when sea level was stabilized at the present position. Therefore, most of the limestones on the island would have been in several different hydrologic environments since deposition.

The study area is situated on an active island-arc setting and has undergone substantial tectonic movement during the late Pleistocene. Obata and Tsuji (1992) and Honda et al. (1994) pointed out that the Irabu Island area subsided tectonically throughout the deposition of the Ryukyu Group. This tectonic subsidence continued at least until 0.39 Ma, and, after that, Irabu Island and surrounding area changed abruptly from a site of subsidence to a site of uplift. The Pleistocene is characterized by high-amplitude sea level changes of approximately 100 m. Thus, the island would have been repeatedly exposed and submerged causing repeated diagenetic alteration in a variety of diagenetic environments, resulting in a complicated diagenetic history.

The limestones on Irabu Island include a variety of diagenetic products, as mentioned before. Equant blocky cements are dominant in all units regardless of present-day diagenetic environment. These cements occupy primary and secondary porosity and are sometimes cut by vuggy pores. Meniscus cements are also observed in limestones within the present marine phreatic zone. These observations support repeated diagenetic alteration in a variety of diagenetic environments. Table 3 shows porosity data of the cored limestones with each present diagenetic zone based on conventional core analyses. The data indicate that porosity is not significantly different in limestones present in different diagenetic zones. This fact supports the movement of diagenetic environments through time. Therefore, rates of porosity change calculated from hydrogeochemical data are reasonable.

SUMMARY

In conclusion, analyses of the groundwater in Irabu and Kikai Islands are useful for understanding the diagenetic alteration of porosity on carbonate islands. Figure 17 schematically illustrates where porosity changes occur on Irabu Island. Rates of porosity change within each diagenetic zone are summarized in Table 4. In the vadose zone, the concentration of Ca^{2+} ions in the groundwater increases gradually by dissolution of host limestones as groundwater flows through the zone. Dissolution of the host limestone is accelerated by elevated P_{CO_2} which was acquired when the groundwaters moved through the soil zone which has high P_{CO_2} levels. Porosity increase is estimated at 1.98% per 10 k.y. in the vadose zone. Excess

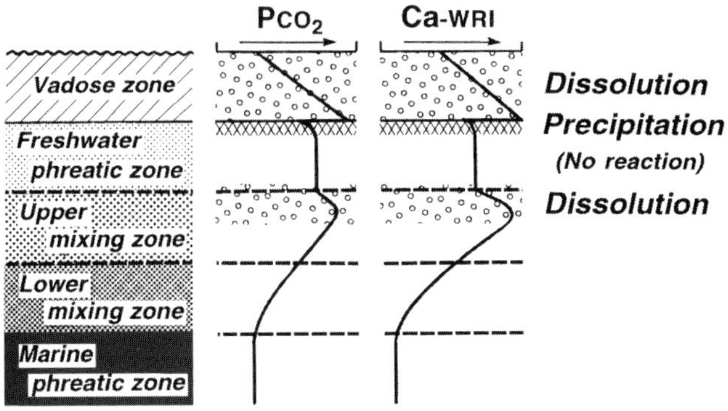

Figure 17. Schematic profile of early diagenetic process in each diagenetic environment.

Ca^{2+} ions decrease from the vadose zone to the top of the freshwater phreatic zone because of calcite precipitation associated with CO_2-degassing across the water table. Porosity decreases in the upper 1.5 m of the freshwater phreatic zone at a rate of 5.36% per 10 k.y. However, P_{CO_2} rises again in the upper mixing zone, probably due to the decay of suspended organic matter. As a result, the *Ca-WRI* value in the upper mixing zone increases due to dissolution of host limestones. Porosity in the coastal mixing zone increases at a rate of 2.10% per 10 k.y. The lower mixing and marine phreatic zones are characterized by a huge depletion of Mg^{2+} ions, caused by modern dolomite formation in Holocene calcareous sediments on an insular shelf. In contrast, the porosity change is very slow in the inland mixing zone with porosity increasing at a rate of 0.0145% per 10 k.y.

Longman (1980) pointed out that diagenetic processes are significantly different depending on whether the diagenetic environment is active or stagnant. According to Longman (1980), little if any dissolution and cementation will occur in a stagnant zone, whereas significant amounts of cementation and dissolution can occur in an active zone. In the Irabu groundwater system, calcite dissolution and precipitation occur in the vadose, upper freshwater phreatic, and upper coastal mixing zones. If significant amounts of cementation and dissolution occur in active zones as pointed out by Longman (1980), calcite cementation should occur not only at the water table but also occur throughout the freshwater phreatic zone, because the entire freshwater phreatic zone is active. Nevertheless, the calcite precipitation is, in fact, limited to the upper few meters of the zone. Therefore the key to diagenetic alteration is not an active hydrologic environment, but the presence CO_2 fluxes. Thus, the diagenetic processes on Irabu Island, which is composed almost entirely of low-Mg calcite, are mainly controlled by CO_2 fluxes.

Table 4. Estimated rates of porosity change associated with early diagenesis.

	Porosity Change Rate (% per 10 k.y.)	
Vadose zone	+1.98	
Freshwater phreatic zone		
Upper 1.5 m thick zone	–5.36	
Middle to lower zone	0	
	Coastal area	Inland area
Upper mixing zone	+2.10	+0.0145

ACKNOWLEDGMENTS

This study was carried out under the research project entitled "Techniques for Interpretation and Reconstruction of Reservoir Development" by Technology Research Center of Japan National Oil Corporation (JNOC-TRC). We thank the numerous colleagues at JNOC-TRC for their discussions and assistance. The Irabu Town Office has provided every convenience for the surveys. Field operations and laboratory analyses were carried out by Okinawa Environmental Analysis Center Co., Ltd. and Japan Oil Engineering Co., Ltd. We thank reviewers Paul Wagner, John Humphrey, Arthur Saller, and David Budd for suggestions that improved this manuscript. Thanks also to JNOC-TRC for permitting the publication of this paper.

REFERENCES CITED

Anthony, S. S., F. L. Peterson, F. T. MacKenzie, and S. N. Hamlin, 1989, Geohydrology of the Laura freshwater lens, Majuro atoll: a hydrogeochemical approach: Geological Society of America Bulletin, v. 101, p. 1066–1075.

Back, W., B. B. Hanshaw, J. S. Herman, and J. N. Van Driel, 1986, Differential dissolution of a Pleistocene reef in the ground-water mixing zone of coastal Yucatan, Mexico: Geology, v. 14, p. 137–140.

Badiozamani, K., 1973, The Dorag dolomitization model—application to the middle Ordovician of Wisconsin: Journal of Sedimentary Petrology, v. 43, p. 965–984.

Berner, R. A., 1980, Early diagenesis. A theoretical approach: Princeton, Princeton University Press, 241 p.

Budd, D. A., 1988a, Aragonite-to-calcite transformation during fresh-water diagenesis of carbonates: Insights from pore-water chemistry: Geological Society of America Bulletin, v. 100, p. 1260–1270.

Budd, D. A., 1988b, Petrographic products of freshwater diagenesis in Holocene ooid sands, Schooner Cays, Bahamas: Carbonates and Evaporites, v. 3, p. 143–163.

Halley, R. B., and P. M. Harris, 1979, Fresh-water cementation of a 1,000 year old oolite: Journal of Sedimentary Petrology, v. 49, p. 969–988.

Hanor, J. S., 1978, Precipitation of beachrock cements: mixing of marine and meteoric waters vs. CO_2-

degassing: Journal of Sedimentary Petrology, v. 48, p. 489–501.

Honda, N., Y. Tsuji, H. Matsuda, and J. Saotome, 1994, Sea level changes and development of the Pleistocene Ryukyu Group in the Irabu Island region, southwest Japan: Journal of the Japanese Association for Petroleum Technology, v. 59, p. 86–98.

Jankowski, J., and G. Jacobson, 1991, Hydrochemistry of groundwater-seawater mixing zone, Nauru Island, central Pacific Ocean: BMR Journal of Australian Geology & Geophysics, v. 12, p. 51–64.

Konishi, K., S. O. Schlanger, and A. Omura, 1970, Neotectonic rates in the central Ryukyu Islands derived from ^{230}Th coral ages: Marine Geology, v. 9, p. 225–240.

Konishi, K., A. Omura, and O. Nakamichi, 1974, Radiometric coral ages and sea level records from the Late Quaternary reef complexes of the Ryukyu Islands: Proceedings of 2nd International Coral Reef Symposium, v. 2, p. 595–613.

Longman, M. W., 1980, Carbonate diagenetic textures from nearshore diagenetic environments: AAPG Bulletin, v. 64, p. 461–487.

Maritime Safety Agency of Japan, Hydrographic Department, 1986, Maritime basic map (1:50,000 scale)—Report of submarine geographic and geological surveys, Miyako Island: 59 p.

Matsuda, H., and A. Iijima, 1986, Trace dolomite in the borehole cores at Irabu Island, Miyako Islands (abs.): Abstracts of Geological Society of Japan 93rd Annual Meeting, p. 314.

Matsuda, H., K. Baba, and Y. Tsuji, 1994, Discovery of modern dolomite in reef-associated shallow-marine shelf carbonates under normal marine condition, off Miyako Islands, southwest Japan (abs.): Abstracts of 14th International Sedimentological Congress, B-10.

McClain, M. E., P. K. Swart, and H. L. Vacher, 1992, The hydrogeochemistry of early meteoric diagenesis in a Holocene deposit of biogenic carbonates: Journal of Sedimentary Petrology, v. 62, p. 1008–1022.

Moore, C. H., 1989, Carbonate diagenesis and porosity: developments in Sedimentology 46: New York, Elsevier, 338 p.

Noma, Y., 1978, Groundwater in Kikai-jima—Hydrogeology of Amami Islands (1): Bulletin of the Geological Survey of Japan, v. 29, p. 145–157.

Obata, M., and Y. Tsuji, 1992, Quaternary geohistory inferred by seismic stratigraphy of a carbonate province in an active margin, off Miyako Islands, South Ryukyus, Japan: Carbonates and Evaporites, v. 7, p. 150–165.

Okinawa General Bureau, 1983, Groundwater in Okinawa Prefecture; hydrogeological map of Okinawa Prefecture: Naha, Okinawa General Bureau, 95 p.

Omura, A., 1988, Geologic history of the Kikai Island, central Ryukyus, Japan; summary of uranium-series dating of fossil corals from the Riukiu Limestone, *in* T. Itaya, N. Imai, A. Omura, M. Suzuki, N. Nakai, K. Ninagawa and K. Hirooka, eds., Quaternary dating methods: Memoirs of the Geological Society of Japan 29, p. 253–268.

Omura, A., Y. Tsuji, K. Ohmura, and Y. Sakuramoto, 1985, New data on uranium-series ages of hermatypic corals from the Pleistocene limestone on Kikai, Ryukyu Islands: Transactions and Proceedings of the Palaeontological Society of Japan, New Series 139, p. 196–205.

Parkhurst, D. L., D. C. Thorstenson, and L. N. Plummer, 1980, PHREEQE—A computer program for geochemical calculations: U.S. Geological Survey, Water-Resources Investigation 80-96, 193 p., revised 1985.

Plummer, L. N., 1975, Mixing of sea water with calcium carbonate ground water: Geological Society of America Memoir, v. 142, p. 219–236.

Plummer, L. N., H. L. Vacher, F. T. MacKenzie, O. P. Bricker, and L. S. Land, 1976, Hydrogeochemistry of Bermuda: a case history of groundwater diagenesis of biocalcarenites: Geological Society of America Bulletin, v. 87, p. 1301–1316.

Sado, K., K. Kameo, K. Konishi, T. Yuki, and Y. Tsuji, 1992, A new age assignment of Riukiu Limestone (Pleistocene) by calcareous nannofossils from the borehole cores at Irabu Island, South Ryukyus, Japan: Journal of Geography, v. 101, p. 127–132.

Sanford, W. E., and L. F. Konikow, 1989a, Simulation of calcite dissolution and porosity changes in saltwater mixing zones in coastal aquifers: Water Resource Research, v. 25, p. 655–667.

Sanford, W. E., and L. F. Konikow, 1989b, Porosity development in coastal carbonate aquifers: Geology, v. 17, p. 249–252.

Smart, P. L., J. M. Dawans, and F. Whitaker, 1988, Carbonate dissolution in a modern mixing zone: Nature, v. 335, p. 811–813.

Stoessell, R. K., W. C. Ward, B. H. Ford and J. D. Schuffert, 1989, Water chemistry and $CaCO_3$ dissolution in the saline part of open-flow mixing zone, coastal Yucatan Peninsula, Mexico: Geological Society of America Bulletin, v. 101, p. 159–169.

Tamai, N., 1989, Hydraulics 2: Tokyo, Baifukan, 207 p.

Tsuji, Y., 1993, Tide influenced high energy environments and rhodolith-associated carbonate deposition on the outer shelf and slope off the Miyako Islands, southern Ryukyu Island Arc, Japan: Marine Geology, v. 113, p. 255–271

Tsuji, Y., H. Matsuda, K. Baba, N. Honda, T. Yuki, and S. Nomoto, 1990, Diagenesis of Quaternary Riukiu Limestone in Irabu Island, Okinawa Prefecture: Journal of the Japanese Association for Petroleum Technology, v. 55, p. 288–289.

Vacher, H. L., 1988, Dupuit-Ghyben-Herzberg analysis of strip-island lenses: Geological Society of America Bulletin, v. 100, p. 580–591.

Wigley, T. M. L., and L. N. Plummer, 1976, Mixing of carbonate waters: Geochimica et Cosmochimica Acta, v. 40, p. 989–995.

Yuki, T., H. Iwamoto, Y. Tsuji, T. Kodato, H. Sunouchi, T. Yamamura, and M. Obata, 1988, Cyclic sedimentary sequences and their lateral variation in the Ryukyu Limestone (Pleistocene), Irabu Island, Okinawa, Japan (abs.): SEPM 1988 Annual Midyear Meeting, p. 59.

Chapter 3

Karst Development on Carbonate Islands

John E. Mylroie
Mississippi State University
Mississippi State, Mississippi, U.S.A.

James L. Carew
University of Charleston
Charleston, South Carolina, U.S.A.

ABSTRACT

Karst development on carbonate platforms occurs continuously on emergent portions of the platform. Surficial karst processes produce an irregular pitted and etched surface, or epikarst. The karst surface becomes mantled with soil, which may eventually result in the production of a resistant micritic paleosol. The epikarst transmits surface water into vadose pit caves, which in turn deliver water to a diffuse-flow aquifer. These pit caves form within a 100,000 yr time frame. On islands with a relatively thin carbonate cover over insoluble rock, vadose flow perched at the contact of carbonate rock with insoluble rock results in the lateral growth of vadose voids along the contact, creating large collapse chambers that may later stope to the surface.

Carbonate islands record successive sequences of paleosols (platform emergence) and carbonate sedimentation (platform submergence). The appropriate interpretation of paleosols as past exposure surfaces is difficult, because carbonate deposition is not distributed uniformly, paleosol material is commonly transported into vadose and phreatic voids at depth, and micritized zones similar in appearance to paleosols can develop within existing carbonates.

On carbonate islands, large dissolution voids called flank margin caves form preferentially in the discharging margin of the freshwater lens from the effects that result from freshwater/saltwater mixing. Similarly, smaller dissolution voids also develop at the top of the lens where vadose and phreatic freshwaters mix. Independent of fluid mixing, oxidation of organic carbon and oxidation/reduction reactions involving sulfur can produce acids that play an important role in phreatic dissolution. This enhanced dissolution can produce caves in freshwater lenses of very small size in less than 15,000 yr. Because dissolution voids develop at discrete horizons, they provide evidence of past sea level positions. The glacio-eustatic sea level changes of the Quaternary have overprinted the dissolutional record of many carbonate islands with multiple episodes of vadose, freshwater phreatic, mixing zone, and marine phreatic conditions. This record is further complicated by collapse of caves, which produces upwardly prograding voids whose current position does not correlate with past sea level positions.

The location and type of porosity developed on emergent carbonate platforms depend on the degree of platform exposure, climate, carbonate lithology, and rate of sea level change. Slow, steady, partial transgression or regression will result in migration of the site of phreatic void production as the freshwater lens changes elevation and moves laterally in response to sea level change. The result can be a continuum of voids that may later lead to development of solution-collapse breccias over an extended area.

INTRODUCTION

Carbonate platforms have become emergent throughout earth history. Karst development on such emergent platforms is a very rapid geological process. Emergent carbonate platforms are likely to develop karst over a broad area and through a significant thickness. Why is paleokarst, as described in this paper, not more common in the geologic record? Do we fail to recognize the evidence, or are these karst features vulnerable to subsequent erosion and alteration that eliminate them? The major purpose of this paper is to motivate others to re-examine ancient carbonates in light of our understanding of karst processes.

For the purposes of this paper, carbonate platforms in the *oceanic* realm, as defined by Smart and Whitaker (1991, 1993), will be the model used to describe carbonate island karst development. Smart and Whitaker (1991, 1993) define the oceanic realm as that in which carbonates are diagenetically immature, sea level fluctuations are of low order, and continental influences are absent. The focus of this paper is on the effects of recharge, climate, degree of platform exposure, lithology, and rate of sea level change upon carbonate platforms and karst development, using Quaternary carbonate settings as examples. The various hydrological environments found in oceanic carbonate islands are illustrated in Figure 1.

Whenever carbonate platforms are subaerially exposed, karst development occurs. The exposure can be the result of eustatic or local (tectonic) sea level change. In some cases, shoaling may occur and eventually tidal flat, beach, and eolian carbonates may be deposited, producing an exposed land surface without the need of a sea level change. Karst features may develop very rapidly, even as deposition continues, a phenomenon called syngenetic karst (Jennings, 1968).

On emergent carbonate platforms meteoric precipitation falls directly onto the carbonate land surface and enters the subsurface vadose zone (Figure 1). Such input is known as autogenic recharge (Mylroie, 1984a). The meteoric water, slightly acidic from incorporation of atmospheric CO_2, gains more dissolutional potential by addition of soil CO_2. Much of this dissolutional potential is expended rapidly by interaction with the exposed carbonate rocks, and, as a result, water enters the subsurface with significantly diminished dissolutional capability. On the scale of hundreds of meters, water infiltration is relatively uniform over the exposed outcrop as diffuse input (Mylroie, 1984a). Exceptions occur where carbonate platforms are on a

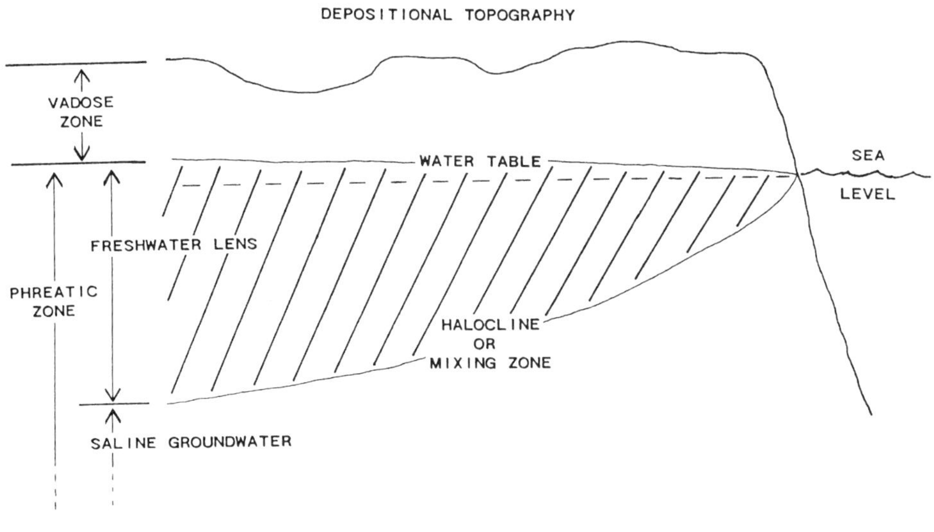

Figure 1. Diagrammatic representation of a freshwater lens in a carbonate island showing the various hydrologic environments found between the land surface and the saline groundwater. The boundary between the freshwater lens and the saline groundwater may be a sharp halocline as pictured here, or a broad region of changing salinity called a mixing zone.

continental margin, or on an island with significant outcrops of insoluble rocks. There, meteoric water collects in streams that flow onto the carbonates. In these cases the input is allogenic, because the water enters the limestone in appreciable volumes at point locations, and its dissolutional potential persists to significant depths in the limestone. Such allogenic inputs are restricted to the contact of the carbonates and insoluble rocks; within the bulk of the carbonate outcrop, autogenic recharge, and the effects of that recharge, dominate.

KARST PROCESSES AND PRODUCTS ON CARBONATE PLATFORMS

Surface Karst

Autogenic recharge to a carbonate platform results in a surficial, irregularly pitted and etched karst surface, with dissolution channels and networks of great complexity that extend through only the upper few meters of rock (Figure 2). The rock contains numerous tubes, holes, and enlarged joints in the size range of centimeters to tens of centimeters. The karst surface is all or partially covered by soil and weathered rock material which may subside into the underlying dissolutional network (Figure 3). This surficial karst and its soil mantle are referred to as epikarst or subcutaneous karst (Williams, 1983). Epikarst follows surface topography and occurs only in the upper few meters of rock. It is, therefore, laterally extensive with potentially great variation in elevation (Figure 4). The epikarst is intimately linked with soil development processes and biological activity. Despite the fact that the epikarst is very permeable, and that the carbonate rocks beneath the epikarst have large porosity, the epikarst is capable of significant water storage. Soil infill and the large surface area provided by the

Figure 2. Photograph of a dissolved and etched carbonate rock surface, San Salvador Island, Bahamas. Camera on right for scale.

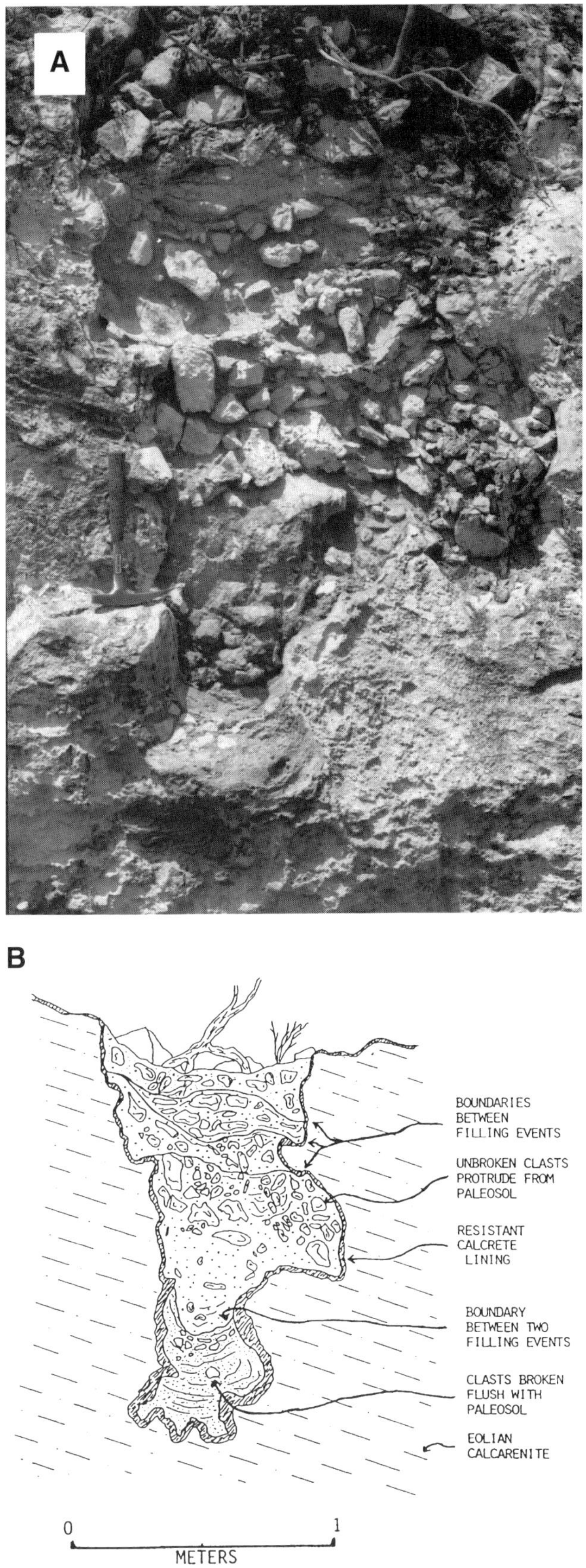

Figure 3. (A) Photograph of a small dissolution pit, filled with weathered rock and paleosol material; (B) diagram of the feature indicates that this infilling may have been episodic.

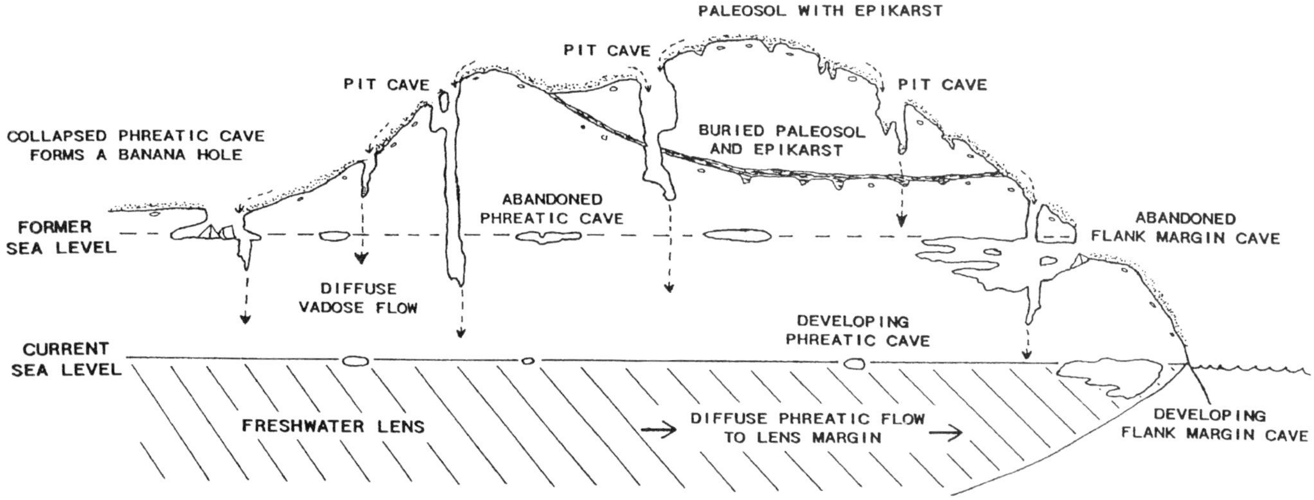

Figure 4. Diagrammatic representation of the main dissolutional features found on carbonate islands: epikarst (with paleosol), pit caves, banana holes and phreatic caves, and flank margin caves. Also shown are the positions of the features relative to a freshwater lens and halocline. Changes in sea level move the position of the karst features. In the Quaternary, these sea level changes led to overprinting of dissolutional environments.

dissolutional network hold water by capillary action. The epikarst should not be confused with coastal karst (commonly referred to as phytokarst), which is the result of modification of carbonate rocks in shoreline settings, where endolithic algae, sea spray, grazing invertebrates, dissolution and precipitation, and wave action combine to produce complex, but spatially restricted, etched carbonate surfaces (Viles, 1988).

Paleosols

Coincident with karst development, the land surface becomes mantled with soil formed by collection of residual insolubles and atmospheric dust. Where the soil development has been pronounced, the underlying epikarst can be completely covered. On Quaternary carbonate islands, resistant and relatively impermeable micritic paleosols (terra-rossa paleosols) have developed (Figure 5). Locally, these micritized layers can be relatively impervious to infiltration, and are capable of diverting meteoric recharge to a limited number of point inputs to the subsurface. While this development of small streams may occur over a lateral scale of meters to tens of meters, nevertheless, the water input to the subsurface on the larger scale (hundreds of meters) remains roughly equal per given area.

The proper interpretation of paleosols and the past exposure surfaces they represent can be difficult in carbonate islands (Carew and Mylroie, 1991). Topography, whether depositional or erosional, can result in poorly developed, thin paleosols on ridge crests, and well-developed, thick paleosols in topographic lows. In the rock record, the difference in elevation and degree of development of the resulting paleosol in cores or widely spaced outcrops may lead an observer to incorrectly interpret multiple exposure events.

Calcarenite protosols must be differentiated from terra-rossa paleosols (Figure 6). Calcarenite protosols are fossiliferous, unstructured paleosols formed during brief emergence events or temporary pauses in carbonate deposition that occur within a single sea level highstand. They represent a minimal exposure time for the carbonate platform. Terra-rossa paleosols are the result of the cumulative weathering effects of long-term exposure largely associated with sea level lowstands, and are underlain by a porous epikarst. Terra-rossa paleosols commonly separate carbonate deposits formed during different sea level highstands and, therefore, represent a substantial time of exposure for the carbonate platform.

Carbonate deposition, especially on platforms that are incompletely flooded, is spatially patchy, so

Figure 5. Typical example of a terra-rossa paleosol from San Salvador Island, Bahamas. Note the resistant micritic crusts above figure, and the complex rhizomorphs beneath those crusts.

paleosols may be merged into composite paleosols at the localities that remained emergent. Yet, the same paleosols may be separated by significant sediment packages at other locations that were flooded and underwent carbonate deposition. On Quaternary carbonate islands such as Bermuda or the Bahamas, the patchy deposition of eolianites commonly produces this result, as shown in Figure 4, where buried and exposed paleosols at the right of the figure merge into a single, composite paleosol on the left. Conversely, apparent terra-rossa paleosol horizons can develop within existing carbonates as a result of shallow vadose flow and weathering, and be misinterpreted as exposure surfaces (Carew and Mylroie, 1991; Rossinsky et al., 1992). In addition, paleosol material may be transported into vadose and phreatic caves at depth (Figure 7). Because phreatic caves may form at common elevations, such as the water table, the collection of soil infill at the specific horizons represented by these caves can be later misinterpreted as a surface paleosol formed at a true exposure surface (Carew and Mylroie, 1991).

Large-Scale Karst Landforms

Quaternary carbonate islands today are found in tropical and subtropical settings. One would expect these islands to have karst landforms of the same type as those found in similar climatic settings on continents; however, this is not the case. Widely recognized tropical karst landforms such as sinking streams, springs, blind valleys, and tower karst are not found in these carbonate-island environments. Their absence is caused by the lack of allogenic recharge, specifically, major streams crossing onto the carbonates from non-carbonate terrain.

Figure 6. Outcrop at The Bluff, San Salvador Island, Bahamas. Vertical exposure is 20 m. Overhanging layer at the top of the outcrop is a resistant terra-rossa paleosol similar to that in Figure 5. In the middle of the outcrop, trending horizontally, is a white, unstructured calcarenite protosol.

Depressions

On exposed carbonate platforms that have only autogenic recharge, the dominant karst landforms are closed depressions and caves. Closed-contour depressions of a variety of sizes are found on Quaternary carbonate islands, but many of these, including the largest, are of depositional origin. Their continued expression as topographic depressions is the result of internal drainage by karst processes. While it is clear that these depressions may undergo significant modification by karst and related weathering processes, the majority of their volume is a result of initial depositional topography. Purdy and Bertram (1993) have argued that the major karst process on oceanic carbonate islands is the dissolutional excavation of large central karst depressions, with an elevated residual island rim. However, they fail to explain why the island center should dissolve more rapidly than the island perimeter. Additionally, in the Quaternary, the rate of carbonate deposition during sea level highstands is equivalent to the lesser amount of carbonate dissolution that takes place during the longer duration sea level lowstands, so no net carbonate is lost (Mylroie, 1993a, b). Depressions should experience

Figure 7. Outcrop on New Providence Island, Bahamas, showing modern soil at A, collecting in shallow depression at B, and being piped into a number of dissolutional voids at C, D, E, F, and G. The vertical dashed line illustrates what would appear in a core drilled along that line. Soils at B, C, D, and G are all the same age, but might be misinterpreted in core as separate exposure horizons.

the longest time for carbonate deposition, elevated rims the least. Mass-balance considerations indicate that these large depressions on carbonate islands must have been created by means other than surface dissolution.

The continued development of depositional closed depressions as major karst landforms is controlled in part by climate. For example, whereas Bermuda and San Salvador Island, Bahamas, are carbonate islands of similar size and geology, they have markedly different depression morphologies that result from the effects of their different climates (Vacher and Mylroie, 1991).

Bermuda has a positive water budget (annual precipitation exceeds evapotranspiration), which results in the depressions being sites of recharge to the freshwater lens. In the depressions, vegetation is lush, soil CO_2 is abundant, and soil acidity is high, so, with time, the depressions deepen and expand by dissolution. Eventually the depression-forming processes may breach topographic barriers and the depressions are invaded by the sea. Subsequent marine bio-erosion further enlarges the depressions.

In contrast, on San Salvador Island, Bahamas, the water budget is negative (annual evapotranspiration exceeds precipitation). Where constructional depressions extend below the water table, lakes occur. Extensive evaporation produces hypersaline lakes by upconing of marine water through the freshwater lens. Significant dissolution of the depression margins is inhibited by the hypersalinity, and the original depression morphology (in this case, swales between eolian calcarenites) persists through time.

Figure 8. Typical pit cave, Isla de Mona, Puerto Rico. Pronounced vertical grooving common to vadose dissolution can be seen adjacent to the person.

Caves

Caves found on carbonate platforms fall into three main categories: vadose, phreatic, and fracture caves. Vadose caves consist of pits, which are dissolutional pathways formed by water descending from the surface toward the water table. Phreatic caves (flank margin caves and "banana holes") develop below the water table in the freshwater lens by the mixing of chemically-distinct waters. Fracture caves develop along the margin of carbonate platforms as a result of mechanical failure of the platform margin.

Pit Caves

On a large scale, carbonate platforms have autogenic diffuse input, whereas on the scale of meters to tens of meters, even without the presence of an impermeable paleosol layer, vadose waters do not infiltrate uniformly. Spatial variation in vadose input can leave a recognizable diagenetic record in the underlying vadose zone (Pelle and Boardman, 1989). The small-scale variation in vadose input is a result of the epikarst gathering meteoric water into discrete point inputs, which produces vadose shafts. These vadose shafts are called pit caves (Figure 4), and they penetrate deeply into the limestone (Pace et al., 1993). Pit caves are characterized by a near vertical, commonly stairstepped pattern; by vertical grooves on the walls; and by the absence of the curvilinear dissolutional surfaces associated with phreatic conditions (Figure 8). Pit caves rarely deliver water directly to the water table, but end above the water table. The water completes its journey to the water table as diffuse flow (Figure 4). The fact that pit caves rarely reach the water table reflects the relatively low dissolutional potential of autogenic water derived from an epikarst. While surface dissolution is active and widespread, continued contact with carbonate rock as the water moves downward through the vadose zone consumes dissolutional capability. The active lifetime of individual pit caves is relatively brief, as their downward development is arrested when a newly formed pit cave develops upstream in the epikarst and pirates their recharge. This process forms pit complexes, where the landscape is studded with numerous pit caves, far more than seem justified based on the available catchment and climate (Pace et al., 1993). Pit complexes represent accumulated pit cave development and abandonment over time. In the Bahamas, pit-cave complexes have developed in eolianites that are only 125,000 yr old (Figure 9); the pit caves themselves must have formed at a faster rate to have produced a complex containing both active and abandoned pit caves (Pace et al., 1993).

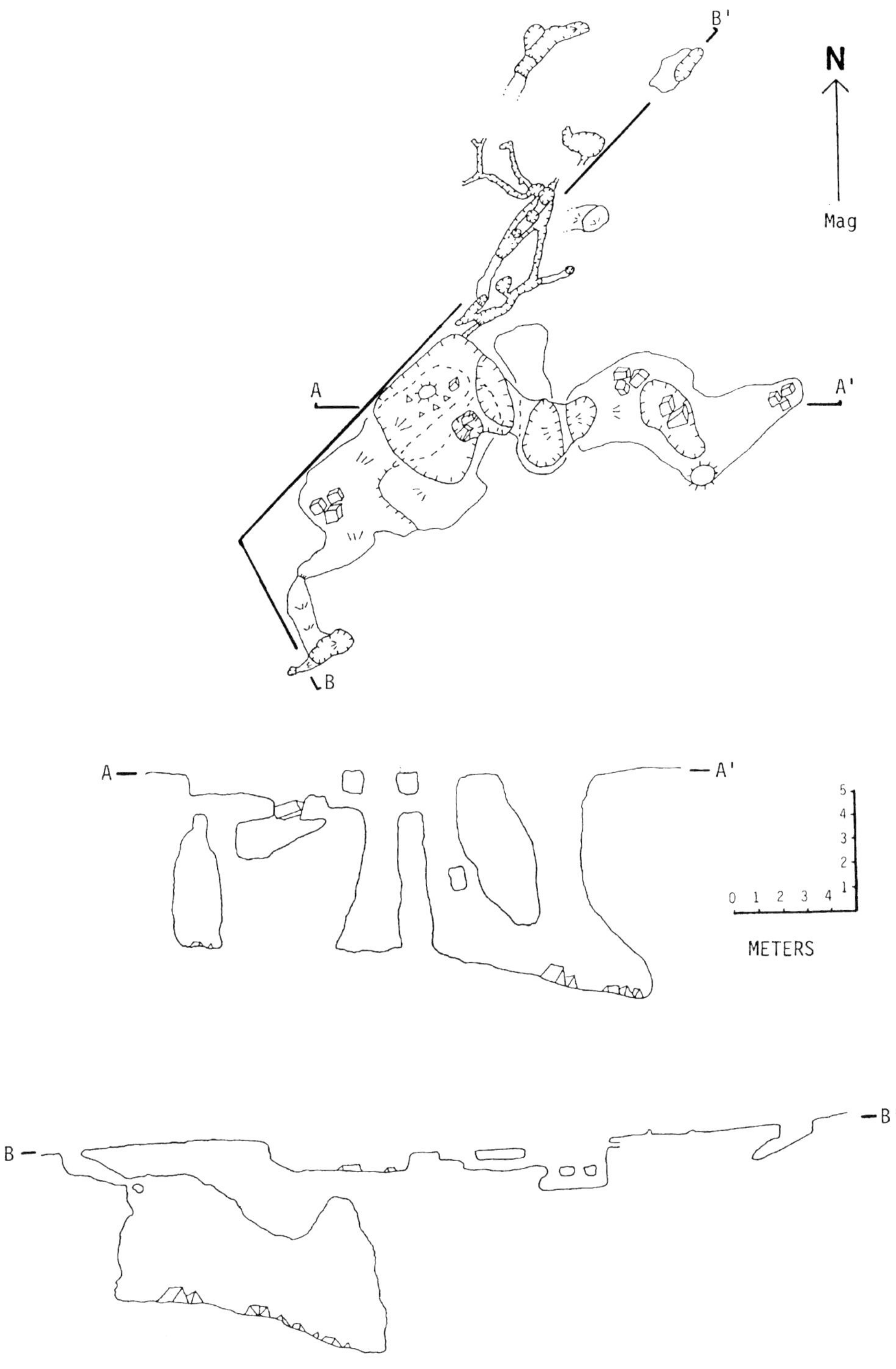

Figure 9. Map and cross section of a typical pit complex at Sandy Point, San Salvador Island, Bahamas. Note the dendritic pattern of shallow channels in the epikarst and the series of pits associated with the shallow flow system.

On islands such as Bermuda, where the carbonates consist of a thin veneer over noncarbonate rocks, vadose flow is channeled along the base of the limestone contact, which results in the lateral growth of vadose voids. Subsequent collapse of these voids, with continued dissolution of the collapse debris, creates large chambers that may stope many tens of meters toward the surface (Figure 10A) (Palmer et al., 1977; Mylroie, 1984b). Though technically not pit caves, these collapse caves owe their origin to dissolutional activity in the vadose zone (Figure 11). A similar phenomenon is reported from Kangaroo Island, Australia (Jennings, 1968).

Flank Margin Caves

In carbonate platforms, large phreatic dissolution voids, called flank margin caves, form preferentially along the margin of the discharging freshwater lens as a result of freshwater/saltwater mixing (Figure 4) (Mylroie and Carew, 1990). The mixing of fresh and marine waters, even if both are saturated with respect to $CaCO_3$, results in a mixture undersaturated with

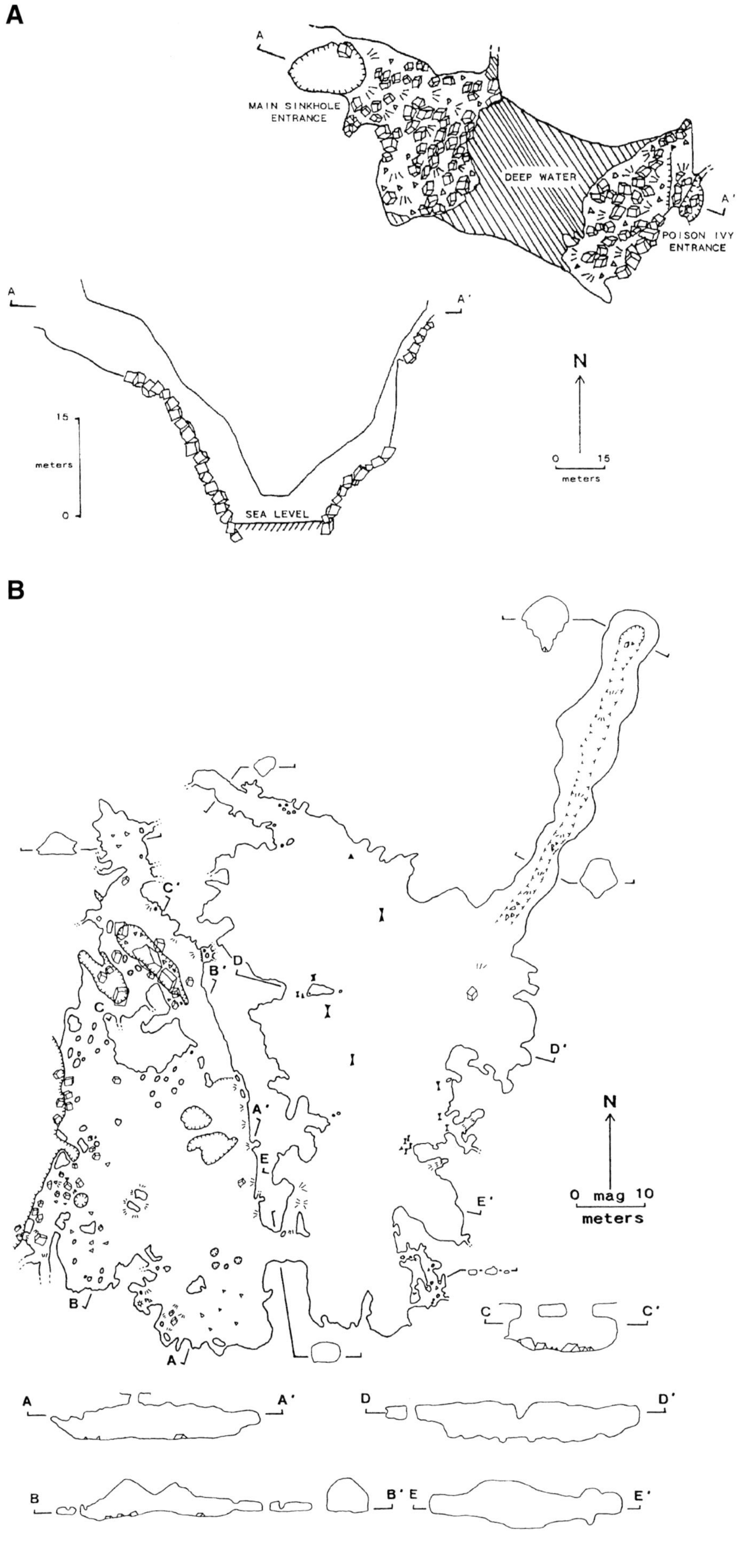

Figure 10. Maps with cross sections of typical caves formed by vadose-induced collapse (A) and flank margin conditions (B). Cave walls shown in bold line, collapse debris shown as blocks, slopes shown as short diverging lines, stalactites and stalagmites as solid triangles, hachures indicate a vertical drop (hachures point to lower elevation), and diagonal lines indicate water. (A) Church Cave, Bermuda, formed by stoping upwards from a deeper void developed by vadose dissolution at the carbonate/igneous rock contact. Note that the floor of the cave is entirely collapse debris, sediment, or seawater. Vertical scale is 2× horizontal scale (redrawn from Mylroie, 1984b). (B) Salt Pond Cave, Long Island, Bahamas, a flank margin cave. The cave has an irregular phreatic morphology as discussed in the text, and is horizontally extensive but vertically restricted.

Figure 11. Photograph of a portion of Admiral's Cave, Bermuda. Note the angular, broken nature of the ceiling, the result of upward progradational collapse of a deep-seated void. Photograph by A. N. Palmer.

respect to $CaCO_3$ (Plummer, 1975). This phenomenon has been recognized as a major means of cave production in the Yucatan of Mexico (Back et al., 1986), where freshwater discharges are high because the Yucatan carbonate platform is fed by allogenic recharge from the North American continent to the west. Based on models from the Yucatan, Sanford and Konikow (1989) determined that large-scale dissolution and cave production resulting from simple physical mixing of fresh and marine water could not operate effectively on small carbonate islands due to limitations of time, catchment, and discharge. Despite this evidence from aquifer modeling, numerous caves are found on small carbonate islands. The dissolution that produces these phreatic voids is only partly the result of physical mixing of fresh and marine waters (Smart et al., 1988a; Mylroie and Carew, 1990). In addition, the presence of organics in the water allows oxidation that produces further CO_2 that drives carbonate dissolution. This process also results in anoxic conditions in the mixing zone of the freshwater lens. Complex oxidation/reduction reactions involving sulfur can occur there, producing acids that lead to further dissolution. Such reactions have been implicated in the development of large phreatic caves in the Bahamas (Bottrell et al., 1991, 1993; Mylroie and Balcerzak, 1992).

Flank margin caves have a morphology that is common to caves formed by mixed-water dissolution in other geologic environments (Figure 10B). That morphology includes large, globular chambers, bedrock spans, thin bedrock partitions between chambers, tubular passages that end abruptly, and curvilinear phreatic dissolutional surfaces (Figures 12 and 13) (Mylroie, 1991; Mylroie et al., 1991). These caves lack ablation scallops, stream channels, and other indications of turbulent conduit flow. Flank margin caves are not conduits, but rather mixing chambers. They receive water from the lens in the island interior as diffuse flow, and discharge that water, after mixing, as diffuse flow to the sea. The caves develop without an external opening to the sea or the land. Entry today is gained as a result of surface erosion breaching the cave (Figure 10B).

Inland from the lens margin, at the top of the lens where vadose and phreatic fresh waters mix, smaller phreatic dissolution voids may also develop (Pace et al., 1993). Collapse of the thin bedrock roofs of these voids forms broad, vertical-walled depressions up to 10 m across, called "banana holes" in the Bahamas (Figure 4). In both banana holes and flank margin caves, the phreatic voids have limited vertical development but are horizontally extensive, the opposite of the pattern seen in pit caves. In the Bahamas, flank

Figure 12. Photograph of a portion of Lighthouse Cave, San Salvador Island, Bahamas. The depositional dip of the eolianites can be seen from upper right to lower left. The configuration of rock pillars and pendants is typical of rock dissolution in the phreatic zone.

margin caves and banana holes are found at elevations of 1 to 6 m above sea level. The Bahamas are tectonically stable, undergoing only isostatic subsidence. Therefore, these caves could only have formed during a glacio-eustatic sea level highstand that reached elevations above modern sea level. Given the subsidence rate of the Bahamas, and the age of these late Quaternary carbonates, only the last interglacial sea level highstand centered approximately 125,000 years ago (oxygen isotope substage 5e) could have produced the caves (Mylroie et al., 1991). This highstand was above modern sea level for only about 15,000 years (Chen et al., 1991); at that time, the Bahamas consisted of islands even smaller than today's, as all land below 6 m elevation was inundated by the sea. These phreatic caves formed in freshwater lenses of very small size in as little as 15,000 years (Mylroie et al., 1991; Carew and Mylroie, 1992, in press). Rapid cave development by the complex geochemistry of the mixing zone leads to a definitive diagenetic record in the cave wall rock that includes trace amounts of dolomite and survival of significant amounts of aragonite in these young rocks (Vogel et al., 1990; Schwabe et al., 1993). Flank margin caves have been described from a number of sites in the Bahamas (Mylroie, 1988; Carew and Mylroie, 1989; Mylroie et al., 1991) and from Isla de Mona, Puerto Rico (Frank, 1993; Mylroie et al., 1993).

Fracture Caves

Many Quaternary carbonate platforms are steep sided. Such steep slopes are prone to mechanical failure, and that may be a major part of how carbonate platform margins are modified through time (Mullins and Hine, 1989). Fractures parallel to bank margins are common in the Bahamas (Daugherty et al., 1987; Carew and Mylroie, 1989; Carew et al., 1992). These fractures are commonly the site of cave systems with extensive vertical and linear development (Palmer, 1986). Such fracture caves may also represent a means by which Neptunian dikes and fissure fills are developed (Smart et al., 1988b). Fracture caves are not dissolutional in origin; however, once they are formed, they act as pathways for water flow into and/or out of the carbonate platform. While they may form without the platform becoming emergent, platform emergence, with subsequent dissolution at the discharging margin of the freshwater lens, may help promote overall rock weakness that leads to bank-margin collapse and fracture development. During glacio-eustatic sea level lowstands, loss of buoyant support that was provided

Figure 13. Photograph of a portion of the main inner chamber of Salt Pond Cave, Long Island, Bahamas (see map, Figure 10B). Floor and ceiling show typical phreatic morphology; dark floor stain is caused by guano, since mined out. The cave is horizontally extensive, but vertically restricted.

by the water during the highstand is probably a contributing factor in the fracture development.

EFFECT OF SEA LEVEL CHANGE ON KARST PROCESSES

Sea level change has an important impact on island karst. Sea level position controls the location of the freshwater lens, and therefore the placement of vadose and phreatic dissolutional environments (Mylroie and Carew, 1988). Whereas the development of vadose voids can occur at a variety of elevations between the surface and the water table, the development of phreatic dissolution voids occurs at discrete horizons that represent past sea level stillstand positions. Flank margin caves are especially indicative of sea level position, because they form in the margin of the lens, where it thins near sea level.

Because flank margin caves develop at the margin of the lens at the edge of islands, they are vulnerable to destruction by scarp and hillslope erosion during subsequent platform emergence. Such erosion appears to be the origin of many long, linear notches found in the sides of hills in the interior of Bahamian islands that were formerly interpreted as fossil bio-erosion notches (Figure 14). Most of these notches are now thought to be the eroded remnants of flank margin caves (Mylroie and Carew, 1991). Such features may be preserved in the rock record, but whether these features originated as bio-erosion notches or flank margin caves makes little difference in their usefulness for determining past sea level position.

In the Bahamas, Quaternary glacio-eustatic sea level changes have repeatedly moved vadose, freshwater phreatic, mixing zone, and marine phreatic environments through a significant vertical range. The top of the freshwater lens has been as high as +6 m, and as low as –125 m (Carew and Mylroie, 1987). The geochemically active mixing zone would have covered an even larger vertical range, as the mixing zone extends well below sea level (to –30 m today on Andros Island). In addition to vertical changes as sea level fluctuated during the Quaternary, the freshwater lenses and their attendant mixing zones changed horizontal position, thickness, and lateral dimension as the amount of exposed platform changed. Caves found at depth in the Bahamas today represent the cumulative dissolutional effects of many sea level events. Many of these caves are accessible through blue holes, which are water-filled shafts that commonly extend downward as much as 100 m. Blue holes are polygenetic; that is, they developed variously as open portions of bank-margin fractures, as prograding collapse (stoping) from voids at depth, or as drowned pit caves formed during a sea level lowstand (Burkeen and Mylroie, 1992). Many contain subaerial speleothems, which indicates that they have been through at least one vadose/phreatic cycle as sea level has changed in response to glacio-eustasy (Palmer, 1985). The overprinting produced by numerous sea level changes, combined with bank-margin fracture development and cavern collapse, makes proper interpretation of the history and origin of blue holes extremely difficult.

KARST PROCESSES AND THE DEVELOPMENT OF POROSITY AND PERMEABILITY

In carbonate islands, dissolution can produce large-scale porosity in a variety of ways over a significant vertical range. This porosity development can occur in a very brief time and requires only a minimal area of exposed platform. It is important to note that the term porosity has come to be casually, but inappropriately, equated to permeability, when in fact the two terms

Figure 14. Altar Cave area, San Salvador Island, Bahamas. The notch was originally interpreted as a fossil bio-erosion notch, but the undulating floor and ceiling of the feature, plus the existence of remnant calcite speleothems such as the column shown here, indicate that it is the back wall of a breached flank margin cave.

identify distinctly different rock properties (Lucia, 1993). Over time, the redistribution of carbonate material by diagenesis usually produces a decrease in bulk porosity, but if karst processes occur there is a consequent increase in permeability as continuous dissolutional flow paths develop.

For karst processes to function, some portion of the carbonate bank must be emergent to collect meteoric input, or there must be an adjacent landmass that delivers allogenic recharge to the carbonates. Research from the Bahamas, discussed earlier, indicates that karst processes produce significant features within tens of thousands of years. Karst features can, therefore, be expected to be produced in response to all but the fastest of sea level changes. To determine how much and where permeability will develop within the carbonate platform requires that four basic criteria be taken into account: (1) position of the lens relative to the exposed bank surface, (2) climate, (3) lithology, and (4) rate and magnitude of sea level change.

FRESHWATER LENS CONFIGURATIONS

Sea Level Change

The starting point for discussions of island karst processes and permeability development is a shallow carbonate platform undergoing carbonate deposition. If the platform is submerged, there is no autogenic freshwater input. If the platform is isolated, there is no allogenic freshwater input from the subsurface and the carbonates are uniformly under marine phreatic conditions. Depositional shoaling, wind, and wave action can produce local areas of emergence without sea level change, and these shoals can develop their own freshwater lenses and begin to undergo meteoric diagenesis (Budd and Land, 1990). Such land phenomena, however, tend to be very small (< 1 km^2). A local (tectonic) or eustatic sea level change would be necessary for a large portion of a platform to become emergent. Platforms that become emergent for whatever reason are subject to autogenic input of meteoric fresh water (Figure 15A).

Allogenic Recharge

Emergent carbonate platforms that grade landward into substantial noncarbonate land masses receive allogenic recharge in the subsurface. The resultant groundwater body may have characteristics that are not in agreement with local conditions (Figure 15B). An excellent example is the Yucatan, an emergent carbonate platform that extends westward into the landmass of Mexico (Back et al., 1986). The carbonate aquifer receives substantial allogenic recharge from highlands to the west. Despite the local semi-arid conditions in the carbonate terrain, the freshwater lens is well developed and has a high discharge.

If the carbonate bank shares an aquifer with a distant meteoric catchment (Figure 15C), it is possible that this aquifer can recharge the carbonate bank from below. Mixing-zone geochemical conditions could then occur within the carbonate platform. If the aquifer discharges through the carbonates of the bank to the ocean, mixing-zone dissolution could occur at the discharge point, and produce an unique in situ submarine karst. Such a scenario has been hypothesized for the Florida–Hatteras Slope (am Ende and Paull, 1991).

Amplitude of Emergence

The degree to which a bank is emergent will control many aspects of karst development, and, therefore, significant aspects of permeability formation. If the emergence is slight, then the water table is just below the surface, and the epikarst communicates directly with the phreatic zone. The surface ecosystem has ample water supply and significant biological productivity, which results in high soil CO_2 that drives karst processes in the epikarst and vadose zone. Interior water bodies are likely to occur if topographic depressions intersect the top of the lens. If the climate is

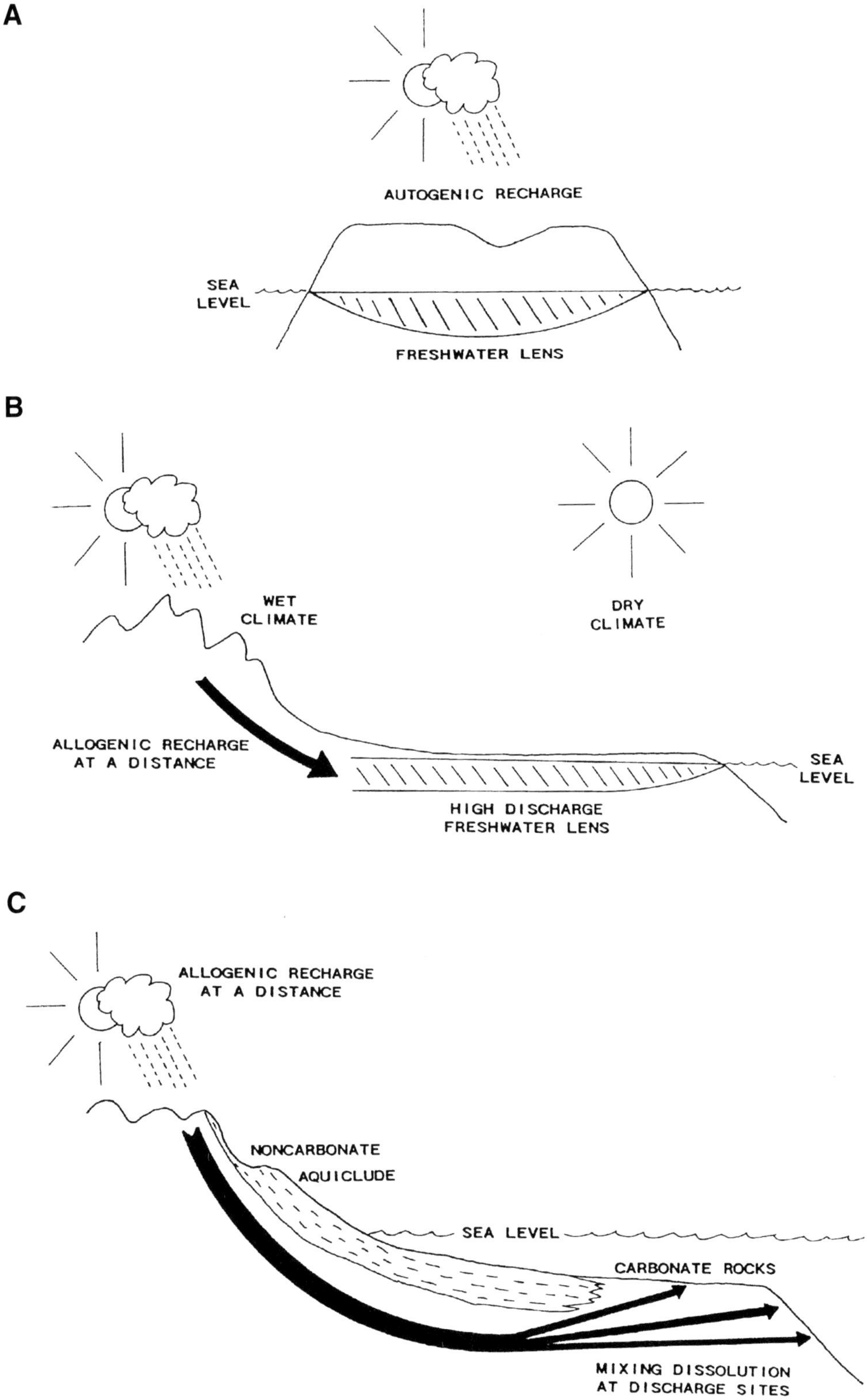

Figure 15. Diagrammatic representation of various carbonate platform recharge situations. (A) Simple autogenic recharge. (B) Allogenic recharge at a distance, with a high discharge freshwater lens in carbonate rocks under conditions that are locally arid (similar to Yucatan, Mexico). (C) Allogenic recharge at a distance that produces submarine freshwater discharge. Mixing-zone dissolution can occur at the platform surface and within the platform, even though the carbonates themselves were never subaerially exposed.

humid, these lakes will recharge the aquifer, and the water-filled depressions will enlarge by dissolution (Figure 16A). If the climate is arid, the lakes will experience excessive evaporation and the underlying marine groundwater will be upconed, thereby partitioning the freshwater lens, as reported for San Salvador Island (Davis and Johnson, 1989) and Exuma Island (Vacher and Wallis, 1992), Bahamas (Figure 16B). In low-lying areas where upconing has occurred, or in coastal lowlands where saltwater intrusion is extensive, the mixing of meteoric water with saline water in the epikarst results in an extremely jagged

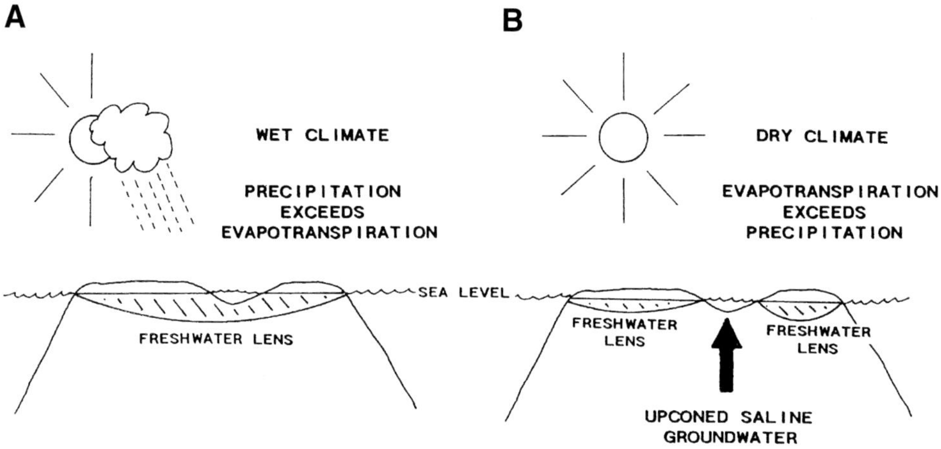

Figure 16. Diagrammatic representation of carbonate islands where the lens is shallow and in communication with the epikarst. For simplification, the diagrams assume uniform lithology and uniform permeability, conditions unlikely in the real world. (A) In a wet climate, where precipitation exceeds evapotranspiration and where interior depressions intersect the lens, they form freshwater lakes and marshes, which recharge the lens. (B) In a dry climate, where evapotranspiration exceeds precipitation and where interior depressions intersect the lens, evaporation removes the exposed lens and saline groundwater is upconed to the surface. Interior water bodies are saline or hypersaline.

and fretted rock surface (Figure 2). In the Bahamas, such surfaces are called "moonrock" (Davis and Johnson, 1989) because of the cratered and chaotic appearance of the rock surface. If this epikarst can be buried and preserved, it has the potential to form a laterally extensive high-permeability zone with limited vertical extent.

If the platform is significantly emergent, then the epikarst is separated from the freshwater lens by a substantial vadose zone (Figure 17). If the topographic depressions do not penetrate down to the top of the lens, evaporative upconing of underlying marine water does not occur, moonrock does not form, and the epikarst is dominated by meteoric karst forms. If the emergence is significant, then the surface ecology is diminished because only some plants will be able to reach deep enough to tap the freshwater lens. Lower organic productivity will mean less soil CO_2 and subsequently lesser amounts of karst development in the epikarst and the vadose zone.

Slight emergence will produce a better developed epikarst than significant emergence. Additionally, slight emergence will tend to superimpose the epikarst onto the dissolutional zone that occurs where vadose and phreatic waters mix at the top of the water table. In the rock record, the proximity of these dissolutional environments may lead to important permeability. When the platform is significantly emergent, two permeability zones develop with respect to the vadose zone: one at the surface epikarst and one at the top of the freshwater lens. Neither horizon is likely to be as permeable as the composite one developed when emergence is slight.

Effects of Climate

Given a carbonate platform of uniform lithology with a substantial amount of emergence, climate differences will change the position and nature of karst processes at depth. If the climate is humid, the lens will have ample recharge, will be thick, and discharge will be high, relative to a lens formed under arid conditions in the same physical setting (Figure 17A). Because the humid environment generates a thick lens, the base of the lens will be well below sea level. The greater discharge from the lens will entrain more of the underlying marine water, and the mixing zone will be thick as well. The resulting dissolutional permeability will not correlate with sea level except at the discharging lens margin.

On the other hand, in the same physical setting, if the climate is dry, the lens will have less recharge, will be thinner, and discharge will be lower (Figure 17B). Because the lens is thin, the base of the lens will be relatively close to sea level, and the mixing-zone dissolutional permeability will develop near sea level. If a thick vadose zone exists, it will prevent the arid climate from causing evaporative upconing of marine water into the lens.

Effects of Lithology

Two carbonate platforms with identical amounts of emergence and identical climates may still exhibit different lens configurations depending on the permeability of the rock. For simplicity, the discussion here assumes single lithologies for the sample platforms; in reality, significant lithologic differences, and hence

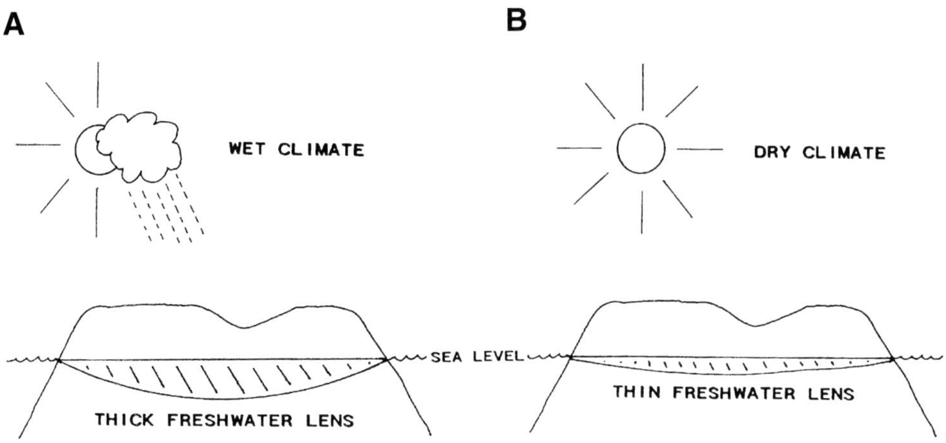

Figure 17. Diagrammatic representation of the effect of climate on a freshwater lens separated from the epikarst by a substantial vadose zone. For simplification, the diagrams assume uniform lithology and uniform permeability, conditions unlikely in the real world. (A) The large amount of water recharge from the wet climate produces a thick freshwater lens. (B) The small amount of water recharge from the dry climate produces a thin freshwater lens.

lens differences, occur in short distances on carbonate platforms. A platform composed of micritic material is likely to have low primary porosity and permeability, so the lens will discharge slowly, thereby causing the water to "pile up" (Figure 18A) (Vacher, 1988). A platform with the same conditions except that it is composed of coarse bioclastic material is likely to have high primary porosity and permeability, so the lens will discharge readily and will be thin (Figure 18B). In the latter case, the mixing zone will be shallower relative to sea level, than in the former case. In the rock record, the resulting permeability zone (if preserved) produced by mixing-zone dissolution will be closer, in vertical section, to the platform exposure surface for rocks with high primary porosity than for rocks with low primary porosity.

RATE AND MAGNITUDE OF SEA LEVEL CHANGE

Carbonate Platforms

The rate at which sea level changes and the duration of stillstands at any given sea level position will control the distribution of karst porosity and permeability produced by the freshwater lens. Time alone is significant; the more time a carbonate rock has spent in a freshwater lens, the more permeable it will become as dissolutional pathways develop. Such a relationship has been demonstrated for Bermuda (Vacher, 1988). With increased time, the differences in lens shape produced initially by primary porosity will diminish as karstic permeability becomes fully developed in the lens, and the lens thins (Figure 18C).

If a carbonate platform experiences slow, steady emergence through time, initially the freshwater lens will be a shallow lens (i.e., the epikarst will be in contact with the freshwater lens), and with time the lens will migrate downward through the platform as sea level falls. If the emergence rate exceeds the denudation rate, then the epikarst will lose contact with the freshwater lens, and the vadose zone will enlarge with time (Figure 19A). The mixing zone and its dissolutional effects will migrate downward through the carbonate platform. The degree of alteration of the rock at any given level will depend on how long the lens is at that level (all other influences being equal).

If the rate of emergence is slow and does not exceed the karst denudation rate, the landscape will be lowered as well, and the platform will maintain a shallow lens configuration with the epikarst in contact with the freshwater lens (Figure 19B). Because the denudation rate is controlled by climate, variations in climate could fortuitously combine with a variety of platform emergence rates to produce the same end result. Rapid emergence will disperse the development of both phreatic and vadose dissolutional permeability through a large section of the platform thickness, whereas slow emergence or a stillstand will concentrate such dissolutional permeability at a single zone approximating the elevation of the sea level stillstand.

A steady, continual, and rapid emergence of a carbonate island will result in a relatively uniform but low level of porosity and permeability production (Figure 20A). If sea level stands at specific levels, then the mixing zone and its attendant effects (physical mixing, organic carbon oxidation, and sulfur oxidation/reduction processes) will be focused on specific zones in the rock (Figure 20B). The development of large, phreatic dissolution voids (flank margin caves and banana holes) in carbonate platforms can depend on such sea level stillstands. Flank margin caves and banana holes from the Bahamas clearly record the high sea level stillstand of the last interglacial, and

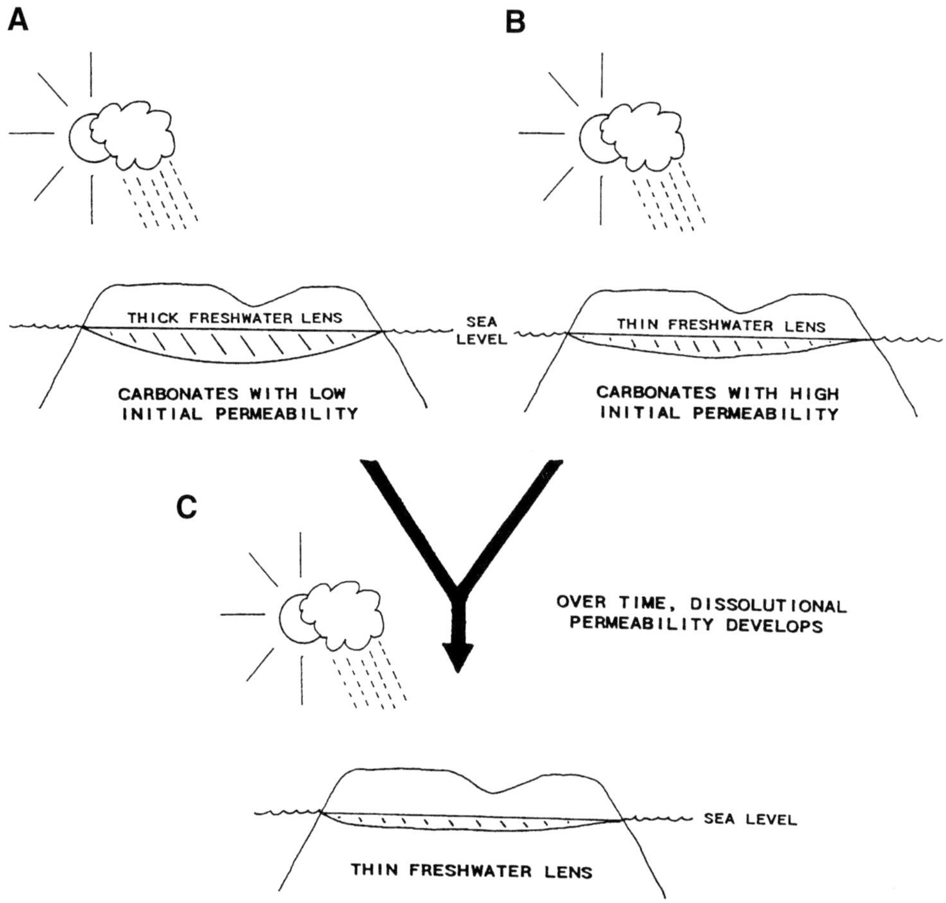

Figure 18. Diagrammatic representation of carbonate islands with different primary permeabilities. For simplification, the initial and subsequent permeabilities are assumed to be uniform over the entire platform, an unlikely occurrence. (A) With low primary permeability, groundwater transmission is impeded and the lens is thick. (B) With high primary permeability, groundwater discharge is facilitated and the lens is thin. (C) Through time, the development of secondary dissolutional permeability makes both lenses efficient in the transmission of water, and the lenses become thinner.

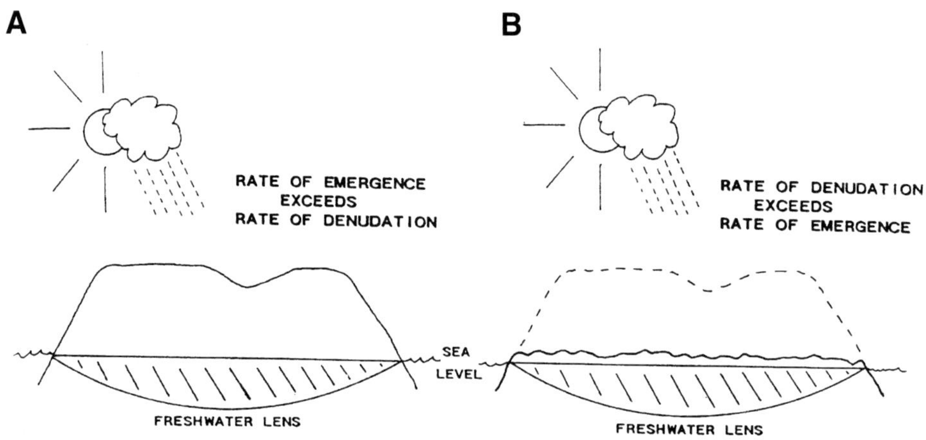

Figure 19. Effects of rate of emergence versus denudation rate. (A) If the rate of emergence greatly exceeds the denudation rate, the epikarst loses contact with the lens, and the vadose zone enlarges with time. (B) If the denudation rate exceeds the rate of emergence, the epikarst stays in contact with the lens and the vadose zone is always minimal.

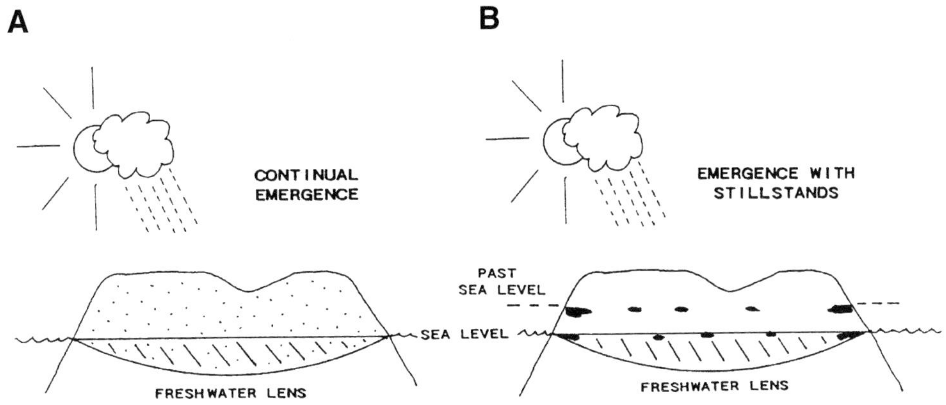

Figure 20. Effects of steady emergence versus episodic emergence with sea level stillstands. (A) When emergence is rapid and uniform, the development of minor dissolutional porosity and permeability is also uniform throughout the emergent portion and lens of the carbonate island (shown as stippling). Large-scale dissolutional voids may not develop. (B) When emergence is punctuated with times of no emergence, the dissolutional potential of the freshwater lens is concentrated and numerous large dissolutional voids develop in the lens (shown as solid patterns).

observations from submersibles has demonstrated that submerged flank margin caves can be found at specific levels that relate to past glacio-eustatic low sea level stillstands (Carew and Mylroie, 1987). The transition between these various stillstand events by glacioeustasy occurred too fast to allow mixing-zone dissolution processes the time to make large caves at intervening elevations.

As noted earlier, not only is the rate of sea level change important, but reversals in the direction of sea level change have a significant impact. The glacioeustatic sea level fluctuations of the Quaternary repeatedly subjected the carbonate rocks of the Bahamas to a variety of subaerial, vadose, freshwater phreatic, mixing zone, and marine phreatic conditions. The overprinting of the rock with multiple karst and diagenetic features from these fluctuations has introduced a bewildering variety and complexity into the Bahamian rock record. These complexities can be very difficult to resolve after millions of years, especially if geologists are looking at scattered outcrops and well records.

Carbonate Ramps

In contrast to a bank setting, when a carbonate ramp undergoes emergence, successively larger areas of carbonate rock are exposed through time. If the rate of emergence is slow, then the degree of epikarst and soil development on the platform will vary laterally; that is, the part of the ramp that becomes emergent first will have the greatest degree of karst and soil development, and the part exposed last will have the least development (Figure 21A). On the other hand, if the emergence is rapid, relative to the rate of karst and soil development, then the emergent ramp will have similar karst and soil development across its entire breadth. If the platform is subsequently submerged, burying the exposure surface in new sediment, the last exposed outcrops will be the first covered, and their karst and soil development will be much less than that of the outcrops farther up the ramp, which will not be buried until submergence of the ramp is almost complete. All other things being equal, the differences in degree of karst and soil (paleosol) development on the ramp will be due solely to duration of exposure.

In the subsurface, the migration of sea level across a carbonate ramp will cause the freshwater lens to migrate as well, and successively different parts of the subsurface will be in the appropriate position for flank margin cave development (Figure 21B). In platform settings, flank margin cave development can be continuous along the lateral margin of the lens, as noted on Isla de Mona (Frank, 1993; Mylroie et al., 1993) and in the Yucatan (Back et al., 1986). In a ramp situation, this laterally continuous zone of cave development will migrate with the migration of the lens. One of the possible end results of this void production is subsequent collapse of these voids (Figure 22). On carbonate ramps, if the migration of the flank margin conditions up (transgression) and down (regression) the carbonate ramp is slow enough to allow full dissolutional expression at each elevation, the result may be to produce successive arrays of cave chambers that subsequently collapse. Over the long term, this will produce (in the subsurface) an extensive solution-collapse and breccia facies. Conceptually, the process is similar to the long-wall coal mining process, in which the working face is advanced into the seam, and the mined void is left behind to collapse.

The characteristics of the collapse features produced by the processes described above depend on many factors. Unlike stream caves formed in continental settings, flank margin caves are not conduits, but mixing-zone chambers. Flank margin caves do not

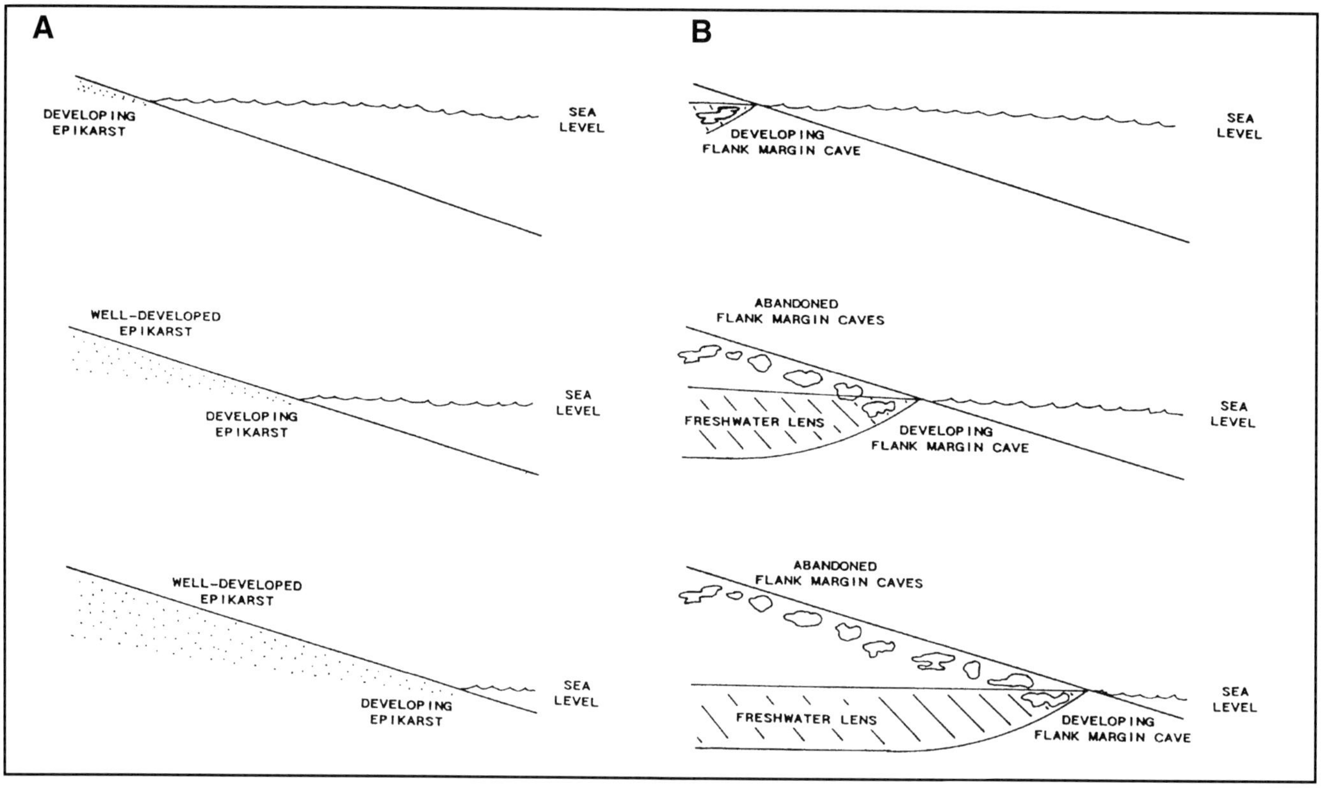

Figure 21. (A) Epikarst development during emergence of a carbonate ramp. Because of exposure time (all other things being equal), the epikarst at the highest position is the best developed, that at the lowest position is least developed. Were sea level to reverse and transgress the platform, epikarst development high on the ramp would still continue the longest. (B) Flank margin cave development during emergence of a carbonate ramp. If the rate of emergence is less than or equal to the time necessary for flank margin cave development, a continuous series of caves would be expected on the ramp. Subsequent erosional degradation and collapse of these caves would produce a continuous horizon of solution-collapse breccia. This figure has been separated into parts A and B, and pit caves omitted, for clarity.

contain fluvial deposits unless the voids collapse to the surface and allow surficial waters to wash material into the caves. If the overburden above the caves were thick, collapse processes would stabilize before the surface was reached, and the cave chambers and their associated collapse debris would be free of material washed in from the surface. On the other hand, if the overburden is thin, collapse would open cave chambers to the surface (especially during a regression), allowing fluvial sediments to work their way into the collapsed voids. In any case, carbonate platforms with autogenic diffuse recharge do not have significant surface streams and fluvial transport is a minor and very local occurrence. Therefore, caves developed in flank margin settings should not have significant volumes of fluvial debris. Mass wasting of soil material into caves opened by collapse or penetrated by pit caves can be significant, however, and cave fills due to this process are recognized in the rock record (Jones and Smith, 1988) (Figure 23). The presence or absence of surface sediments in paleokarst caves is not sufficient, by itself, to determine whether the caves developed as mixing-zone dissolution chambers or as through-flowing conduits. In oceanic settings, carbonate platforms receiving autogenic recharge should develop phreatic caves notably free of washed-in fluvial material.

KARST TO PALEOKARST: THE ROCK RECORD

The material presented to this point has been based on examples from Quaternary carbonate islands. If "the present is the key to the past," then the rock record should contain karst preserved as paleokarst. Successful identification of paleokarst in cores and outcrop has important implications for sequence stratigraphy, paleo-environments, and hydrocarbon reservoirs. Caves, whether conduits in continental settings or mixing chambers in island settings, are regions of very high porosity and permeability. Can paleokarst caves be easily located, and if located, easily identified? In modern karst environments, a major part of the research effort is to locate and characterize caves. Caves are difficult to find, and determining their water flow paths requires sophisticated techniques such as quantitative dye tracing and humanly rigorous exploration. To locate and characterize

Figure 22. Edge of Isla de Mona, Puerto Rico, showing breached flank margin caves receding into the distance. The talus in the background represents the former outer walls of these chambers.

Figure 23. Infilling paleosol breccia in Cueva de Pajaros, Isla de Mona, Puerto Rico; lens cap at left for scale. This material piped into the cave from the surface, lithified, and has been subsequently etched by phreatic dissolution, most clearly seen at lower left and right.

unknown and unenterable caves millions of years old and thousands of meters beneath the surface is a daunting task.

The first question is, do cave and karst features recognizably survive time and burial? The literature has many references to paleokarst (James and Choquette, 1988; Bosak, 1989), so clearly the answer is yes; in most cases it has been recognized in surface outcrop, or after the fact in cores and mines. Less commonly has there been deliberate prospecting for paleokarst in the subsurface.

Do dissolution voids survive deep burial in carbonate island settings? In the Bahamas, cavernous porosity has been penetrated by wells at depths ranging from 21 to 4082 m, the deepest void being large enough to accept 2430 m of drill pipe (Meyerhoff and Hatten, 1974). On Isla de Mona, Puerto Rico, caves with volumes in excess of 100,000 m^3 have survived since initial uplift over 1 Ma (Briggs and Seiders, 1972; Mylroie et al., 1993). Dean's Blue Hole, on Long Island, Bahamas, is a upwardly prograding collapse feature 202 m deep, with a volume of 1.15×10^6 m^3 (Wilson, 1994). The initial void, which accepted collapse material as the blue hole prograded upward, formed below any past Quaternary sea level lowstand, and thus represents a cave of significant age (pre-Quaternary). Clearly, large voids can survive long periods of time and deep burial.

Can paleokarst cave locations be predicted in the subsurface? Based on the flank margin model presented in this paper, the margin of carbonate platforms would be the most likely spot for caves to have developed if the platforms were ever emergent. The classic work of Craig (1988) from the Yates field of west Texas demonstrates that, with large amounts of data from a producing field, good correlations can be made with models of an island freshwater lens, cave development, and cave location by borehole. However, cave height was used by Craig (1988) to determine lens thickness and position. Cave height is a poor measure of these characteristics, because adjacent cave levels formed on successive but closely spaced sea level stillstands can merge by dissolution and collapse. If a borehole intersects a pit cave, it will yield a large "cave height" but be almost impossible to differentiate from a flank margin cave. The extremely large flank margin caves of Isla de Mona, Puerto Rico, the largest currently known in a subaerial position, generally have original dissolutional ceiling heights of less than 8 m. Flank margin caves, as their name suggests, tend to form on the margin of carbonate platforms, making them vulnerable to destruction by bank-margin fracture and other weathering processes. Flank margin caves formed in the Bahamas during the last interglacial (circa 125,000 years ago) can be found in various stages of degradation (Mylroie and Carew, 1991), suggesting that their life span is brief. Conversely, those on Isla de Mona have survived a million years of exposure, primarily because they were so large that despite the loss of exterior chambers to cliff retreat, major portions of these caves survive (Figure 22).

To successfully predict the occurrence of paleokarst caves in ancient carbonate platforms requires use of additional information (Kerans, 1988). Evidence of emergence is a critical factor (Saller et al., 1994). If the platforms were never emergent, then the caves as described in this paper are unlikely to have ever formed, although special cases, as noted in Figure 15C, are possible. Subaerial unconformities or evidence of mixing-zone diagenesis would be an important clue that pit caves and flank margin caves were possible. If subaerial unconformities in the carbonate section can be identified, the degree to which the epikarst has been developed can provide clues as to the climate at the time of karstification, and the degree of separation of the surface from the water table (Webb, 1994). While a paleo-epikarst may itself be a region of high permeability, the existence of an epikarst is an indication of emergence and the potential for pit caves and flank margin caves.

CONCLUSIONS

Carbonate platforms that receive only autogenic recharge may appear to be a special case, but many areas in the world, both today and during the Pleistocene, fit this case. The Bahamas are the most notable example. With only regionally diffuse meteoric input, macroscopic karstic porosity and cave permeability occur in three basic areas: the epikarst of the surface, the pit caves of the vadose zone, and the phreatic caves (flank margin caves and banana holes) of the freshwater lens.

Through time, emergent carbonate platforms may develop karst features faster than other diagenetic processes can alter the carbonate rock. The groundwater geochemistry that develops carbonate-platform karst features is affected by atmospheric and soil CO_2, subsurface oxidation of organic carbon to CO_2, inorganic mixing of fresh and saline waters, and oxidation/reduction reactions involving sulfur. A possible area of future work is to study these diverse and complex chemical reactions to see if they introduce unexpected variations in stable isotopes that may confuse results aimed at assessing the occurrence of exposure events.

Climate and lithology are obvious controls of karst processes. The amount of emergence of carbonate terrains, and the rate at which this emergence occurs, interacts with climate and lithology to place karstic porosity and permeability, especially that developed within the freshwater lens, at a variety of locations within the carbonate section. Because macroscopic dissolutional voids that are meters to tens of meters in size can form in as few as tens of thousands of years, even minor variations in the rate of sea level change can greatly affect the size and location of the caves. The primary site of flank margin cave formation is at the margin of the freshwater lens. The freshwater lens margin occupies the exposure margins of carbonate platforms. The caves are therefore developed in a location in which minimal erosion will disrupt and collapse the caves. This collapse may produce breccia zones in the carbonate section. If subaerial erosion is significant, it can remove all evidence of the caves, obscuring the important role of cave formation.

ACKNOWLEDGMENTS

The authors are indebted to many people who over the years have exchanged ideas with us, gone into the field with us, and critically reviewed our work. Special thanks go to Bill Back, Art Palmer, Rob Palmer, Pete Smart, Joe Troester, Len Vacher, Fiona Whitaker, Will White, and numerous students. David Budd, Art Palmer, and an anonymous reviewer made many helpful suggestions regarding the manuscript. Much of our paper is based on the results of research conducted over the last two decades in the Bahamas. That work has been supported by the Bahamian Field Station, Dr. Donald T. Gerace, Chief Executive Officer; the University of Charleston; and Mississippi State University.

REFERENCES CITED

am Ende, B. A., and C. K. Paull, 1991, Submarine karst on the Florida-Hatteras slope? (abs.): Program of the National Speleological Society Annual Convention, p. 55.

Back, W., B. B. Hanshaw, J. S. Herman, and J. N. Van Driel, 1986, Differential dissolution of a Pleistocene reef in the ground-water mixing zone of coastal Yucatan, Mexico: Geology, v. 14, p. 137–140.

Bosak, P., 1989, ed., Paleokarst a systematic and regional review: Amsterdam, Elsevier, 725 p.

Bottrell, S. H., P. L. Smart, F. Whitaker, and R. Raiswell, 1991, Geochemistry and isotope systematics of sulphur in the mixing zone of Bahamian blue holes: Applied Geochemistry, v. 6, p. 97–103.

Bottrell, S. H., J. L. Carew, and J. E. Mylroie, 1993, Bacterial sulphate reduction in flank margin environments: evidence from sulphur isotopes, *in* B. White, ed., Proceedings of the Sixth Symposium on the Geology of the Bahamas, Port Charlotte, Florida, Bahamian Field Station, p. 17–21.

Briggs, R. P., and V. M. Seiders, 1972, Geologic map of Isla de Mona quadrangle, Puerto Rico: U.S. Geological Survey Miscellaneous Investigations, Map I-718.

Budd, D. A., and L. S. Land, 1990, Geochemical imprint of meteoric diagenesis of carbonates: insights from pore-water chemistry: Geological Society of America Bulletin, v. 100, p. 1260–1270.

Burkeen, B., and J. E. Mylroie, 1992, Bahamian blue holes: description and definition (abs.): National Speleological Society Bulletin, v. 54, p. 92–93.

Carew, J. L., and J. E. Mylroie, 1987, Submerged evidence of Pleistocene low sea levels on San Salvador, Bahamas, *in* R. A. Cooper, and A. N. Shepard, eds., National Oceanographic and Atmospheric Administration Undersea Program Symposium Series for Undersea Research, v. 2, p. 167–175.

Carew, J. L., and J. E. Mylroie, 1989, The geology of eastern South Andros Island, Bahamas: a preliminary report, *in* J. E. Mylroie, ed., Proceedings of the Fourth Symposium on the Geology of the Bahamas, Port Charlotte, Florida, Bahamian Field Station, p. 313–321.

Carew, J. L., and J. E. Mylroie, 1991, Some pitfalls in paleosol interpretation in carbonate sequences: Carbonates and Evaporites, v. 6, p. 69–74.

Carew, J. L., and J. E. Mylroie, 1992, Subaerial fossil reefs and phreatic dissolution caves: indicators of Late Quaternary sea level and the tectonic stability of the Bahamas: Geological Society of America Abstracts with Programs, v. 24, no. 1, p. 6.

Carew, J. L., and J. E. Mylroie, in press, Quaternary

tectonic stability of the Bahamian Archipelago: evidence from fossil coral reefs and flank margin caves: Quaternary Science Reviews, v. 13.

Carew, J. L., J. E. Mylroie, and N. E. Sealey, 1992, Field guide to sites of geological interest, western New Providence Island, Bahamas: Field Trip Guidebook—The Sixth Symposium on the Geology of the Bahamas: Port Charlotte, Florida, Bahamian Field Station, p. 1–23.

Chen, J. H., H. A. Curran, B. White, and G. J. Wasserburg, 1991, Precise chronology of the last interglacial period: ^{234}U–^{230}Th data from fossil coral reefs in the Bahamas: Geological Society of America Bulletin, v. 103, p. 82–97.

Craig, D. H., 1988, Caves and other features of Permian karst in San Andres Dolomite, Yates field reservoir, west Texas, *in* N. P. James and P. W. Choquette, eds., Paleokarst: New York, Springer-Verlag, 416 p.

Daugherty, D. R., M. R. Boardman, and C. V. Metzler, 1987, Characteristics of joints and sedimentary dikes of the Bahama Islands, *in* H. A. Curran, ed., Proceedings of the Third Symposium on the Geology of the Bahamas, Fort Lauderdale, Florida, CCFL Bahamian Field Station, p. 45–56.

Davis, R. L., and C. R. Johnson, Jr., 1989, Karst hydrology of San Salvador, *in* J. E. Mylroie, ed., Proceedings of the Fourth Symposium on the Geology of the Bahamas, Port Charlotte, Florida, Bahamian Field Station, p. 118–135.

Frank, E. F., 1993, Aspects of karst development and speleogenesis Isla de Mona, Puerto Rico: an analogue for Pleistocene speleogenesis in the Bahamas: MSc. thesis, Mississippi State University, Mississippi State, Mississippi, 349 p.

James, N. P., and P. W. Choquette, 1988, eds., Paleokarst: New York, Springer-Verlag, 416 p.

Jennings, J. N., 1968, Syngenetic karst in Australia, *in* P. W. Williams, and J. N. Jennings, eds., Contributions to the study of karst: Canberra, Australia, Australian National University Research School of Pacific Studies Publication G5, p. 41–110.

Jones, B., and D. S. Smith, 1988, Open and filled karst features on the Cayman Islands: Implications for the recognition of paleokarst: Canadian Journal of Earth Sciences, v. 25, p. 1277–1291.

Kerans, C., 1988, Karst-controlled reservoir heterogeneity in Ellenburger Group carbonates of west Texas: AAPG Bulletin, v. 72, p. 1160–1183.

Lucia, F. J., 1993, Myths and fantasies about porosity and unconformities (abs.): AAPG Program of the Hedberg Conference on Unconformities and Porosity Development in Carbonate Strata: Recognition, Controls, and Predictive Strategies, p. 36.

Meyerhoff, A. A., and Hatten, C. W., 1974, Bahamas salient of North America: Tectonic framework, stratigraphy, and petroleum potential: AAPG Bulletin, v. 58, p. 1201–1239.

Mullins, H. T., and A. C. Hine, 1989, Scalloped bank margins: beginning of the end for carbonate platforms?: Geology, v. 17, p. 30–33.

Mylroie, J. E., 1984a, Hydrologic classification of caves and karst, *in* R. G. LaFleur, ed., Groundwater as a geomorphic agent: Boston, Allen and Unwin, p. 157–172.

Mylroie J. E., 1984b, Speleogenetic contrast between the Bermuda and Bahama Islands, *in* J. W. Teeter, ed., Proceedings of the Second Symposium on the Geology of the Bahamas, Fort Lauderdale, Florida, CCFL Bahamian Field Station, p. 113–128.

Mylroie, J. E., 1988, Karst of San Salvador, *in* J. E. Mylroie, ed., Field guide to the karst geology of San Salvador Island, Bahamas: Fort Lauderdale, Florida, Bahamian Field Station, p. 17–44.

Mylroie, J. E., 1991, Cave development in the glaciated Appalachian karst of New York: surface-coupled or saline-freshwater mixing hydrology?, *in* E. H. Kastning, ed., Proceedings of the Appalachian Karst Symposium, Huntsville, Alabama, National Speleological Society, p. 85–90.

Mylroie, J. E., 1993a, Return of the coral reef hypothesis: basin to shelf partitioning of $CaCO_3$ and its effect on atmospheric CO_2 (Comment): Geology, v. 21, p. 475.

Mylroie, J. E., 1993b, Carbonate deposition/dissolution cycles and carbon dioxide flux in the Pleistocene, *in* B. White, ed., Proceedings of the Sixth Symposium on the Geology of the Bahamas, Port Charlotte, Florida, Bahamian Field Station, p. 103–107.

Mylroie, J. E., and W. J. Balcerzak, 1992, Interaction of microbiology and karst processes in Quaternary carbonate island aquifers: *in* J. A. Stanford and J. J. Simons, eds., Proceedings of the First International Conference on Ground Water Ecology, Bethesda, Maryland, American Water Resources Association, p. 37–46.

Mylroie, J. E., and J. L. Carew, 1988, Solution conduits as indicators of late Quaternary sea level position: Quaternary Science Reviews, v. 7, p. 55–64.

Mylroie, J. E., and J. L. Carew, 1990, The flank margin model for dissolution cave development in carbonate platforms: Earth Surface Processes and Landforms, v. 15, p. 413–424.

Mylroie, J. E., and J. L. Carew, 1991, Erosional notches in Bahamian Carbonates: bioerosion or groundwater dissolution?, *in* R. J. Bain, ed., Proceedings of the Fifth Symposium on the Geology of the Bahamas, Port Charlotte, Florida, Bahamian Field Station, p. 85–90.

Mylroie, J. E., J. L. Carew, N. E. Sealey, and J. R. Mylroie, 1991, Cave development on New Providence Island and Long Island, Bahamas: Cave Science, v. 18, p. 139–151.

Mylroie, J. E., E. F. Frank, R. Carrasquillo, B. E. Taggart, J. W. Troester, and J. L. Carew, 1993, Flank margin cave development, Isla de Mona, Puerto Rico (abs.): Program of the National Speleological Society Annual Convention, Huntsville, Alabama, National Speleological Society, p. 39–40.

Pace, M. C., J. E. Mylroie, and J. L. Carew, 1993, Petrographic analysis of vertical dissolution features on

San Salvador Island, Bahamas, *in* B. White, ed., Proceedings of the Sixth Symposium on the Geology of the Bahamas, Port Charlotte, Florida, Bahamian Field Station, p. 109–123.

Palmer, A. N., M. V. Palmer, and J. M. Queen, 1977, Geology and the origin of caves in Bermuda, *in* T. D. Ford, ed., Proceedings of the 7th International Speleological Congress, Sheffield, British Cave Research Association, p. 336–338.

Palmer, R. J., 1985, The blue holes of the Bahamas: London, Jonathan Cape, 184 p.

Palmer, R. J., 1986, The blue holes of South Andros, Bahamas: Cave Science, v. 13, p. 3–6.

Pelle, R. C., and M. R. Boardman, 1989, Stratigraphic distribution and associations of trace elements in vadose-altered multicomponent carbonate assemblages, *in* J. E. Mylroie, ed., Proceedings of the Fourth Symposium on the Geology of the Bahamas, Port Charlotte, Florida, Bahamian Field Station, p. 118–135.

Plummer, L. N., 1975, Mixing of sea water with calcium carbonate ground water, *in* E. H. T. Whitten, ed., Quantitative studies in geological sciences: Geological Society of America Memoir v. 142, p. 219–236.

Purdy, E. G., and G. T. Bertram, 1993, Carbonate concepts from the Maldives, Indian Ocean: AAPG Studies in Geology 34, 56 p.

Rossinsky, V. Jr., H. R. Wanless, and P. K. Swart, 1992, Penetrative calcretes and their stratigraphic implications: Geology, v. 20, p. 331–334.

Saller, A. H., D. A. Budd, and P. M. Harris, 1994, Unconformities and porosity development in carbonate strata: ideas from a Hedberg Conference: AAPG Bulletin, v. 78, p. 857–871.

Sanford, W. E., and Konikow, L. F., 1989, Porosity development in coastal carbonate aquifers: Geology, v. 17, p. 249–252.

Schwabe, S. J., J. L. Carew, and J. E. Mylroie, 1993, The petrology of Bahamian Pleistocene eolianites and flank margin caves: implications for late Quaternary island development, *in* B. White, ed., Proceedings of the Sixth Symposium on the Geology of the Bahamas, Port Charlotte, Florida, Bahamian Field Station, p. 149–164.

Smart, P. L., and F. F. Whitaker, 1991, Karst processes, hydrology, and porosity evolution, *in* V. P. Wright, M. Esteban, and P. L. Smart, eds., Palaeokarsts and palaeokarstic reservoirs: Reading England, Postgraduate Research Institute for Sedimentology, University of Reading, p. 1–55.

Smart, P. L., and F. F. Whitaker, 1993, Controls on the distribution and extent of porosity development in paleokarst terrains: AAPG Program of the Hedberg Conference on Unconformities and Porosity Development in Carbonate Strata: Recognition, Controls, and Predictive Strategies, p. 69.

Smart, P. L., J. M. Dawans, and F. Whitaker, 1988a, Carbonate dissolution in a modern mixing zone: Nature, v. 335, p. 811–813.

Smart, P. L., R. J. Palmer, F. Whitaker, and V. P. Wright, 1988b, Neptunian dikes and fissure fills: An overview and account of some modern examples, *in* N. P. James and P. W. Choquette, eds., Paleokarst: New York, Springer-Verlag, p. 149–163.

Vacher, H. L., 1988, Dupuit-Ghyben-Herzberg analysis of strip-island lenses: Geological Society of America Bulletin, v. 100, p. 580–591.

Vacher, H. L., and J. E. Mylroie, 1991, Geomorphic evolution of topographic lows in Bermudian and Bahamian Islands, effect of climate, *in* R. J. Bain, ed., Proceedings of the Fifth Symposium on the Geology of the Bahamas, Port Charlotte, Florida, Bahamian Field Station, p. 221–234.

Vacher, H. L., and T. N. Wallis, 1992, Comparative hydrology of Bermuda and Great Exuma Island, Bahamas: Groundwater, v. 30, p. 15–20.

Viles, H. A., 1988, ed., Biogeomorphology: New York, Basil Blackwell, 365 p.

Vogel, P. N., J. E. Mylroie, and J. L. Carew, 1990, Limestone petrology and cave morphology on San Salvador Island, Bahamas: Cave Science, v. 17, p. 19–30.

Webb, G. E., 1994, Paleokarst, paleosol, and rocky-shore deposits at the Mississippian–Pennsylvanian unconformity, northwestern Arkansas: Geological Society of America Bulletin, v. 106, p. 634–648.

Williams, P. W., 1983, The role of the subcutaneous zone in karst hydrology: Journal of Hydrology, v. 61, p. 45–67.

Wilson, W. L., 1994, Morphology and hydrology of the deepest known cave in the Bahamas: Dean's Blue Hole, Long Island (abs): Abstracts and Program of the 7th Symposium on the Geology of the Bahamas, p. 21.

Chapter 4

Geochemical Models for the Origin of Macroscopic Solution Porosity in Carbonate Rocks

Arthur N. Palmer
State University of New York
Oneonta, New York, U.S.A.

ABSTRACT

Any single geologic setting may include a variety of geochemical environments, each capable of producing a different type of carbonate solution porosity. Also, many types of porosity can form in more than one geologic setting. Thus, the interpretation of solution porosity is best approached by first delineating the geochemical processes necessary to form the observed pattern of porosity, and then using these insights to assess the broader geologic context. Throughout most of any carbonate formation the solution process is highly selective, and only those openings of maximum groundwater flow are enlarged, while surrounding openings undergo little or no enlargement. Pervasive macroscopic porosity, in which nearly all initial openings are enlarged by solution, is formed by: (1) meteoric water with high discharge and/or low flow distance, (2) mixing of waters of disparate chemistry, (3) oxidation of hydrogen sulfide, or (4) production of acids by redox reactions involving carbon compounds in reducing environments. Areally extensive solution porosity within a narrow stratigraphic range usually indicates solution or reduction of sulfates. Cavernous solution porosity is negligible where aggressive infiltration is lacking, in deep zones where groundwater chemistry is uniform, and in low-flow areas of diagenetically mature carbonate rocks far from sources of groundwater recharge.

INTRODUCTION

Although the origin and hydrologic function of macroscopic solution pores are well understood in the context of present-day karst (see Jennings, 1985; Ford, 1988; White, 1988; Ford and Williams, 1989), the relation of porosity to unconformities is inconsistent and unpredictable. One would expect porosity to be most intense just beneath unconformities and to diminish with depth, but this is rarely the case. In parts of some carbonate strata, virtually every available presolutional opening has been enlarged, but elsewhere solution pores may be entirely absent. Most commonly only a minority of the initial openings has been enlarged significantly by solution, while the rest are left in nearly their original state. Interpretation is clouded by the fact that solution porosity can form in several dissimilar ways.

This apparent disorder can be resolved by relating the pattern of solution pores to specific geochemical processes, which may be only indirectly controlled by local geology. Once a specific geochemical model is ascertained, the distribution of solution porosity is more easily interpreted in a broad geologic context.

This paper focuses on cavernous solution porosity (and other macroscopic porosity) associated with near-surface or deep-seated karst processes, with particular emphasis on its recognition in petroleum reservoirs. Porosity resulting from diagenesis is mentioned only briefly for contrast.

Scale is an important criterion in distinguishing among the various pore types. Macroscopic pores resulting from karst processes range up to several tens of meters (rarely more than 100 m) in mean diameter and comprise integrated systems, defined by paths of greatest meteoric groundwater flow or by the generation of deep-seated acids. Many of these serve, or once served, as conduits for turbulent groundwater flow connecting recharge areas to spring outlets. The term cave (or cavern) should be limited to solution conduits or vugs large enough for human access (Choquette and Pray, 1970). This usage is not as anthropocentric as it might seem, because direct observation and continuous mapping are what distinguish the results of cave studies from other types of subsurface data, which are discontinuous or indirect. Although cave-sized openings represent but a small fraction of the total solution porosity in petroleum reservoirs (see references in Roehl and Choquette, 1985; Candelaria and Reed, 1992; and Fritz et al., 1993), they provide insight into nearly all types of carbonate solution porosity. This advantage is illustrated by Loucks and Handford (1992) in a comparison between features observed in accessible caves and those in paleo-cave reservoirs. However, much of the porosity in carbonate rocks, particularly microporosity, is diagenetic and has little relation to cave-forming processes. Most diagenetic pore sizes are adjusted to the scale of individual grains or to that of the intergranular matrix, and tend to be distributed more uniformly than macroscopic solution pores.

ORIGINS OF SOLUTION POROSITY

A few well known but important concepts should first be reviewed. Solution porosity is formed by the movement of undersaturated water through a preexisting network of intergranular pores, fractures, or partings in soluble bedrock or semi-lithified sediments. The hydraulic gradient necessary to drive the flow is usually controlled by the pattern of recharge and discharge in response to topographic relief, although thermal gradients or differential pore pressures can account for local and usually short-lived water movement at depth. The flow rate of aggressive water is an important variable, because diffusion of chemical species alone is unable to produce significant amounts of solution porosity. The volume of solution porosity also depends on the duration of solutionally aggressive water flow through a carbonate aquifer.

Although most evaporite minerals are highly soluble in pure water, carbonate rocks dissolve readily only where there is a source of acidity more potent than the dissociation of water alone. The most common is carbonic acid produced by carbon dioxide in the atmosphere and soil. The effect of organic acids is generally overwhelmed in near-surface environments by carbonic acid, but can be of local importance in the vicinity of hydrocarbons at depth. Of importance in certain areas is sulfuric acid produced by the oxidation of hydrogen sulfide, which is released by the reduction of sulfates in contact with organic carbon compounds. Carbonate solution processes are discussed in detail by Dreybrodt (1988), Lohmann (1988), White (1988), and Ford and Williams (1989).

Solution rates are greatest where acid is generated or where solutional aggressiveness is rejuvenated. In most karst areas this is the soil/bedrock contact, where infiltrating water first encounters carbonate rock (Figure 1). As the dissolved load increases, the solution rate diminishes, as does the volume of solution porosity generated by the solvent water. The greater the discharge and velocity, the farther the solution process extends in the direction of flow. Aggressiveness may also be generated along the flow route, for example by the oxidation of organic compounds to produce carbon dioxide. Cooling of water produces a similar effect, owing to the inverse relationship between carbonate solubility and temperature, even if the P_{CO_2} remains constant. Aggressiveness can also increase abruptly by mixing of waters of different chemistry or by acquisition of new sources of acid (e.g., by oxidation of hydrogen sulfide), creating local concentrations of solution porosity.

Vuggy porosity consists of enlarged intergranular, moldic, and other "matrix" pores resembling those in a sponge. In contrast, discrete tunnels formed along major flow paths (solution conduits) require that the water remains aggressive over the entire distance from the recharge point to the surface outlet. Most solution conduits represent the selective enlargement of only a few of the myriad interconnecting fractures and partings, but rarely matrix porosity. In prominently fractured rock they have angular patterns consisting of fissure-like segments (Figure 2). In well-bedded rock with few prominent fractures, solution conduits are mainly sinuous, curvilinear tubes and canyon-like passages (Figures 3 and 4).

Laminar flow prevails within the presolutional network, but turbulent flow develops if there is sufficient enlargement and flow rate. Depending on the hydraulic gradient and temperature, this transition takes place at fissure widths of about 0.1–2 cm. Turbulent flow is commonly used as the criterion for whether the interconnected openings that define a flow path have enlarged enough to constitute a solution conduit.

Most of the water in any carbonate aquifer is at or near saturation with the dominant mineral species, but perfect saturation is rarely achieved or maintained (Thrailkill, 1968; Thrailkill and Robl, 1981). Changes in temperature, P_{CO_2}, and dissolved solids cause the water to hover around, but not exactly at, saturation. Furthermore, certain reactions (e.g., solution or precipitation of dolomite) are so sluggish that considerable

Figure 1. Epikarst in the sawed face of a dimension-stone quarry in the Salem Limestone, Indiana. Note solutional enlargement of all major fractures and downward narrowing of fissures.

undersaturation or supersaturation can persist for lengthy periods. Slight supersaturation of minerals is common because of the kinetic threshold that must be exceeded to achieve nucleation (Drever, 1982, p. 118–120).

The vadose zone behaves generally as an open system with respect to CO_2, with P_{CO_2} values similar to those of the overlying soil (roughly 0.01–0.1 atm), except in openings that exchange air with the surface, in which the P_{CO_2} is typically about one-tenth as great. Atkinson (1977) describes the CO_2 distribution in a typical karst vadose zone. Phreatic solution takes place under approximately closed-system conditions with respect to CO_2, as there is no abundant gas phase present, and, therefore, the CO_2 content of the water may decrease in the downflow direction as carbonates are dissolved. However, mixing and redox processes prohibit truly closed conditions.

SOLUTION RATES IN THE CARBONIC ACID SYSTEM

In humid climates, where infiltrating water is charged with carbon dioxide, the volume of carbonate rock that can be dissolved depends on the equilibrium P_{CO2}, temperature, and presence of other acids and dissolved solids. Figure 5 shows the solubility of calcite, aragonite, and well-ordered dolomite as a function of P_{CO_2}. In keeping with the goals of this paper, solubility is expressed as maximum pore volume (cm^3) produced by each liter of solvent water. These values were obtained by iterative calculation of all relevant equilibria, including those for ion pairs, while maintaining the charge balance and adjusting activity coefficients with the extended Debye-Hückel equation. Equilibrium constants were those recommended by Plummer and Busenburg (1982) for the H_2O-CO_2-$CaCO_3$ system, and others were calculated from thermodynamic data listed by Woods and Garrels (1987).

The curves in Figure 5 do not represent true pore volume because much of the solutional capacity of water is "wasted" in degrading the bedrock surface. Nor do they indicate the rates at which porosity is produced or how it is distributed. These must be determined by examining the mass balance, flow rates, and solution kinetics. These relationships are complex enough that exact trends can be revealed only by finite-difference computations (James and Kirkpatrick, 1980; Palmer, 1984; Dreybrodt, 1990; Groves and Howard, 1994). This approach requires a great deal of trust on the part of the reader, for whom the exact method and computer code are rarely available for scrutiny. With a few simplifications, the same results

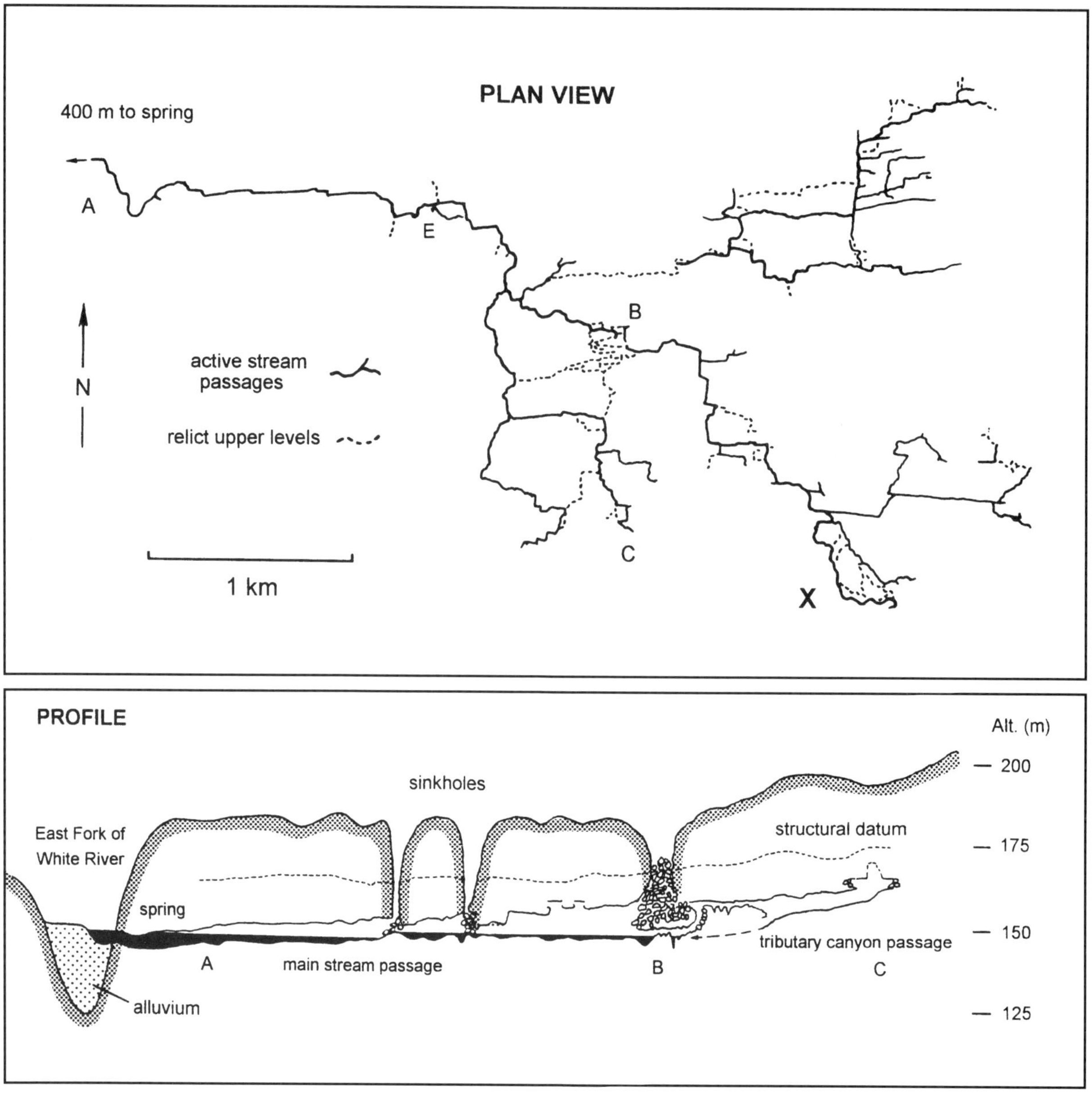

Figure 2. A branchwork cave consisting of vadose and phreatic conduits fed by sinkholes (Blue Spring Cave, Indiana, in the Salem Limestone). The branchwork pattern is especially clear because only a single major level is present, with only short fragments of a relict upper level. Profile A-B-C shows the relatively steep gradient of vadose tributaries compared to the main passage, which originally formed at and just below the water table. Bedding structure is illustrated by an arbitrary structural datum derived from mapping of bedding planes in the cave. Prominent jointing not only imparts an angular pattern to the cave but allows local discordance to the strata. The river valley has undergone late-Pleistocene alluviation, reflooding the lower reaches of the cave (depth of alluvium determined by refraction seismology). E = entrance. Map and profile by A. and M. Palmer.

can be achieved by pure mathematical analysis (Palmer, 1988), but the most satisfactory approach is to develop each concept in an intuitive, conceptual manner. Numerical and analytical solutions can stand as guideposts along the way, but if they make no intuitive sense their utility is limited.

The mass balance dictates that the mass removed from pore walls by solution must equal that which is carried away by the solvent water. Dividing by the bulk density of the bedrock and by time gives the rate of pore-volume increase (rock volume / time). In aqueous solution, the mass being removed is expressed as the product of dissolved load (mass of dissolved solids / water volume) and discharge (water volume / time). Mass transfer is therefore limited by the discharge rate and by the carbonate solution kinetics.

Figure 3. A curvilinear solution conduit in the prominently bedded Ste. Genevieve Limestone (Marengo Cave, Indiana). Note stream and banks of insoluble sediment.

In most karst systems, groundwater discharge depends on the rate of infiltration from the land surface. Flow is distributed unevenly among the various flow paths according to their efficiency in transmitting water and their orientation with respect to recharge and discharge areas. The hydraulic gradient along each flow path adjusts to the discharge and to the overall resistance to flow (which increases with path length and decreases greatly with fracture width or conduit diameter).

Solution rates can be quantified only with the aid of empirical data. Experiments with calcite show that at the moderate pH of typical karst water, the rate-limiting effect is the reaction between solvent groundwater and bedrock, rather than mass-transfer rates within the fluid (Plummer and Wigley, 1976). The transition from laminar to turbulent flow causes a sharp jump in solution rate at low pH but has little effect at the pH values of karst groundwater (typically about 6.5–8.5). In this respect limestone contrasts with many other soluble materials, such as gypsum, for which the solution rate increases with turbulence, regardless of pH. Empirical data on solution rates are given by Berner and Morse (1974), Plummer and Wigley (1976), and Plummer et al. (1978) for calcite and by Busenburg and Plummer (1982) and Herman and White (1985) for dolomite. The mass balance and kinetic relationships can be arranged into a general equation for the rate of wall retreat within pores in calcitic limestone (Palmer, 1991):

$$S = 31.56\,(k/\rho)\,(1 - C/C_s)^n \text{ cm/yr} \qquad (1)$$

where S = rate of wall retreat (cm/yr), k = reaction coefficient (mg-cm/L-sec), ρ = bulk bedrock density (g/cm^3), n = reaction order, and C/C_s = saturation ratio (actual concentration/saturation concentration), which is commonly given the symbol Ω (omega). C_s in g/L can be obtained from Figure 5 by multiplying volume/liter by the appropriate mineral density (e.g., 2.71 g/cm^3 for calcite). The constant 31.56 is a conversion factor that provides the desired units. The equation is valid for any flow type, pore or conduit configuration, or location within the groundwater system.

Equation 1 reflects the fact that S decreases as C/C_s increases—i.e., with time and flow distance. At C/C_s less than about 0.7, k is approximately 0.01 and n is approximately 2.0. Values of n decrease slightly with P_{CO_2}, and k increases with temperature and P_{CO2} but decreases with grain size, dolomite content, and insoluble percentage (Rauch and White, 1977; Palmer, 1991). At higher C/C_s values, however, n rises to about 4 (Plummer and Wigley, 1976) and k to about 0.1 (Palmer, 1991). Because $(1 - C/C_s)$ is less than one, the increase in n *decreases* the solution rate, causing it to drop sharply while the water is still far from saturation and allowing the remaining aggressiveness to extend over a much longer distance than would otherwise be possible.

This change in kinetics represents an important geochemical threshold (White, 1977). Palmer (1984, 1991) has shown with finite-difference modeling and field examples that if the low values of k and n were to persist all the way to saturation, the average solution

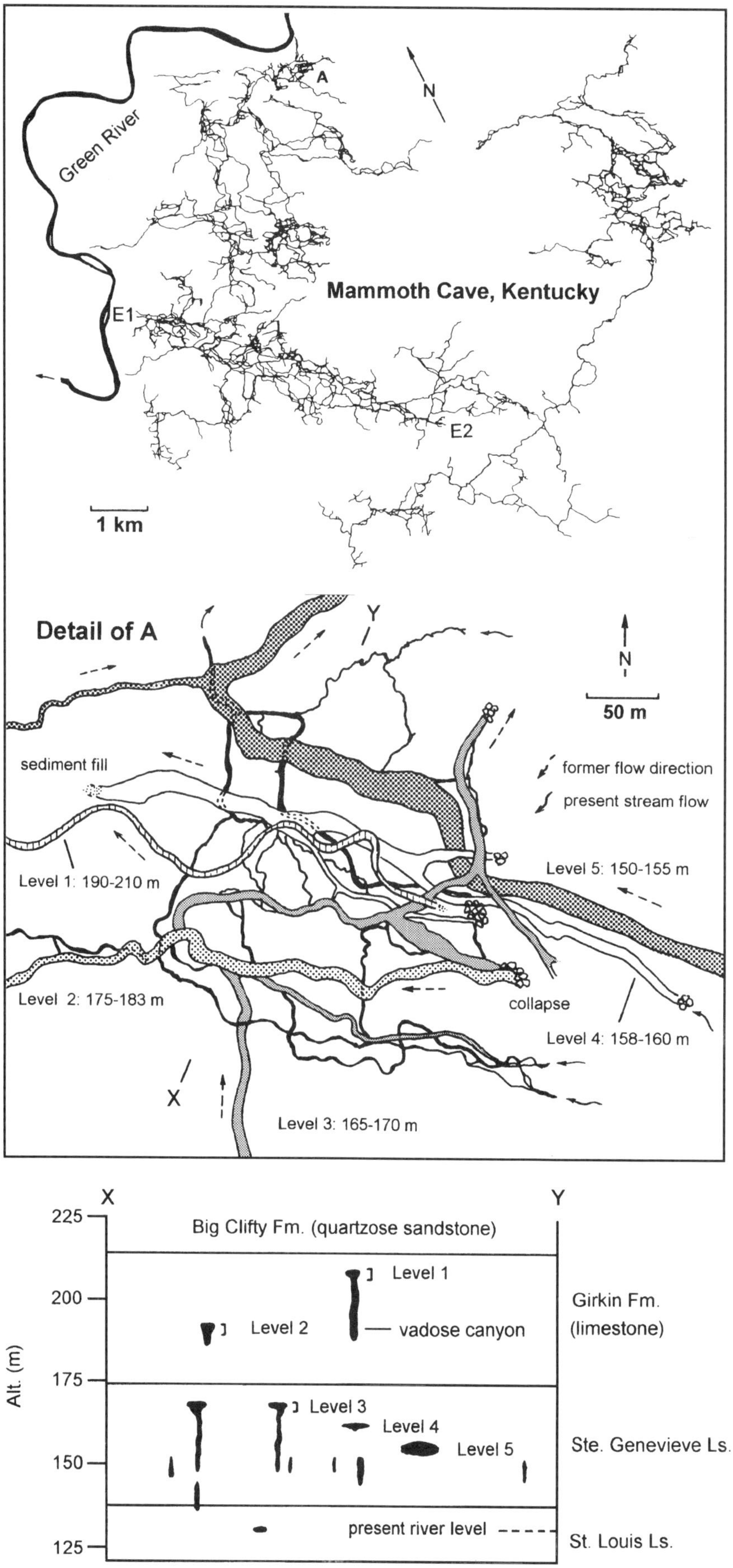

Figure 4. Mammoth Cave, Kentucky, is the longest known cave in the world, with more than 545 km of mapped passages. Prominent bedding in the Mississippian limestone promotes curvilinear patterns. The limestone is capped by detrital strata that limit recharge to certain areas, causing the irregular distribution of passages. Groundwater flows northwest from where the limestone is exposed updip, and from karst valleys that have breached the insoluble cap rock, discharging into the Green River. Variations in local dip and strike determine passage sinuosity (average dip is 0.3°NW). The branchwork pattern is obscured by relict upper levels. The detailed map shows an unusually dense cluster of passages with several major levels of tubular conduits (originally phreatic) formed at successive erosion levels of the Green River. Some tubes show later vadose entrenchment. The overall cavernous porosity of the limestone is only about 4%, as shown in cross section X-Y. Most phreatic conduits are highly concordant with the strata. Vadose passages (black) descend in stairstep patterns as alternating canyons and vertical shafts. E1 = Historic Entrance; E2 = Frozen Niagara Entrance. Map courtesy of Cave Research Foundation; geology and cross section by A. and M. Palmer.

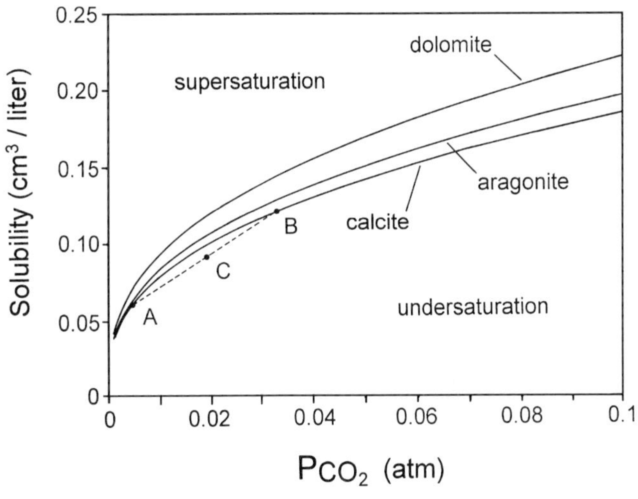

Figure 5. Volumetric solubility of calcite, aragonite, and dolomite vs. CO_2 partial pressure at 10°C. Mixing of two calcite-saturated solutions (A and B) with contrasting P_{CO_2} produces an undersaturated mixture [(C), for a 1:1 mixing ratio].

rate would be so rapid that nearly all aggressiveness would be squandered in the first few tens of meters of flow through the initial system of pores and narrow fissures. It is the abrupt rise in k and n as the dissolved load exceeds approximately 70% saturation that allows the solution process to extend for great distances through the aquifer, provided the flow rate is sufficient. On the other hand, these low solution rates are not sufficient to produce large conduits within geologically feasible time periods. Both solution rates are essential to the origin of solution conduits: slow solution enlarges the entire flow route rather uniformly, and then, when the water is able to pass through the entire system at less than the critical C/C_s, the solution rate jumps and large caverns form rather quickly. Were it not for this change in kinetics, limestone solution conduits could form only in the most ideal circumstances of steep hydraulic gradient and large initial fissure width (e.g., along steep escarpments). Instead, caves are common in virtually every present-day carbonate terrane in humid regions.

The time, t, required for the onset of maximum solution rate in a planar fissure (fracture or parting) in calcitic limestone can be determined analytically or numerically to be approximately

$$t = \alpha w_o^{-3} P_{CO_{2o}}^{-1} (i/L)^{-1.4} \text{ yr} \quad (2)$$

(Palmer, 1988, 1991; Dreybrodt, 1990), where w_o = initial fissure width (cm), $P_{CO_{2o}}$ = carbon dioxide partial pressure at the upstream end (atm), i = hydraulic gradient (dimensionless), L = fissure length (cm), and α is a constant that varies between 1×10^{-12} for open systems and 5×10^{-12} for systems closed to further addition of carbon dioxide. Hydraulic gradient (i) and flow distance (L) have exactly opposite effects. These relationships show the importance of discharge—proportional to (w^3) (i) in laminar flow—and of the mass balance, in which the amount of solution necessary to enlarge the fissure depends on the fissure length and the amount of mass that the water can dissolve. The time required to reach the maximum solution rate is typically about 10^5 yr in long conduits, as verified by geomorphic and paleomagnetic evidence in which cave passages at different levels have formed at intervals of several hundred thousand years (Miotke and Palmer, 1972; Schmidt, 1982).

COMPETITION BETWEEN FLOW PATHS

Figure 6 shows how the mean rate of wall retreat (S) along flow paths in limestone varies with mean conduit radius, discharge (Q), and flow length (L) from the input point. It is assumed that carbonic acid is the source of aggressiveness and initial $C/C_s = 0$. Deviations from a circular cross section have little effect on the relationships. This graph was constructed from finite-difference calculations (Palmer, 1991), but its pattern can be explained intuitively in the following way. At low discharge and/or high flow distance, the water becomes nearly saturated within the flow distance L, and so the average S depends simply on the mass balance, i.e., how much bedrock can be dissolved by the available amount of water. This relationship accounts for the sloping lines in the center of the graph (Zone A), which show greatly disparate S values for different values of r and Q/L. However, if the water passes through the system rapidly enough to retain very low saturation ratios, the solution rate is fixed at or near a maximum value, regardless of discharge. This condition is represented by the convergence of all the individual sloping lines into a single horizontal line at the top of the graph (Zone B).

For any given conduit radius (r), the rate of enlargement increases with the Q/L ratio, i.e., with increasing discharge or decreasing flow distance from the upstream end. However, S eventually reaches the upper limit beyond which further Q/L increase has virtually no effect (Zone B in Figure 6). This limit is reached at approximately $Q/L > 0.001\ r$ (cgs units). The maximum S depends on temperature, P_{CO_2}, lithology, and initial dissolved load, and is usually about 0.05–0.1 cm/yr in fresh water infiltrating through soil directly into limestone. This value fluctuates with seasonal P_{CO_2} levels and is best considered an annual mean. Other factors such as turbulence, abrasion by sediment, and presence of solution-retarding agents such as phosphates make the S values in Figure 6 only approximate, but the general concepts still hold.

Low Q/L ratios (Zone A in Figure 6) are typical of the early stages of flow through a carbonate aquifer before turbulent-flow conduits have developed. There is a great disparity in mean enlargement rates, as illustrated by the scattered dots in Figure 6. Mean growth rate in a given conduit can increase only if the discharge increases, which is usually accomplished by piracy from less favorable routes, as there is only a limited amount of recharge to the aquifer. Those paths

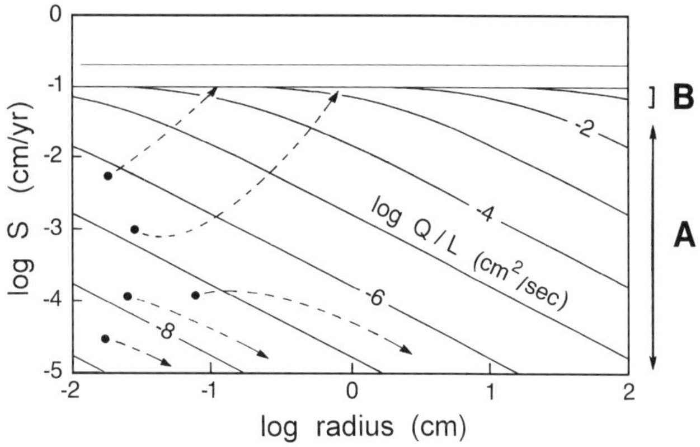

Figure 6. Mean rate of solutional wall retreat (*S*) in a solution conduit vs. discharge (*Q*), flow distance (*L*), and effective conduit radius, at 10°C and P_{CO_2} = 0.01 atm. Zone A = long flow paths with disparate *S* values. Zone B = short or high-*Q* paths that grow simultaneously at comparable rates. Dashed lines show growth histories of a few typical flow paths, where • = conditions at any given time early in the history of meteoric water circulation. Two reach high growth rates and become conduits, while the others stagnate.

with relatively high Q/L tend to enlarge at accelerating rates (ascending arrows in Figure 6) and eventually reach the maximum possible enlargement rate at the top of the graph.

Only those few openings grow to large size, while all others languish with low and perhaps diminishing enlargement rates, as illustrated by the descending arrows in Figure 6. The result is a sparse branchwork of stream conduits with confluent tributaries (Figures 2 and 7). The branchwork pattern is commonly obscured on cave maps by the presence of multiple levels (the upper ones relict), by structural control of conduit orientation, and by segmentation of conduits by collapse (Figures 2 and 4). Such a system is usually fed by infiltration through a karst surface with discrete points of recharge, such as sinkholes (Figures 2 and 7). Conduits have well-defined walls representing a sharp demarcation from the surrounding bedrock, rather than gradational boundaries with spongelike zones of smaller openings (Figure 3). Conduits tend to diminish in number with depth below the water table, because the presolutional openings are narrower and sparser, and flow routes are longer (Ford and Ewers, 1978). One of the most accessible examples of this kind of conduit system is Mammoth Cave in Kentucky (White and White, 1989).

PERVASIVE ENLARGEMENT OF INITIAL OPENINGS BY METEORIC GROUNDWATER

Where *Q/L* is simultaneously large throughout many competing flow paths (Zone B, at the top of the graph in Figure 6), nearly all openings grow at comparable rates regardless of size or discharge, and a maze of interconnected solution voids is formed. Only those openings narrower than a few tens of microns escape solutional enlargement within geologically feasible times. This is the opposite of the selective, competitive growth of conduits fed by recharge from sinkholes, whose development is initiated in Zone A in Figure 6, as described in the previous section. Several geologic settings provide the necessary conditions for pervasive solution porosity.

Most common is the epikarst, immediately below the soil, where water is highly aggressive and flow distances are short (Figure 1). All but the narrowest openings enlarge simultaneously, forming a network of interconnected fissures and irregular voids within the top few meters or tens of meters of the bedrock. At greater distances from the surface (larger *L*) the water approaches saturation, and the *Q/L* ratio drops enough that the various flow paths begin to differ in enlargement rate. Water passes through the epikarst in a dispersed fashion, but most is gradually focused into relatively few major conduits that penetrate deeply into and through the aquifer (Williams, 1983). These are the few that win the competition among the widely varied *S* values by emerging from Zone A into Zone B in Figure 6.

A similar situation prevails in those parts of a carbonate aquifer subject to the sudden influx of floodwater, e.g., where surface runoff furnishes rapid recharge to carbonate rocks, either as sinking streams or as episodic bank storage adjacent to entrenched rivers. Aggressive water is forced into all openings under steep hydraulic gradients, enlarging them all at similar rates (Zone A in Figure 6). This process can occur deep inside a carbonate aquifer where pre-existing air-filled conduits just above the water table are subject to sudden flooding, especially in the vicinity of constrictions formed by collapse or sediment fill. Pervasive solution can take place despite the great distance from the recharge source, because water is delivered rapidly to the site by the conduits while still far from calcite saturation. Conduits that receive such flow become surrounded by a maze of fissures, vugs, and secondary passages localized in areas of great water-table fluctuation. In prominently jointed rock these openings form fissure networks. An example accessible to the public is Mystery Cave, Forestville State Park, in southeastern Minnesota, which is a network maze still forming as a subsurface meander cutoff of an entrenching river (Milske et al., 1983; Figure 8). In well-bedded rocks they tend to comprise anastomotic mazes of braided, sinuous, interconnected tubes along partings (Figure 9). In rocks with high matrix porosity, interconnected voids like those of a sponge are formed ("spongework"), but these are rare. Palmer (1975, 1991) shows further examples of maze caves.

Where recharge enters carbonate rock through permeable but insoluble material, such as quartzose sandstone, all significant fractures in the soluble rock enlarge at comparable rates, because discharge is held nearly uniform by the insoluble rock and flow paths are short (Figure 7). A dense network of intersecting

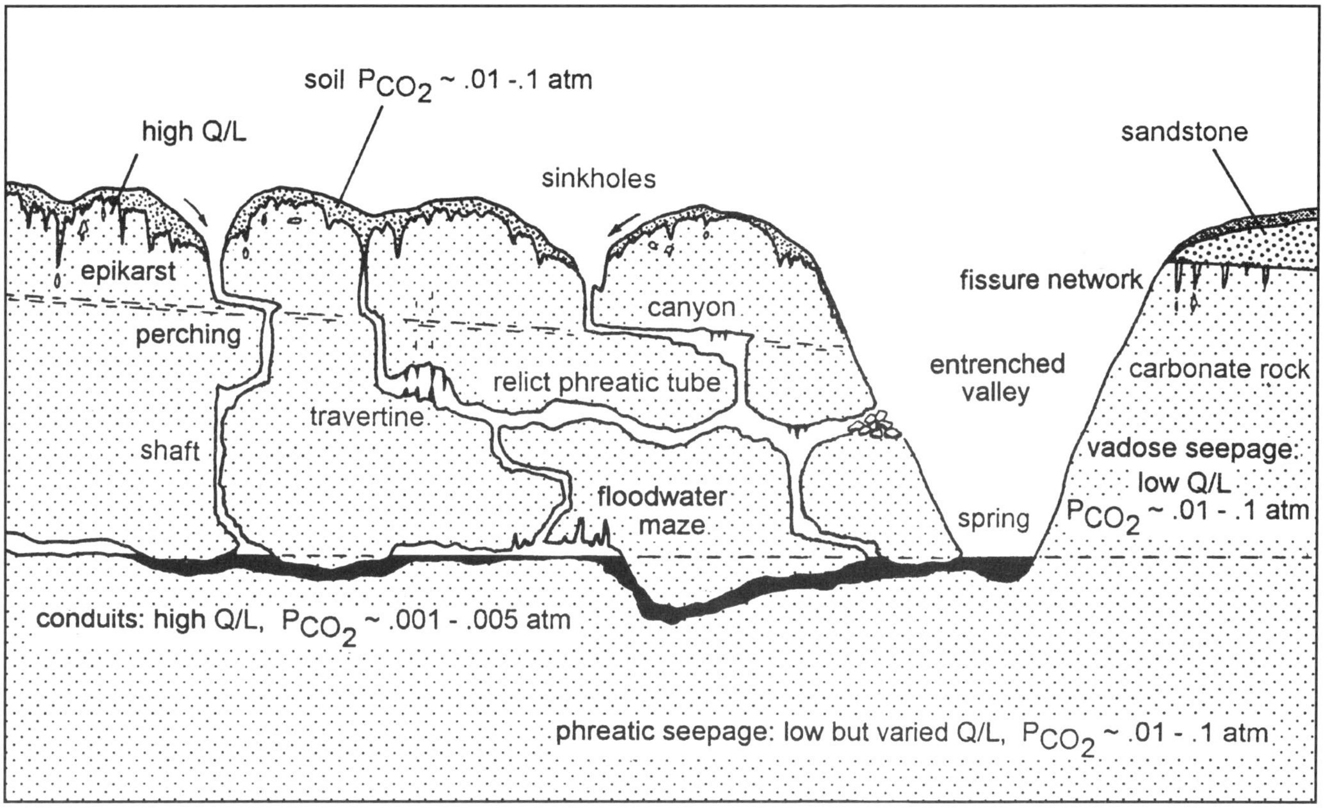

Figure 7. Geochemical setting and distribution of porosity in diagenetically mature karst regions. See text and Figure 6 for an explanation of terms.

fissures is produced, which is the subsurface equivalent of epikarst. Examples are shown in Figure 10 and in Palmer (1975).

Secondary microporosity can also be pervasive in carbonate rocks, but it originates from diagenesis or from selective solution of relatively soluble grains, rather than by the processes described above. In artesian basins with low-permeability outlets there can be a gradation in the style of solution porosity along the paths of flow. In the Lincolnshire Limestone of eastern England, for example, most karst voids are produced within the upper few tens of meters of the recharge surface, whereas at depth, up to tens of kilometers from the recharge source, widespread microporosity is produced by selective solution of ooids, micrite, and fossils (Smalley et al., 1994).

MIXING ZONES

Mixing of waters of contrasting chemistry can enhance or rejuvenate solutional aggressiveness. Because of the concave-downward saturation curves for carbonate minerals as a function of P_{CO_2} (Figure 5), mixing of two waters saturated with a carbonate mineral at different P_{CO_2} values will produce an undersaturated solution (Bögli, 1964, 1980). Even if the initial solutions are not at saturation, the saturation ratio (C/C_s) of the mixture will be lower than that of either source. The effect resembles a local boost in acidity (or increase in Q/L), and seemingly isolated zones of solution porosity may be produced. The same effect is caused by mixing of waters having different salinity, owing to the diminution of activity coefficients with increasing ionic strength (Runnells, 1969). Mixing of fresh groundwater with seawater at $P_{CO_2} > 0.01$ atm often causes calcite undersaturation at low seawater percentages, but calcite supersaturation at high seawater percentages (Plummer, 1975; Wigley and Plummer, 1976). Dolomite solubility varies in a similar way but is influenced by the degree of order within the dolomite lattice and by which of several possible solubility constants is selected (Hardie, 1987). A third mixing effect, caused by differences in H_2S concentration, is discussed in a later section.

Mixing is accomplished by hydrodynamic dispersion (branching and convergence of flow lines), ionic diffusion, and, in large voids, by turbulent eddies. In diffuse-flow systems, where mixing is the greatest source of solution porosity, flow rates are usually low. As a result, most of the dissolving is localized rather than drawn out in the downflow direction. The volume and rate of porosity production are governed mainly by the rate of inflow and mixing rather than by solution kinetics, and usually can be represented simply by the mass balance (rate of flux of solvents and solutes).

Mixing-zone solution is most conspicuous in young seacoast carbonates with high primary porosity (Vacher, 1978; Back et al., 1979, 1984; Mylroie and Carew, 1990; see Figure 11). There are two zones of greatest mixing: one at the top of the freshwater lens,

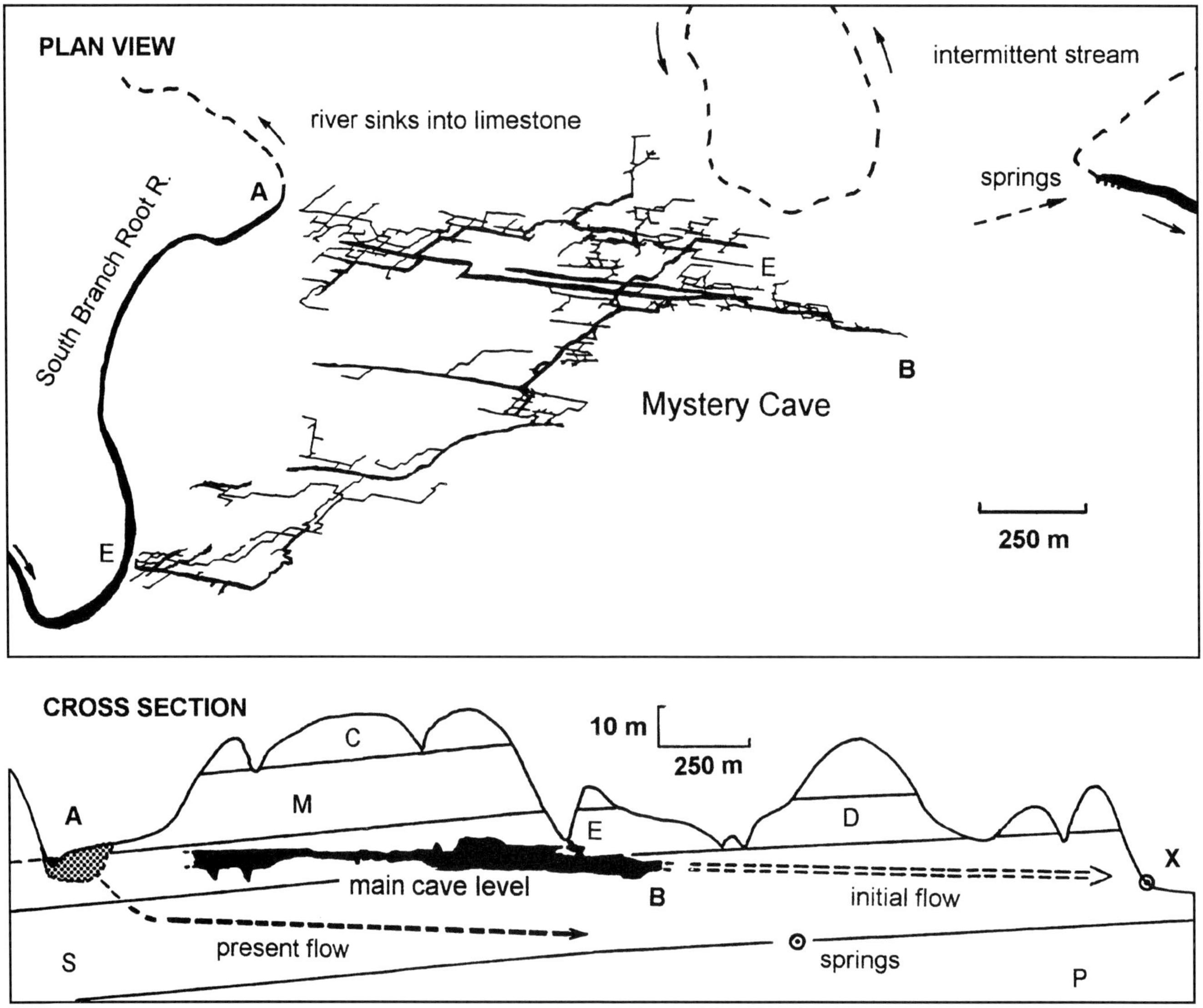

Figure 8. A floodwater network formed by subterranean piracy of the South Branch of Root River, Minnesota (Mystery Cave). The cross section shows the original and present passage gradients. X = original spring location. Few of the presently active lower-level passages are of explorable size. Prominent jointing allows strong discordance to the strata. Depth of alluvium (shaded) determined by refraction seismology. C = Cedar Valley Ls.; M = Maquoketa Fm. (limy dolomite); D = Dubuque Fm. (shaly limestone); S = Stewartville Fm. (limy dolomite); P = Prosser Fm. (cherty limestone); E = entrance. Map courtesy of the Minnesota Speleological Survey; profile and geology by A. and M. Palmer.

where high-CO_2 infiltrating water meets lower-CO_2 phreatic water; and another at the freshwater/saltwater interface. Data from wells and caves in Bermuda (Plummer et al., 1976) show a crude positive correlation between aggressiveness and P_{CO_2}, but no systematic relationship between saturation levels and salinity. Solution rates may be increased by reduction of sulfate in the seawater and oxidation of the resulting H_2S to sulfuric acid (Bottrell et al., 1991; Stoessel, 1992). Flow rates and mixing at the freshwater/saltwater interface depend partly upon interactions among hydraulic gradient, buoyant circulation driven by meteoric recharge, reflux, and thermal convection, which may augment or partly counteract each other (Whitaker and Smart, 1993).

Porosity in seacoast mixing zones varies from tiny matrix voids to large caverns, predominantly with an irregular vug-like geometry (Figures 12 and 13). Caves are concentrated just inland from the coastline, where mixing rates are greatest (Mylroie and Carew, 1990). In caves of the Yucatan peninsula, Stoessel et al. (1989) found the highest solution rates in areas of steepest vertical salinity gradient. As the porosity and hydraulic conductivity increase, the freshwater lens along the seacoast may dwindle to a thin layer of brackish water, especially where flow rates are low.

Mixing has little effect at the water table in continental karst, because phreatic water tends to equilibrate with the CO_2 levels of the vadose water that feeds it. Within conduits there is little geochemical

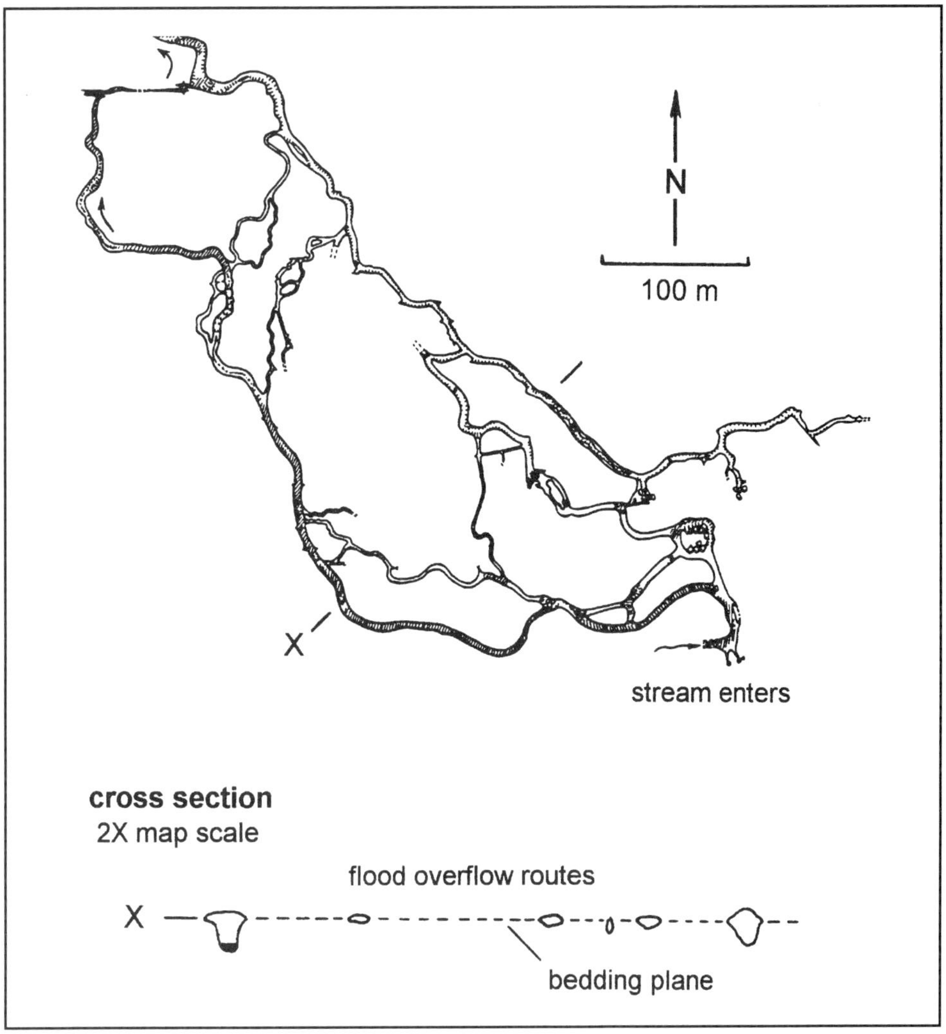

Figure 9. Anastomotic floodwater maze at the upstream end of Blue Spring Cave, Indiana, in bedded upper Salem Limestone (see Figure 2, X, for general setting). Note concentration of passages at the same stratigraphic level. During high discharge, the entire maze fills to the ceiling with water.

change between the vadose and phreatic zones, and solution rates show little or no increase at confluences.

HYPOGENETIC PROCESSES

Many of the concepts that govern solution by carbonic-acid–rich meteoric water also apply to deep-seated processes, although the origin and distribution of aggressiveness are quite different and solution rates are poorly known. Instead of the rather predictable conditions of humid continental karst, where carbonic acid is generated at the recharge site and the solution process is attenuated in the downflow direction, solution porosity deep beneath the surface is usually created by bursts of aggressiveness that are spatially and temporally limited. Average flow rates are comparatively low, and solution porosity tends to be localized rather than distributed over large distances. As in mixing zones, the rate of porosity generation is governed more by the mass balance than by solution kinetics.

A common origin for deep-seated porosity begins with the bacterial or thermal reduction of sulfates in anoxic zones by organic carbon compounds (Machel, 1987, 1989; Hill, 1987, 1990; Mazzullo and Harris, 1991). Calcium and bicarbonate ions are produced, which have the potential to precipitate calcite. The smaller molar volume of calcite, compared to that of the original sulfate minerals, can cause increased porosity. In closed systems, replacement of gypsum or anhydrite by calcite can produce up to 50% and 20% porosity, respectively, but these percentages are rarely achieved because exact mole-for-mole replacement is rare. Diagnostic calcite textures include pseudomorphs after sulfates, nodules, and doubly terminated crystals. Large negative oxygen and carbon isotope ratios are typical. Examples are given by Pierre and Rouchy (1988) and A. N. Palmer and M. V. Palmer (1989, 1991).

Hydrogen sulfide is another product of sulfate reduction. Some or all is retained in solution, while some may be released as gas bubbles. Aqueous

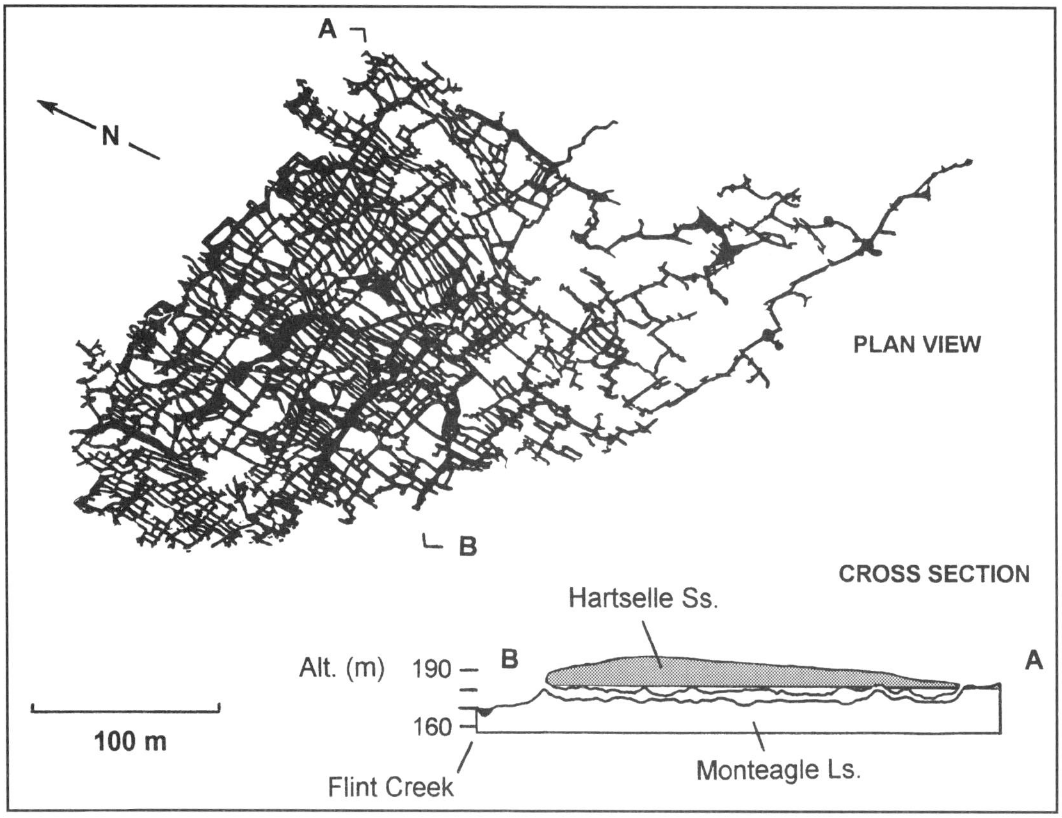

Figure 10. An exceptionally large network cave formed by water infiltrating through a permeable cap of Hartselle Sandstone (Anvil Cave, Alabama). In places, the spatial density in plan view approaches 40%, but the single level and small vertical range greatly limit the overall porosity of the limestone. Bank flooding from the adjacent stream apparently caused much of the enlargement, but the network pattern was controlled by infiltration through the sandstone, since all networks in the area lie directly beneath the Hartselle. (Modified from Varnedoe, 1964.)

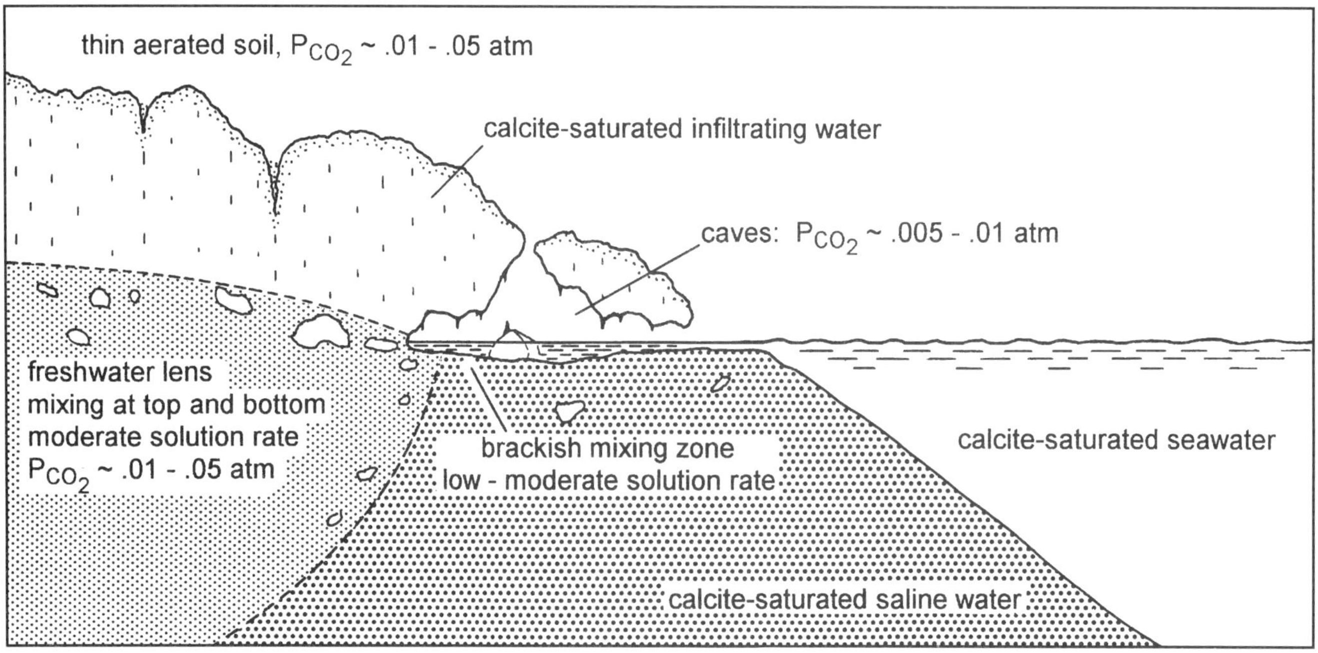

Figure 11. Geochemical setting and distribution of karst-related porosity in seacoast mixing zones. Reduction of sulfate from seawater and oxidation of the resulting H_2S can enhance solution rates considerably.

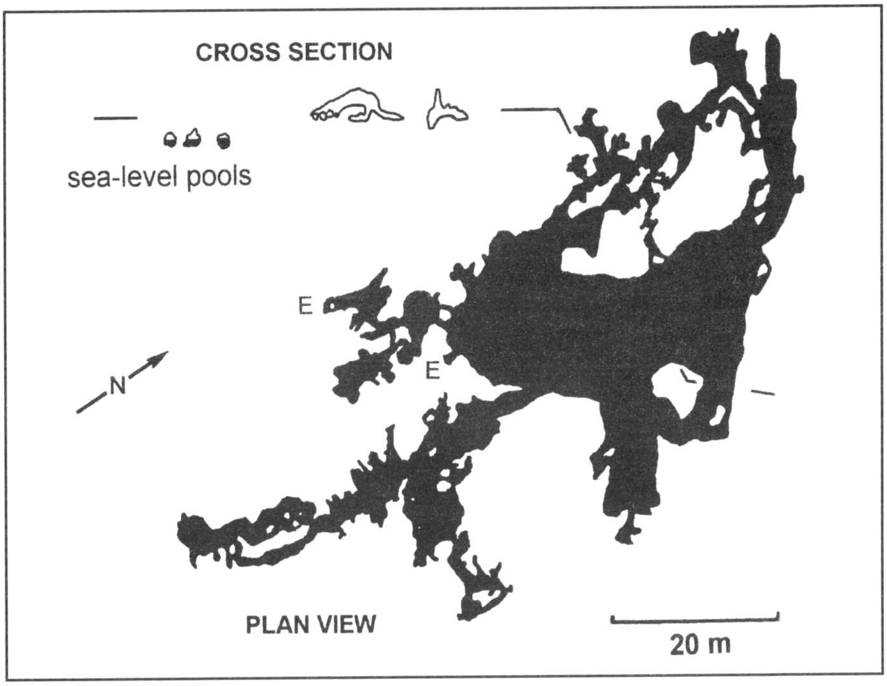

Figure 12. Cave formed by seacoast mixing (Light House Cave, San Salvador, Bahamas). Note the irregular pattern and nearly horizontal cross section, which is discordant to the strata. The cave is located in dune eolianites only 85,000–125,000 years old. Water in the cave is brackish and has a 1 m average tidal range. E = entrance. (Modified from Mylroie, 1988.)

Figure 13. Cave formed by mixing at the freshwater/saltwater contact of a seacoast aquifer, Walsingham Formation, Bermuda. People to right of opening show scale. Infiltrating water is dispersed among many small pores and loses its aggressiveness within a few meters of the surface. The warm, shallow seawater is supersaturated with calcite. The only solutionally aggressive water available to form caves is in mixing zones (see Plummer et al., 1976). This cave correlates with zones of spongework in nearby caves at the same elevation. The freshwater lens has degenerated to a thin zone of brackish water because of the high permeability of the cavernous limestone. The sea-level niche is erosional and biogenic.

hydrogen sulfide is by itself a weak acid with almost the same ability as carbonic acid to dissolve carbonate rock. Yet, the fluids in sulfate-reducing environments are usually at or near saturation with respect to calcite and have little tendency to dissolve more carbonates. If this water migrates from the site of sulfate reduction it can dissolve further carbonate rock in either of two ways:

1. Mixing of waters of contrasting H_2S content can produce considerable undersaturation. The saturation curves for carbonate minerals vs. H_2S concentration resemble those for carbonic acid solutions (compare with Figure 5) and show a similar ability to renew solutional aggressiveness, where mixing of waters of differing H_2S content takes place (Palmer, 1991). This process is most potent if one of the initial solutions has a very low H_2S concentration, which is a common occurrence. Porosity zones can be produced at any depth with virtually no relation to the overlying land surface.

2. If aqueous or gaseous hydrogen sulfide comes in contact with oxygen-rich water, the H_2S oxidizes to sulfuric acid, either directly or indirectly through the intermediate step of native sulfur. This produces a burst of solutional capacity in which one equivalent of hydrogen sulfide is capable of dissolving two of calcite or one of dolomite. The effect is the same as a sharp increase in Q/L (see Figure 6) and results in maze-like porosity in which most initial pores, fractures, or partings enlarge simultaneously (Figures 14 and 15). The oxygen requirement tends to limit the depth to which this process can take place, although certain caves in the Guadalupe Mountains of New Mexico show evidence of H_2S oxidation over a vertical span of several hundred meters, culminating upward at the former water table (Hill, 1987; Figure 15). The resulting porosity volume depends on the ambient P_{CO_2}, since CO_2 is generated by this solution process. If CO_2 escapes, the solutional capacity of the water diminishes (Palmer, 1991). Solution of carbonate bedrock by sulfuric acid may drive the ion activity product $(Ca^{++})(SO_4^{=})$ to supersaturation with respect to gypsum or, less commonly, anhydrite. Where there is mole-for-mole ionic exchange, as in a closed system, the volume of gypsum produced can almost exactly equal the volume of limestone dissolved. However, most of the calcium and sulfate are removed by flowing groundwater, either at the same time as the sulfuric acid reaction or later by the invasion of meteoric water, resulting in a large net increase in porosity. Caves and pore systems formed by rising and oxidizing hydrogen sulfide commonly have ramifying patterns, in which irregular rooms and

Figure 14. Pervasive enlargement of initial pores in limestone by sulfuric acid, forming a spongework cave pattern (Capitan Formation, Carlsbad Caverns National Park, New Mexico).

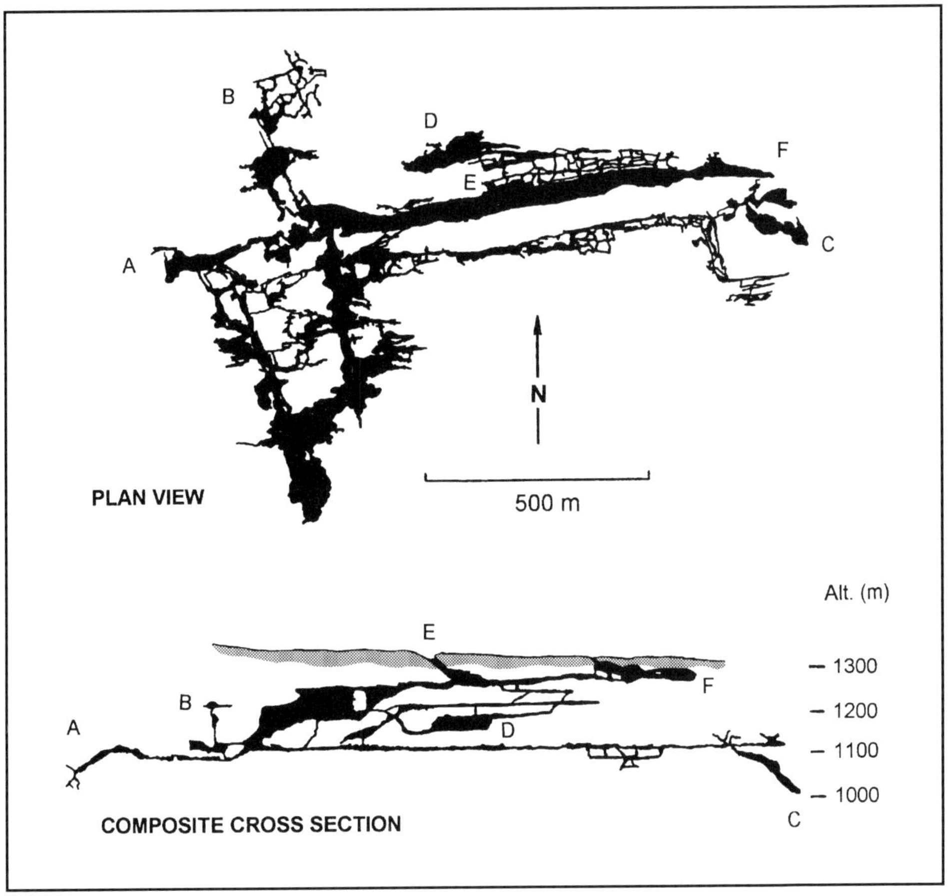

Figure 15. Carlsbad Cavern, New Mexico, is an exceptional example of solution by oxidation of rising H_2S to sulfuric acid. Note the large rooms with ramifying and network patterns, ascending passage segments, and prominent levels at former water-table elevations. The cave is located mainly in the massive Capitan reef (Permian), although the southeastern areas are in fore-reef talus and upper levels in the northwestern parts are in the bedded back-reef Tansill and Yates formations. The cave is highly discordant to the strata. E = entrance. Map and profile courtesy of Cave Research Foundation.

maze-like galleries wander in three dimensions with branches exiting from the main areas of development at various levels (Hill, 1987; Palmer, 1991). Where inflow to the carbonate rock is dispersed among many fractures, a network of intersecting solutionally enlarged fissures is formed. Isolated vugs in sulfide ore zones also appear to be the result of H_2S oxidation and/or mixing (Palmer and Palmer, 1991; Furman, in press).

Hydrocarbon maturation, thermal degradation, and reaction with mineral oxidants can produce organic acids capable of dissolving carbonates (Meshri, 1986; Moore, 1989, p. 267; Surdam et al., 1993). Some solution porosity in oil fields has been attributed to processes of this type.

Solution of carbonates can also be achieved by the cooling of thermal waters rising from depth. Accessible field examples include Wind and Jewel Caves in South Dakota (Figure 16), which are thought to have been formed in part by rising thermal water (Bakalowicz et al., 1987). This process is slow but quantitatively feasible (Palmer, 1991), although its recognition and significance are clouded by the fact that mixing with shallow meteoric water of contrasting CO_2 content usually accounts for most of the undersaturation. Mixing of cold and warm waters is not by itself a viable mechanism for renewing aggressiveness, because the saturation curve has a negative slope that diminishes with temperature (Figure 17). Such mixing would tend to produce supersaturation.

In arid and semi-arid karst regions, solution by shallow meteoric water is minimal, and the epikarst is poorly developed. Solution sinkholes and vadose solution conduits are extremely rare. Soil is thin and calcareous or entirely absent, resulting in bare bedrock with solution pockets and runnels (Esteban and Klappa, 1983; Ford and Williams, 1989, p. 467–472). Porosity formed by hypogenetic acids (produced by deep-seated processes rather than by gases or organic processes in the atmosphere or soil) is much more prominent in dry climates because in humid regions the effects of these acids are easily overwhelmed by those of epigenetic carbonic acid.

SOLUTION POROSITY IN AREAS OF INTERBEDDED SULFATES AND CARBONATES

Interbedded sulfates have an immense impact on carbonate rocks, owing to their mobility and chemical instability. The reduction of sulfates to hydrogen sulfide, described earlier, is only one of several related phenomena. Solution, hydration, and dehydration of sulfates cause fracturing and collapse of surrounding strata. Reprecipitation of sulfates in fractures, either by evaporation or by crystallization during the sulfuric acid reaction with carbonates, wedges clasts apart to produce mosaic breccias. Precipitation of gypsum to form breccias may be accomplished by cooling of

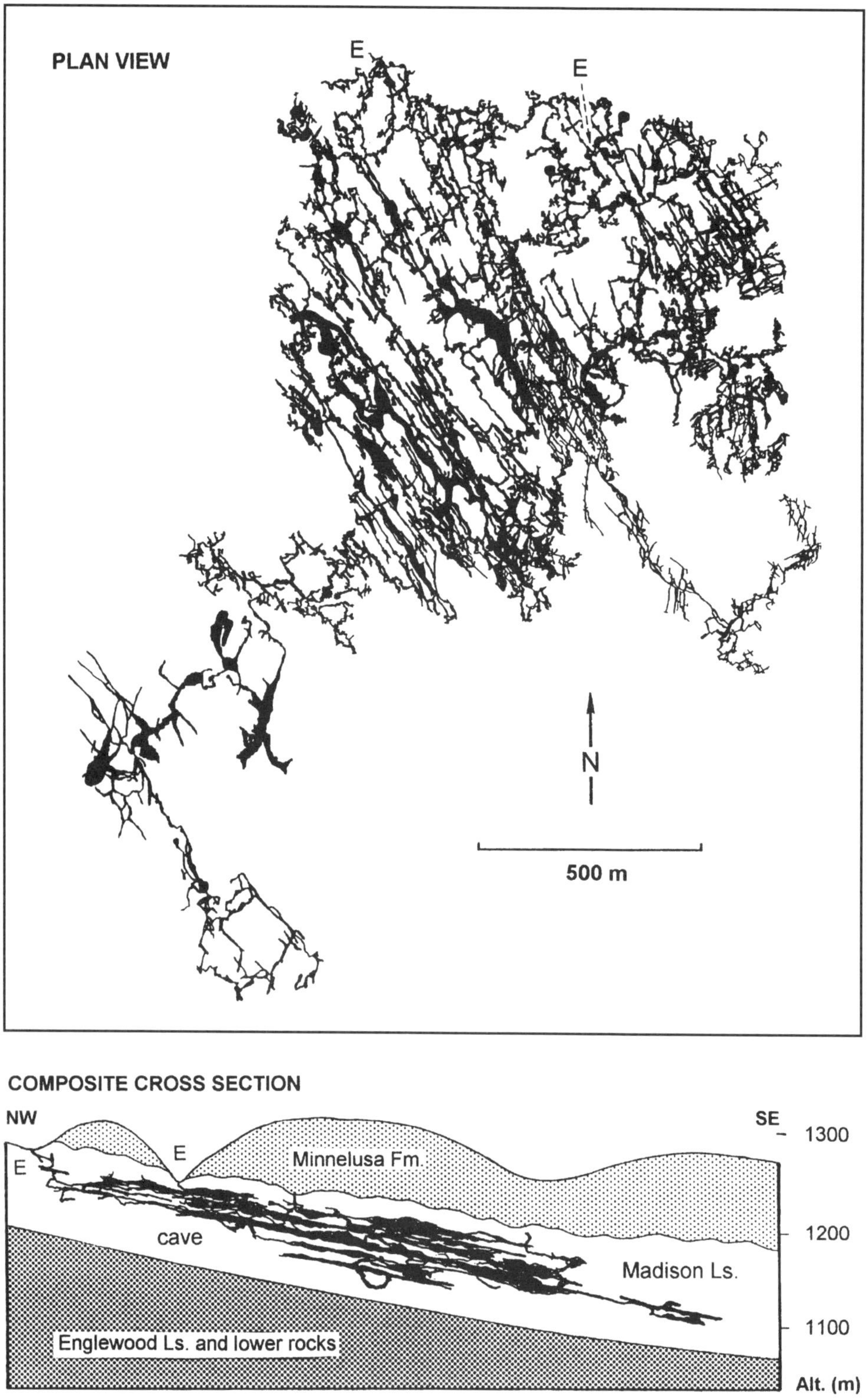

Figure 16. Wind Cave is a network occupying a former sulfate zone of the Madison (Pahasapa) Limestone, South Dakota. Tertiary groundwater flow has enlarged the initial Mississippian caves. Note the strong stratal influence, lack of horizontal levels determined by past water tables, and concentration of passages in certain beds, which are typical of porosity in sulfate-carbonate zones. The cross section is viewed in the direction of the strike from the southwest. E = entrance. Map and profile courtesy of National Park Service; geology by A. and M. Palmer.

ascending water saturated with anhydrite. Anhydrite in contact with water is unstable at low pressures and temperatures below about 40°C, and as the water cools, less-soluble gypsum becomes the stable phase and is forced to precipitate.

If gypsum or anhydrite is dissolved by water rich in calcium bicarbonate, calcite is forced to precipitate because of the common-ion effect, in which the shared Ca^{++} increases the saturation ratio of each mineral. This process leaves only indirect evidence for the former sulfates, such as pseudomorphs of sulfate crystals, doubly terminated calcite crystals, anastomotic (braided) veining, mosaic breccias, and dedolomitization (A. N. Palmer and M. V. Palmer, 1989). Widespread porosity zones limited to narrow stratigraphic intervals are common, as are nearly vertical breccia

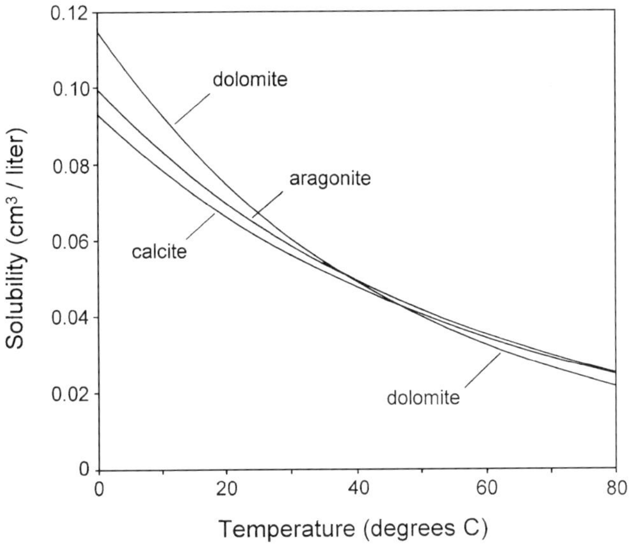

Figure 17. Solubility of calcite, aragonite, and dolomite vs. temperature at P_{CO_2} = 0.01 atm.

pipes (Roberts, 1966; Sando, 1974, 1988; Loucks and Anderson, 1985; M. V. Palmer and A. V. Palmer, 1989; Dravis and Muir, 1993; Demiralin et al., 1993; see Figure 18). Where the initial sulfates are now absent, such zones may be erroneously interpreted as the result of cavernous solution and collapse within carbonate rocks. The potential role of now-vanished sulfates should be considered in the interpretation of the origin of areally widespread interstratal porosity in karst reservoirs, such as the Ellenburger Group in Texas, which is generally interpreted in terms of meteoric processes (Kerans, 1988; Loucks and Handford, 1992; Canter et al., 1993). Worthington (1994) has shown that solution conduits in carbonate aquifers can be initiated by solution of sulfates by deeply circulating meteoric water.

The common-ion effect also controls the relative solubility of calcite and dolomite. Dissolved gypsum or anhydrite diminishes the solubility of both limestone and dolomite, but the effect on limestone is greater. As a result, dolomite becomes far more soluble than calcite or aragonite, and selective solution of dolomite can occur (Figure 19). Dedolomitization and the selective solution and incongruent solution of dolomite have been noted in such areas (Evamy, 1967; A. N. Palmer and M. V. Palmer, 1989, 1991) and observed experimentally in sulfate-rich solutions (DeGroodt, 1967). The geochemistry of groundwater in the Madison aquifer of Wyoming and South Dakota, in the vicinity of the Black Hills, indicates simultaneous dolomite solution and calcite precipitation in the presence of sulfates (Back et al., 1983).

DIAGENETIC SOLUTION POROSITY

Diagenesis is considered here in the rather strict sense of broad-scale changes in mineralogy and fabric, although many geologists consider all solution porosity to be diagenetic (Choquette and James, 1988, p. 2). This topic has been treated at length by other workers (e.g., Choquette and Pray, 1970; Bathurst, 1971; James and Choquette, 1984; Budd, 1988; Moore, 1989) and is briefly reviewed here only to place it in context with the preceding sections.

Where infiltrating water first comes in contact with carbonate rock it is undersaturated with all carbonate species. In diagenetically immature carbonates, calcite is the first to approach saturation, while aragonite and high-Mg calcite continue to dissolve. As a result, low-Mg calcite is precipitated, filling much of the new and pre-existing porosity. In the vadose zone, diagenetic boundaries are highly irregular, with the most advanced diagenesis localized along major infiltration paths. Aragonite and high-Mg calcite can persist in isolated zones of low moisture content even after the carbonates along the major flow routes have been converted entirely to low-Mg calcite. Diagenesis proceeds rapidly in the phreatic zone because of the persistent availability of water, and where waters of varied composition are able to mix (James and Choquette, 1984). As the mineralogy along a given flow path stabilizes, solution becomes more fabric-selective, with micrite, fossils, and ooids preferentially dissolved.

Interstitial pores in dolomite account for much of the porosity in certain petroleum reservoirs (Roehl and Choquette, 1985). Dolomitization of calcite to form well-ordered dolomite in the ideal 2:1 ratio of a closed system has the potential to increase porosity by 13% because of the decrease in molar volume. Such molar balance is rarely achieved, although the volume change must still be accounted for in interpreting the porosity (Choquette et al., 1992).

POROSITY PRESERVATION

The most extensive karst is formed during lengthy continental exposure, and one would expect its relict porosity to be concentrated below major sequence boundaries. However, such openings are highly susceptible to sediment filling or erosional destruction. Deep pores and conduits survive easily, but they are either sparse, having originated under highly competitive conditions (Zone A of Figure 6), or consist of scattered, irregular zones of hypogenetic porosity that have little relation to the overlying erosion surface. Nevertheless, relict caves and isolated solution vugs are common beneath many unconformities in carbonates, and former surface features such as sinkholes, fissures (cutters or grikes), weathering breccias, and paleosols are present in some areas. Voids are most effectively preserved through burial and filling by continental deposits such as fluvial, glacial, lacustrine, or volcanic materials, which also afford protection from erosion during later marine transgression (James and Choquette, 1988; Bosák et al., 1989).

Direct preservation by marine deposits, though documented, usually follows considerable erosional destruction of karst features. The most widespread continental paleokarst zones in the United States,

Figure 18. Some features typical of former sulfate zones: breccia, vuggy porosity, "zebraic" texture, and calcite cement (Madison Formation, Custer County, South Dakota). Knurled section of drafting pencil = 3.25 cm long.

those of the post-Sauk and post-Kaskaskia sequence boundaries, have discontinuous and poorly preserved paleosols and epikarst features (Sando, 1974; Mussman et al., 1988; M. V. Palmer and A. N. Palmer, 1989). Much of the porosity in these zones appears to have resulted from early sulfate-related processes and mixing (Palmer and Palmer, 1988). Solution voids, even those of cavern size, can survive many kilometers of burial, usually acquiring only a thin coating of euhedral calcite. For example, numerous intact caves and vugs in the Black Hills of South Dakota are of Mississippian age and have been enlarged further during Laramide uplift following sedimentary burial of about 2 km (A. N. Palmer and M. V. Palmer, 1989; Figure 16).

ABSENCE AND OCCLUSION OF SOLUTION POROSITY

Interpretation of porosity distribution must include reasons for its absence. In purely geochemical terms the explanation is simple: there has been no solutionally aggressive groundwater flow. Either there was insufficient hydraulic gradient, as is common at great depth beneath the surface, or (far less likely) a lack of initial openings for water to follow. Saturation may have been reached before the water penetrated to the zone in question, as is typical for diffuse infiltration through small pores. In arid and semi-arid regions, most infiltrating water becomes saturated at or just below the surface. Porosity may also appear to be absent deep within karst aquifers, in the large expanses between solution conduits, where the ratio of discharge to flow distance has been low.

Porosity can be partly or completely occluded in several ways. Mineral replacement involving a density decrease (e.g., calcite to gypsum) can diminish porosity in systems that maintain an approximate molar balance. Heating drives groundwater toward calcite supersaturation, producing widespread pore linings of spar. Mixing of waters of contrasting temperature increases the saturation ratios of all carbonate minerals, although this process is usually accompanied by

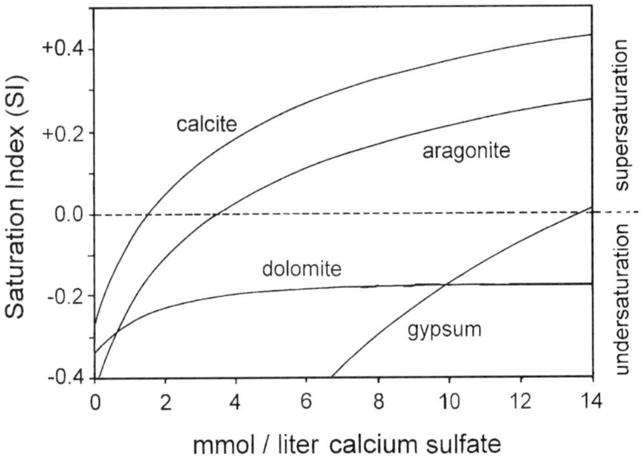

Figure 19. Effect of dissolved gypsum or anhydrite on the saturation index (*SI*) of calcite, aragonite, and dolomite at 10°C and P_{CO_2} = 0.01 atm. If calcium sulfate is added to an initial solution with 75% dolomite (shown here), calcite and aragonite rapidly become supersaturated (SI>0), but dolomite remains undersaturated. Additional $CaCO_3$, beyond what is contributed by dolomite, causes even greater disparity in *SI*. $SI = (2/n)\log(IAP/K)$, where *n* = number of ions released by solution, *IAP* = ion activity product, and *K* = solubility product. The term (2/*n*) makes the *SI* values numerically compatible for all minerals, regardless of the number of ions produced.

more potent chemical differences that decrease the saturation ratio.

CO_2 degassing reduces the solubility of carbonate minerals, generally causing calcite to precipitate as travertine or pore-lining cement. This process is most common where vadose seepage through narrow openings drips or flows into aerated caves. Travertine is concentrated in large openings and fissures that communicate with the surface, as they have the lowest P_{CO_2}. Local accumulations of travertine can be massive, although they are rarely extensive enough to reduce total cavernous porosity by more than a few percent. Less commonly, widespread cementation can take place where rising high-CO_2 water degasses because of decreasing hydrostatic pressure (Bakalowicz et al., 1987). Ford and Williams (1989, p. 346–347) discuss the spatial distribution of carbonate precipitates in caves, although quantitative data are sparse. In most actively forming karst the volume of reprecipitation is small compared to that of the dissolved load removed to springs.

SUMMARY: DISTINCTIONS AMONG TYPES OF SOLUTION POROSITY

Quantitative data on the distribution of solution voids are available from accessible caves, but comparisons with porosity data from petroleum reservoirs must be made cautiously. The effective percentage of macroscopic solution porosity varies greatly over short distances and depends upon the volume of rock considered (Ford and Williams, 1989, p. 134). Porosity may appear to be great if the boundaries are drawn tightly around the cavernous zone, but the percentage decreases considerably if larger volumes are considered. Porosity measurements from cave maps are biased by the fact that not all openings are accessible to mappers. Furthermore, much carbonate porosity is not cavernous (e.g., initial porosity, diagenetic porosity, fractures). From the statistical standpoint it is appropriate to refer to plan-view spatial density, which is equivalent to the probability of a drill hole encountering a solution cavity, since the limited vertical range of most caves gives deceptively low values if porosity is expressed as a percentage of the total rock volume. The following summaries may help in distinguishing the type and pattern of solutional porosity from drill hole and geophysical data.

Porosity Beneath Continental Karst Surfaces

Much solution porosity emanates from present or relict karst surfaces. Active examples are found in any humid region where relatively pure carbonate rocks are exposed to rapid groundwater flow. Paleokarst voids, mostly sediment filled, are abundant directly beneath the post-Sauk and post-Kaskaskia erosion surfaces in the southeastern United States and Rocky Mountain region respectively (M. V. Palmer and A. N. Palmer, 1989).

The *epikarst* consists of pervasive porosity formed at small distances from the surface (meters to several tens of meters) and is best developed in humid climates (Figure 1). Porosity decreases sharply with depth. In most paleokarst beneath widespread continental erosion surfaces, the solution features at and immediately below the original surface have been destroyed prior to burial, and only the deepest sinkholes and fissures are preserved. Paleosols and bedrock clasts commonly fill their lower parts and exhibit little grading or sorting. Solution porosity in intact epikarst is typically about 10-20%, but detrital fill reduces the net porosity by about half. Examples of epikarst porosity are given by Williams (1983), Smart and Friederich (1986), and Smart and Hobbs (1986).

Solution conduits descend from the erosion surface and represent only the selectively enlarged major flow paths (Figure 7). By volume, approximately 65–70% of all known solution caves are of this type, but they are least likely to be preserved beneath unconformities. Conduits are bounded by discrete walls, with little solutional enlargement of surrounding openings. Their most distinctive characteristics (branchwork pattern and huge length/diameter ratio) are difficult to identify by drilling or geophysical surveys. *Vadose conduits* have continuously downward profiles relative to the original horizontal datum and exhibit considerable stratigraphic perching interrupted by abrupt stratal discordance along fractures. Shaft, canyon, and fissure morphologies are typical. *Phreatic conduits* consist of tubes or fissures with irregular profiles and overall low gradients relative to the original horizontal datum. In prominently bedded rocks, most phreatic conduits are fairly concordant with the strata, even in steeply dipping rocks where they commonly follow strike-oriented trends. Discordance to the strata is greater in prominently fractured rocks, but even they show a tendency for stratal confinement of conduits. The vertical distribution of phreatic conduits may show several sharp peaks caused by stratigraphically or geomorphically controlled tiers or levels (Palmer, 1987; Ford, 1988; White, 1988, p. 85; Ford and Williams, 1989, p. 274). Conduits of vadose origin are generally as numerous as those of phreatic origin, and so the arrangement of levels may be obscure in drill-hole data. Lomando et al. (1993) interpret multiple paleokarst levels in the Casablanca field off the Spanish coast to represent cavernous porosity formed at progressively lower base levels.

Conduits diminish greatly in number and size below the lowest preburial base level, although this tendency is less evident in regions that were tectonically deformed prior to karst development, where deep flow paths are more common (Moneymaker, 1941). Borehole evidence from the Balkans shows a roughly exponential decrease in permeability with depth below present-day karst surfaces (Milanovic, 1981). Roof collapse and sinkhole development during cave enlargement produce many local breccia zones rarely more than a few hundred meters in horizontal or vertical extent (see examples in White and White, 1969; White, 1988, p. 229-237; and Ford and Williams, 1989, p. 309–314). Bedded siliciclastic sediments are common, including coarse-grained deposits such as sand and gravel, as well as silt and clay (White and White, 1968; Ford and Williams, 1989, p. 318–330). Travertine can be locally extensive, especially in segmented upper-level conduit fragments, but the overall percentage of travertine fill is small. The spatial density of solution conduits of any single conduit level is low, only about 1–5% in plan view, but is much greater in multilevel caves (see examples in Figures 2 and 4). The total cavernous porosity is rarely more than 4–5% (Figure 4).

Floodwater mazes consist of fissure networks in which every major fracture has been enlarged by solution, or anastomotic bands of tubular conduits that are usually guided by a few dominant bedding-plane partings or low-angle faults (Figures 8 and 9). Approximately 10–15% of the volume of all known solution caves (including parts of caves) is of this type. They are rarely more than a few hundred meters in lateral extent. Spatial density in plan view can reach 15–20% in local areas, with rather sharp outlines beyond which there is little or no solutional enlargement of openings. Vertical extent relative to the original horizontal datum rarely exceeds a few tens of meters. Examples are given by Palmer (1975, 1991) and Ford and Williams (1989, p. 273–274). Such mazes lie in close proximity to present or former river valleys or preexisting solution conduits, which furnished their floodwater recharge. Siliciclastic sediments in solution cavities range from coarse gravels and cobbles (if sources were available) to silt and clay (Milske et al., 1983). Travertine is very sparse and in many places entirely absent.

Fissure networks formed by diffuse meteoric water are clustered below (and rarely above) contacts with insoluble permeable rock through which the aggressive recharge entered the carbonates. Approximately 5% of the volume of all known solution caves is of this type. They represent rather uniform solution at small distances from where the water first encountered the carbonate rock. Porosity diminishes rapidly away from the contact. Such porosity is most abundant where the insoluble rock is thin and topographically suited to transmitting diffuse recharge, and on a regional scale these networks vary greatly in spatial density. Detrital bank-flooding sediment from nearby rivers is common. Travertine is sparse. The typical spatial density of solution networks is about 15–20%, with local maxima of nearly 40% (see examples in Palmer, 1975; White, 1988, p. 78–84; and Figure 10).

Solution porosity formed by meteoric water beneath continental erosion surfaces is least commonly preserved, and what does survive consists mainly of scattered conduit fragments. The porosity types described in the following paragraphs are more likely to be preserved and to form petroleum reservoirs.

Seacoast Mixing Zones

In seacoast mixing zones, solution porosity ranges greatly in size, with large voids surrounded by numerous smaller ones. Only about 1% of the volume of known solution caves is of this type, but inaccessible solution pores represent a much greater volume. Travertine is common in relict caves. Collapse breccia is abundant in caves in poorly indurated rocks, such as those of Bermuda, but is nearly absent in caves in more competent rocks (Mylroie, 1988). Porosity is distributed irregularly and in proportion to the local infiltration rate and proximity to the shore (Figure 11). Discrete levels are present where solution has concentrated at various sea-level stands. Modern examples are given by Back et al. (1979) and Mylroie and Carew (1990), and an inferred paleokarst example of solution porosity produced by meteoric recharge and mixing in former carbonate islands is described by Craig (1988) in the San Andres Dolomite in the Yates field of west Texas. A variety of detrital sediment types is typical of karst voids in seacoast areas, including shell material, carbonate breccias, soil, and indurated limestone (Jones, 1992).

Hypogenetic Porosity

Solution voids formed by deep-seated processes unrelated to aggressive meteoric infiltration (e.g., by redox or thermal processes) are summarized in Figure 20. They can be recognized by their great variation in spatial density, almost complete lack of coarse, bedded internal siliciclastic sediment, and absence of vadose perching on low-permeability beds (Figures 14–16). This type represents about 10–15% of the volume of known solution caves, but this figure must be greatly underestimated, because many related solution voids are too small for human access. Such caves can extend to considerable depth below erosion

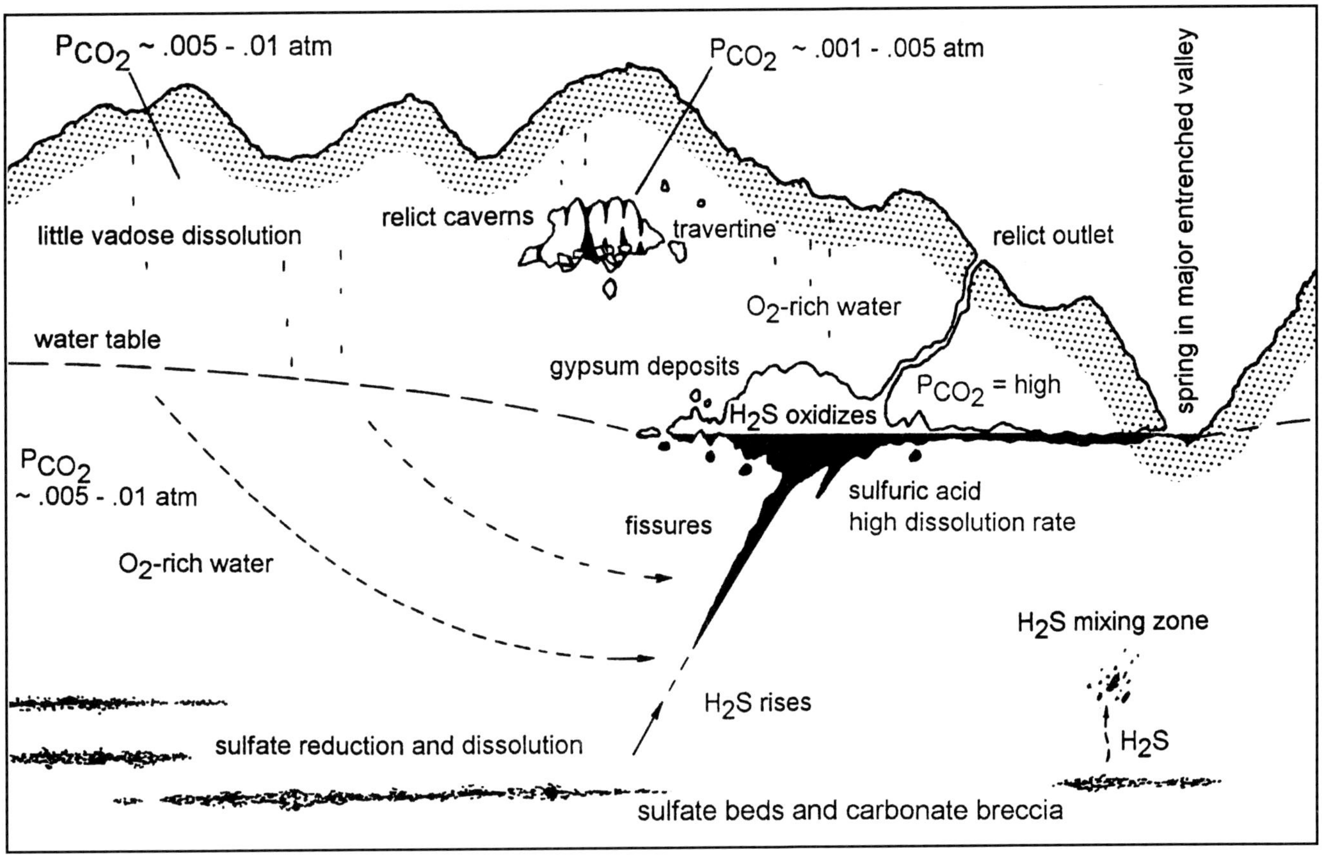

Figure 20. Geochemical setting and distribution of solution porosity unrelated to aggressive meteoric infiltration. See text and Figure 6 for an explanation of terms. Solution by rising thermal waters resembles that of rising H_2S, except that CO_2 is the usual source of acidity, and mixing with local meteoric water accounts for most of the solution at and near the water table.

surfaces, with little apparent genetic relationship to the surface. Highly porous, weathered, or mineralized zones surround the main solution conduits in some areas. In plan view, the spatial density of pores can reach 25–30% throughout areas as large as a square kilometer but is commonly far less (see examples in Hill, 1987; Ford, 1988; Ford and Williams, 1989; and Palmer, 1991). Travertine may be very abundant in local areas but is sparse overall.

Mixing of waters of contrasting H_2S content produces pervasive vuggy or fissure porosity localized within areas of convergent flow. Redox reactions that produce organic acids can produce solution porosity of similar appearance. Surdam et al. (1993) cite examples of petroleum reservoirs in which the carbonate matrix has been dissolved from ferruginous sandstones as a result of reactions between hydrocarbons, iron oxide, and mineral oxidants.

Oxidation of H_2S to sulfuric acid produces intense concentrations of pervasive porosity, typified by large voids surrounded by lesser ones (Egemeier, 1981; Hill, 1987, 1990; Palmer, 1991). Sponge-like patterns and irregular fissure networks are common. Partial or complete filling of solution pores with gypsum is diagnostic, although it has been entirely removed from many areas by meteoric groundwater. Native sulfur and clays such as halloysite, dickite, and alunite are also diagnostic, though rare. Oxidation of minerals in pore walls may produce bleached or multicolored halos. Caves have ramifying patterns in which irregular rooms and maze-like galleries wander in three dimensions with branches exiting from the main areas of development at various levels. Travertine is common in voids that have received vadose seepage. An example is the cavernous porosity in the Guadalupe Mountains of New Mexico (Hill, 1987, 1990; Figures 14 and 15).

Solution caused by the cooling of ascending water increases upward along many intersecting fissures and culminates at or near the former water table, where mixing with shallow meteoric water has taken place (Bakalowicz et al., 1987; Palmer, 1991). Such systems cluster around former groundwater outlets. An example of cavernous porosity formed in this way, both active and relict, is located at Manitou Springs, Colorado (Luiszer, 1994), although mixing of upward-flowing and near-surface water contributes to most of the solutional aggressiveness. Upward-moving H_2S-rich water is suggested as a brecciating agent in Devonian carbonates in Alberta by Dravis and Muir (1993).

Porosity formed by the solution, reduction, and replacement of gypsum and anhydrite commonly consists of widespread, stratally limited zones of vuggy pores that vary greatly in size and shape. Even where there are no remnant sulfates, diagnostic features can be recognized, such as calcite veins (typically red or yellow as the result of iron oxide impurities), doubly terminated or tapered calcite crystals, widespread chaotic and mosaic breccias, and great variety of pore sizes. Good examples are located below and in association with paleokarst features near the top of the Madison Limestone of the Northern Rocky Mountains (Sando, 1988). Much of the cavernous porosity of the Madison in South Dakota represents Mississippian porosity formed in sulfate zones and enlarged by post-Laramide solution (A. N. Palmer and M. V. Palmer, 1989; Figure 16).

CONCLUSIONS

Solution porosity is not a random phenomenon, but instead is rigidly controlled by the chemical and hydrologic mass balance, flow equations, and chemical kinetics. By examining the geochemical constraints under which a given porosity type must have formed, one can more easily fit field observations into the regional geologic picture. Even where the regional interpretation is not clear, geochemical models provide a basis with which to explain the occurrence of solution porosity and perhaps to extrapolate its distribution elsewhere.

REFERENCES CITED

Atkinson, T.C., 1977, Carbon dioxide in the atmosphere of the unsaturated zone: an important control of ground-water hardness in limestones: Journal of Hydrology, v. 35, p. 111–123.

Back, W., B.B. Hanshaw, T.E. Pyle, L.N. Plummer, and A.E. Weidie, 1979, Geochemical significance of groundwater discharge and carbonate solution to the formation of Caleta Xel Ha, Qintana Roo, Mexico: Water Resources Research, v. 15, no. 6, p. 1521–1535.

Back, W., B.B. Hanshaw, L.N. Plummer, P.H. Rahn, C.T. Rightmire, and M. Rubin, 1983, Process and rate of dedolomitization: mass transfer and ^{14}C dating in a regional carbonate aquifer: Geological Society of America Bulletin, v. 94, p. 1415–1429.

Back, W., B. Hanshaw, and J.N. Van Driel, 1984, Role of groundwater in shaping the eastern coastline of the Yucatan Peninsula, Mexico, *in* LaFleur, R.G., ed., Groundwater as a geomorphic agent: Boston, Allen and Unwin, p. 281–293.

Bakalowicz, M.J., D.C. Ford, T.E. Miller, A.N. Palmer, and M.V. Palmer, 1987, Thermal genesis of dissolution caves in the Black Hills, South Dakota: Geological Society of America Bulletin, v. 99, p. 72–99.

Bathurst, R.G.C., 1971, Carbonate sediments and their diagenesis: Amsterdam, Elsevier, 620 p.

Berner, R.A., and J.W. Morse, 1974, Dissolution kinetics of calcium carbonate in seawater; IV: theory of calcite dissolution: American Journal of Science, v. 274, p. 108–134.

Bögli, A., 1964, Mischungskorrosion, ein Beitrag zur Verkarstungsproblem: Erdkunde, v. 18, p. 83–92.

Bögli, A., 1980, Karst hydrology and physical speleology: Berlin, Springer-Verlag, 284 p.

Bosák, P., D.C. Ford, J. Glazek, and I. Horácek, eds., 1989, Paleokarst: Prague and Amsterdam, Academia and Elsevier, 725 p.

Bottrell, S.H., P.L. Smart, F. Whitaker, and R. Raiswell, 1991, Geochemistry and isotope systematics of sulphur in the mixing zone of Bahamian blue holes: Applied Geochemistry, v. 6, p. 97–103.

Budd, D.A., 1988, Aragonite-to-calcite transformation during fresh-water diagenesis of carbonates: insights from pore-water chemistry: Geological Society of America Bulletin, v. 100, p. 1260–1270.

Busenberg, E., and L.N. Plummer, 1982, The kinetics of dissolution of dolomite in CO_2–H_2O systems at 1.5 to 65°C and 0 to 1 atm P_{CO2}: American Journal of Science, v. 282, p. 45–78.

Candelaria, M.P., and C.L. Reed, eds., 1992, Paleokarst related hydrocarbon reservoirs: Field Trip Guidebook, Permian Basin Section Society of Economic Paleontologists and Mineralogists, Publication 92–33, 202 p.

Canter, K.L., D.B. Stearns, R.C. Greesaman, and J.L. Wilson, 1993, Paleostructural and related paleokarst controls on reservoir development in the lower Ordovician Ellenburger Group, Val Verde Basin, Texas, *in* R.D. Fritz, J.L. Wilson, and D.L. Yurewicz, eds., Paleokarst related hydrocarbon reservoirs: Society for Sedimentary Geology Core Workshop No. 18, p. 61–99.

Choquette, P.W., and N.P. James, 1988, Introduction, *in* N.P. James and P.W. Choquette, eds., Paleokarst: New York, Springer-Verlag, p. 1–21.

Choquette, P.W., and L.C. Pray, 1970, Geologic nomenclature and classification of porosity in sedimentary carbonates: AAPG Bulletin, v. 54, p. 207–250.

Choquette, P.W., A. Cox, and W.J. Meyers, 1992, Characteristics, distribution and origin of porosity in shelf dolostones: Burlington-Keokuk Formation (Mississippian), U.S. Mid-Continent: Journal of Sedimentary Petrology, v. 62, no. 2, p. 167–189.

Craig, D.H., 1988, Caves and other features of Permian karst in San Andres Dolomite, Yates field reservoir, west Texas, *in* N.P. James and P.W. Choquette, eds., Paleokarst: New York, Springer-Verlag, p. 342–363.

DeGroodt, K., 1967, Experimental dedolomitization: Journal of Sedimentary Petrology, v. 37, p. 1216–1220.

Demiralin, A.S., N.F. Hurley, and T.W. Oesleby, 1993, Karst breccias in the Madison Limestone (Mississippian), Garland Field, Wyoming, *in* R.D. Fritz, J.L. Wilson, and D.L. Yurewicz, eds., Paleokarst related hydrocarbon reservoirs: Society for Sedimentary Geology Core Workshop 18, p. 101–118.

Dravis, J.J., and I.D. Muir, 1993, Deep-burial brecciation in the Devonian Upper Elk Point Group, Rainbow basin, Alberta, western Canada, *in* R.D. Fritz, J.L. Wilson, and D.L. Yurewicz, eds., Paleokarst related hydrocarbon reservoirs: Society for Sedimentary Geology Core Workshop 18, p. 119–166.

Drever, J.I., 1982, The geochemistry of natural waters: Englewood Cliffs, Prentice Hall, 388 p.

Dreybrodt, W., 1988, Processes in karst systems—physics, chemistry and geology: Berlin, Springer-Verlag, 288 p.

Dreybrodt, W., 1990, The role of dissolution kinetics in the development of karst aquifers in limestone: a model simulation of karst evolution: Journal of Geology, v. 98, no. 5, p. 639–655.

Egemeier, S.J., 1981, Cave development by thermal waters: National Speleological Society Bulletin, v. 43, p. 31–51.

Esteban, M., and C.F. Klappa, 1983, Subaerial exposure environment, *in* P.A. Scholle, D.G. Bebout, and C.H. Moore, eds., Carbonate Depositional Environments: AAPG Memoir 33, p. 1–54.

Evamy, B.D., 1967, Dedolomitization and the development of rhombohedral pores in limestone: Journal of Sedimentary Petrology, v. 37, p. 1204–1215.

Ford, D.C., 1988, Dissolutional cave systems in carbonate rocks, *in* N.P. James and P.W. Choquette, eds., Paleokarst: New York, Springer-Verlag, p. 25–57.

Ford, D.C., and R.W. Ewers, 1978, The development of limestone cave systems in the dimensions of length and depth: Canadian Journal of Earth Sciences, v. 15, p. 1783–1798.

Ford, D.C., and P.W. Williams, 1989, Karst geomorphology and hydrology: London, Unwin Hyman, 601 p.

Fritz, R.D., J.L. Wilson, and D.L. Yurewicz, eds., 1993, Paleokarst related hydrocarbon reservoirs: Society for Sedimentary Geology Core Workshop 18, 275 p.

Furman, F.C., in press, Formation of east Tennessee Knox MVT bodies by hypogenetic-interstratal-evaporite-TSR-sulfuric acid karstification, *in* K.L. Shelton and R.D. Hagni, Geology and geochemistry of Mississippi Valley-type ore deposits: Rolla, Missouri, University of Missouri at Rolla Press.

Groves, C.G., and A.D. Howard, 1994, Minimum conditions for limestone cave development: Water Resources Research, v. 30, no. 10, p. 2837–2846.

Hardie, L.A., 1987, Dolomitization: a critical review of some current views: Journal of Sedimentary Petrology, v. 57, p. 166–183.

Herman, J.S., and W.B. White, 1985, Dissolution kinetics of dolomite: effects of lithology and fluid flow velocity: Geochimica et Cosmochimica Acta, v. 49, p. 2017–2026.

Hill, C.A., 1987, Geology of Carlsbad Caverns and other caves in the Guadalupe Mountains, New Mexico and Texas: New Mexico Bureau of Mines and Mineral Resources Bulletin 117, 150 p.

Hill, C.A., 1990, Sulfuric acid speleogenesis of Carlsbad Cavern and its relationship to hydrocarbons, Delaware Basin, New Mexico and Texas: AAPG Bulletin, v. 74, p. 1685–1694.

James, N.P., and P.W. Choquette, eds., 1984, Diagenesis 9—Limestones. The meteoric diagenetic environment: Geoscience Canada, v. 11, no. 4, p. 161–194.

James, N.P., and P.W. Choquette, eds., 1988, Paleokarst: New York, Springer-Verlag, 416 p.

James, A.N., and I.M. Kirkpatrick, 1980, Design of foundations of dams containing soluble rocks and soils: Quarterly Journal of Engineering Geology, v. 13, p. 189–198.

Jennings, J.N., 1985, Karst geomorphology: Oxford, Basil Blackwell, 293 p.

Jones, B., 1992, Void-filling deposits in karst terrains of isolated oceanic islands: a case study from Tertiary carbonates of the Cayman Islands: Sedimentology, v. 39, p. 877–903.

Kerans, C., 1988, Karst-controlled reservoir heterogeneity in Ellenburger Group carbonates of west Texas: AAPG Bulletin, v. 72, p. 1160–1183.

Lohmann, K.C., 1988, Geochemical patterns of meteoric diagenetic systems and their application to studies of paleokarst, *in* N.P. James and P.W. Choquette, eds., Paleokarst: New York, Springer-Verlag, p. 58–80.

Lomando, A.J., P.M. Harris, and D.E. Orlopp, 1993, Casablanca field, Tarragona Basin, offshore Spain: a karsted carbonate reservoir, *in* R.D. Fritz, J.L. Wilson, and D.L. Yurewicz, eds., Paleokarst related hydrocarbon reservoirs: Society for Sedimentary Geology Core Workshop 18, p. 201–225.

Loucks, R.G., and J.H. Anderson, 1985, Depositional facies, diagenetic terranes, and porosity development in lower Ordovician Ellenburger dolomite, Puckett field, west Texas, *in* P.O. Roehl and P.W. Choquette, eds., Carbonate petroleum reservoirs: New York, Springer-Verlag, p. 19–37.

Loucks, R.G., and C.R. Handford, 1992, Origin and recognition of fractures, breccias, and sediment fills in paleocave-reservoir networks, *in* M.P. Candelaria and C.L. Reed, eds., Paleokarst related hydrocarbon reservoirs: Field Trip Guidebook, Permian Basin Section, Society of Economic Paleontologists and Mineralogists Publication 92-33, p. 31–44.

Luiszer, F.G., 1994, Speleogenesis of Cave of the Winds, Manitou Springs, Colorado, *in* I.D. Sasowsky and M.V. Palmer, eds., Breakthroughs in karst geomicrobiology and redox geochemistry: Charleston, West Virginia, Karst Waters Institute Special Publication 1, p. 91–109.

Machel, H.G., 1987, Some aspects of diagenetic sulphate-hydrocarbon redox reactions, *in* J.D. Marshall, ed., Diagenesis of sedimentary sequences: Geological Society of America Special Publication 36, p. 15–28.

Machel, H.G., 1989, Relationships between sulphate reduction and oxidation of organic compounds to carbonate diagenesis, hydrocarbon accumulations, salt domes, and metal sulphide deposits: Carbonates and Evaporites, v. 4, p. 137–151.

Mazzullo, S.J., and P.M. Harris, 1991, An overview of solution porosity deveopment in the deep-burial environment, with examples from carbonate reservoirs in the Permian Basin, *in* M.P. Candelaria, ed., Permian Basin plays—tomorrow's technology today: West Texas Geological Society Symposium Publication 91-89, p. 125–138.

Meshri, I.D., 1986, On the reactivity of carbonic and organic acids and generation of secondary porosity, *in* D.L. Gautier, ed., Roles of organic matter in sediment diagenesis: Society of Economic Paleontologists and Mineralogists Special Publication 38, p. 123–128.

Milanovic, P.T., 1981, Karst hydrogeology: Littleton, Colorado, Water Resources Publications, 434 p.

Milske, J.A., C.A. Alexander, and R.S. Lively, 1983, Clastic sediments in Mystery Cave, southeastern Minnesota: National Speleological Society Bulletin, v. 45, p. 55–75.

Miotke, F.-D., and A.N. Palmer, 1972, Genetic relationship between caves and landforms in the Mammoth Cave National Park area: Geographic Institute, Technical University of Hannover, Germany, Böhler Verlag, 69 p.

Moneymaker, B.C., 1941, Subriver solution cavities in the Tennessee Valley: Journal of Geology, v. 49, p. 74–86.

Moore, C.H., 1989, Carbonate diagenesis and porosity: New York, Elsevier, 338 p.

Mussman, W.J., I.P. Montanez, and J.F. Read, 1988, Ordovician Knox paleokarst unconformity, Appalachians, *in* N.P. James and P.W. Choquette, eds., Paleokarst: New York, Springer-Verlag, p. 211–228.

Mylroie, J.E., ed., 1988, Field guide to the karst geology of San Salvador Island, Bahamas: Mississippi State University, Department of Geology and Geography, Proceedings of 10th Friends of Karst Meeting, 108 p.

Mylroie, J.E., and J.L. Carew, 1990, The flank margin model for dissolution cave development in carbonate platforms: Earth Surface Processes and Landforms, v. 15, p. 413–424.

Palmer, A.N., 1975, The origin of maze caves: National Speleological Society Bulletin, v. 37, p. 56–76.

Palmer, A.N., 1984, Recent trends in karst geomorphology: Journal of Geological Education, v. 32, p. 247–253.

Palmer, A.N., 1987, Cave levels and their interpretation: National Speleological Society Bulletin, v. 49, p. 50–66.

Palmer, A.N., 1988, Solutional enlargement of openings in the vicinity of hydraulic structures in karst regions: Dublin, Ohio, Proceedings of 2nd Conference on Environmental Problems in Karst Terranes, Association of Ground Water Scientists and Engineers, p. 3–13.

Palmer, A.N., 1991, Origin and morphology of limestone caves: Geological Society of America Bulletin, v. 103, p. 1–21.

Palmer, A.N., and M.V. Palmer, 1989, Geologic history of the Black Hills caves, South Dakota: National Speleological Society Bulletin, v. 51, p. 72–99.

Palmer, A.N., and M.V. Palmer, 1991, Replacement mechanisms among carbonates, sulfates, and silica in karst regions: some Appalachian examples, *in* E.H. Kastning and K.M. Kastning, eds., Proceedings of Appalachian Karst Symposium, Radford University, Radford, Virginia, p. 109–115.

Palmer, M.V., and A.N. Palmer, 1989, Paleokarst of the United States, *in* P. Bosák, D.C. Ford, J. Glazek, and I. Horácek, eds., Paleokarst: Prague and Amsterdam, Academia and Elsevier, p. 337–363.

Pierre, C., and J.M. Rouchy, 1988, Carbonate replacements after sulfate evaporites in the middle Miocene of Egypt: Journal of Sedimentary Petrology, v. 58, p. 446–456.

Plummer, L.N., 1975, Mixing of seawater with calcium carbonate ground water: Geological Society of America Memoir 142, p. 219–236.

Plummer, L.N., and E. Busenberg, 1982, The solubilities of calcite, aragonite, and vaterite in CO_2-H_2O solutions between 0° and 90°C and an evaluation of the aqueous model for the system $CaCO_3$–CO_2–H_2O: Geochimica et Cosmochimica Acta, v. 46, p. 1011–1040.

Plummer, L.N., and T.M.L. Wigley, 1976, The dissolution of calcite in CO_2-saturated solutions at 25°C and 1 atmosphere total pressure: Geochimica et Cosmochimica Acta, v. 40, p. 191–202.

Plummer, L.N., H.L. Vacher, F.T. Mackenzie, O.P. Bricker, and L.S. Land, 1976, Hydrochemistry of Bermuda: a case history of groundwater diagenesis of biocalcarenites: Geological Society of America Bulletin, v. 87, p. 1301–1316.

Plummer, L.N., T.M.L. Wigley, and D.L. Parkhurst, 1978, The kinetics of calcite dissolution in CO_2-water systems at 5° to 60°C and 0.0 to 1.0 atm CO_2: American Journal of Science, v. 278, p. 179–216.

Rauch, H.W., and W.B. White, 1977, Dissolution kinetics of carbonate rocks. 1. Effects of lithology on dissolution rate: Water Resources Research, v. 13, p. 381–394.

Roberts, A.E., 1966, Stratigraphy of the Madison Group near Livingston, Montana, and discussion of karst and solution-breccia features: U.S. Geological Survey, Professional Paper 52B, p. B1–B22.

Roehl, P.O., and P.W. Choquette, eds., 1985, Carbonate petroleum reservoirs: New York, Springer-Verlag, 622 p.

Runnells, D.D., 1969, Diagenesis, chemical sediments, and mixing of natural waters: Journal of Sedimentary Petrology, v. 39, p. 1188–1201.

Sando, W.J., 1974, Ancient solution phenomena in the Madison Limestone (Mississippian) of north-central Wyoming: U.S. Geological Survey Journal of Research, v. 4, no. 2, p. 133–141.

Sando, W.J., 1988, Madison Limestone (Mississippian) paleokarst: a geologic synthesis, *in* N.P. James and P.W. Choquette, eds., Paleokarst: New York, Springer-Verlag, p. 256–277.

Schmidt, V.A., 1982, Magnetostratigraphy of sediments in Mammoth Cave, Kentucky: Science, v. 217, p. 827–829.

Smalley, P.C., P.K. Bishop, J.A.D. Dickson, and D. Emery, 1994, Water-rock interaction during meteoric flushing of a limestone: implications for porosity development in karstified petroleum reservoirs: Journal of Sedimentary Research, v. A64, no. 2, p. 180–189.

Smart, P.L., and H. Friederich, 1986, Water movement and storage in the unsaturated zone of a maturely karstified carbonate aquifer, Mendip Hills,

England: Dublin, Ohio, National Water Well Association, Proceedings of Conference on Environmental Problems in Karst Terranes and their Solutions, p. 59–87.

Smart, P.L., and S.L. Hobbs, 1986, Characterisation of carbonate aquifers: a conceptual base: Dublin, Ohio, National Water Well Association Proceedings of Conference on Environmental Problems in Karst Terranes and their Solutions, p. 1–14.

Stoessel, R.K., 1992, Effects of sulfate reduction on $CaCO_3$ dissolution and precipitation in mixing-zone fluids: Journal of Sedimentary Petrology, v. 62, p. 873–880.

Stoessel, R.K., W.C. Ward, B.H. Ford, and J.D. Schuffert, 1989, Water chemistry and $CaCO_3$ dissolution in the saline part of an open-flow mixing zone, coastal Yucatan Peninsula, Mexico: Geological Society of America Bulletin, v. 101, p. 159–169.

Surdam, R.C., Z.S. Jiao, and D.B. MacGowan, 1993, Redox reactions involving hydrocarbons and mineral oxidants: a mechanism for significant porosity enhancement in sandstones: AAPG Bulletin, v. 77, no. 9, p. 1509–1518.

Thrailkill, J., 1968, Chemical and hydrologic factors in the excavation of limestone caves: Geological Society of America Bulletin, v. 79, p. 19–46.

Thrailkill, J., and T.L. Robl, 1981, Carbonate geochemistry of vadose water recharging limestone aquifers: Journal of Hydrology, v. 54, p. 195–208.

Vacher, H.L., 1978, Hydrogeology of Bermuda—significance of across-the-island variation on permeability: Journal of Hydrology, v. 39, p. 207–226.

Varnedoe, W.W., 1964, The formation of an extensive maze cave in Alabama: Alabama Academy of Science Journal, v. 35, no. 4, p. 143–148.

Whitaker, F.F., and P.L. Smart, 1993, Circulation of saline ground water in carbonate platforms—a review and case study from the Bahamas, *in* A.D. Horbury and A.G. Robinson, eds., Diagenesis and basin development: AAPG Studies in Geology 36, p. 113–132.

White, E.L., and W.B. White, 1968, Dynamics of sediment transport in limestone caves: National Speleological Society Bulletin, v. 30, p. 115–129.

White, E.L., and W.B. White, 1969, Processes of cavern breakdown: National Speleological Society Bulletin, v. 31, p. 83–96.

White, W.B., 1977, Role of solution kinetics in the development of karst aquifers, *in* J.S. Tolson and F.L. Doyle, eds., Karst hydrogeology: International Association of Hydrogeologists 12th Memoirs, p. 503–517.

White, W.B., 1988, Geomorphology and hydrology of karst terrains: New York, Oxford University Press, 464 p.

White, W.B., and E.L. White, eds., 1989, Karst hydrology—concepts from the Mammoth Cave region: New York, Van Nostrand Reinhold, 346 p.

Wigley, T.M.L., and L.N. Plummer, 1976, Mixing of carbonate waters: Geochimica et Cosmochimica Acta, v. 40, p. 989–995.

Williams, P.W., 1983, The role of the subcutaneous zone in karst hydrology: Journal of Hydrology, v. 61, p. 45–67.

Woods, T.L., and R.M. Garrels, 1987, Thermodynamic values at low temperature for natural inorganic materials: an uncritical summary: New York, Oxford University Press, 242 p.

Worthington, S.R.H., 1994, The possible importance of sulfur minerals in initiating epigenic caves, *in* I.D. Sasowsky and M.V. Palmer, eds., Breakthroughs in karst geomicrobiology and redox geochemistry: Charleston, West Virginia, Karst Waters Institute Special Publication 1, p. 80–82.

Chapter 5

Interplay of Water-Rock Interaction Efficiency, Unconformities, and Fluid Flow in a Carbonate Aquifer: Floridan Aquifer System

Harris Cander
Amoco Production Company
Houston, Texas, U.S.A.

ABSTRACT

The middle Eocene Avon Park Formation comprises shallow subtidal skeletal limestones and dolomitized peritidal limestones that underwent several periods of unconformity-related exposure during the Cenozoic. The porous limestones are typical of the Tertiary Floridan aquifer system, which has both high interparticle, matrix porosity and a conduit flow system comprising karst zones, caves, vugs, channels, and bedding planes. Avon Park limestones have retained most primary porosity (ϕ = 20 to 30%) and Eocene marine-like geochemical compositions, despite being exposed to flushing by meteoric groundwater during these long-lived unconformities. The marine-like geochemical compositions indicate low water/rock ratios during mineralogical stabilization to calcite. The most common diagenetic product in the limestones is isopachous bladed calcite cement that precipitated during intraformational unconformities or immediately after deposition. The limestones are in oxygen, carbon, and strontium isotopic disequilibrium with modern Floridan aquifer groundwater (limestone: $\delta^{18}O$ = –1.0 to +1.0‰, PDB; $\delta^{13}C$ = 0 to +2.0‰; $^{87}Sr/^{86}Sr$ = 0.70777 ; Sr = 400 ppm; dilute groundwater: $\delta^{18}O$ = –0.5‰, SMOW; $\delta^{13}C$ = –5 to –14‰; $^{87}Sr/^{86}Sr$ = 0.7081 to 0.7089). Based on geochemical modeling, quantitative estimates of the number of pore volumes that have reacted with Avon Park limestone compared to the number of pore volumes that have flowed through the rocks indicate that the long-term efficiency of water-rock interaction is less than 0.002%.

In contrast to the matrix limestone, late-stage, conduit-lining coarse calcite cements in the Avon Park are in isotopic and elemental equilibrium with modern Floridan aquifer groundwater, indicating precipitation at extremely high water/rock ratios and interaction efficiency (late calcite: $\delta^{18}O$ = –3.3‰, PDB; $\delta^{13}C$ = –7.0‰; $^{87}Sr/^{86}Sr$ = 0.70875; Sr = 15–20 ppm). The radiogenic $^{87}Sr/^{86}Sr$ composition of these calcite cements indicates that they contain Sr of Middle Miocene age or younger. The contrasting data from the matrix limestones and the conduit-lining cements indicate that the two fluid-flow systems give rise to two different diagenetic systems in the same aquifer. The matrix system is characterized by low efficiency with products precipitated at low water/rock ratios; the conduit system is characterized by high water-rock interaction efficiency and products precipitated at extremely high water/rock ratios. The conduit system has active diagenesis, where large mass transfer of calcium carbonate is occurring and the matrix system is relatively inert. The response of the Avon Park Formation to unconformity-related diagenesis can be interpreted based on the Eocene age of matrix cements, the post-Middle Eocene age of conduit-lining cements, and the timing of long-lived regional unconformities. During periods of subaerial exposure asociated with intraformational and early postdepositional unconformities, the conduit system was poorly developed and the matrix system was the locus of water-rock interaction; the dominant product was intra- and interparticle calcite cement precipitated in near-equilibrium with the host limestone. During the later-stage, long-lived exposure associated with regional unconformities (Late Oligocene, Late Miocene, and throughout the Pliocene–Pleistocene), the conduit fluid-flow system developed and focused both fluid flow and water-rock interaction out of the matrix and into the conduits; the dominant product became coarse cavity-lining calcite cement, precipitated in equilibrium with the groundwater. Today, the conduit system has active diagenesis where large mass transfer of calcium carbonate is occurring and the matrix system is relatively inert.

The history of water-rock interaction in the Avon Park Formation suggests that as diagenesis in carbonate platform limestones evolves, a conduit fluid-flow system may develop in response to meteoric diagenesis during long-lived unconformity-related exposure. In these systems, the conduit porosity system overtakes the matrix porosity system as the locus of diagenesis and carbonate mass transfer. In so doing, the conduit system serves to limit diagenesis in the matrix and preserve matrix porosity. Results of this study indicate that the type of fluid-flow system(s) must be considered, as well as the fluids and rocks, when interpreting carbonate rock-water interaction and porosity modification below unconformities.

INTRODUCTION—LIMESTONE RECRYSTALLIZATION AND CEMENTATION

Defining the conditions that cause destruction or favor survival of primary porosity in platform limestones has remained a fundamental problem in carbonate petrology. The importance of freshwater diagenesis at unconformities as a mechanism of porosity modification through limestone recrystallization, dissolution, and cementation has been documented via petrography and geochemistry of aquifer rocks and pore fluids in many studies of Quaternary carbonate aquifers (Matthews, 1968, 1971, 1974; Harris and Matthews, 1968; Halley and Harris, 1979; Allan and Matthews, 1982; Budd and Land, 1990). In these systems, significant porosity destruction and resetting of original rock chemistry have occurred within tens of thousands of years after deposition. In much larger Paleozoic systems, limestone recrystallization and calcite cementation have also been interpreted as resulting from early meteoric phreatic diagenesis in paleoaquifers (Meyers, 1974; Grover and Read, 1983; Meyers and Lohmann, 1985; Dorobek, 1987; Kaufman et al., 1988).

Relatively few petrographic and geochemical studies have concentrated on limestone recrystallization and cementation in active freshwater hydrologic systems comparable in scale to ancient Paleozoic platform carbonates for which early meteoric diagenesis has been invoked as the agent of porosity destruction. The Floridan aquifer system, one of the world's largest carbonate aquifers, is an example of such an active sytem. Studies of this extant system have stressed both the preservation of porosity during shallow burial of Florida platform Tertiary carbonates (Halley and Schmoker, 1983; Cander, 1991; Budd et al., 1993) and addressed the role of the groundwater system in dolomitization of Eocene strata (Hanshaw and Back, 1972; Randazzo et al., 1977; Randazzo and Cook, 1987; Cander, 1991, 1994). This paper evaluates the efficiency of limestone recrystallization and calcite cementation during subaerial unconformities in the Floridan aquifer system. The theme of this study is that as a carbonate platform aquifer evolves two pore systems may develop, an intergranular/intercrystalline matrix system and a conduit system, with the efficiency of water-rock interaction in these two flow systems being completely different. This study attempts to show how the relative importance of the two flow systems may change over time, resulting in changes in the products and occurrences of porosity-modifying reactions. Petrographic (transmitted and cathodoluminescent microscopy), isotopic (C, O, and Sr), and elemental (Ca, Mg, Sr, Fe, and Mn) data for middle Eocene Avon Park Formation limestones and their pore fluids are integrated and quantitatively modeled and compared to data from the underlying Oldsmar Formation limestones (lower Eocene). This study also proposes conditions under which primary porosity and primary geochemical compositions can be preserved in a marine limestone subjected to long-lived (albeit cold) freshwater diagenesis during numerous exposure events associated with unconformities.

GEOLOGIC AND HYDROLOGIC SETTING

The Floridan aquifer system system is a continuous succession of Paleocene to Miocene carbonates that underlies all of Florida and extends northward into Alabama, Georgia, and South Carolina (Figure 1). The upper and lower bounds of the aquifer are, respectively, the phosphatic, clastic-rich sections of the Miocene Hawthorn Group and the anhydritic Paleocene Cedar Keys Formation (Figure 2). The system is subdivided into the upper and lower Floridan aquifer system (Miller, 1986), separated by middle confining units composed of gypsiferous dolomite and low-permeability dolomite, primarily of middle Eocene age (Figure 2). The upper Avon Park Formation is part of

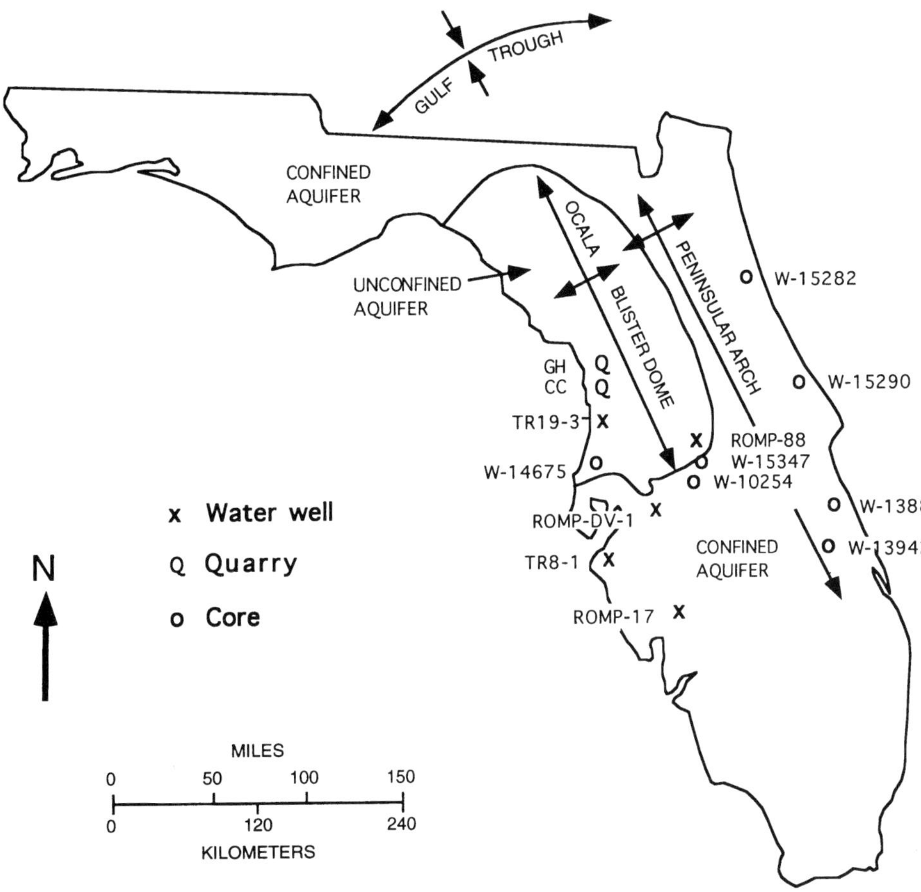

Figure 1. Study area in peninsular Florida, showing structural highs, and locations of cores (o), quarries (Q), and groundwater wells (x) used in this study. The Floridan aquifer system underlies all of Florida and extends north into Alabama and Georgia.

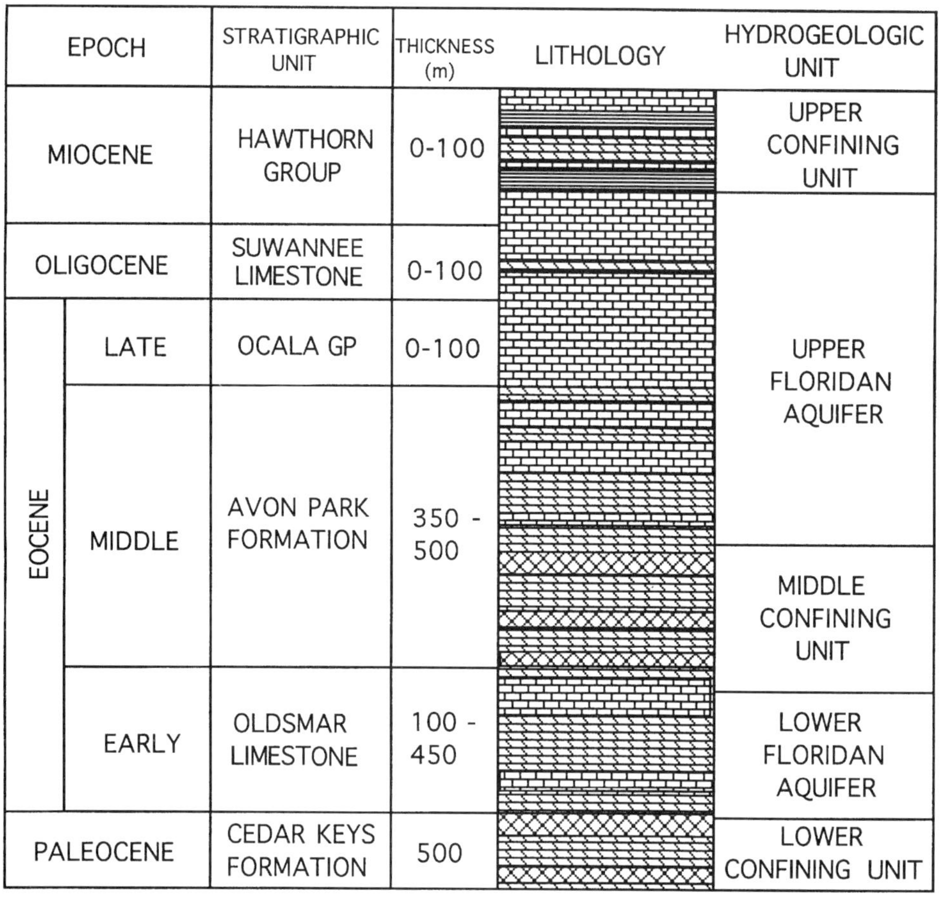

Figure 2. Cenozoic stratigraphy and schematic hydrogeology of Floridan aquifer system in peninsular Florida. Brick pattern is limestone, slanted brick is dolomite, thin lines are silty shales, and diamond pattern is gypsum and anhydrite.

the upper Floridan aquifer system, and gypsiferous intervals of the lower Avon Park Formation can comprise parts of the middle confining layers (Figure 2). The top of the Floridan aquifer system system does not coincide with a specific lithologic unit, but is defined by permeability (Miller, 1986). In the study area, the top of the Floridan aquifer system occurs in the Oligocene Suwannee Limestone or in the upper Eocene Ocala Group (Miller, 1986). The top of the aquifer is slightly above sea level in central Florida and deepens to about –60 m on the east coast and –30 m on the west coast in the study area (Miller, 1986). Thickness of the Floridan aquifer system in peninsular Florida ranges from 500 to 1000 m. The aquifer thickens southward from north-central Florida (Alachua County) (Miller, 1986). The Floridan aquifer system is unconfined in parts of central and west-central Florida where Miocene strata have been thinned or removed by erosion (Figure 1).

The Peninsular Arch and the Ocala dome are structural highs that influence the potentiometric surface of the Floridan aquifer system (Figure 3) such that groundwater flows radially away from central peninsular Florida. Recharge to the aquifer ranges from 2 to 30 cm/yr (Ryder, 1985). Estimates of the average linear velocity of groundwater in the strata in the study area range from 5 to 30 m/yr (Meyer, 1989). However, estimating flow velocity is complicated by the flow network in the Floridan aquifer system. In addition to the high matrix porosity of the Tertiary limestones and dolomite, the Floridan aquifer is riddled with caves and karst systems that serve as important fluid conduits (Stringfield, 1966; Miller, 1986; Meyer, 1989; Sprinkle, 1989). In effect, there are two types of porosity and permeability in the Floridan aquifer and, therefore, two fluid-flow systems. It is unclear what percentage of the total fluid flow occurs in the conduit network versus the matrix porosity. However, groundwater flow velocities in karst systems in other limestone aquifers are often greater than 100 m/hr (Bögli, 1980). The estimates of average linear velocity of Floridan aquifer system groundwater probably represent an average of relatively slow fluid flow through the matrix and relatively rapid fluid flow through the conduit system.

STRATIGRAPHY AND TIMING OF UNCONFORMITIES

The Paleocene Cedar Keys Formation and lower Eocene Oldsmar Limestone unconformably underlie the middle Eocene Avon Park Formation throughout the study area (Figure 2; Miller, 1986). The Paleocene Cedar Keys Formation comprises pervasively dolomitized peritidal carbonates with extensive bedded and intergranular anhydrite and gypsum that form the lower bound of the Floridan aquifer system in peninsular Florida (Applin and Applin, 1944; Miller, 1986).

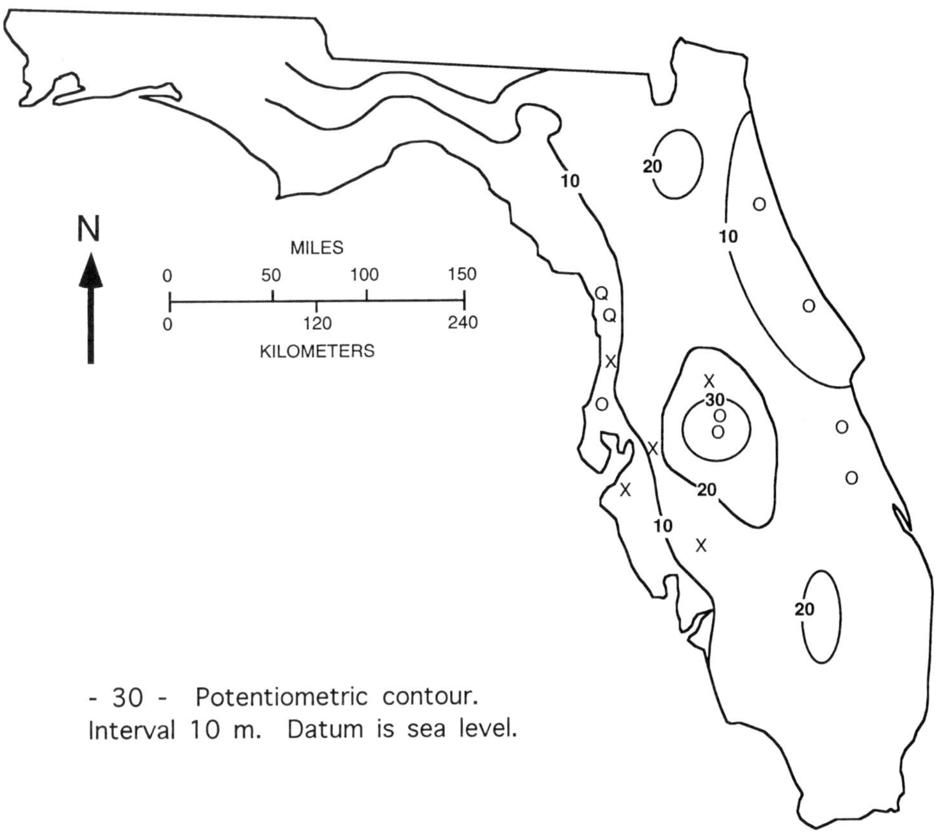

Figure 3. Potentiometric surface of the Floridan aquifer (after Miller, 1986). Groundwater recharges in central Florida and flows radially toward both coasts.

The lower Eocene Oldsmar Limestone comprises shallow subtidal to supratidal carbonates with fewer evaporites and less dolomite than the underlying Cedar Keys Formation (Applin and Applin, 1944; Miller, 1986; Thayer and Miller, 1984). In a core from central Florida (core W-15347, Figure 1), the upper Oldsmar Limestone is not extensively dolomitized and is dominated by benthonic foraminiferal grainstone indicating shallow open-marine deposition. The upper Oldsmar comprises poorly to well-sorted foraminiferal grainstones. Porosity is commonly between 20 and 30% and the rocks are only slightly more indurated than the upper Avon Park limestones.

The middle Eocene Avon Park Formation unconformably overlies the Oldsmar Limestone in central Florida (Figure 4) (Chen, 1965; Miller, 1986). The Avon Park comprises over 400 m of partially dolomitized, cyclic, shallow, open-marine to tidal-flat carbonates deposited on the stable Florida platform (Randazzo and Saroop, 1976). Allochems consist mostly of benthic foraminifers, echinoderms, algal grains, and pellets. In the cores observed in this study, undolomitized limestone (Figure 5A) is common only in the upper 100 m of the formation, where it is interbedded with dolomitized packstone, wackestone, and mudstone. The limestone is typically skeletal rich, with little evidence of cementation in core (Figure 5A). Porosities are commonly greater than 20% and the limestones range from well indurated to friable. Intergranular and interbedded gypsum and anhydrite are common in the lower two-thirds of the formation. The evaporites serve to reduce porosity such that the upper one-third of the Avon Park is more porous (total porosity averages about 20%) than the lower two-thirds of the formation in two deep cores observed in this study (cores W-10254 and W-15347; see Figure 1 for locations).

The Avon Park Formation is unconformably overlain by the upper Eocene Ocala Group and the Oligocene Suwannee Limestone (Figure 2). The Ocala Group comprises gray to white, coarse- to medium-grained foraminiferal grainstone to chalky foraminiferal wackestone with little to no dolomite (Figure 5B). The Ocala Group is overlain by the Suwannee Limestone, a porous, skeletal-rich to pelletal, white to cream limestone. Thin intervals of the Suwannee are pervasively dolomitized (Miller, 1986). Both the upper Eocene Ocala Group (Randazzo and Saroop, 1976; Miller 1986) and the Oligocene Suwannee Limestone (Miller, 1986; Hammes, 1992) were deposited in shallow open-marine water on the broad carbonate bank of the Florida platform.

Oldsmar deposition was followed by an unconformity (Figure 4) that exposed the formation to meteoric diagenesis, at least in central Florida. Avon Park deposition was punctuated by numerous intraformational unconformities in central Florida (Randazzo and Saroop, 1976; Miller, 1986) and followed by an unconformity (Figure 4; Miller, 1986). The first significant regional exposure event occurred in the late Oligocene, after Suwannee deposition (Miller, 1986). The modern Floridan aquifer system was established in the late Miocene, during regional subaerial exposure following deposition of the Hawthorn Group (Miller, 1986; Scott, 1989). Exposure during the

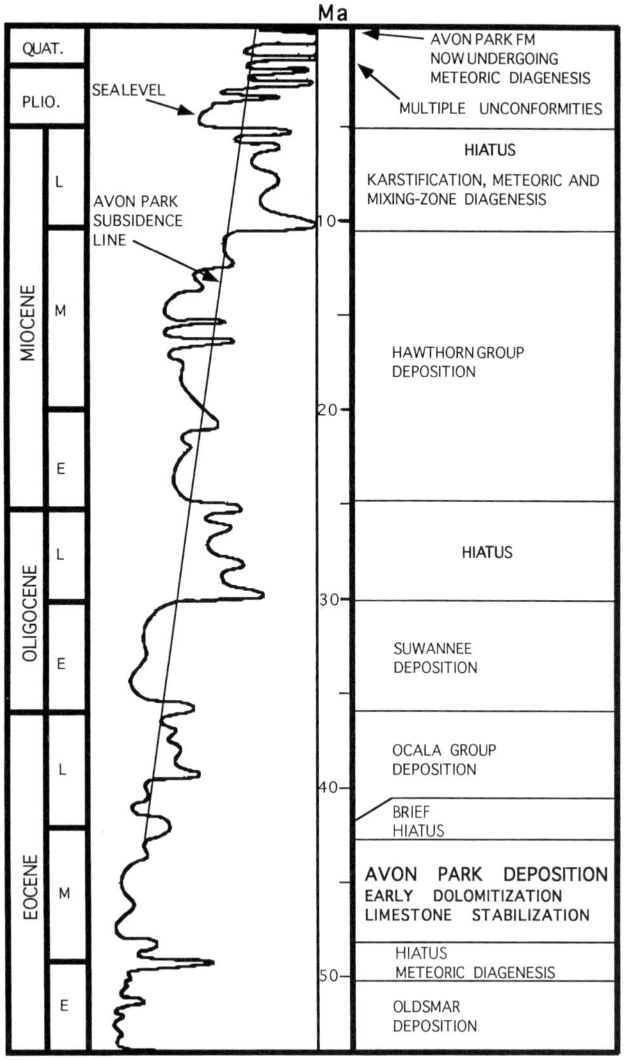

Figure 4. Schematic representation of the post-Paleocene stratigraphy of the Floridan aquifer system and timing of hiatuses as compared to the Cenozoic global eustatic sea level curve from Haq et al. (1988). The locations of boundaries between formations are approximately correlated with time on the sea level curve. The straight line drawn from the top of the Avon Park Formation to present sea level is an approximate subsidence line that tracks the position of the top of this formation relative to Cenozoic sea level. Those periods when the sea level curve drops below the subsidence line indicate hiatuses during which the Avon Park was saturated with meteoric groundwater.

late Oligocene and late Miocene unconformities lasted for up to 5 m.y., thereby allowing meteoric groundwater to circulate through Eocene strata for several million years (Figure 4).

METHODS

The study area and core, water well, and outcrop locations are shown in Figure 1. Avon Park and Oldsmar limestone and dolomite samples were obtained from ten subsurface water well and gypsum exploration cores (3–6 cm diameter) and two surface quarries. Five groundwater wells were sampled to compare isotopic and elemental data from current Avon Park Formation fluids with the rocks (Figure 1).

Approximately 150 standard-size, uncovered, epoxy-impregnated, polished thin sections were examined in transmitted and cathodoluminescent illumination. For cathodoluminescence petrography, a Technosyn MK 11 Cold Cathode Luminescence device was mounted on a Nikon Labophot microscope. Operating conditions were 15–20 kv and 300–600 microamp beam current. Photomicroscopy was performed with a Nikon UFX automatic camera system using high-speed film.

All geochemical analyses were performed at the Department of Geological Sciences, University of Texas at Austin. More than 120 limestone, dolomite, and calcite cement samples were analyzed for stable carbon and oxygen isotopic composition. For these isotope analyses, 3–10 mg of powder were filled from thin section heels and core pieces using a hand-held or vice-mounted drill with variably sized carbide drill bits. In some cases, dolomite was purified of coexisting calcite by reaction in 8% acetic acid. X-ray diffraction analyses confirmed at least 98% mineralogic purity of samples. All samples were reacted off-line at 25°C (calcite) or 50°C (dolomite) in anhydrous H_3PO_4 for 24 to 36 hr. Carbon and oxygen isotopes were measured on the evolved CO_2 gas on a Nuclide gas source mass spectrometer. Dolomite stable isotope analyses were corrected to 25°C and all values normalized to NBS 20 ($\delta^{13}C = -4.14‰$; $\delta^{18}O = -0.96‰$, PDB). Precision on stable isotopic analyses is ± 0.04 for individual runs and ± 0.2 for total procedural duplicates.

Trace (Sr, Mn, and Fe) and major (Ca and Mg) element analyses of rocks were done by inductively coupled argon plasma atomic emission spectroscopy (ICAP-AES) and electron microprobe analyses. For ICAP-AES, approximately 10 mg of drilled powder were dissolved in 2 N HCl, filtered through 0.22 µm nucleopore filters, dried, and dissolved in 10% HCl for analysis. Insoluble residue comprised <1% by weight of most powders, with an upper limit of 3%. Accuracy of ICAP-AES was determined by analyses of NBS 88b dolomitic limestone and variable concentrations of an in-house secondary standard with a carbonate matrix and similar cation concentrations to the unknowns. For groundwaters, ICAP-AES measurements were also made to determine the concentration of Sr in each groundwater sample, which ranged from <1 ppm to 25 ppm. Electron microprobe analyses were performed on a JEOL microprobe using a 10 µm spot size, 10 kv beam, with a 12 nannoamp current on brass. A Bence-Albee reduction scheme was employed. Detection limits were 150 ppm Fe, 140 ppm Mn, and 110 ppm Sr.

For strontium isotopic analyses of limestones, dolomites, and sulfates, 10–20 mg of powder were drilled from thin section heels and washed three times over a 24 hour period with 0.2 N ammonium acetate, then rinsed three times with triply distilled water

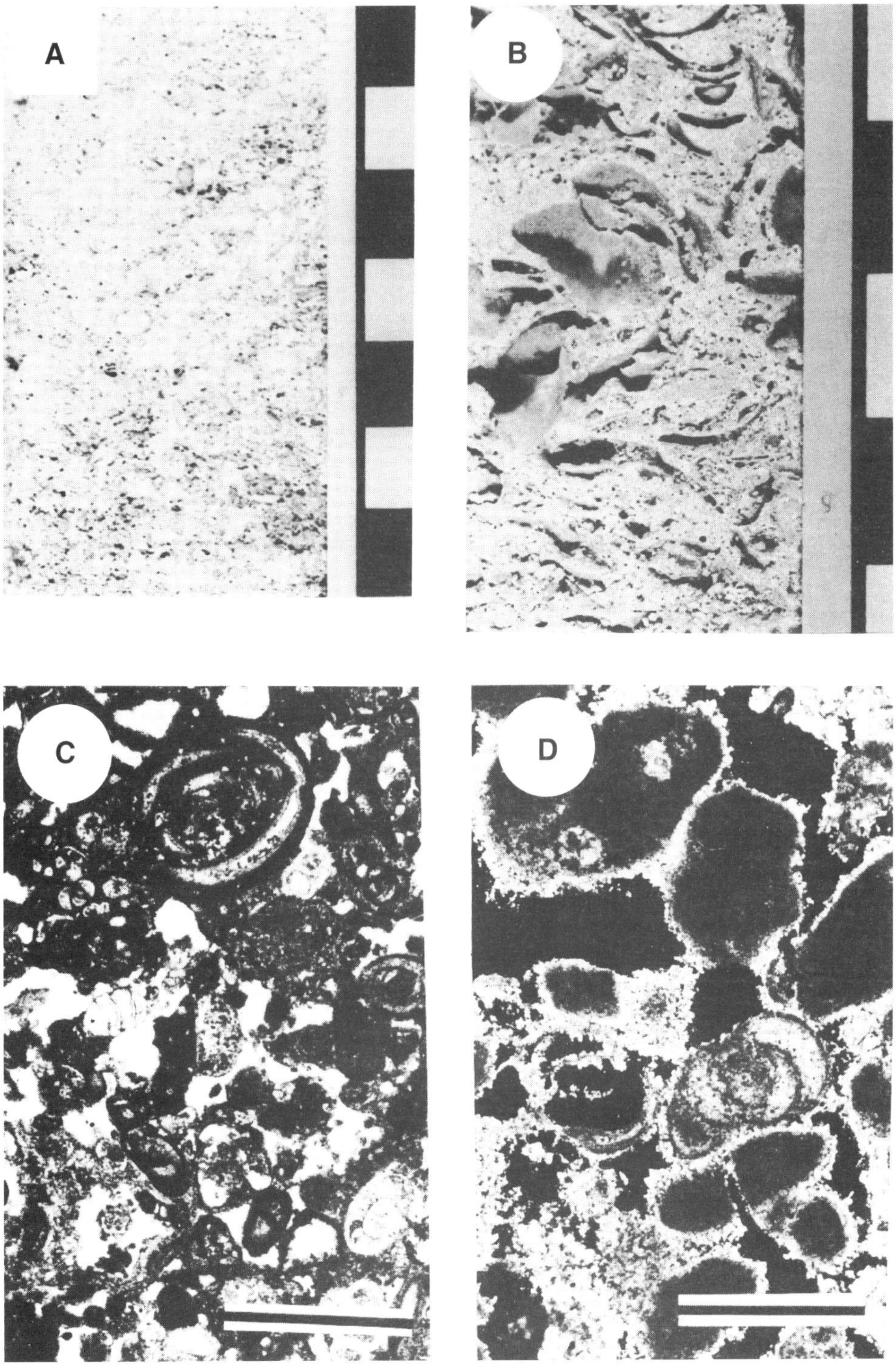

Figure 5. (A) Foraminiferal grainstone, upper Avon Park Formation, core W-14675. Porosity is greater than 20% and is mostly primary interparticle with some mollusk molds. Lithification is moderate; skeletal grains can be abraded off the rock sample by hand. Scale is in centimeters. (B) Skeletal grainstone, lower Ocala Group, core W-15282. Rock sample is 100% low-Mg calcite. Note high primary and moldic porosity. Scale is in centimeters. (C) Transmitted light photomicrograph of foram-skeletal grainstone, upper Avon Park Formation, core W-14675. Note that nearly all foraminifers are micritized. Also note minor intraparticle blocky calcite cement. Scale bar = 1.0 mm. (D) Transmitted light photomicrograph of upper Avon Park foraminiferal grainstone (core W-15290) showing granular to blocky calcite cement lining and filling intraparticle porosity in foraminifers. Also note that most grains are micritized. Scale bar = 0.5 mm.

following the procedure of Morton and Long (1983). This procedure attempts to leach exchangeable Sr from clay sites. This loosely bound Sr is otherwise leached during later chemical procedures, and, because it is not contained in the dolomite, calcite, or sulfate, is a potential contaminant. Powders were then leached in 1.4 N (8%) distilled acetic acid for five minutes (calcite) or ten minutes (dolomite). For both rock and water samples approximately 1 µg of Sr in solution was loaded onto cation exchange columns. Sr was eluted using distilled 2 N HCl and the appropriate interval (3 ml total volume) was collected in Teflon beakers and dried. Evaporations and cation exchange chemistry were conducted under laminar flow. Total procedural blank was less than 1 ng Sr.

Solid source thermal ionization mass spectrometry was performed by simultaneous multicollection on a Finnegan MAT 261 thermal ionization mass spectrometer under static collection mode. Fractionation in the mass spectrometer was corrected using $^{86}Sr/^{88}Sr$ = 0.1194. Sr was loaded in H_3PO_4 on Ta filaments. For each 13-sample turret that was run, at least two samples of NBS 987 Sr were analyzed. Analyses of the standard NBS 987 Sr yielded 0.710225 ± 19 (n = 21).

Five groundwater wells were chosen for sampling of Avon Park pore fluids (see Figure 1 for locations). All five wells are cased with PVC to the upper Avon Park Formation. Groundwater was sampled from these five wells by the Southwest Florida Water Management District. Wells were purged until pH, temperature, and conductivity had stabilized (ranging from ten minutes to five hours purge time). Both acidified (HNO_3) and unacidified samples were taken. Cations were analyzed by ICAP-AES. Anions were analyzed twice by liquid ion chromatography (IC). Precision for analyses of duplicate unknowns is within 10% for trace elements. Cl concentrations were also measured by titration with $AgCl_2$, and a duplicate SO_4 measurement was made on sample TR8-1 by precipitation with $BaCl_2$. Difference between IC and Cl titration is <1.5% and difference between IC and $BaCl_2$ precipitation for SO_4 concentration is less than 4%.

LIMESTONE PETROGRAPHY

The petrographic characteristics of upper Avon Park Formation limestones and upper Oldsmar limestones are similar and will be described together. Most of the undolomitized limestone in the Avon Park and the upper Oldsmar is foraminiferal grainstone (Figure 5A). Subordinate allochems include echinoderm fragments, peloids, and algal grains. Most mollusks have been leached from the grainstones and occur as molds. The most striking petrographic feature of these limestones is the high primary porosity (Figure 5C, D). Avon Park and upper Oldsmar limestones average between 20 and 30% porosity (Halley and Schmoker, 1983; Thayer and Miller, 1984; Cander, 1990, 1991). There has been very little porosity reduction through diagenesis. Most porosity in the limestone is primary interparticle porosity (Figure 5C). Moldic porosity can account for as much as half of the total porosity in mollusk-rich limestone samples. Due to high porosities and minimal cementation, samples are generally rather friable with Oldsmar rocks being only slightly more indurated than upper Avon Park rocks.

MARINE DIAGENESIS

Nearly all grains have micritic envelopes and many grains have been completely micritized (Figure 5C, D), suggesting extensive marine boring processes that presumably contributed to some early lithification. Radial fibrous low-Mg calcite cement lines the intraparticle porosity of a few foraminifers. This cement is commonly about 5 µm in length and is interpreted to be a replacement of marine high-Mg calcite or aragonite.

Phreatic Calcite Cement

The most common cement is 5 to 100 µm, isopachous, anhedral, blocky to bladed low-Mg calcite (Figures 5D and 6A). This cement occurs on the outer and inner walls of foraminifers and lines mollusk and peloid molds (Figures 5D and 6A). Crystals can be equigranular or, more commonly, elongate normal to grain surfaces (Figure 6A). This cement postdates leaching of mollusks and peloids, yet, in partially dolomitized rocks, predates precipitation of at least some dolomite (Figure 6D). This is concluded because dolomite rhombs overgrow and replace the calcite cement (Figure 6D). This dolomite is interpreted to have formed from normal to hypersaline middle Eocene seawater (Cander, 1991, 1994). Based on its anhedral, blocky to bladed morphology, equal distribution around most grains (Halley and Harris, 1979), and post-leaching and pre-marine dolomite timing, the calcite cement is interpreted as a phreatic cement precipitated nearly contemporaneously with deposition.

It is problematic whether this low-Mg calcite cement formed from marine water as either a direct precipitate or replacement of aragonite or high-Mg calcite, or whether this cement precipitated in freshwater phreatic lenses formed during brief syndepositional exposure of the Avon Park to meteoric water. The occurrence of this cement lining molds of aragonitic allochems is consistent with an origin from fresh water that leached the aragonite. Alternatively, the leaching of aragonitic allochems could have occurred in fresh water during brief exposure, and subsequent calcite cementation could have occurred in the marine phreatic zone upon resubmergence. One example was found where a boring cuts through partially cemented allochems (Figure 6C), indicating cementation during Avon Park deposition. Similar marine calcite cementation processes are occurring in Holocene submarine Persian Gulf limestones (Shinn, 1969).

Syntaxial Calcite Cement

Inclusion-free syntaxial calcite cement is present on most echinoderm grains, and ranges in thickness from 200 to 500 µm. Where echinoderm grains are common, this cement occludes much primary porosity (Figure 6B). However, because echinoderm grains rarely account for more than 10% of all allochems in a hand sample or thin section, this cement is volumetrically minor. The presence of coarse calcite cement syntaxially growing over echinoderm grains, and absence of large calcite crystals on foraminifers, peloids, and mollusk molds, indicates the importance of substrate selectivity in calcite cement development (cf. Cullis, 1904; Murray, 1960; Taylor and Illing, 1969; Meyers, 1974; Kaufman et al., 1988). Based on cross-cutting petrographic relationships, the syntaxial calcite cement predates grain-to-grain compaction (Figure 6A). Although the petrographic relationships are equivocal, syntaxial calcite cement appears to be overgrown and replaced by dolomite of Eocene seawater origin. Therefore, the syntaxial calcite cement may be a very early diagenetic phase, contemporaneous with the blocky to bladed calcite cement.

Cavity-Lining Calcite Cement

At surface quarry locations, the Avon Park has common cavities and dissolution conduits (Figure 7A).

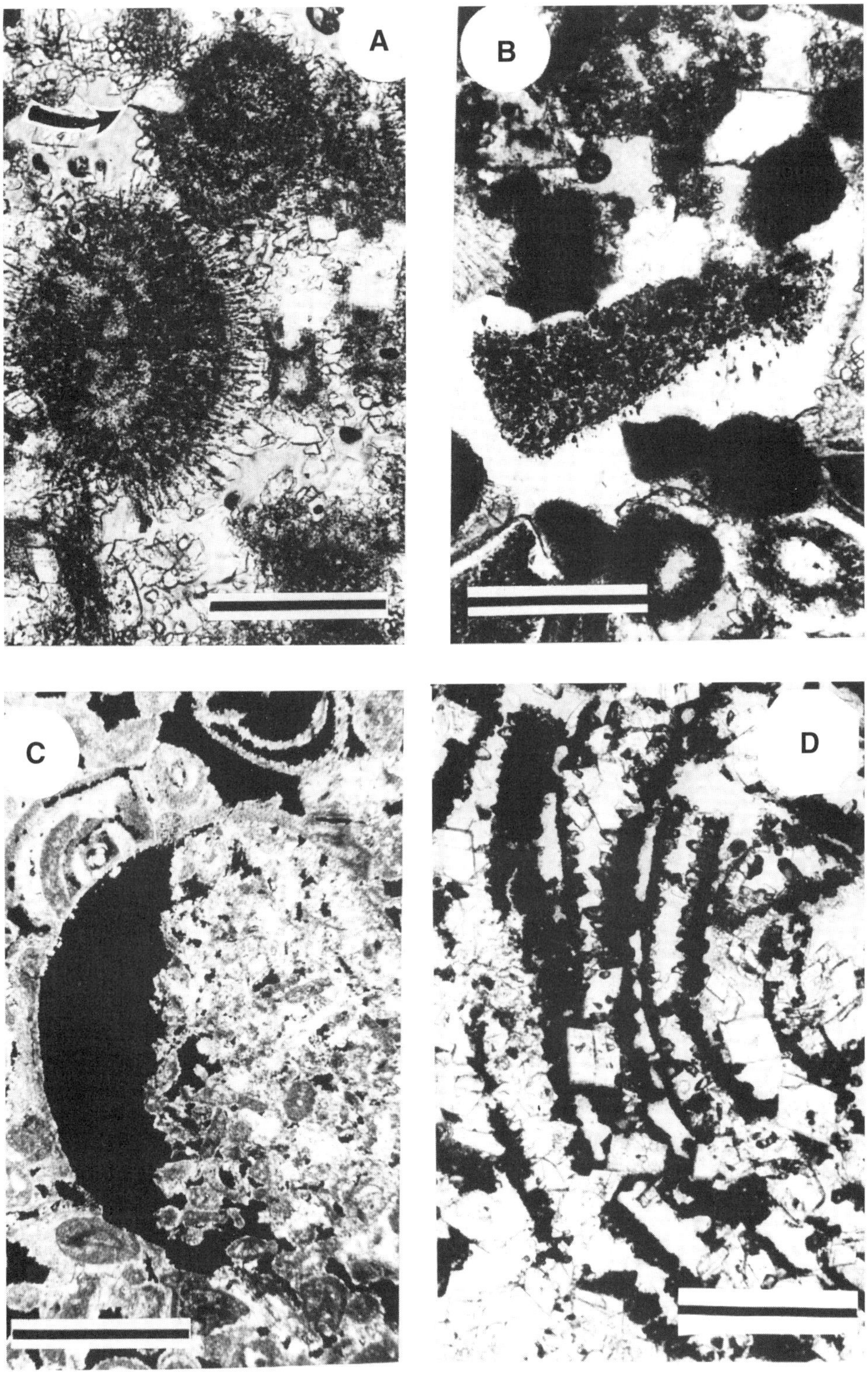

Figure 6. Dolomite-calcite cement timing, syntaxial cement, and late-stage diagenesis. (A) Transmitted light photomicrograph of upper Avon Park foraminiferal grainstone (core W-15290) showing bladed to blocky calcite cement lining and filling intraparticle porosity in foraminifers. Also note euhedral dolomite rhombs partially replacing calcite allochems and cement. Scale bar = 0.5 mm. (B) Transmitted light photomicrograph of Avon Park skeletal grainstone. Syntaxial calcite cement is well developed on crinoid grains, but calcite cement is volumetrically minor on other allochems. Scale bar = 0.3 mm. (C) Transmitted light photomicrograph of upper Avon Park skeletal grainstone (core W-15290) showing a marine boring that crosscuts grains with bladed and blocky calcite cement. At least some calcite cementation and mineral stabilization occurred during Avon Park deposition. Scale bar = 0.3 mm. (D) Transmitted light photomicrograph, upper Avon Park Formation, core W-10254. Euhedral dolomite rhomb overgrows and includes bladed to blocky calcite cement, indicating post-calcite cement timing. Scale bar = 0.3 mm.

Several samples of very coarse cavity-lining calcite cement (Figure 7B) were taken from quarries in northwestern peninsular Florida. Based on geochemical data discussed below, the cavity-lining cements formed during late-stage precipitation from dilute meteoric phreatic Floridan aquifer groundwater. The volumetric importance of these cements is difficult to estimate, but may be significant. The calcite can form layers several centimeters thick in many cavities, suggesting that the caves and conduits are important sinks for calcium carbonate in the Floridan aquifer system. Previous studies of Floridan aquifer cementation have noted that there is little evidence for calcite cementation in the matrix porosity (Cander, 1991, 1992; Budd et al., 1993). It is possible that the main sink for calcite carbonate is calcite lining the cavities and vugs that exist in the extensive conduit flow system.

Figure 7. Late-stage cavities and calcite cement. (A) Example of cavity in the Avon Park Formation formed by meteoric dissolution from quarry location CC. (B) Coarse calcite cement found on floor of 1 m diameter cavity in quarry CC in western Florida (Figure 1). The calcite cements can exhibit a variety of crystal morphologies, such as the bladed crystals in this sample. Scale is in centimeters.

GEOCHEMISTRY

Stable Isotopes

Figure 8 is a plot of carbon versus oxygen isotopes for 33 whole rock and microsampled separates of Avon Park limestone, five Oldsmar limestones, and eight microsamples of vug and cavity-lining coarse calcite cements in Avon Park host rock. Oxygen isotopic compositions for all samples range from $\delta^{18}O$ = –3.2 to +0.9‰, PDB, and carbon isotopic compositions range from $\delta^{13}C$ = –7.8 to +3.1‰, PDB. All Avon Park limestones have carbon isotope values greater than +0.5‰, PDB. The lightest Avon Park limestone has a

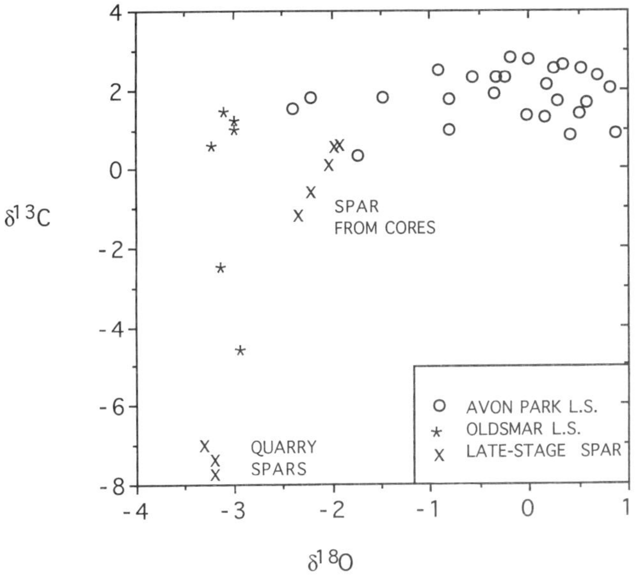

Figure 8. Carbon versus oxygen isotopic compositions of Avon Park limestones (o), Oldsmar limestones (*), and coarse calcite cements that line cavities (X). All rock data are presented relative to PDB standard.

$\delta^{18}O = -2.8‰$, and over half of the Avon Park limestones have oxygen isotopic values greater than −0.5‰. None of the Avon Park samples exhibit a significant depletion in $\delta^{13}C$. In contrast, the Oldsmar limestone samples have a very narrow range of $\delta^{18}O$ compositions (Figure 8), from −2.9 to −3.2‰, and a wide range of $\delta^{13}C$ values, from +0.8 to −4.5‰. All of the Oldsmar limestone samples are depleted in ^{18}O relative to Avon Park limestones, and most Oldsmar samples have depleted $\delta^{13}C$ compositions relative to Avon Park samples. Data occur in an L-shaped pattern extending from limestones with original marine compositions ($\delta^{18}O > 0‰$) to the coarse calcite cements reflecting precipitation from the groundwater system with little buffering from the host rock ($\delta^{18}O = -3.0$ to −3.3‰). Note that some Avon Park limestones have retained their original marine isotopic compositions and that, in contrast, all Oldsmar limestone samples have been reset by the meteoric groundwater system. The L-shaped array of Avon Park Formation carbon and oxygen isotope data is characteristic of freshwater recrystallization of marine limestone (Meyers and Lohmann, 1985; Banner et al., 1988; Banner and Hanson, 1990), in which the oxygen isotopic composition of the limestone is completely equilibrated with freshwater prior to perturbation of the $\delta^{13}C$ composition.

The near-surface cave-lining calcite cements ($\delta^{18}O = -3.1‰$, $\delta^{13}C = -7.5‰$, PDB) (Figure 8) have stable isotopic compositions that are close to being in equilibrium with typical, shallow Floridan aquifer groundwater (average Floridan aquifer water: $\delta^{18}O = -1.7‰$, SMOW; $\delta^{13}C = 10.0‰$, PDB; based on data from Sprinkle, 1989) and, thus, are end-member groundwater phases. The vug-lining calcite cements from subsurface Avon Park host rock have stable isotopic compositions that lie on a mixing line between the freshwater calcite cements and slightly recrystallized (^{18}O-depleted) Avon Park marine limestone. It is possible that various amounts of host limestone could have been sampled during micro-drilling of the vug-lining cements.

Strontium Isotopic Compositions

The $^{87}Sr/^{86}Sr$ compositions of Avon Park and Oldsmar limestones are presented in Table 1 and graphically presented in relation to $\delta^{18}O$ compositions in Figure 9. The composition of middle Eocene seawater is estimated to range from 0.70774 to 0.70780 (DePaolo and Ingram, 1985). The $^{87}Sr/^{86}Sr$ compositions of Avon Park limestone samples range from 0.707717 ± 15 to 0.707812 ± 15 (n = 24). Oldsmar limestone samples have a smaller $^{87}Sr/^{86}Sr$ range, from 0.707730 ± 16 to 0.707761 ± 13 (n = 6, with 320 to 461 ppm Sr).

Coarse calcite cements in limestones of the Avon Park Formation have a much wider range of $^{87}Sr/^{86}Sr$ compositions, from 0.708700 ±14 to 0.707790 ± 8 (n = 6) (Figure 9). These $^{87}Sr/^{86}Sr$ compositions are minimum values because any incorporation of Avon Park host rock ($^{87}Sr/^{86}Sr$ = 0.70772–0.70780) during sampling would lower the measured $^{87}Sr/^{86}Sr$ composition of the coarse calcite cements. Because their $^{87}Sr/^{86}Sr$ compositions are often considerably more radiogenic than Avon Park host rock, these calcite cements must have derived Sr from above middle Eocene strata. It is unlikely that radiogenic Sr for these cements was derived from within the Avon Park for the following reasons: (1) The Avon Park is siliciclastic-poor, and (2) based on their depleted stable isotopic compositions and low Sr concentrations, the coarse calcite cements precipitated at very high water-rock ratios, with little incorporation of host rock material. The $^{87}Sr/^{86}Sr$ compositions of the cavity-lining cements are equivalent to estimates of late Miocene seawater, indicating that Sr was likely derived from Miocene or younger strata in the Florida platform. Because host Avon Park limestone and dolomite would only serve as a source of nonradiogenic Sr for groundwater precipitating calcite, the radiogenic compositions of the cavity-lining calcite cements (>0.7086), coupled with their depleted oxygen isotope compositions, are evidence of post-middle Miocene groundwater diagenesis, but do not document the occurrence of older groundwater systems in the Avon Park Formation.

The $^{87}Sr/^{86}Sr$ compositions of Avon Park limestones are unmodified from coeval middle Eocene seawater (Figure 9). There has been no detectable exchange of Sr between Avon Park limestone and modern infiltrating groundwaters, even though fresh (dilute) Avon Park groundwaters have $^{87}Sr/^{86}Sr$ ratios (0.7080–0.7085) far more radiogenic than the host rock (0.70772–0.70780) (Table 2). This is consistent with the interpretation of limited water-rock interaction between Floridan aquifer groundwater and Avon Park limestone based on the retention of most original oxygen and carbon in the limestones. As well, the radiogenic $^{87}Sr/^{86}Sr$ compostions of cavity-lining cements indicate very little buffering by Avon Park

Table 1. Limestone and cement geochemistry.

Sample	Description	$\delta^{13}C$	$\delta^{18}O$	$^{87}Sr/^{86}Sr$	Sr (ppm)	Mn (ppm)	Fe (ppm)
W-1 3942							
326	Ocala Ls	2.12	0.17	0.707754	323	6	77
339	Packstone	1.73	0.29	0.707750	400	7	127
349	Packstone	1.32	0.16	0.707812	373	8	202
366	Packstone	1.38	0.50	0.707809	709	9	458
367	Packstone	0.86	0.41	0.707717	350	7	194
412	Mudstone	1.65	0.58	0.707787	488	9	263
426	Grainstone	2.34	0.69	0.707789	452	6	64
W-14675							
417	Grainstone	1.53	−2.39				
438	Grainstone	1.80	−1.48	0.707807	409	16	105
464	Grainstone	1.92	−2.73	0.707786	477	30	145
497	Grainstone	1.78	−2.22	0.707791	480	18	201
W-15282							
295	Grainstone	2.80	−0.18	0.707802	366	25	284
301	Grainstone	2.30	−0.33	0.707786	356	12	104
329	Packstone	2.02	0.81		470	8	110
W-15290							
108.5	Grainstone	1.90	−0.35				
111	Grainstone	0.90	0.88				
W-13881							
1025	Packstone	2.31	−0.25				
1740	Packstone	2.32	−0.57				
2005		0.71	−1.85				
W-15347							
907	Spar			0.707790		136	522
1795	Oldsmar G.S.	0.82	−3.00	0.707741	346	40	222
1812.5	Oldsmar G.S.	0.58	−3.18	0.707778	412	33	582
1826	Spar	0.78	−1.90				
1892.5	Oldsmar G.S.	−2.50	−3.09	0.707730	322	15	201
1896.5	Oldsmar G.S.	−4.48	−2.86	0.707740	461	11	198
1984	Oldsmar G.S.	1.26	−3.00	0.707750	320	14	314
1990	Oldsmar G.S.	1.12	−2.92	0.707761	402	8	104
W-10254							
367	Ocala Ls	3.06	0.10	0.707781	460	11	188
370	Ocala Ls	2.80	0.40	0.707736	463	19	145
412	Mudstone	1.34	−0.03	0.707787	435	24	231
420	Grainstone	1.77	−0.81	0.707749	530	12	101
465	Grainstone	2.50	−0.92	0.707774	333	7	238
485	Grainstone	2.64	0.34	0.707750	450	22	566
493	Mudstone	2.77	0.00	0.707740	400	8	132
493B	Grainstone	2.53	0.52	0.707740	430	13	201
504.5	Grainstone	2.17	0.12		365	15	226
953A	Cavity Spar	0.14	−1.53			7	709
953B	Wall Rock	0.97	−0.80				
954	Cavity Spar	−0.64	−2.24	0.707799			
955.5	Cavity Spar	0.86	−1.80	0.707870	790	214	548
956	Cavity Spar	−1.15	−2.27	0.707802			
957	Wall Rock	0.35	−1.74			12	402
Cavity Calcite							
CC-1	Coarse Spar	−7.21	−3.17	0.708700	16	3	67
CC-2	Coarse Spar	−7.68	−3.07	0.708671	17	4	84
CC-2C	Coarse Spar	−7.83	−3.08	0.708710	11	6	77
Average Limestone		1.95	−0.41	0.707773	432	19	224
Average Avon Park Dolomite (n = 57)		2.03	3.65	0.707827	167	13	192

host rock (Figure 9). These data reveal that there has been very limited interaction between dilute Floridan aquifer groundwater and Avon Park host marine carbonate rocks.

The $^{87}Sr/^{86}Sr$ compositions, Sr concentrations, and chlorinity of five Avon Park groundwaters are consistent with these interpretations (Table 2). The $^{87}Sr/^{86}Sr$ compositions of the dilute groundwaters are all significantly more radiogenic than Avon Park host limestone and dolomite. The groundwater data are consistent with the rock data, indicating that water-rock interaction is very limited today between dilute Avon Park pore fluids and host rock.

Sr Concentrations

Strontium concentrations for Avon Park and Oldsmar limestones are comparable. The range of Sr concentrations in Avon Park limestones is from 320 to 530 ppm, and from 320 to 460 ppm in the Oldsmar samples. In stark contrast, coarse calcite cements lining cavity walls in the Avon Park Formation have very low Sr concentrations (10 to 20 ppm).

The low Sr concentrations in coarse calcite cements are consistent with the very low Sr/Ca ratios in Avon Park groundwater samples (Table 2) and, as is the case with $^{87}Sr/^{86}Sr$ data reported above, indicate limited leaching of Sr from Avon Park host rock. The dilute groundwaters today have less than 1 ppm Sr, very low Sr/Ca {ratios, and $^{87}Sr/^{86}Sr$ compositions more radiogenic than host Avon Park rock (Table 2). The two mixing-zone groundwater samples have enriched Sr concentrations (>20 ppm) and $^{87}Sr/^{86}Sr$ compositions in equilibrium with Avon Park host rock. The excess Sr correlates with excess SO_4. Thus, the nonradiogenic Sr is derived from dissolution of Eocene gypsum from deeper in the Floridan aquifer, not from Avon Park carbonate dissolution (Cander, 1991, 1992). Virtually all near-recharge Avon Park groundwater is calcite saturated; only dilute mixing-zone groundwater is calcite undersaturated (Table 2).

The Sr concentrations of Avon Park limestones, Oldsmar limestones, and coarse late-stage calcite cements are plotted against their $\delta^{18}O$ compositions in Figure 10. Many of the Avon Park limestones have heavy $\delta^{18}O$ compositions (>0.5‰) and high Sr concentrations (> 400 ppm). As discussed previously, the $\delta^{18}O$ values are considered to be near the original marine isotopic composition. Avon Park Sr concentrations range from 323 to 530 ppm (one sample has 709 ppm Sr). Oldsmar limestone samples have both a smaller range of Sr concentrations (320 to 461 ppm) and, as discussed previously, a narrower range of $^{87}Sr/^{86}Sr$ compositions. Most of the allochems that comprise the limestone samples are benthic foraminifers. The relatively lower Sr concentrations of the Avon Park and Oldsmar samples compared to modern foraminifers (greater than 1000 ppm Sr; Milliman, 1974) may be the result of recrystallization during burial in Eocene seawater; this process would expel Sr from the limestones without altering their $\delta^{18}O$ or $^{87}Sr/^{86}Sr$ compositions.

A few Avon Park limestone and virtually all Oldsmar limestone samples lie along a pathway of recrystallization (Figure 10). That is, the covariant changes in Sr and $\delta^{18}O$ composition of these samples can be

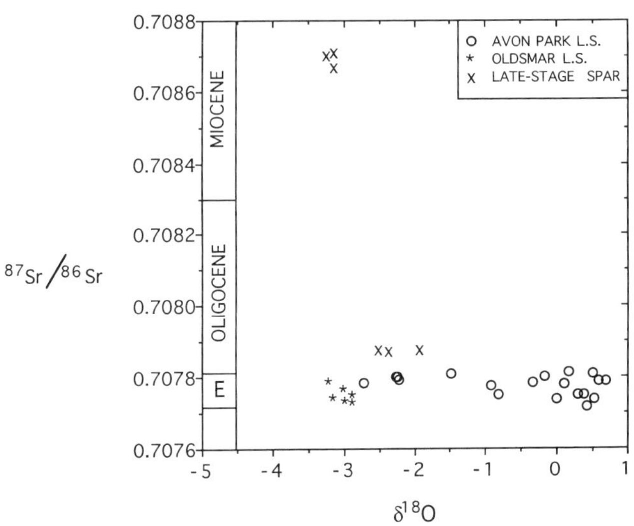

Figure 9. $\delta^{18}O$ versus $^{87}Sr/^{86}Sr$ for Avon Park limestones (o), Oldsmar limestones (*), and cavity-fill calcite cements (X). Also shown are estimated ranges for seawater since the Cenozoic.

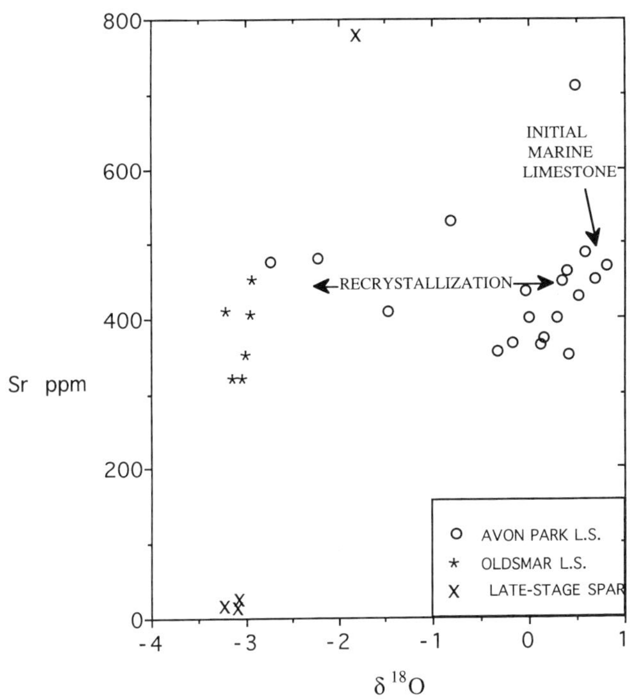

Figure 10. $\delta^{18}O$ versus Sr concentration for Avon Park limestones (o), Oldsmar limestones (*), and late-stage cavity-fill calcite cements (X). Also shown are pathways of meteoric recrystallization of marine limestone and physical mixing between marine limestones and meteoric calcite cements.

Table 2. Geochemistry of Avon Park groundwater samples.

Sample	Water Type	Depth (m)	TDS	pH	Ca	Mg	Na
TR8-1	Mixing Zone	274–287	3400	7.35	340	150	651
ROMP-17	Mixing Zone	340-436	913.66	7.25	123	63	63
ROMP 88	Fresh	59–117	460.93	7.3	87	32.4	39.7
ROMP DV-I	Fresh	162–259	327.49	7.35	65.5	9.5	7
TR19-3	Fresh	134-184	208.12	7.65	43.3	5.8	2
	Modern Seawater		34,700	8.2	410	1,287	10,685

accounted for by recrystallization, as opposed to cementation processes. The covariant trends in Avon Park and Oldsmar limestone compositions are consistent with the paucity of calcite cement in these rocks.

DISCUSSION

Water-Rock Interaction Modeling

Based on the above reasoning that the isotopic compositions of Avon Park limestones can be interpreted within the hydrogeologic framework of the modern Floridan aquifer system, the carbon and oxygen isotope data can be combined for water-rock interaction modeling. Several authors have theoretically modeled water-rock interaction in carbonates using quantitative models (Taylor, 1977; Land, 1980; Banner et al., 1988; Banner et al., 1989; Banner and Hanson, 1990). When applied to actual geologic systems with geochemical data on rocks and/or pore fluids, these models can place quantitative constraints on extents of water-rock interaction between carbonate aquifers and pore fluids. This study applies the water-rock interaction model of Banner and Hanson (1990) to estimate the number of pore volumes of Floridan aquifer groundwater that have reacted with Avon Park limestone since deposition. In this model, successive pore volumes of fluid are reacted with a fixed volume of rock until isotopic equilibrium is achieved. Figure 11 shows the results of interacting Floridan aquifer groundwater with a starting composition $\delta^{18}O$ = –0.5‰, $\delta^{13}C$= –14.0‰, PDB, with Avon Park marine limestone having starting composition $\delta^{18}O$ = +0.8‰, $\delta^{13}C$ = +2.0‰, PDB. The starting compositions of both the initial marine Avon Park limestone and the diagenetic Floridan aquifer groundwaters are based on actual data of this study, as well as supporting data from a regional study of Floridan aquifer groundwater chemistry (Sprinkle, 1989). In other words, the calculated water-rock interaction pathway is tightly constrained by existing data.

A comparison of the calculated water-rock interaction model (Figure 11) and the actual Avon Park stable isotope data (Figure 8) illustrates several points. First, both the calculated model and the actual data have L-shaped water-rock interaction paths whereby oxygen isotopes in the rock are completely equilibrated with Floridan aquifer groundwater prior to detectable resetting of the $\delta^{13}C$ composition of the rock. Second, all Avon Park limestone samples fall along the first branch of the interaction pathway, indicating that

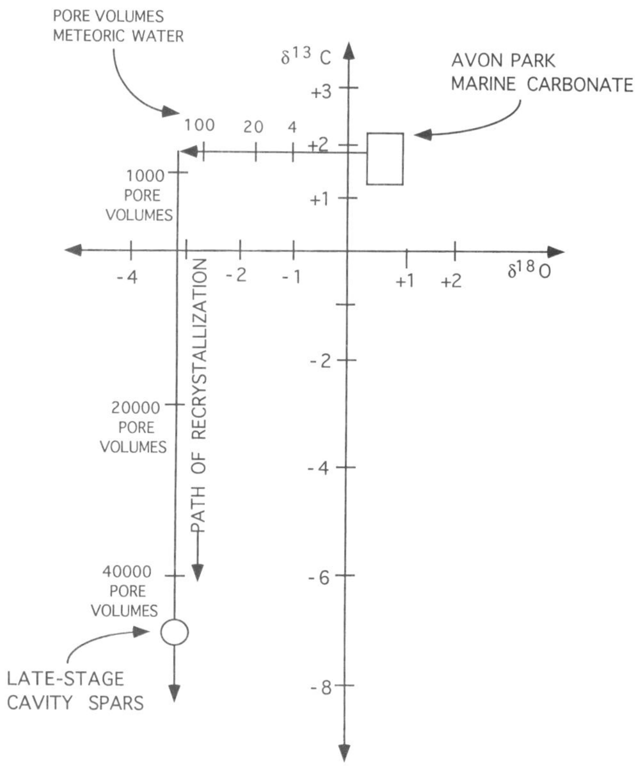

Figure 11. Path of limestone recystallization during calculated water-rock interaction between Floridan aquifer groundwater ($\delta^{18}O$ = –0.5 ‰, SMOW; $\delta^{13}C$ = –10‰, PDB) and Avon Park marine limestone ($\delta^{18}O$ = +0.8‰, PDB; $\delta^{13}C$ = +2.0‰, PDB). End-member compositions are based on actual data from this study and Sprinkle (1989). Water-rock interaction calculations and computer program used are from Banner and Hanson (1990). For the calculations, porosity = 25% and efficiency of reaction = 100%. Numbers along pathway indicate pore volumes of groundwater that have interacted with the rock.

Table 2. (*continued*).

Sample	Sr	Sr/Ca	$^{87}Sr/^{86}Sr$	Bicarbonate	Cl	Sulfate	Nitrate	Br	F
TR8-1	24	0.0706	0.707814	105.1	1370.5	755.5	0	3.93	0
ROMP-17	25	0.2033	0.707781	151.45	117.8	368.8	1.33	0.33	0
ROMP 88	0.6	0.0069	0.708540	278.8	22	0	0.34	0.058	0.03
ROMP DV-1	0.1	0.0015	0.708095	233.03	10.5	0.47	0.38	0.04	0.97
TR19-3	0.2	0.0046	0.708214	142.9	5.62	6	0.22	0.033	1.27
	8	0.0195	0.709164	142	19,215	2511	0	67	0.1

Avon Park limestones interacted with less than 100 pore volumes of Floridan aquifer groundwater. Third, only the vug-lining and cavity-lining coarse calcite cements in the Avon Park Formation precipitated at high water-rock ratios, in probable equilibrium with Floridan aquifer groundwater.

At least two factors contribute to the excellent agreement between the theoretically modeled L-shaped water-rock interaction pattern and the L-shaped pattern of the actual data of this study. First, there is very little calcite cement in the limestones, which would cause data points to fall along a straight line connecting end-member marine and freshwater calcite. Secondly, the Avon Park Formation has had a relatively simple two-component hydrologic history involving marine water and fresh water. Elevated temperature fluids or saline basin-derived fluids do not appear to have interacted with Avon Park limestones.

Water-Rock Interaction Efficiency

The Avon Park Formation has experienced multiple episodes of exposure during which long duration, active circulation of meteoric groundwater has probably occurred under hydrologic conditions similar to present (Stringfield, 1966; Thayer and Miller, 1984; Miller, 1986; Randazzo and Cook, 1987; Scott, 1989). Karst features in the Avon Park Formation have been ascribed to the late Miocene eustatic fall, as well as to post-Miocene eustatic falls (Stringfield, 1966). The probable periods of freshwater saturation of the Avon Park Formation are illustrated in Figure 4 and were estimated by correlating the Floridan stratigraphy with the sea level curve from Haq et al. (1988). Assuming linear subsidence of the Florida platform and based on the top of the Avon Park Formation being slightly above present sea level, it can be deduced that Floridan aquifer-like conditions would have developed during the late Oligocene, late Miocene, and intermittently throughout the Pliocene–Quaternary (Figure 4). These conclusions agree with interpretations of paleohydrogeology in central Florida by Scott (1989). A high amplitude pre-late Eocene sea level fall in Florida is discounted because: (1) upper Eocene strata cover all middle Eocene strata in Florida except where they have been removed by erosion (Chen, 1965; Miller, 1986), and (2) the upper Eocene shoreline was landward relative to the middle Eocene shoreline (Chen, 1965; Miller, 1986). Since the Floridan aquifer system comprises Paleocene through middle Miocene strata, groundwater circulation in the system during post-middle Miocene sea level falls would have been similar to its configuration today. Based on the above reasoning, the Avon Park Formation must have experienced one or more prolonged periods of saturation with meteoric phreatic groundwater.

Given quantitative constraints on the extents of water-rock interaction in Avon Park limestones, it is possible to estimate the efficiency of Floridan aquifer groundwater in recrystallizing Avon Park limestone. The water-rock interaction efficiency is herein defined as the ratio of the number of reacted pore volumes of groundwater to the number of total pore volumes of groundwater. That is, the efficiency is the number of pore volumes that have reacted with Avon Park Formation limestones (<100 pore volumes) divided by the total number of pore volumes that have flowed through the limestones. The total number of pore volumes of groundwater that have flowed through the Avon Park Formation can be estimated with the following data: (1) average linear velocity of Floridan aquifer groundwater, (2) porosity of the rock, and (3) duration of freshwater flow. A conservative estimate of the average linear velocity of Floridan aquifer groundwater in the study area is approximately 5 m/yr (Sprinkle, 1989). The average porosity of Avon Park Formation limestone is 25% (this study; Halley and Schmoker, 1983; Thayer and Miller, 1984). The estimated duration of freshwater flow is more difficult to assess. Based on the eustatic sea level curve and relative position of the top of the Avon Park Formation (Figure 4), there have been several periods of prolonged exposure to meteoric infiltration. In particular, a 5 m.y. late Oligocene duration and a 5 m.y. late Miocene duration appear to be likely times for establishment of meteoric groundwater systems (Figure 4).

The geochemical data on coarse vug-lining and cavity-lining calcite cements support the occurrence of a post-middle Miocene groundwater system, but neither confirm nor deny the occurrence of a late Oligocene groundwater system. Hammes (1992) indicates that the late Oligocene hiatus lasted between 2.5 and 6.5 m.y. For conservative estimates, taking only one of these times of exposure into account, the duration of freshwater flow was up to 5 m.y. It is possible that the duration of freshwater flow to at least the upper Floridan aquifer was much longer.

Consider a 5 m.y. groundwater system flowing through 1 m^3 of Avon Park limestone with homogeneous and isotropic permeability, porosity of 0.25, and average linear velocity of $v = 5$ m/yr. The following equations relate the given variables to determine the total number of pore volumes of meteoric groundwater that have flowed through the upper Avon Park limestones during one of the long durations of exposure:

$$Q = vA \quad (1)$$

$$\text{Total Pore Volumes} = Q \times (\text{Time}) \quad (2)$$

where Q is specific discharge (m^3/s), v is linear discharge (m/s), and A is cross-sectional area (m^2). A comparison of the number of pore volumes that have reacted (PV_r) with Avon Park limestone with the total number of pore volumes that have flowed through the Avon Park Formation (PV_t) during one of the Tertiary meteoric lenses yields the efficiency of Floridan aquifer recrystallization

$$\text{Efficiency (\%)} = PV_r/PV_t\,(100) \quad (3)$$

Solving these equations for just 5 m.y. of freshwater flow yields an efficiency of 0.0016%. In other words, only one out of every 62,500 pore volumes of Floridan aquifer groundwater reacted with Avon Park limestone.

It is important to note that all of the numbers used in these calculations are approximations and that the number of pore volumes of reacted water (<100) could be slightly different for a different water-rock interaction modeling program. Specifically, the water-rock interaction program of Banner and Hanson (1990) assumes 100% reactivity between each pore volume of fluid and the host rock. This ideal condition is probably not met in most natural systems due to differential reactivity of carbonates based on crystal morphology, size, or surface contaminants, among other factors. The estimate for the water-rock efficiency of the system is merely intended to provide a perspective on interaction between Floridan aquifer groundwater and Avon Park limestone. Banner and Hanson (1990) state that uncertainties in mineral-water reactivity will affect the absolute water-rock ratio at a given point on the calculated pathway, but not the diagnostic pathway. Results of the calculations presented here may provide important constraints from a natural system on the efficiency of reaction between each pore volume of fluid and the host rock. In the case of Avon Park limestone, the efficiency of reaction is less than 0.002%.

The inefficiency of diagenesis of the Floridan aquifer system with respect to Avon Park limestone is also shown by the $^{87}Sr/^{86}Sr$ disequilibrium between Avon Park host rock limestone (and dolomite) and late-stage cavity-lining calcite cements. The late-stage calcite cements have $^{87}Sr/^{86}Sr$ compositions as radiogenic or more radiogenic than dilute Avon Park groundwaters. There is apparently little contribution of Sr to dilute fresh water from middle Eocene host rock, even though the groundwaters have very low Sr concentrations (<1 ppm) (Table 2) and the rocks are relatively Sr-rich (100–450 ppm) (Table 1). Due to the relatively high concentrations of Sr in the rock compared to the groundwater, rock buffering of the groundwater $^{87}Sr/^{86}Sr$ composition should occur at very low rock/water ratios.

Interpretation of Water-Rock Interaction Efficiency

The above calculations demonstrate the long-term inefficiency of water-rock interaction in Avon Park limestone over the past 40 m.y. However, interpretations of diagenetic efficiency must consider the fluid-flow system in the Eocene strata of the Floridan aquifer system and, in particular, how the fluid-flow system evolved through time. Also, it is important to consider the episodic nature of diagenesis. That is, a long-term diagenetic inefficiency may result from one or two highly efficient episodes followed by longer episodes of limited water-rock interaction. The following section discusses the role of the fluid-flow system and the timing of unconformity-related diagenesis that result in a carbonate water-rock interaction system with long-term inefficiency.

As stated previously, the Eocene Floridan aquifer system comprises both a matrix permeability system and a conduit flow system of karst-related caves, fractures, bedding planes, and solution vugs. In addition, analogous karst systems suggest the velocity of groundwater in the matrix is orders of magnitude slower than in the conduits. The 5 m/yr estimate of average linear velocity of groundwater in the Floridan aquifer probably represents a value between the slow velocity of groundwater in the matrix and the rapid velocity of groundwater in the conduits. The above water-rock interaction modeling utilizes stable isotopic compositions of limestone samples representing the matrix pore network and thus reflects only the water-rock interaction efficiency in the matrix. In order to interpret water-rock interaction efficiency in the conduits, one must analyze diagenetic phases from the conduits. The cavity-lining coarse calcite cements have stable and radiogenic isotopic compositions in equilibrium with Floridan aquifer groundwater, suggesting extremely high water-rock ratios (>40,000 pore volumes Floridan aquifer groundwater) and, as a consequence, much higher water-rock interaction efficiency. These data suggest that the Floridan aquifer has two diagenetic systems (Figure 12): (1) an inert matrix system in which marine isotopic compositions are preserved and calcite cement is rare, and (2) the efficient conduit network in which significant calcite dissolution and precipitation at high water-rock ratios has occurred.

It is proposed that the development of the conduit flow system contributes to inefficiency of diagenesis in the matrix. That is, the development of the conduit

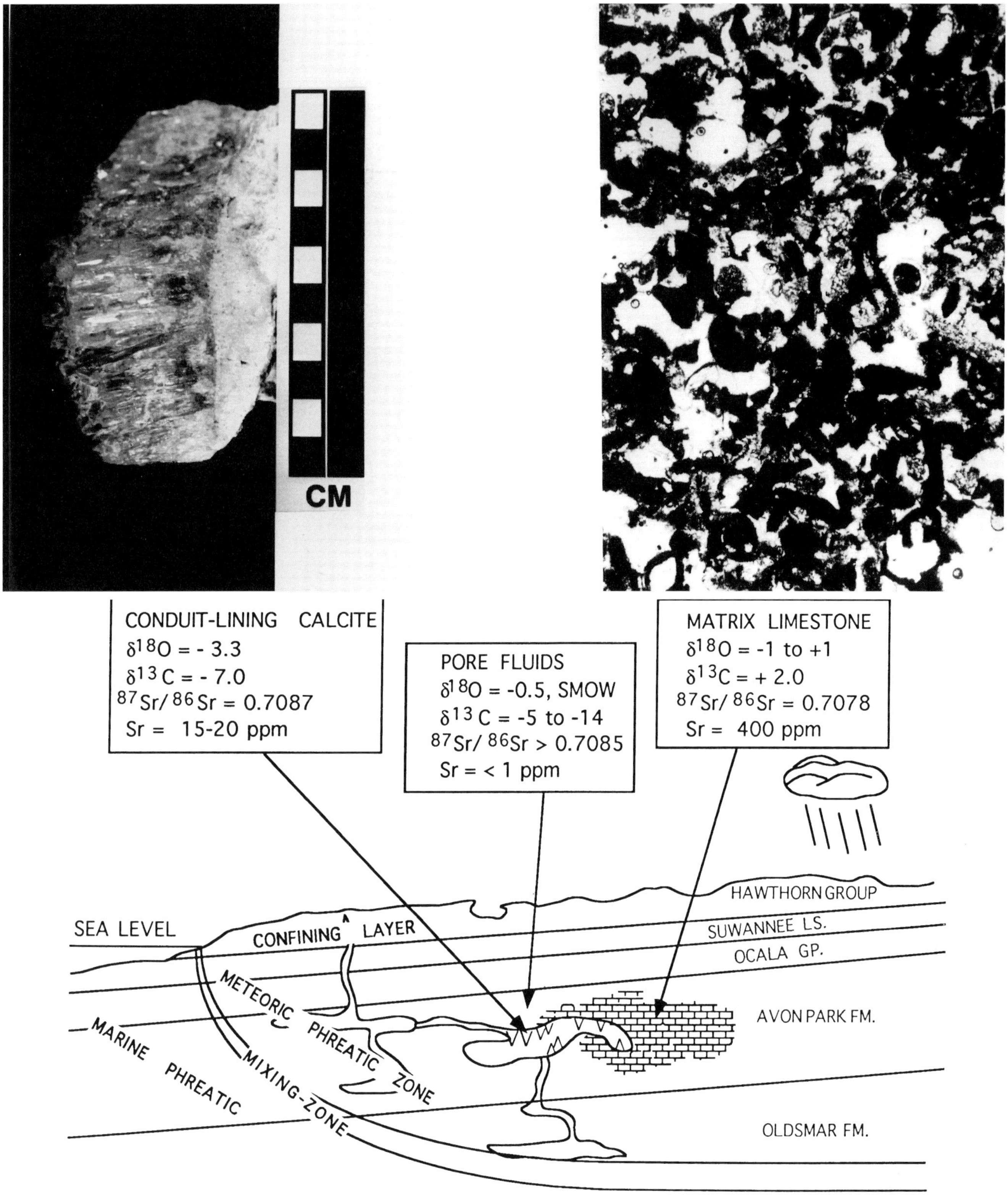

Figure 12. Schematic model of the two water-rock interaction systems in the Avon Park Formation, Floridan aquifer system. Model depicts the confined part of the Floridan aquifer system. Matrix limestones have completely different carbon, oxygen, and $^{87}Sr/^{86}Sr$ isotopic compositions and Sr concentrations than late diagenetic cavity-lining calcite or extant pore waters. The matrix system was active during intraformational and early post-depositional hiatuses, but has been relatively inert since the Eocene. In contrast, the conduit water-rock interaction system has been dominant since establishment of the conduit fluid-flow system during later, regional unconformities of the late Oligocene and late Miocene. This model shows how interpretations of carbonate rock-fluid interaction should carefully consider the fluid-flow system(s), as well as the compositions of the fluids and rocks.

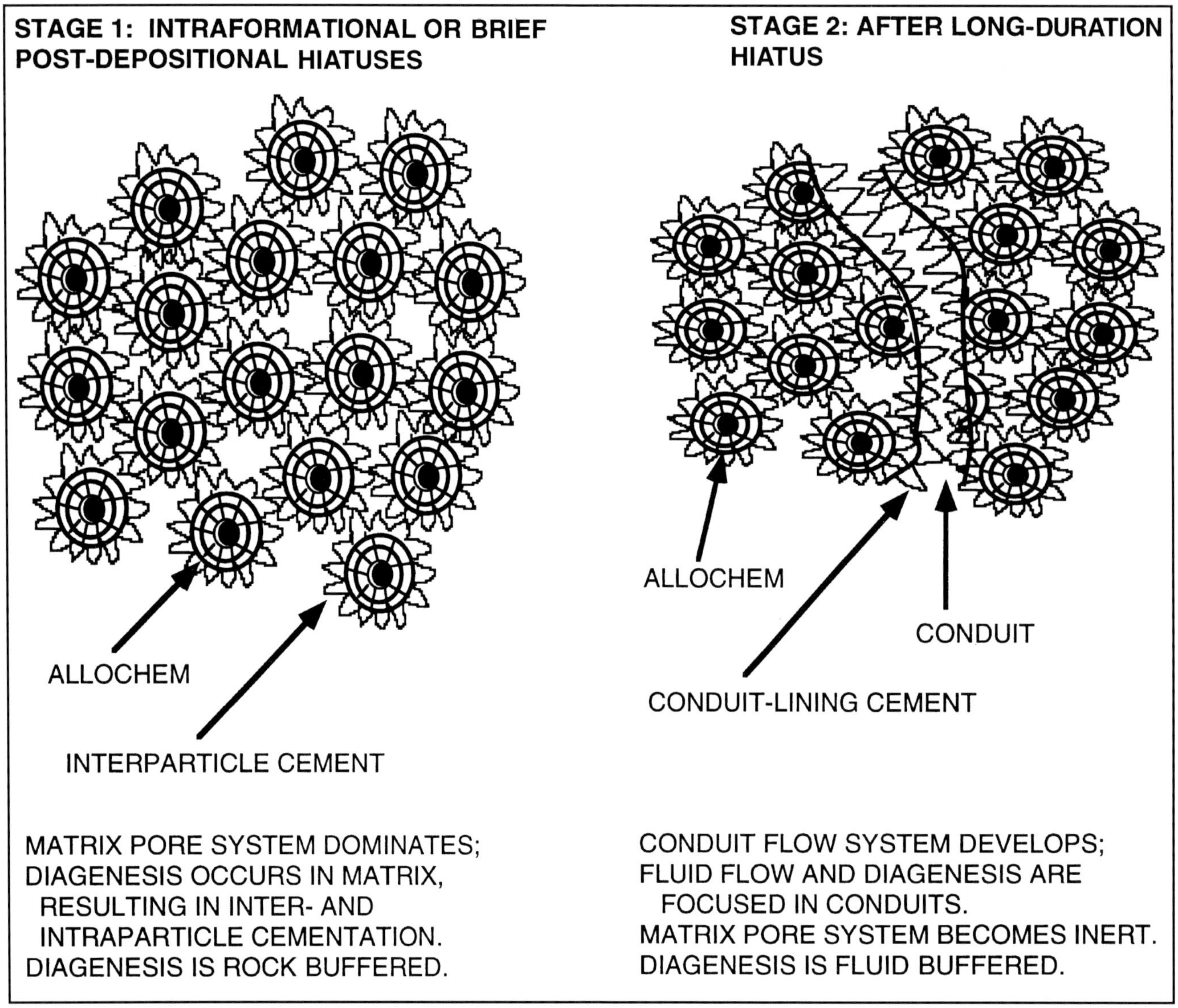

Figure 13. Schematic depiction of the interplay between hiatuses, fluid flow, and carbonate diagenesis. During intraformational and early postdepositional hiatuses (Stage 1), the matrix-porosity system dominates. Diagenesis is concentrated in the matrix pores, resulting in interparticle cementation as well as generation of secondary porosity during allochem dissolution. After a long-duration hiatus (Stage 2), a conduit pore system can develop via dissolution and karstification. Fluid flow is focused, and, thus, diagenesis is focused. The result is calcite cementation along conduits, as opposed to between particles, or enlargement of conduits through dissolution. Diagenesis during Stage 1 is more likely rock buffered. Diagenesis during Stage 2 is more likely fluid buffered.

system focuses fluid transmission out of the matrix and into the conduits. As fluid flow is focused, water-rock interaction becomes focused in the conduits and decreases in the matrix system. Where go the fluids, so go the reactions and products of diagenesis. The Avon Park data suggest the intimate interplay of the type of fluid-flow system and the efficiency of water-rock interaction. There is, in effect, a feedback mechanism between diagenesis and the flow system. Just as diagenesis influences the flow network, so, too, does the resulting flow network influence the products and occurrences of diagenesis (Figure 13).

The Avon Park data also suggest that the timing of diagenesis associated with unconformities can be related to both the fluid-flow system and water-rock interaction efficiency. As discussed earlier, the most common diagenetic product in Avon Park limestones is isopachous bladed calcite that rims allochems, as well as skeletal moldic porosity. Petrographic and geochemical data indicate that this ubiquitous calcite precipitated during subaerial exposure associated with intraformational or early post-depositional unconformities. The timing and equilibrium isotopic composition of this calcite suggest that during early exposure events the matrix pores were the loci of diagenesis. In effect, during early exposure events, the matrix fluid flow was dominant and matrix water-rock interaction was far more efficient than at present.

There is, however, little petrographic or geochemical record of diagenesis in the matrix pores of Avon Park limestones since the Eocene. Rather, the major, late diagenetic phase is coarse calcite in vugs, caves, and conduits. The $^{87}Sr/^{86}Sr$ composition of this calcite suggests a post-middle Miocene timing, subsequent to two significant sea level falls that exposed the Eocene to long-duration meteoric groundwater systems (Figure 4). Apparently, the conduit fluid-flow system developed from meteoric or mixing-zone dissolution during post-middle Eocene exposure events and became the dominant water-rock interaction system no later than the late Miocene. As water-rock interaction increased in the conduit system, it decreased in the matrix pore system. If the Floridan aquifer system is analogous to other regional platform carbonate aquifers, past and present, then it may be a general rule that diagenesis occurs mainly in the matrix (inter- and intraparticle) pores during early, brief-duration exposure events and becomes progressively focused out of the matrix as the conduit system develops during later, long-duration exposure associated with regional unconformities. As well, diagenesis in the matrix pore system is more likely to be rock buffered, and diagenesis in the conduit flow system is more likely to be fluid buffered (Figure 13).

In addition to the development of a conduit fluid-flow system, there are other factors that contribute to the inefficiency of the Floridan aquifer groundwater system as an agent of limestone recrystallization in the Avon Park Formation. First, the strata overlying the Avon Park allow for calcite saturation to be achieved prior to the entrance of the pore waters into the Avon Park. As a result, there is a reduced potential for water-rock interaction within the Avon Park. Such a situation has probably persisted throughout most of the diagenetic history of the Avon Park, since the first long-duration freshwater lens probably did not develop until the late Oligocene or post-middle Miocene when the Avon Park was everywhere buried under the upper Eocene Ocala Group, Oligocene Suwannee Limestone, or Miocene Hawthorn Group (Figure 4). Because the groundwater reaches calcite saturation during interaction with post-middle Eocene limestones, it loses its thermodynamic drive to recrystallize and cement Avon Park limestones. Further, the state of near isothermal chemical equilibrium in calcitic portions of the Avon Park Formation allows for maintenance of isotopic disequilibrium between Floridan aquifer groundwater and host Avon Park limestone. This conclusion is supported by other data, including: (1) the low concentration and radiogenic composition of Sr in late-stage calcite cements in the Avon Park, (2) preservation of marine isotopic compositions in most dolomite (Cander, 1991, 1994), and (3) groundwater chemistry that shows isotopic disequilibrium between Avon Park pore fluids and host rock (Table 2). Furthermore, Budd et al. (1993) found a similar lack of matrix calcite cementation in the shallower Oligocene section of the Floridan aquifer system. Budd et al. (1993) suggest that possible drives for cementation in the upper Floridan aquifer include CO_2 fluxes (degassing) or common-ion effects associated with dissolution of calcium sulfate. The occurrence of either of these processes in the Avon Park could shift the system from equilibrium and result in precipitation of calcite cement.

The second reason for inefficient water-rock interaction in the Avon Park Formation is that the Avon Park underwent significant syndepositional mineralogic stabilization that reduced potential for later recrystallization by meteoric water, but did not greatly reduce interparticle porosity. The effect of early recrystallization on subsequent meteoric water diagenesis has been discussed for other carbonate aquifers by Harris and Matthews (1968) and Banner et al. (1989). In Avon Park samples, there is little difference in $\delta^{18}O$ composition between limestones with abundant isopachous calcite cement and limestones with minor isopachous calcite cement. This may be due to the relatively small volume of calcite cement or because the isopachous calcite cement may have precipitated from an isotopically enriched fluid, such as seawater. Micritization, conversion of high-Mg calcite to low-Mg calcite, dissolution of aragonite allochems, most dolomitization (Cander, 1991), and possibly some isopachous calcite cementation occurred during the middle Eocene. Precipitation of isopachous calcite cement around most inter- and intraparticle pores may have reduced surface reaction potential with nonequilibrium meteoric pore fluids.

Third, most of the allochems that comprise Avon Park Formation limestones are poor substrates for rapid growth of pore-filling calcite cement. Previous workers have demonstrated the substrate selectivity of calcite cement (cf. Cullis, 1904; Murray, 1960; Taylor and Illing, 1969; Meyers, 1974; Bathurst, 1975; Kaufman et al., 1988). The dominant middle Eocene allochems in the Floridan aquifer system, benthonic foraminifers, are multicrystalline and apparently poor substrates for calcite cement nucleation and growth. Avon Park allochems that provide better substrates, such as crinoid grains, are often coated with large cement crystals.

Relation of Oldsmar Diagenesis to Avon Park Diagenesis

The geochemical data for underlying upper Oldsmar limestones in central Florida suggest that the Oldsmar Limestone experienced diagenetic alteration prior to deposition of the Avon Park Formation. All Oldsmar limestone samples have $\delta^{18}O$ compositions more depleted than Avon Park limestone samples (Figure 7). The most ^{18}O-depleted Oldsmar limestone samples ($\delta^{18}O$ = –3.0 to –3.2‰, PDB) have the same oxygen isotopic composition as the cavity-lining calcite cements precipitated in equilibrium with pure meteoric Floridan aquifer groundwater (Figure 8). The $^{87}Sr/^{86}Sr$ compositions of Oldsmar limestones are all equivalent to estimates of coeval middle Eocene seawater (Figure 9), indicating that the meteoric groundwater that caused complete resetting of the oxygen isotopic composition did not introduce new Sr into the rocks. As described above, at least some, if not all, of the groundwater recrystallizing and interacting with the Avon Park Formation introduced allochthonous Sr into the limestones. The extent of water-rock interaction is thus very different for the underlying Oldsmar Limestone.

Because lower Eocene strata are not exposed on peninsular Florida and because the middle Eocene Avon Park Formation overlies the Oldsmar throughout the study area, present-day meteoric groundwater recharging the Oldsmar, as well as any post-middle Eocene recharge into the Oldsmar, would flow through and interact with Avon Park strata first. It is highly unlikely that such groundwaters would flow inertly through Avon Park limestones, then react efficiently with Oldsmar Formation strata. Meteoric recrystallization of the Oldsmar most probably occurred during intraformational subaerial exposure or immediately after Oldsmar deposition, when the Florida platform was exposed during the pre-middle Eocene eustatic fall (Figure 4). Infiltrating meteoric groundwaters would not have introduced allochthonous, more radiogenic Sr to the Oldsmar Formation because the Oldsmar limestones would have been the first and dominant Sr source that meteoric waters would have encountered.

CONCLUSIONS

1. Limestones in the middle Eocene Avon Park Formation, Floridan aquifer system, have experienced limited water-rock interaction by meteoric phreatic groundwater systems, despite being exposed to meteoric flushing multiple times during the Cenozoic. The limestones have retained most primary interparticle porosity and a large percentage of the limestones sampled have retained their marine $\delta^{18}O$ composition. Virtually all of the limestones sampled have also retained original middle Eocene marine $\delta^{13}C$ and $^{87}Sr/^{86}Sr$ compositions.

2. The limestones are in oxygen, carbon, and strontium isotopic disequilibrium with modern freshwater pore fluids throughout most of the study area. In contrast, late-stage cavity-lining calcite cements are in equilibrium with meteoric groundwater having depleted $\delta^{18}O$ and $\delta^{13}C$ compositions, radiogenic $^{87}Sr/^{86}Sr$ compositions, and depleted Sr concentrations.

3. Quantitative estimates of the number of pore volumes of groundwater that have reacted with Avon Park Formation limestone with the number of pore volumes of groundwater that have flowed through the Avon Park Formation indicate that less than one out of every 62,500 pore volumes of groundwater has reacted to recrystallize the limestone. The long-term efficiency of reaction between Avon Park limestones and Floridan aquifer system groundwater is less than 0.002%. Conversely, water-rock interaction efficiency in Avon Park cavities and vugs is far more efficient and yields phases in isotopic and trace element equilibrium with meteoric groundwater (high water-rock ratios).

4. It is proposed that the inefficiency of water-rock interaction in the limestone results, in part, from the existence of the extensive conduit fluid-flow system that developed during subaerial exposure associated with later regional unconformities. The conduit flow system focuses fluid, and thus diagenesis, out of the matrix and into the conduits. The matrix remains relatively inert. At the time of intraformational unconformities and immediately after deposition of the Avon Park, when the conduit flow system was absent or poorly developed, the matrix fluid-flow system was dominant and water-rock interaction was more efficient, resulting in mineralogic stabilization and ubiquitous bladed calcite cements on matrix grains and allochems.

5. The inefficiency of this regional meteoric phreatic carbonate aquifer in the Avon Park Formation is aided by the deposition of limestone formations above the Avon Park prior to the onset of the first long-duration, regionally extensive meteoric groundwater system. Groundwater infiltrates and reacts with younger limestone and achieves calcite saturation prior to significant interaction with Avon Park limestone. This hydrogeologic system has preserved most primary porosity and has allowed for maintenance, over the past 40 m.y., of C, O, and Sr isotopic disequilibrium between Avon Park limestones and Floridan aquifer groundwater.

6. Results of this study indicate that there is a feedback mechanism whereby diagenesis influences the pore system, and the resulting pore system, in turn, influences subsequent reaction products and their occurrences in the carbonate aquifer. Interpretations of carbonate rock-water interaction must carefully consider not only the compositions of the fluids and rocks involved in diagenesis, but also the fluid-flow system(s) and its evolution over time.

ACKNOWLEDGMENTS

This study is part of a dissertation on water-rock interaction and dolomitization in the Floridan aquifer system completed in 1991 at the University of Texas at Austin, and supervised by Lynton Land. The research was supported by fellowships from Texaco and Conoco, and a grant from the American Chemical Society Petroleum Research Fund, no. 19316AC2, held by Lynton Land. The manuscript benefited from detailed reviews by David Budd and Eva Moldovanyi. Careful reviews of this study were also provided by Lynton Land, Jay Banner, Don Bebout, and Jack Sharp, from the University of Texas, and Tony Randazzo, from the University of Florida. Tom Scott, Ken Campbell, and Jon Arthur of the Florida Geological Survey generously assisted core and field work. Doug Dewitt, Eric Dehaven, and Lee Clarke of the Southwest Florida Water Management District graciously supplied water samples.

REFERENCES CITED

Allan, J.R., and R.K. Matthews, 1982, Isotopic signatures associated with early meteoric diagenesis: Sedimentology, v. 29, p. 797–817.

Applin, P.L., and E.R. Applin, 1944, Regional subsurface stratigraphy and structure of Florida and southern Georgia: AAPG Bulletin, v. 28,p. 1673–1753.

Banner, J.L., and G.N. Hanson, 1990, Calculation of simultaneous isotopic and trace element variations during water-rock interaction with applications to carbonate diagenesis: Geochimica et Cosmochimica Acta, v. 54, p. 3123–3137.

Banner, J.L., G.N. Hanson, and W.J. Meyers, 1988, Water-rock interaction history of regionally extensive dolomites of the Burlington-Keokuk Formation (Mississippian): Isotopic evidence, *in* V. Shukla and P.A. Baker, eds., Sedimentology and Geochemistry of Dolostones: Society of Economic Paleontologists and Mineralogists Special Publication 43, p. 97–113.

Banner, J.L., G.J. Wasserburg, P.F. Dobson, A.B. Carpenter, and C.H. Moore, 1989, Isotopic and trace element constraints on the origin and evolution of saline groundwaters from central Missouri: Geochimica et Cosmochimica Acta, v. 53, p. 383–398.

Bathurst, R.G.C., 1975, Carbonate sediments and their diagenesis: New York, Elsevier, 658 p.

Bögli, A., 1980, Karst hydrology and physical speleology: New York, Springer-Verlag, 284 p.

Budd, D.A., and L.S. Land, 1990, Geochemical imprint of meteoric diagenesis in Holocene ooid sands, Schooner Cays, Bahamas: correlation of calcite cement geochemistry with extant groundwaters: Journal of Sedimentary Petrology, v. 60, p. 361–378.

Budd, D.A., U. Hammes, and H.L. Vacher, 1993, Calcite cementation in the upper Floridan aquifer: a modern example for confined-aquifer cementation models?: Geology, v. 21, p. 33–36 .

Cander, H.S., 1990, Dolomitization and water-rock interaction efficiency in the middle Eocene Avon Park Formation, Floridan aquifer (abs.): AAPG Annual Meeting, p. 33.

Cander, H.S., 1991, Dolomitization and water-rock interaction in the middle Eocene Avon Park Formation, Floridan aquifer: Ph.D. dissertation, The University of Texas, Austin, Texas, 172 p.

Cander, H.S., 1992, Late-stage diagenesis in the Floridan aquifer, middle Eocene Avon Park Formation (abs.): AAPG Annual Meeting, p. 16.

Cander, H.S., 1994, An example of mixing-zone dolomite, Middle Eocene Avon Park Formation, Floridan aquifer system: Journal of Sedimentary Research, v. 64A, p. 615–629.

Chen, C.S., 1965, The regional lithostratigraphic analysis of Paleocene and Eocene rocks of Florida: Florida Geological Survey Bulletin 45, 105 p.

Cullis, C.G., 1904, The mineralogical changes observed in the cores of Funafuti borings, *in* T.G. Bonney, ed., The Atoll of Funafuti: London, Royal Society, p. 392–420.

DePaolo, D.J., and B.L. Ingram, 1985, High resolution stratigraphy with strontium isotopes: Science, v. 227, p. 938–941.

Dorobek, S.L., 1987, Petrograph, geochemistry, and origin of burial diagenetic facies, Siluro-Devonian Helderberg Group (carbonate rocks), Central Appalachians: AAPG Bulletin, v. 71, p. 492–514.

Grover, G., Jr., and J.F. Read, 1983, Paleoaquifer and deep-burial related cements defined by regional cathodoluminescent patterns, Middle Ordovician carbonates, Virginia: AAPG Bulletin, v. 67, p. 1275–1303.

Halley, R.B., and P.M. Harris, 1979, Freshwater cementation of a 1,000-year-old oolite: Journal of Sedimentary Petrology, v. 49, p. 969–988 .

Halley, R.B., and J.W. Schmoker, 1983, High-porosity Cenozoic carbonate rocks of south Florida: Progressive loss of porosity with depth: AAPG Bulletin, v. 67, p. 191–200.

Hammes, U., 1992, Sedimentation patterns, sequence stratigraphy, cyclicity, and diagenesis of early Oligocene carbonate ramp deposits, Suwannee Formation, Southwest Florida, U.S.A: Ph.D. dissertation, University of Colorado, 344 p.

Hanshaw, B.B., and W. Back, 1972, On the origin of dolomites in the Tertiary aquifer of Florida, *in* H.S. Puri, ed., Proceedings of the Seventh Forum on Geology of Industrial Minerals Special Publication 17, p. 139–153.

Haq, B.U., J. Hardenbol, and P.R. Vail, 1988, Mesozoic and Cenozoic chronostratigraphy and cycles of sea-level change, *in* C.K. Wilgus, B.S. Hastings, C.G.St.C. Kendall, C.A. Posamentier, C.A. Ross, and J.C. Van Wagoner, eds., Sea-level Changes: An Integrated Approach: Society of Economic Paleontologists and Mineralogists Special Publication 42, p. 71–108.

Harris, W.H., and R.K. Matthews, 1968, Subareal diagenesis of carbonate sediments: efficiency of the solution-reprecipitation process: Science, v. 160, p. 77–79.

Kaufman, J., H.S. Cander, L.D. Daniels, and W.J. Meyers, 1988, Calcite cement stratigraphy and cementation history of the Burlington-Keokuk Formation (Mississippian), Illinois and Missouri: Journal of Sedimentary Petrology, v. 58, p. 312–326.

Land, L.S., 1980, The isotopic and trace element geochemistry of dolomite: the state of the art, *in* D. H. Zenger, J.B. Dunham, and R.L. Ethington, eds., Concepts and Models of Dolomitization: SEPM Special Publication 28, p. 87–110.

Matthews, R.K., 1968, Carbonate diagenesis: equilibration of sedimentary mineralology to the subaerial environment: coral cap of Barbados, West Indies: Journal of Sedimentary Petrology, v. 38, p. 1110–1119.

Matthews, R.K., 1971, Diagenetic environment of possible importance to the explanation of cementation fabric in subaerially exposed carbonate sediments, *in* O. P. Bricker, ed., Carbonate Cements: Johns Hopkins University Studies in Geology 19, p. 127–132.

Matthews, R.K., 1974, A process approach to diagenesis of reefs and reef associated limestones, *in* L.F. Laporte, ed., Reefs in Time and Space: Society of Economic Paleontologists and Mineralogists Special Publication 18, p. 234–256.

Meyer, F.W., 1989, Hydrogeology, ground-water movement, and subsurface storage in the Floridan aquifer system in southern Florida: U.S. Geological Survey Professional Paper 1403-G, 59 p.

Meyers, W.J., 1974, Carbonate cement stratigraphy of the Lake Valley Formation (Mississippian) Sacramento Mountains, New Mexico: Journal of Sedimentary Petrology, v. 44, p. 837–861.

Meyers, W.J., and K.C. Lohmann, 1985, Isotope geochemistry of regionally extensive calcite cement zones and marine components in Mississippian limestones, New Mexico, *in* N. Schneiderman and P.M. Harris, eds., Carbonate Cements: Society of Economic Paleontologists and Mineralogists Special Publication 36, p. 223–240.

Miller, J.A., 1986, Hydrogeologic framework of the Floridan aquifer system in Florida and in parts of Georgia, Alabama, and South Carolina: U.S. Geological Survey Professional Paper 1403-B, 91 p.

Milliman, J.D., 1974, Marine carbonates: Berlin, Springer-Verlag, 375 p.

Morton, J.P., and L.E. Long, 1984, Rb-Sr ages of glauconite recrystallization: dating times of regional emergence above sea level: Journal of Sedimentary Petrology, v. 54, p. 495–506.

Murray, R.C., 1960, Origin of porosity in carbonate rocks: Journal of Sedimentary Petrology, v. 30, p. 59–84.

Randazzo, A.F., and D.J. Cook, 1987, Characterization of dolomitic rocks from the coastal mixing zone of the Floridan aquifer, U.S.A.: Sedimentary Geology, v. 54, p.169–192.

Randazzo, A.F., and H.C. Saroop, 1976, Sedimentology and paleoecology of middle and upper Eocene carbonate shoreline sequences, Crystal River, Florida, U.S.A.: Sedimentary Geology, v. 15, p. 259–291.

Randazzo, A.F., G.C. Stone, and H.C. Saroop, 1977, Diagenesis of middle and upper Eocene carbonate shoreline sequences, central Florida: AAPG Bulletin, v. 61, p. 492–503.

Ryder, P.D., 1985, Hydrology of the Floridan aquifer system in west-central Florida, U.S. Geological Survey Professional Paper 1403-F, 63 p.

Scott, T.M., 1989, The lithostratigraphy of the Hawthorn Group (Miocene) of Florida: Florida Geological Survey Bulletin No. 59, 148 p.

Shinn, E.A., 1969, Submarine lithification of Holocene carbonate sediments in the Persian Gulf: Sedimentology, v. 12, p. 109–144.

Sprinkle, C.L., 1989, Geochemistry of the Floridan aquifer system in Florida and in parts of Georgia, South Carolina, and Alabama: U.S. Geological Survey Professional Paper 1403-1, 105 p.

Stringfield, V.T., 1966, Artesian water in Tertiary limestone in the Southeastern United States: U.S. Geological Survey Professional Paper 517, 226 p.

Taylor, H.P., Jr., 1977, Water/rock interactions and the origin of H_2O in granitic batholiths: Journal of Geological Society of London, v. 133, p. 509–558.

Taylor, J.M.C., and L.V. Illing, 1969, Holocene intertidal calcium carbonate cementation, Qatar, Persian Gulf, *in* O.P. Bricker, ed., Carbonate Cements: Baltimore, Johns Hopkins University Press, p. 36–39.

Thayer, P.A., and J.A. Miller, 1984, Petrology of lower and middle Eocene carbonate rocks, Floridan aquifer, central Florida: Gulf Coast Association of Geological Societies Transactions, v. 34, p. 421–434.

Chapter 6

Regional Exposure Events and Platform Evolution of Zhujiang Formation Carbonates, Pearl River Mouth Basin: Evidence from Primary and Diagenetic Seismic Facies

Eva P. Moldovanyi
Amoco Production Company
Houston, Texas, U.S.A.

F. M. Wall
Amoco Norway Oil Company
Stavanger, Norway

Zhang Jun Yan
China Offshore Oil Nanhai East Corporation
Guangzhou, People's Republic of China

ABSTRACT

The concept of chaotic or "diagenetic" seismic facies is introduced as a tool for understanding factors controlling the evolution of Miocene Zhujiang Formation carbonate platforms in the Pearl River Mouth Basin, South China Sea. Seismic data from this basin contain a continuum of chaotic seismic facies, which are present only within the carbonate interval and range in appearance from simple concave-shaped reflectors to highly irregular, high-amplitude chaotic reflectors. Diagenetic seismic facies are significant because they are overprinted on depositional (primary) seismic facies. The common association between chaotic seismic facies and other features, such as truncated zones and hummocky carbonate surfaces, supports a model involving karstification of Zhujiang Formation carbonates.

Two contrasting stages of carbonate platform development are recognized from seismic stratigraphic relationships within the Zhujiang Formation interval: (1) the "lower" platform, a broad and areally extensive low-relief ramp, and (2) the "upper" platform, a narrow and high-relief feature. A regional episode of subaerial diagenesis separating the two stages of platform development is inferred from the distribution of chaotic seismic facies and from the presence on seismic of erosional carbonate surfaces. The magnitude of sea level fluctuation is thought to have been on the order of a few 100 m and was likely eustatically controlled. Numerous backstepping events in the upper Zhujiang carbonate suggest this stage of platform development was characterized by continually rising sea level. Smaller-scale exposure events, not resolvable on seismic, are recognized in cores of upper Zhujiang carbonates in the Liuhua 11-1 field and are seemingly at odds with the drowning model. These exposure events are interpreted to be sudden brief excursions in an otherwise continuously drowning sequence.

INTRODUCTION

Exposure surfaces are a common phenomenon in many carbonate accumulations in Southeast Asia. In some places, such as China and Vietnam, prolonged subaerial diagenesis has resulted in a very mature karsted upland, while in others, such as in the subsurface Natuna-L complex, Indonesia (Rudolph and Lehmann, 1989), the effects are much less dramatic. In either case, the process of karstification has been of some consequence to the carbonate texture, but just how much the reservoir potential (e.g., porosity and permeability) of a carbonate is affected by exposure is a point of contention. There are many who argue that exposure leads to a significant enhancement of reservoir quality, whereas others believe just the opposite and argue that the net effect is negligible.

Seismic data from the study area in the Pearl River Mouth Basin, offshore the People's Republic of China (Figure 1), reveal that Miocene Zhujiang Formation carbonates within this basin were subjected to periods of regionally extensive subaerial exposure. Although a pulse of brief and localized exposure was already known from core data in Liuhua field (Turner and Hu, 1991), regionally extensive events had not been previously documented in this part of the basin. The relevance of recognizing exposure events in the Pearl River Mouth Basin cannot be overstated. From a sedimentological perspective, their recognition has added to the understanding of regional processes controlling carbonate platform development in the basin. Moreover, this knowledge has helped put diagenetic processes into their proper context and has aided in the understanding and prediction of reservoir-scale heterogeneities in the Liuhua 11-1 field. But also, from a more global and academic perspective, recognition of these exposure events has provided additional fuel for the ongoing debate concerning the role of unconformities in the porosity evolution of any given carbonate succession.

Seismic data from Amoco Orient Petroleum Company's acreage in Contract Area 29/04 in the Pearl River Mouth Basin form the basis of this study. Because of their superb quality, these seismic data are ideally suited for stratigraphic and structural interpretation. Depositional seismic facies are easily interpreted and provide a clear window into the internal geometry of the platform. The seismic data are unique in that they also provide a window into early diagenetic processes. In select portions of Contract Area 29/04, the clear seismic response, which is so helpful for the identification of depositional seismic facies, changes to highly irregular and chaotic reflectors. This chaotic seismic response generally occurs only within the Zhujiang Formation carbonate interval and is not an artifact of processing. Although depositional facies cannot be inferred in areas of seismic chaos, these areas are equally important for understanding the processes that control platform development. Chaotic seismic facies occur in association with large concave-shaped reflectors, truncated reflectors, and with what are interpreted to be zones of collapse and scalloped or hummocky carbonate surfaces. It is thought that these features are a result of alteration of the carbonate interval during subaerial exposure.

It is the purpose of this paper to present the results of a seismic stratigraphic study of the Miocene Zhujiang Formation carbonates in Contract Area 29/04 and to put forth a sedimentological model for their evolution. Also introduced is the concept of chaotic or diagenetic seismic facies, which in the case of Zhujiang Formation carbonates, has been instrumental for piecing together their depositional and diagenetic history. Although our model relies heavily on regional seismic data, it should be noted that details from our petrophysical and geoscientific evaluation of the Liuhua 11-1 field are integrated where necessary.

The proposed model for Zhujiang Formation carbonate platform evolution involves two distinct phases of platform development separated by a major episode of exposure and erosion. This major exposure event is thought to be eustatically controlled and is likely responsible for the demise of the first, or "lower," phase of carbonate development. A notable reorganization of platform geometry occurred after exposure of the lower platform and is thought to be a consequence of rapid drowning following exposure. Subsequent carbonate development during the later, or "upper," phase is punctuated by numerous backstepping events, suggestive of a system striving to keep up with continuously rising sea level. Sedimentological evidence in cores, too fine to be resolved by seismic, argues against such a continuous drowning model and suggests instead that sea level rise was interrupted on numerous occasions by times of subaerial exposure. However, these events appear to have been caused by local tectonics, had only limited regional effect, and did not shut down carbonate production.

Importantly, the integration of seismic and lithologic data suggests that local tectonic effects were as important as eustatically controlled sea level fluctuations in

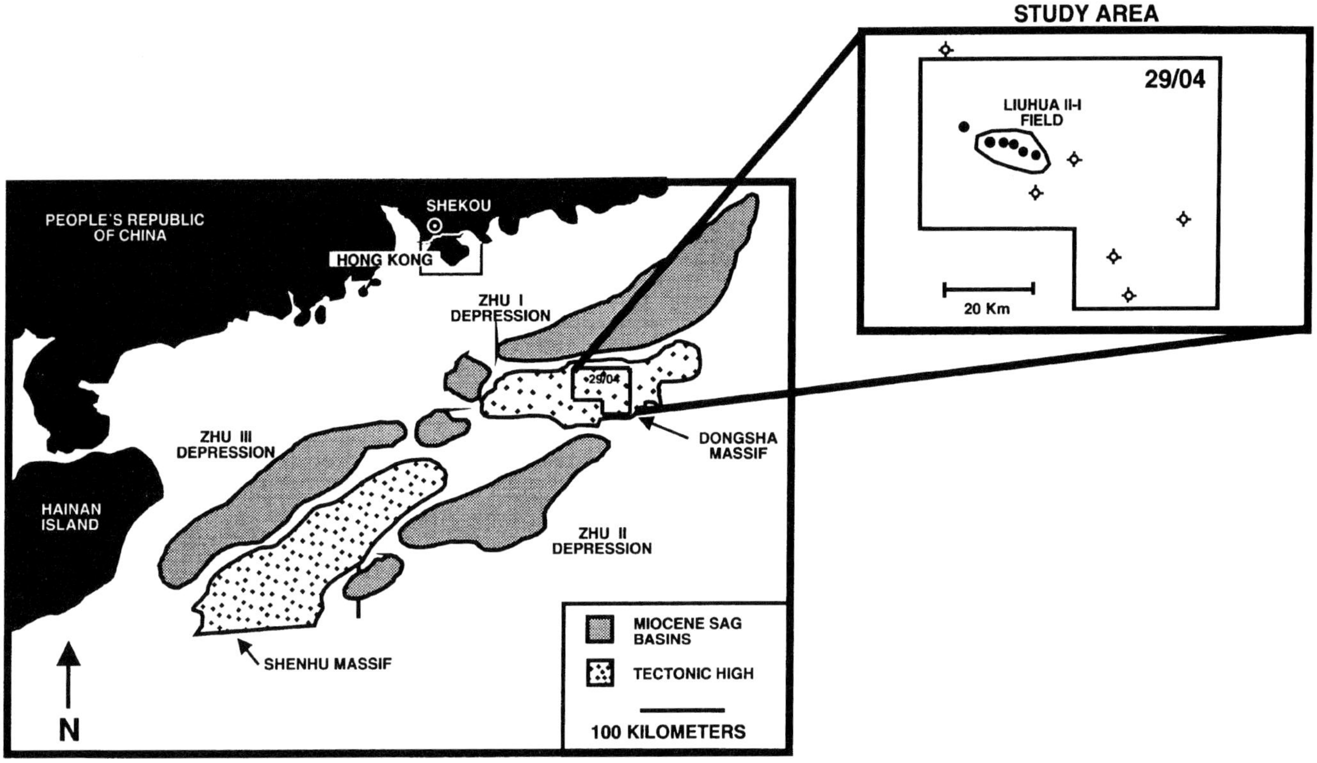

Figure 1. Map of the study area showing location of Contract Area 29/04 and various tectonic elements of the Pearl River Mouth Basin, South China Sea. Carbonate platforms are only developed atop the Dongsha and Shenhu massifs. Thick accumulations of Paleogene lacustrine sediments underlying Miocene sag basins (Zhu I, Zhu II, and Zhu III depressions) are thought to be a major source of hydrocarbons (Guo, 1989; Tyrrell and Christian, 1992). Inset shows the location of Liuhua 11-1 field and other wells used in this study.

governing carbonate platform development. The integrated data set also indicates that there was virtually no net enhancement of reservoir quality of Zhujiang Formation carbonates as a result of subaerial exposure.

REGIONAL GEOLOGY

The Pearl River Mouth Basin is a block-faulted and subsident linear basin of the passive-rifted margin type (Taylor and Hayes, 1980; Roberts, 1988; Ru and Pigott, 1988). This basin, along with the East China Sea Basin, make up the rifted Atlantic-type continental shelf of China. A linear uplift trend, consisting of faulted Mesozoic granites of the Dongsha and Shenhu massifs, extends from Hainan Island to Taiwan and forms the backbone of the basin. Thick accumulations of Paleogene lacustrine sediments filled grabens and half grabens surrounding these massifs. These Paleogene sediments are thought to be the dominant source of hydrocarbons in the Pearl River Mouth Basin (Guo, 1989; Turner and Hu, 1991; Tyrrell and Christian, 1992). Carbonate platforms of early Miocene age are commonly developed on top of the Dongsha and Shenhu massifs but are absent in other areas of the basin. The massifs and associated sag basins are shown in Figure 1. The study area, including the boundaries of Contract Area 29/04 and Liuhua 11-1 field, is also shown in this figure.

Stratigraphy

The general stratigraphy for Contract Area 29/04 is shown in Figure 2. Throughout most of the contract area, basement consists of Mesozoic igneous rocks (Roberts, 1988). In areas away from the massifs, presence of a middle Oligocene regional unconformity marks the end of rifting (Roberts, 1988). At this time, rapid sedimentation and thermal contraction resulted in a change in the depositional environment, as evidenced by the onset of marine conditions in sediments overlying the granite massifs. At the base of the transgressive cycle are the deltaic and paralic sandstones and shales of the late Oligocene Zhuhai Formation.

Marine carbonates and shales of the lower Miocene Zhujiang Formation directly overlie the Zhuhai Formation. Zhujiang Formation carbonates form the primary reservoir unit in the study area and attain a maximum thickness of 560 m in Contract Area 29/04. Two distinct phases of carbonate platform development are distinguished on seismic data and are referred to in this report as the lower Zhujiang carbonate and the upper Zhujiang carbonate. Regional trends in seismic facies suggest that the lower Zhujiang platform was an areally extensive, low-relief, ramplike feature (see for example, Read, 1985). In contrast, the upper Zhujiang platform was a much smaller and relatively narrow, high-relief feature. Figure 3 shows the relative position and geographic extent of the lower

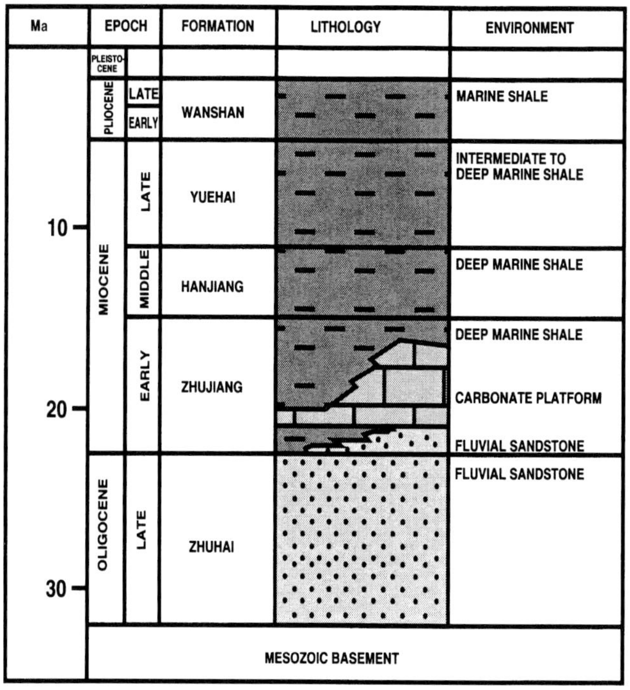

Figure 2. Post-rift stratigraphic associations for sediments overlying the Dongsha Massif in Contract Area 29/04.

and upper Zhujiang platforms. For reference, the approximate position of the early Miocene shoreline and the outer limit of lower Miocene sands are also shown. Prevailing winds are thought to have come from the southwest (Erlich et al., 1990).

Overlying the Zhujiang Formation carbonates are marine shales and siltstones of the Miocene Hanjiang and Yuehai formations and the Pliocene Wanshan Formation. These shales provide the primary seal for Zhujiang Formation reservoirs in the Pearl River Mouth Basin.

SEISMIC DATABASE AND REPROCESSING

During 1986 and 1987, Amoco Orient Petroleum Company acquired a large, regional, seismic database in the South China Sea. Although these data were collected in several phases, they were identically processed by Digicon Geophysical Corporation to zero phase using a far field signature recording. With an upper frequency limit of approximately 80 Hz, these data are of excellent quality. These sections formed the foundation of the regional seismic stratigraphic study.

In order to do a detailed evaluation of the Zhujiang Formation reservoir, a subset of the regional survey over the carbonate buildup in the Liuhua 11-1 field area was selected for reprocessing. Resolution of the original data was improved by expanding the frequency content to a high of 150 Hz, far exceeding the normal range of seismic resolution. This was achieved through traditional whitening steps supplemented by L-1 norm deconvolution. Additional reprocessing steps were applied to address the problems of multiples and short-period reverberations, while residual statics and residual NMO steps were added just prior to stack to prevent deterioration in the stack. The resulting resolution of the reprocessed seismic data is approximately 6 m.

SEISMIC FACIES

A northeast–southwest-trending seismic line through the central part of the study area is displayed

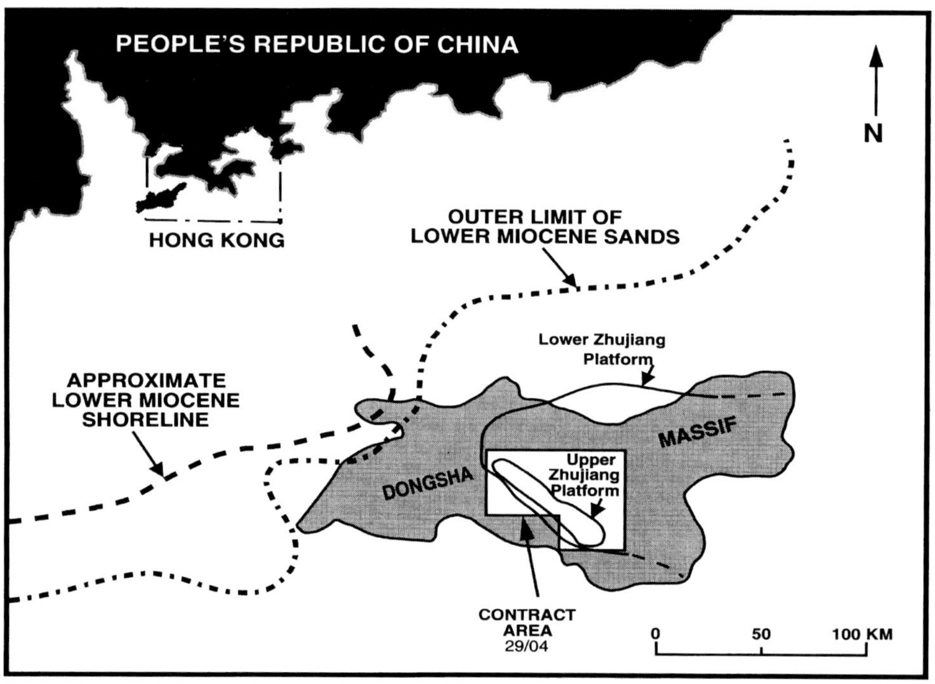

Figure 3. Position and geographic extent of lower and upper Zhujiang carbonate platforms relative to the Dongsha Massif. Approximate positions of lower Miocene shoreline and outer limit of Miocene sand deposition are also shown. Outline of Contract Area 29/04 is provided for reference.

in Figure 4. For reference, a map showing the location of this and all other seismic lines used in this report is provided in Figure 5. On seismic lines, the top of the Zhujiang Formation carbonate and the platform margin are easily identified because of the strong reflector at the shale/carbonate interface. Other depositional features relevant to the carbonate system, such as foreslope carbonate fans and backstepped margins, are also evident in the data and are indicated on Figure 4. Within the study area, front margins of both lower and upper Zhujiang platforms have a northwest-to-southeast orientation.

Depositional Facies

Four depositional seismic facies patterns characteristic of carbonate terrains are evident on seismic lines from Contract Area 29/04 and are referred to here as mounded, prograding, parallel-continuous, and parallel-discontinuous facies. The distribution of these seismic facies in lower and upper Zhujiang carbonate platforms is shown in Figures 6 and 7, respectively. It should be noted that both platforms, and especially the upper one, were continuously evolving in response to changing environmental conditions. Facies did not remain fixed in time, so the distribution shown in Figures 6 and 7 represents the dominant facies within the interval. Figure 7 shows the facies distribution and geometry of the upper Zhujiang carbonate at its maximum geographic extent during earliest stages of upper platform development.

Mounded Seismic Facies

Mounded facies are characterized by divergent downlapping reflectors and likely represent coral and coralline algae topographic buildups (Mitchum et al., 1977; Fontaine et al., 1987). In a regional sense, these facies are relatively rare in both lower and upper Zhujiang carbonate platforms (Figures 6 and 7, respectively), but the presence of coralgal boundstones in cores cut through what is thought to be a latest-stage high-energy reef rim suggests that these facies likely became more common in final stages of upper platform growth (Erlich et al., 1990).

Prograding Seismic Facies

Prograding facies are recognized by their tangential or shingled downlapping reflectors (Mitchum et al., 1977). These facies are interpreted to represent prograding bioclastic carbonate sands. Within the study

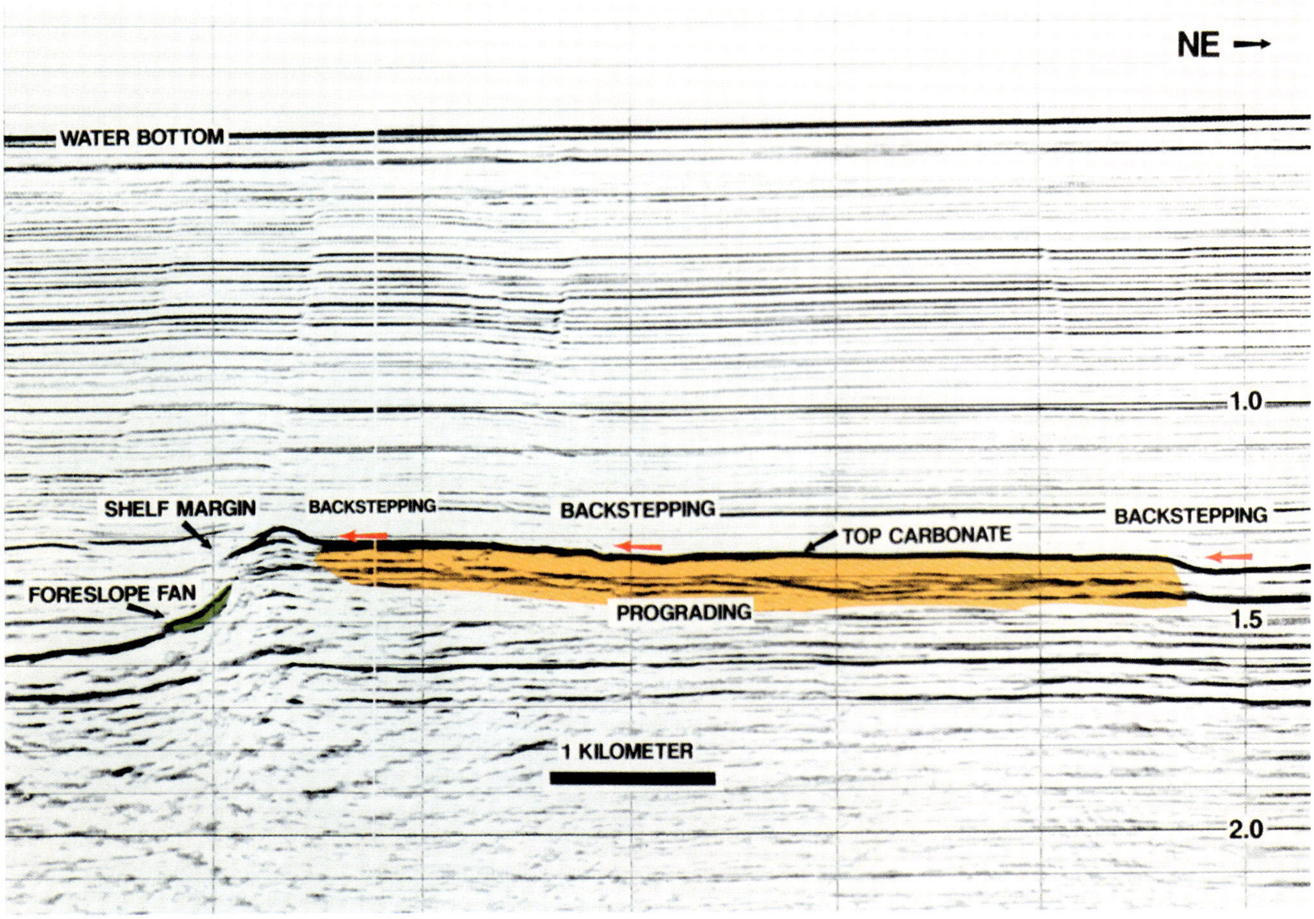

Figure 4. Northeast–southwest-trending seismic line through the study area. Presence of a strong reflector at the shale/carbonate interface facilitates the identification of sedimentological features such as the platform margin, foreslope fans, and backstepped margins. An interval characterized by prograding facies is highlighted in yellow. Red arrows point to stages of backstepping. See Figure 5 for location of seismic line.

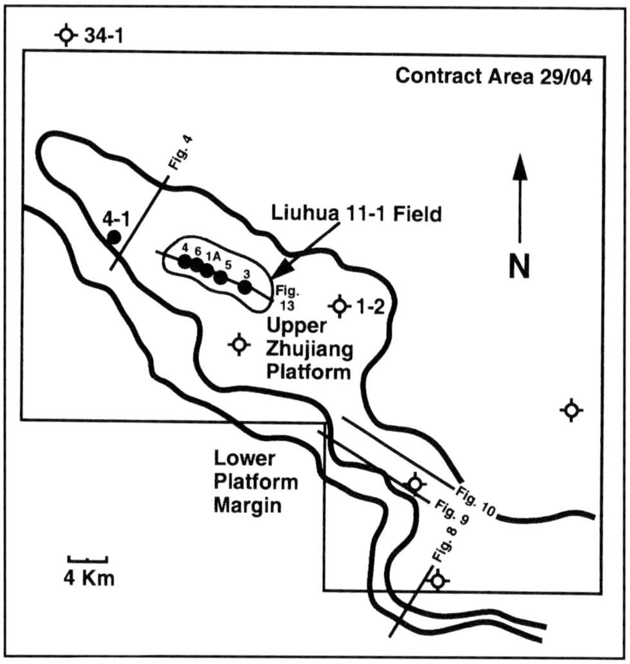

Figure 5. Map of Contract Area 29/04, South China Sea, indicating the approximate location of seismic lines shown in Figures 4, 8, 9, and 10 and the location of the cross section shown in Figure 13. The positions of lower and upper Zhujiang platform margins are also provided for reference.

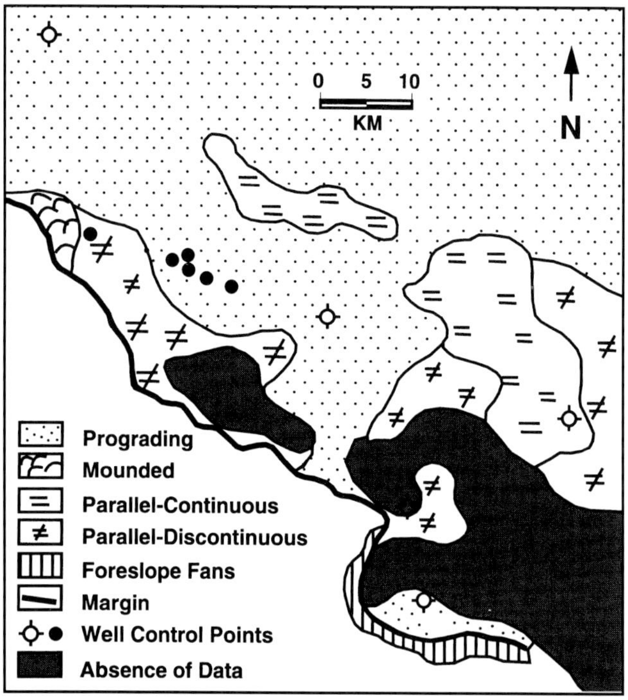

Figure 6. Regional distribution of depositional seismic facies in the lower Zhujiang carbonate platform. Note the predominance of prograding facies across most of the platform. Areas designated as having an "absence of data" refer to areas where the depositional seismic facies are masked by a diagenetic overprint. The cluster of five wells located in the center of the figure makes up the Liuhua 11-1 field.

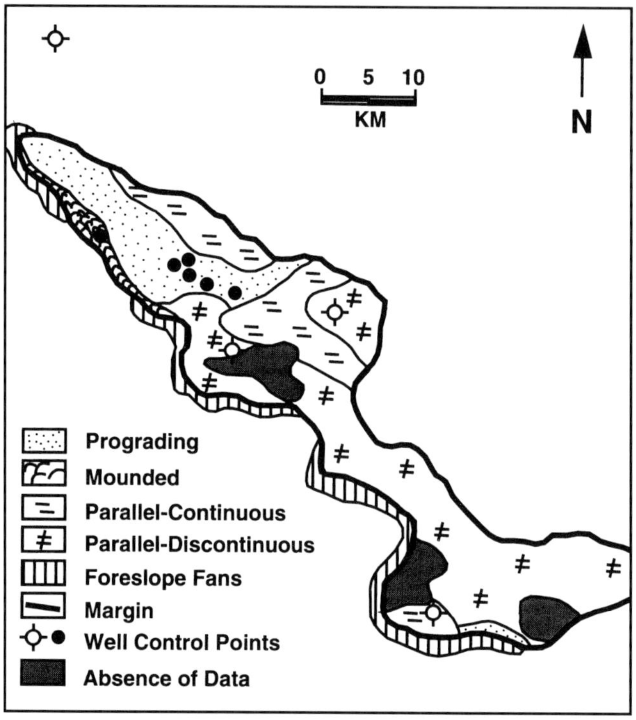

Figure 7. Regional distribution of depositional seismic facies in the upper Zhujiang carbonate platform. Parallel-discontinuous facies dominate in the southern half of the platform, where as prograding facies are most common in the northern half. As in Figure 6, areas designated with "absence of data" are those where a diagenetic overprint obscures the depositional seismic facies.

area, these facies mostly comprise very broadly sweeping and low-angle downlapping surfaces (Figure 4). Locally, in what appear to be areas of high-energy cross-bedded carbonate sands, the foresets exhibit somewhat higher angles. Large portions of both upper and lower carbonate platforms comprise prograding seismic facies (Figures 6 and 7).

Parallel-Continuous and Parallel-Discontinuous Facies

Parallel-continuous and parallel-discontinuous facies (Figures 8 and 9) are characterized by horizontal to subhorizontal reflectors of varying amplitude. These facies are thought to represent beds deposited under relatively uniform conditions in quieter parts of the inner shelf or in lagoonal settings (Mitchum et al., 1977; Fontaine et al., 1987). However, the widespread occurrence of these facies and the absence of lithological control make it difficult to offer a single unequivocal sedimentological interpretation. It seems likely that parallel-discontinuous facies represent a slightly lower energy environment than that which produces the higher amplitude parallel-continuous facies. One possibility is that these facies represent a continuum of facies within inter- to supratidal environments.

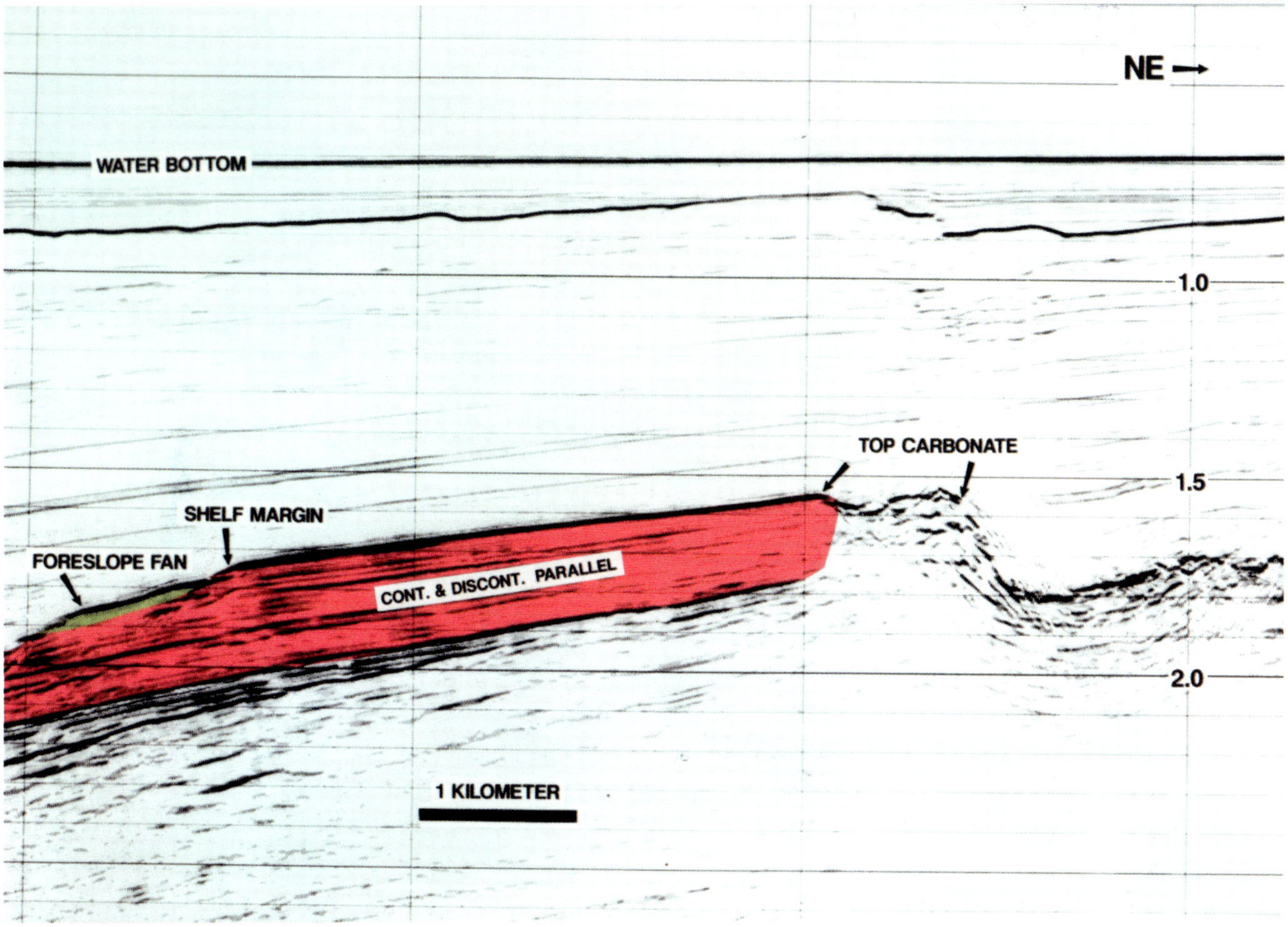

Figure 8. Northeast–southwest-trending seismic line through southeast portion of Contract Area 29/04. Interval of parallel-continuous and parallel-discontinuous seismic facies is highlighted in red. The shelf margin and foreslope fan are also indicated.

Diagenetic Seismic Facies

The highly regular character of seismic data evident over most of the study area changes to an irregular or "chaotic" character in certain parts of Contract Area 29/04. In these areas, sedimentological interpretations become very tenuous because the original depositional signature is obscured by a chaotic seismic overprint. All seismic data in Contract Area 29/04 were acquired and processed equally, so we are certain that this chaotic seismic character is *not* an artifact of processing, and that it has geological significance. Concave or dish-shaped reflectors, similar to those reported by Popenoe et al. (1984) in Eocene and Oligocene strata of Florida, are common in areas of seismic chaos. The arcuate geometry of these reflectors indicates disruption of the carbonate surface and likely represents collapse features or sinkholes. The superposition of chaotic facies and collapse reflectors on what are considered to be original depositional facies suggests that the chaotic signature is a secondary phenomenon likely resulting from the diagenetic alteration of Zhujiang Formation carbonates. Hence, the introduction of the term "diagenetic seismic facies."

A continuum of diagenetic seismic facies is recognized throughout the study area. A few examples are provided in Figures 9 and 10. At one end of the spectrum are concave or dish-shaped reflectors that cut across adjacent seismic facies (Figure 9). At the opposite end of the spectrum are "eroded-chaotic facies," which are associated with an apparent disruption of the top carbonate surface (Figure 10). Between these two extremes are what we refer to as "chaotic seismic facies." These have the characteristic chaotic internal response, but do not involve a disruption of the upper carbonate surface (Figure 9). Maps showing the regional distribution of diagenetic seismic facies in lower and upper Zhujiang carbonates are provided in Figures 11 and 12, respectively.

Concave Collapse Features

Dish-shaped reflectors are present in both upper and lower Zhujiang platform carbonates. They commonly occur as isolated dips in the carbonate surface, but a few also develop along a linear trend, thus giving the appearance of a channel. Importantly, they are never present in the latest stage of upper carbonate development, consistent with the drowning model of

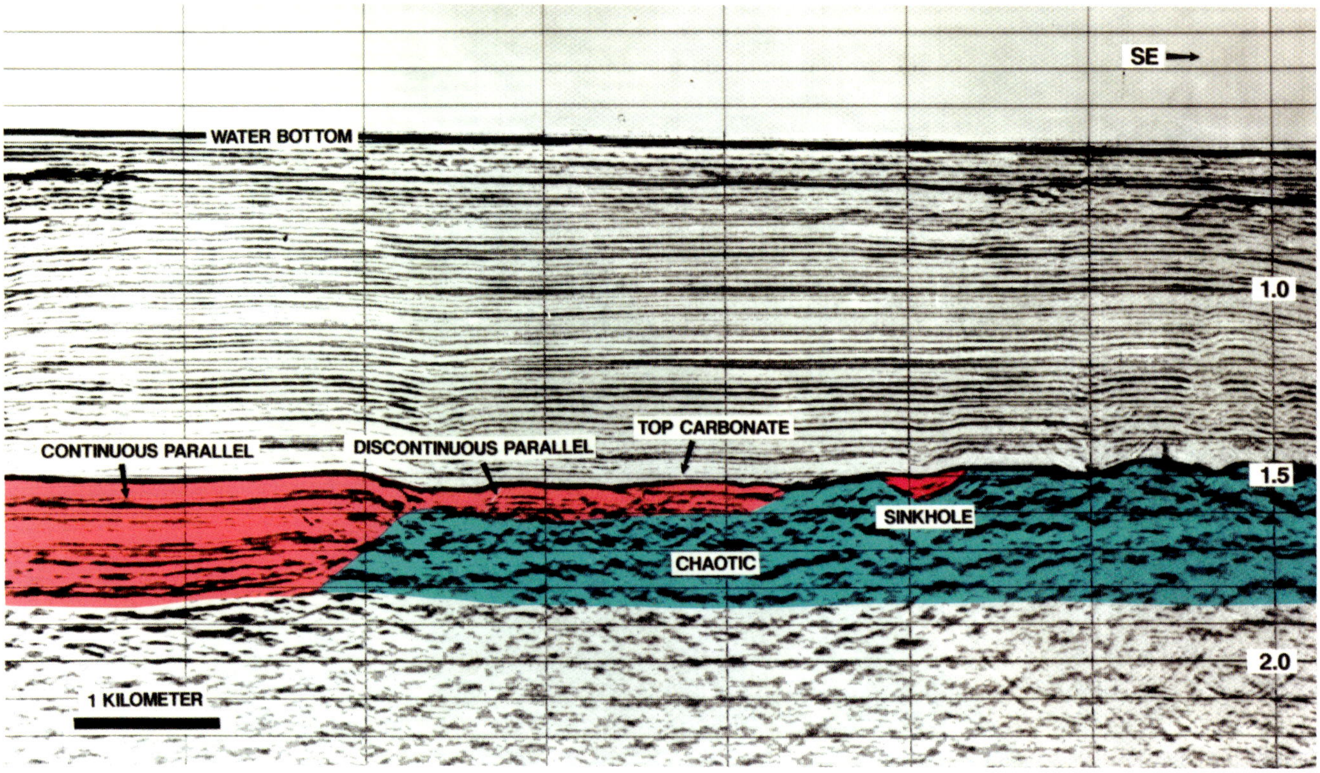

Figure 9. Northwest–southeast-trending seismic line from the southern part of Contract Area 29/04 contrasting depositional and diagenetic seismic facies. Parallel discontinuous facies (red) on the left are obscured by the diagenetic seismic facies (blue). Concave-shaped reflectors are thought to represent sinkholes. Note the concave shape of reflectors in the overlying shale package conforming to the top carbonate surface.

Erlich et al. (1990). Dish-shaped reflectors are generally small (<0.25 km in width), although some may reach up to 0.50–0.75 km in width. Carbonate beds immediately below these concave features are often chaotic, but the concave geometry is also observed in areas void of "seismic chaos." In those cases where the top carbonate surface is affected, overlying beds (i.e., basinal shales) commonly show evidence of soft-sediment deformation or collapse, conforming to the concave shape (Figure 9). As mentioned, concave reflectors displaying similar geometry to those found on the Zhujiang platforms have also been observed on seismic surveys acquired over sinkhole terrains in northern Florida (Popenoe et al., 1984), hence the preference for the sinkhole analogy.

Chaotic Seismic Facies

Plain chaotic seismic facies consist of irregular, internal seismic reflectors and short discontinuous high-amplitude events (Figure 9). They may be but are not always associated with concave reflectors. Commonly, chaotic facies are present on the upthrown side of fault blocks, but they do not appear to be genetically linked to faults. Figures 11 and 12 show the regional distribution of chaotic seismic facies in lower and upper Zhujiang platforms, respectively. As with depositional seismic facies distribution, these maps represent a single instance in time and by no means capture the entire vertical succession of chaotic facies.

Eroded-Chaotic Facies

Severe chaotic seismic character is differentiated from plain chaotic facies by the occurrence of truncated reflectors and irregular to hummocky carbonate surfaces (Figure 10). On some seismic lines, the shales directly overlying zones of eroded chaotic facies appear to be heavily faulted. It is here that it becomes difficult to discern whether the overlying sequence is faulted as a consequence of the eroded and irregular carbonate surface or whether the carbonate surface only appears irregular because it is severely faulted. However, as shown in Figure 10, the overlying shale section does not always appear to be faulted in areas of eroded chaotic seismic facies. Examples of eroded chaotic seismic facies are known from areas of intense paleokarst development, such as Casablanca field in the western Mediterranean (e.g., Watson, 1982) and the Bresse basin, France (Fontaine et al., 1987). Therefore, a model involving subaerial exposure and erosion is favored.

A large part of the lower platform exhibits this type of extreme diagenetic seismic facies (Figure 11). In contrast, there is little evidence of eroded-chaotic facies within the upper Zhujiang platform (Figure 12). Only two small patches of this facies are recognized in the upper carbonate, and these occur in the southeastern part of the platform. It is postulated that these patches of eroded-chaotic facies represent a transition zone between an area of complete erosion and an area unaffected by subaerial processes.

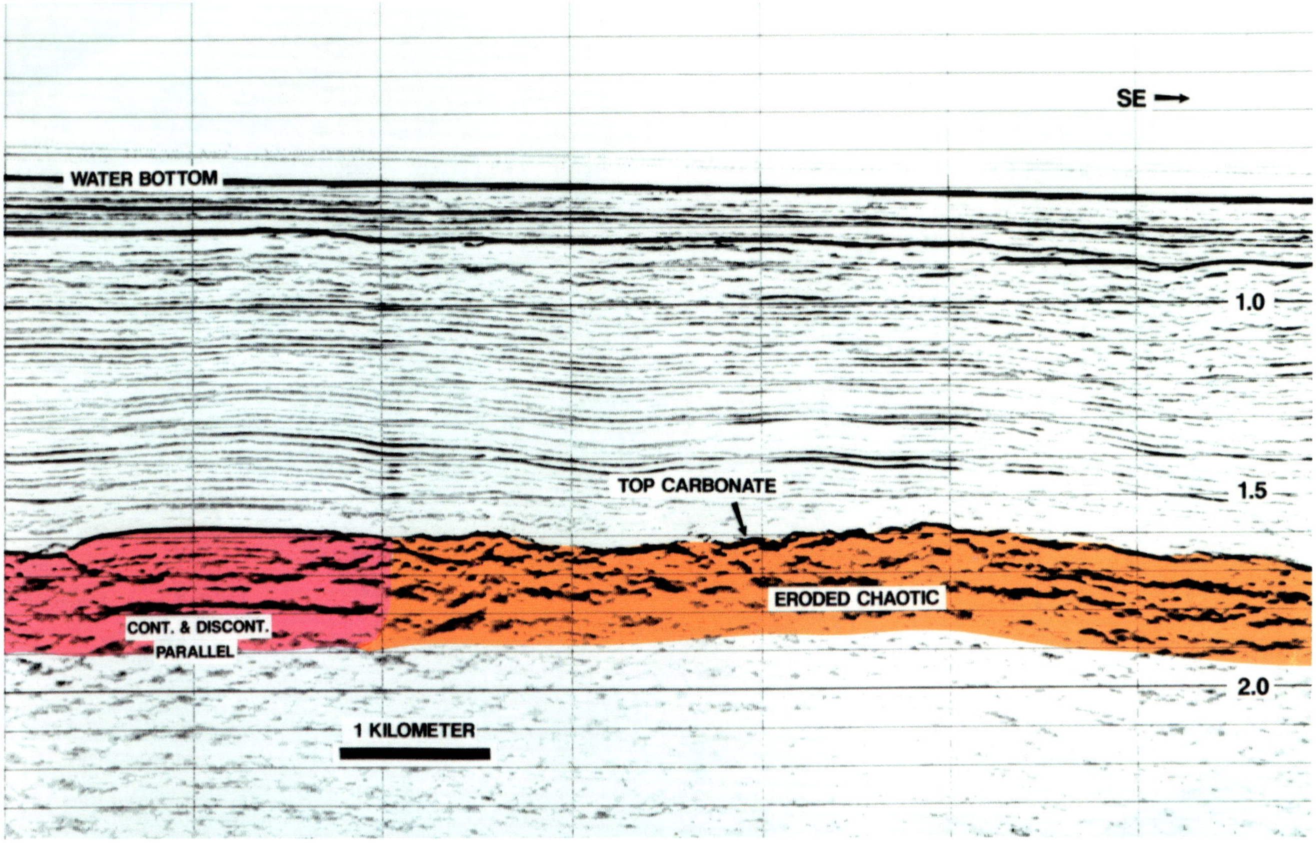

Figure 10. Northwest–southeast-trending seismic line through an area of eroded chaotic seismic facies (orange highlight). Contrast the smooth carbonate reflector on the left of figure (and also in Figure 9) with the irregular surface in the chaotic zone. The thickness of the carbonate interval appears to decrease in the right-hand side of the seismic line.

EVIDENCE FROM THE ROCK RECORD

Miocene carbonate buildups in other areas of Southeast Asia consist of "reefy" and bioclastic textures, very similar to the textures evident in Zhujiang Formation carbonates. They are typically vuggy and have abundant "chalky" microporosity, and although they may not have penetrated large caverns, some cavernous porosity is inferred from bit drops (May and Eyles, 1985; Rudolph and Lehmann, 1989). Regional sequence stratigraphic studies have shown that many Miocene carbonate buildups in Southeast Asia and the South China Sea have been subjected to multiple episodes of subaerial exposure during sea level lowstands (e.g., May and Eyles, 1985; Rudolph and Lehmann, 1989). Our current work suggests that the carbonate complex on the Dongsha Massif may have been similarly affected.

Contract Area 29/04

During 1991, Amoco Orient Petroleum Company drilled two wells in the southeastern part of the study area within areas of chaotic seismic character. These wells encountered thick intervals of vuggy carbonates having up to 25% porosity and abundant clay-filled fractures. While drilling through the Zhujiang Formation, complete loss of circulation and bit drops of 1–2 m were common occurrences. Core recovery was relatively poor because much of the carbonate consisted of loose rubble. Karstification of carbonate terrains often results in brecciated textures (e.g., James and Choquette, 1987). In light of these drilling results, and given the similarity in seismic response in known karsted reservoirs, the preferred model for the Zhujiang carbonate platform is one involving karstification.

Liuhua 11-1 Field

Liuhua 11-1 field is located approximately 220 km offshore the People's Republic of China (Figure 1) and is the largest oil-bearing accumulation found to date in the South China Sea (e.g., Tyrrell and Christian, 1992). The present-day Liuhua 11-1 structure consists of a northwest–southeast-trending elongate anticline bounded by faults on its northern and southern flanks. Results of our current seismic stratigraphic analysis suggest that this "structure" is, in fact, a carbonate bank or reef that developed on a subtle basement high within the interior of the upper Zhujiang platform. A thicker carbonate package characterized by prograding and mounded seismic reflectors in the area of a basement high is clearly evident on seismic and forms the basis of this interpretation. Faults bounding the

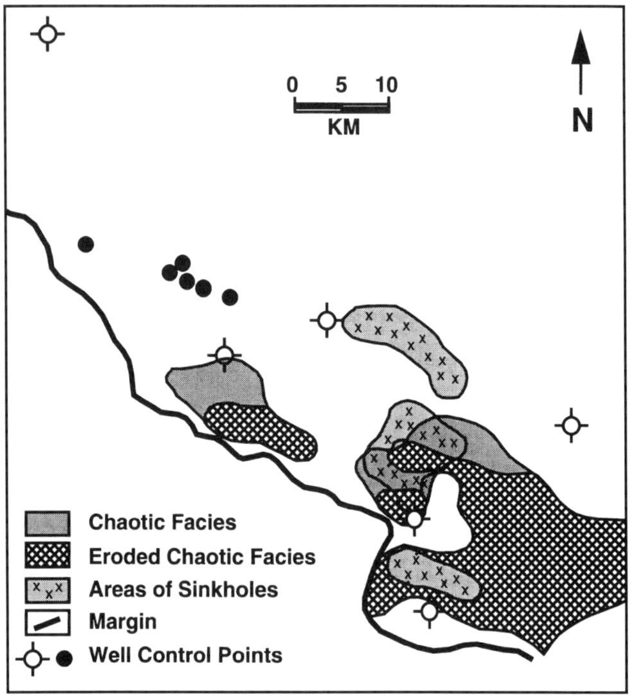

Figure 11. Regional distribution of diagenetic seismic facies in the lower Zhujiang carbonate platform. Eroded chaotic facies are best developed in the southeastern part of the platform. The Liuhua 11-1 field (cluster of five wells) is located north of zones of chaotic seismic reflections.

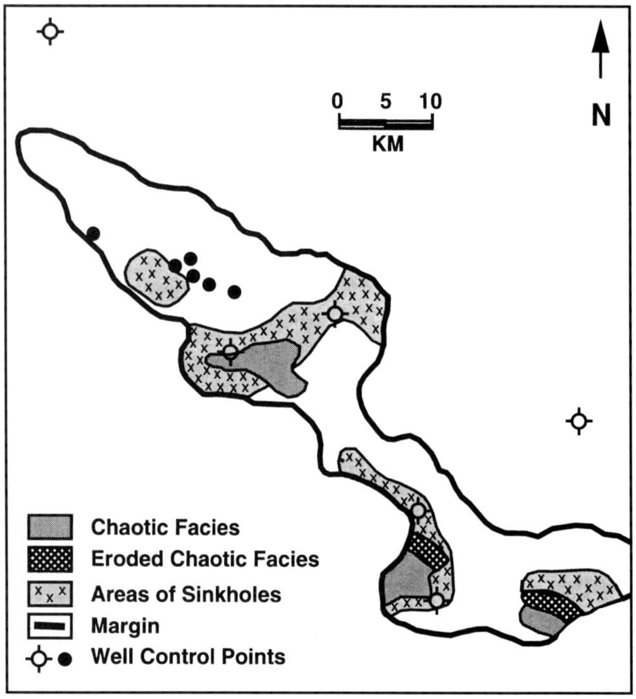

Figure 12. Regional distribution of diagenetic seismic facies in the upper Zhujiang carbonate platform. Sinkholes are the most common type of diagenetic seismic facies in the upper carbonate and these occur in several locations, including in the vicinity of Liuhua 11-1 field (cluster of five wells). Two small patches of eroded-chaotic seismic facies are observed in the southern part of the platform. Note absence of diagenetic seismic facies in the northern nose of the platform.

uplifted basement block die out in the lower Miocene but were apparently active long enough to exert a control on carbonate deposition and early diagenesis. Carbonate development over the basement high continued until the late Miocene, when rising sea level drowned the platform and carbonate production ceased. The shallower faults associated with the carbonate buildup represent a later phase of structural deformation than the basement faulting and are thought to result from differential compaction of sediments overlying the buildup.

To date, five wells have been drilled in the Liuhua 11-1 field: LH 11-1-1A, LH 11-1-3, LH 11-1-4, LH 11-1-5, and LH 11-1-6 (Figure 5). Examination of cores from three of these wells (LH 11-1-1A, LH 11-1-3, and LH 11-1-4) has revealed that the reservoir interval is a white, friable and vuggy limestone consisting of medium- to high-energy packstones, grainstones, and boundstones. Coralline red algae, foraminifera, and Codiacean algae are the dominant components. Within the pay interval, the carbonates have porosity up to 36% and permeability up to 7 d (Tyrrell and Christian, 1992). Although leached pore space is abundant, some of the porosity is considered to be of primary origin. Several cemented intervals as well as highly compacted intervals, both having 10% or less porosity, are interspersed throughout the otherwise highly porous section. The cemented intervals commonly cut across facies boundaries.

It should be noted that the Liuhua 11-1 reservoir comprises only the upper 60 to 100 m of the Zhujiang Formation and that it is, therefore, wholly contained within the upper carbonate platform system. Any conclusions that are drawn from the reservoir data apply only to the upper carbonate platform. It should also be noted that seismic resolution over the field is approximately 6 m, so many of the lithologic features observed in cores and inferred from wireline logs are hardly resolvable on seismic. Still, despite the very different scales, the well data provide critical support for the current sedimentological model.

Preliminary petrophysical evaluations of wireline logs and cores from the Liuhua 11-1 field depict a reservoir comprised of alternating horizontal layers of porous and tight carbonate, with layers truncating at the edges of the structure (Turner and Hu, 1991; Paul Wagner, personal communication). This layering was originally thought to be of diagenetic origin. Current work has revealed that the reservoir is, in fact, layered, but not in the horizontal layer-cake fashion described by earlier workers. Layers are interpreted from seismic reflectors on the high-resolution reprocessed data set, but these are not the same as, nor should they be confused with, the diagenetic layers described by Turner and Hu (1991). Our interpretation of seismic data within the oil-bearing reservoir interval suggests that the layers correspond to discrete lithological packages. Each layer marks a depositional event, and each has a distinct paragenesis, depending on the mineralogy of primary components and on the diagenetic domains that existed within the buildup at any given time. A schematic representation of the reservoir layering prepared directly from seismic data is provided in Figure 13. Layers A, C, and E are characteristically tight.

The Diagenetic Record

Petrographic and geochemical data from Liuhua field cores indicate that the reservoir potential of Zhujiang carbonates has been modified at different times by the following processes: (1) meteoric leaching of metastable grains; (2) precipitation of calcite cements; (3) leaching by deeply circulating cold marine waters; (4) solution-compaction; (5) mechanical compaction; and (6) compaction-water leaching (Turner and Hu, 1991; Paul Wagner, 1992, personal communication).

Core data suggest that earliest phases of diagenesis occurred primarily within the marine environment with occasional episodes of subaerial exposure. Subsequent to these episodes of marine and meteoric diagenesis, upper Zhujiang Formation carbonates underwent a continuum of burial diagenesis. Shallow burial diagenesis, extending to approximately 600 m of burial, was characterized by dissolution of metastable grains, precipitation of rim cements, and by incipient chemical compaction. The deeper burial phase, commencing at about 600 m and continuing to present-day depths of 1200–1500 m, involved processes such as dissolution by compaction waters (whose source was the dewatering of Tertiary shales), and significant mechanical and chemical compaction. Hydrocarbon emplacement is thought to have occurred during this deeper burial phase (Turner and Hu, 1991; Paul Wagner, 1992, personal communication).

Because of their relevance to the regional exposure model, only the effects of earliest marine and meteoric diagenesis are discussed below.

Early Diagenesis

Earliest diagenesis was initiated within the marine depositional environment but scant shallow submarine cements suggest that marine diagenesis was of little consequence to the overall reservoir quality of Zhujiang Formation carbonates. However, the occasional interruption of marine deposition brought about by periodic episodes of subaerial exposure and meteoric diagenesis *was* important to the porosity evolution of Zhujiang carbonates. These periods of exposure and meteoric diagenesis have been inferred from seismic data and are supported on a finer scale by the occurrence of meniscus cements (Figure 14), matrix neomorphism, early moldic porosity (Figure 14), and a light carbon isotope composition of early calcite cements. A cross section of carbon isotope trends with depth across Liuhua field (Figure 15) shows the occurrence of light isotopic peaks at numerous stratigraphic intervals.

Two diagenetic regimes having opposite effects on porosity were established during periods of exposure. In the vadose zone, between the exposure surface and the water table, leaching of metastable grains led to the development of some moldic porosity, a process which somewhat enhanced the early reservoir potential of the section. But this enhancement appears minor in light of the destructive processes that occurred below the water table. In the meteorically stabilized phreatic lens, where cementation actively plugged inter- and intragranular pores, the reservoir potential (porosity and permeability) of the sediments was greatly reduced (Turner and Hu, 1991; Wagner, 1993). This is evidenced by the presence of calcite cements having light carbon isotope composition (Figures 14 and 15). Within the Zhujiang Formation carbonate,

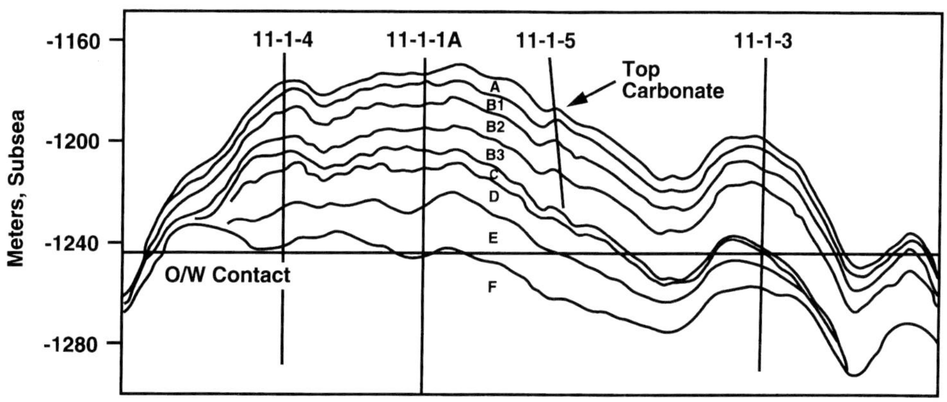

Figure 13. Liuhua 11-1 multisystem reservoir model derived from high-resolution seismic data. Each interval is characterized on seismic by distinct and mappable reflectors representing depositional surfaces. Pinch-out of reflectors at the edges of the Liuhua 11-1 structure suggests that the Liuhua "structure" is in reality a carbonate bank that accumulated on subtle basement topography.

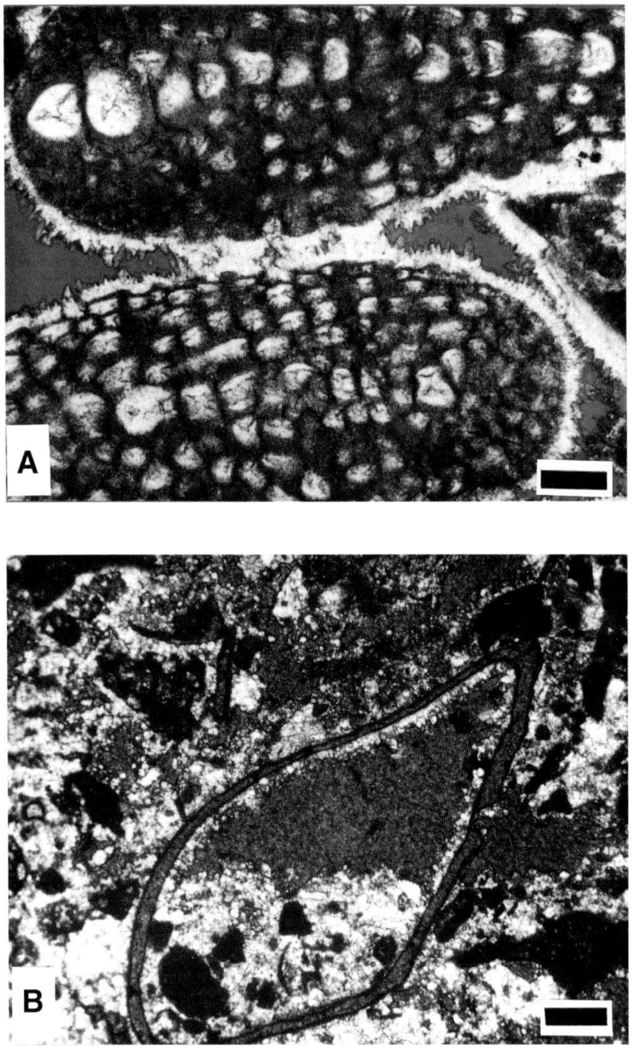

Figure 14. (A) Photomicrograph of meniscus cements found in algal grainstones of the Zhujiang Formation. Scale bar represents 500 μ. Sample is from the Liuhua 11-1-4 core at 1292.5 m. (B) Moldic dissolution cavity partially filled with loose, internal marine sediment and early equant calcite. Scale bar is 500 μ. Sample is from the Liuhua 11-1-3 core at 1250.3 m.

Layer C appears to have been most impacted by meteoric-phreatic cementation. It is the thickest and most continuous of the cemented layers (Figure 16). However, the repeated occurrence of vadose and meteoric-phreatic cement textures and isotopically light early cements suggests that these dichotomous meteoric-vadose and meteoric-phreatic regimes were established at multiple stages in the evolution of the upper Zhujiang platform.

The magnitude of each of the relative sea level drops and the duration of exposure events are relevant to the paragenesis of Zhujiang Formation carbonates, because these factors control both the area affected by leaching as well as the thickness and regional extent of the underlying cemented (tight) zone. Turner and Hu (1991) and Wagner (1993) invoke a single episode of meteoric diagenesis, resulting in a laterally continuous and horizontal tight layer of uniform thickness. Their model, however, did not have the benefit of high-resolution seismic and does not take into account the sedimentological configuration of the Liuhua bank interpreted from these data.

Integration of conventional and reprocessed seismic data with lithological data partially supports the original interpretation put forth by Turner and Hu (1991) and Wagner (1993), but enables a more complete sedimentological depiction of upper Zhujiang carbonate platform development. Based upon this integrated data set, rather than tying all the evidence to one exposure event, we favor a model involving numerous brief and geographically limited episodes of subaerial exposure. We would agree that of these inferred exposure surfaces and associated tight intervals, only Layer C spans the entire Liuhua bank, but we differ in the manner in which the layering is depicted. As shown in Figure 13, based upon the interpretation of high-resolution seismic data, Layer C is thickest at LH 11-1-4 and LH 11-1-1A, and thins toward the edges of the bank. Moreover, Layer C occurs at a greater depth at the edges of the field. Other cemented zones associated with light carbon isotope composition (Figure 16) are relatively thin and not laterally extensive, and represent what are thought to be sites of local and brief exposure. Importantly, light carbon isotope compositions and meniscus cements are absent in wells drilled outside of the bank area, east and southeast of the Liuhua 11-1 field area, thereby supporting the idea that during upper Zhujiang platform development, only small portions of the larger platform were exposed at any given time.

ZHUJIANG PLATFORM EVOLUTION

The lower Zhujiang carbonate accumulated on a vast platform characterized by a low-angle and relatively low-energy margin. No major disruptions in carbonate development can be discriminated on seismic within the lower carbonate interval. This, coupled with the succession of seismic facies and the tremendous lateral continuity of the feature, suggests that lower Zhujiang carbonates developed on a relatively stable platform that was later abruptly terminated.

The break between upper and lower platforms is sharp and is evidenced on seismic by a significant shift in the positions of the margins. Both the front and rear margins migrated toward the interior of the platform, thus resulting in a major downsizing of the platform and formation of a relatively narrow, northwest–southeast-trending feature.

The demise of the lower platform may have been influenced by a number of factors, including climate, ocean chemistry, tilting or local tectonic deformation, and rate of carbonate production (Schlager, 1992). We have concluded, based upon the distribution of chaotic seismic facies in the lower Zhujiang carbonate, that the lower platform experienced a major subaerial exposure event at the end of lower carbonate deposition, just prior to or coincident with its demise.

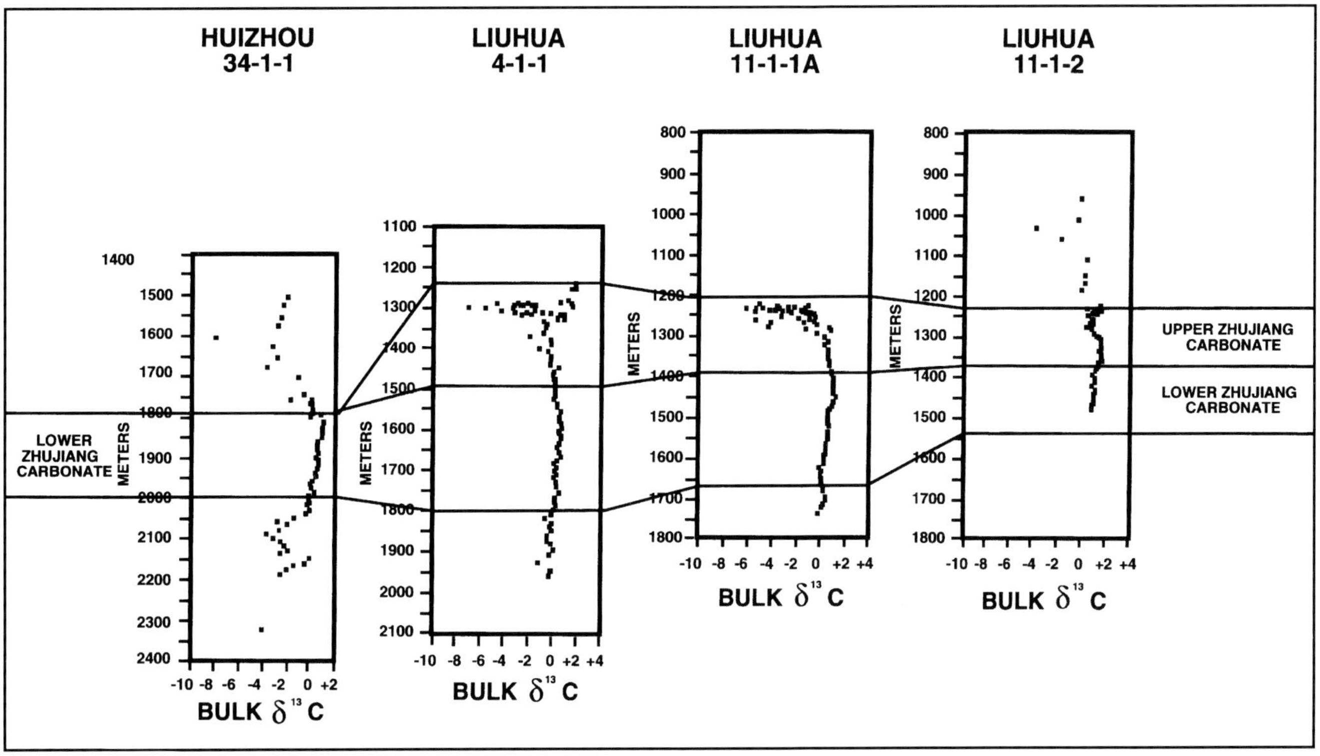

Figure 15. Vertical and lateral trends in carbon isotope composition of equant calcite cements in Zhujiang Formation carbonates. Cross section extends from the northwestern corner of Contract Area 29/04 (Huizhou 34-1-1), beyond the boundaries of the upper carbonate platform, and continues through the Liuhua field area. For reference, the location of wells and the distance between them is shown in Figure 5. Note that light carbon isotope composition is observed in only two wells (Liuhua 4-1-1 and Liuhua 11-1-1A).

Based upon the amount of erosion inferred from seismic data, we postulate a minimum sea level drop on the order of 100 m. Haq et al. (1988) have interpreted two drops in sea level of this magnitude at the beginning of the middle Miocene, one at 16.5 and the other at 15.5 Ma. Li (1984) also interprets a global fall in sea level in this approximate time frame. These events are consistent with the general age bracket determined for the Zhujiang carbonate (Aquitanian to Langhian). Unfortunately, there is no unequivocal method of dating events within the Zhujiang Formation carbonate, so it is difficult to constrain the timing of exposure. Nevertheless, it is reasonable to postulate that it coincided with one (or possibly both) of these global events.

Elements of the Quaternary carbonates of the Caribbean coast of the Yucatan Peninsula, Mexico, provide one possible modern analog for lower Zhujiang Formation subaerial diagenesis. Here, marine-phreatic, mixing-zone, and vadose diagenesis, affected by numerous fluctuations in sea level, have resulted in a rock texture that is similar to that of Zhujiang Formation carbonates. Of special interest to us is the coastal mixing zone where undersaturated meteoric waters mix in varying proportions with normal marine seawater. Circulation of the resultant brackish waters through the Yucatan carbonates is facilitated through fracture-controlled channels and caverns (Ward et al., 1985; Stoessell et al., 1989). Rocky and erosional inlets are very common along the coast and abundant sinkholes dot the interior of the Yucatan Peninsula. Dissolution of aragonite and calcite grains occurs in this mixing-zone environment, even at relatively high salinity ranges (i.e., with up to 90–95 % seawater component) (Stoessell et al., 1989) and produces a vuggy and chalky texture (Back et al., 1986), a texture that is not unlike that of Zhujiang Formation carbonates.

The upper Zhujiang carbonate platform evolved under somewhat different circumstances. Numerous backstepping events are evident throughout upper Zhujiang platform development, each resulting in a net reduction in platform size (Figure 17). This reduction in response to increasing sea level is consistent with the progressive drowning model presented by Erlich et al. (1990). Yet, despite the overall drowning trend, lithological data from Liuhua field wells suggest that the upper platform was also impacted by subaerial diagenesis. Relative to the lower platform, however, the upper platform lacks strong evidence of any large-scale destructive exposure event. The Schooner Cays complex (e.g., Wood Cay) west of Eleuthera Island, Bahamas, is our preferred modern analog for this phase of subaerial exposure. The inferred smaller-scale exposure events likely resulted from sudden, brief regressions caused by local tectonic adjustments of the Dongsha Massif, or, perhaps, they resulted from frequent eustatic drops occurring on the long-term rise. Local tectonic influence seems reasonable, given the

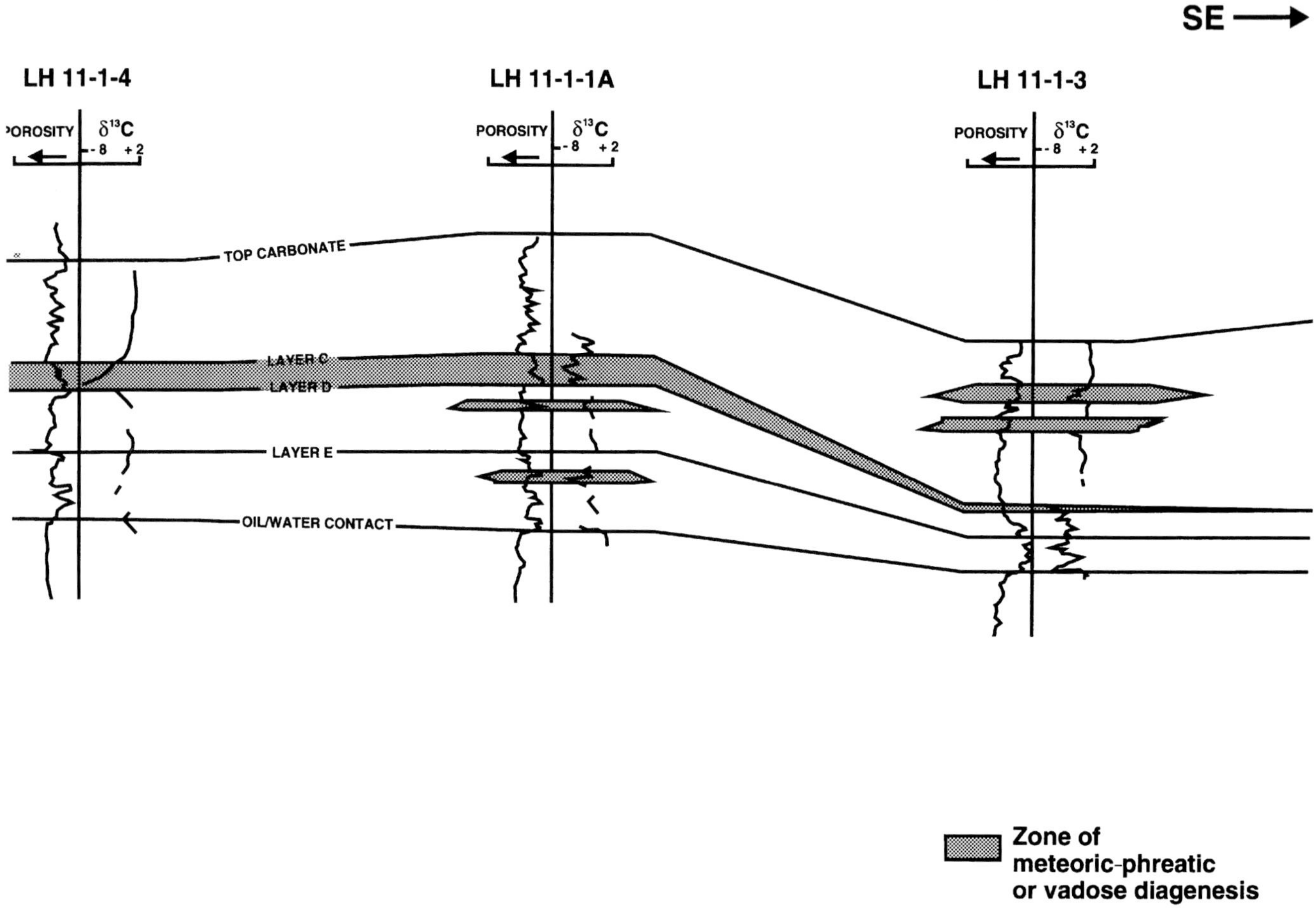

Figure 16. Schematic depiction of carbon isotope stratigraphy and porosity trends in upper Zhujiang Formation carbonates in three Liuhua field wells. Intervals characterized by lower porosity and negative carbon isotope signatures (marked in gray) are thought to result from meteoric-phreatic diagenesis. Repeated occurrence of meteoric-phreatic diagenetic effects supports a model involving multiple exposure events.

inferred structural and basement control on Liuhua Bank development, but it is certainly not a unique interpretation.

Implications for Reservoir Analysis

A broad spectrum of reservoir properties can be expected in karstified terranes. There may be zones of both porosity enhancement and porosity destruction, depending on the volume of fluids flushing through the system, the conduits available for flow, and the time available for geochemical reaction. Additionally, karstification involves both macro- as well as micro-processes, each impacting the reservoir differently. Therefore, different tools must be applied in order to quantify the various micro- and macro-effects.

With respect to Zhujiang Formation carbonates, the effects of subaerial diagenesis appear to have been variable, occurring on both a macroscopic as well as a microscopic scale. Regional karstification resulting from major sea level fluctuations appears to be characteristic of the lower carbonate, whereas more discrete subaerial exposure events are characteristic of the upper carbonate. Large-scale erosional carbonate surfaces and vuggy carbonates are common in the lower Zhujiang Formation. In this section, the seismic data cannot resolve smaller-scale heterogeneities. In contrast, erosional surfaces are lacking in upper Zhujiang carbonates, and, based upon seismic alone, one would be inclined to conclude that the upper system was one of continuous drowning. However, as inferred from core and wireline data, there were several episodes of exposure during upper Zhujiang carbonate platform development. Importantly, the data also suggest that these periods of exposure resulted in both porosity enhancement as well as porosity destruction.

CONCLUSIONS

The distribution of chaotic seismic facies in Zhujiang Formation carbonates supports a model of repeated episodes of subaerial exposure. These events likely occurred on both a large scale, as inferred from regional seismic interpretations, and on a local scale, as evidenced by the rock record at the Liuhua 11-1 field.

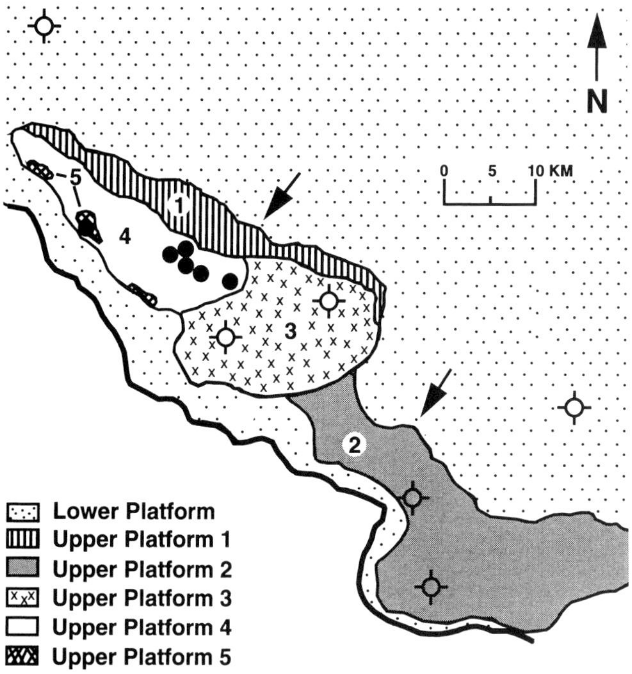

Figure 17. Progression of backstepping during the evolution of the lower and upper Zhujiang carbonate platforms. Black arrows indicate the direction of downsizing that occurred after the demise of the lower system. Numbers and corresponding patterns identify portions of the platform that were successively drowned at the end of any given stage. Upper Platform 1 includes the areas marked with numbers 1–5. Upper Platform 2 includes only the part of the platform marked with numbers 2–5. Upper Platform 3 includes areas marked 3–5; Upper Platform 4 includes only 4 and 5. Final late stage reef rims described by Erlich et al. (1990) are equivalent to stage 5 in this figure.

Before the relationship between primary and chaotic seismic facies had been resolved, little was understood about factors controlling the origin and evolution of the Zhujiang carbonate platform in the Pearl River Mouth Basin. The integration of regional seismic data with the lithologic details from wells in Liuhua field has allowed for a much more complete model, which in turn has been critical for understanding vertical and lateral heterogeneities evident within the Liuhua reservoir. From this integration we conclude:

1. Two major stages of carbonate growth are recognized on seismic data from Contract Area 29/04 in the Pearl River Mouth Basin. These include a lower Zhujiang platform stage and an upper Zhujiang platform stage. Eustasy as well as local tectonic factors influenced platform development.
2. The lower Zhujiang carbonate platform developed as a low-energy, ramp-style platform. The higher-energy, upper Zhujiang carbonate platform was smaller in regional extent. Multiple backstepping and platform downsizing events suggest that the upper Zhujiang carbonate developed in a continuously drowning system.
3. Chaotic seismic facies overprinted on primary seismic character are interpreted as evidence for karstification of Zhujiang Formation carbonates. Miocene carbonate buildups in Indonesia, Malaysia, and the Philippines show similar-staged developments and commonly exhibit paleokarst features. Neither the Miocene of the Pearl River Mouth Basin nor Miocene buildups across Southeast Asia show any major paleocavern development. Carbonates in the present-day Yucatan Peninsula, Mexico, have many commonalities with Zhujiang Formation carbonates.
4. Two scales of exposure events are inferred from seismic and lithologic data. At least one of these events is thought to have had tremendous regional impact and caused the termination of lower carbonate development. This event is inferred from distribution of chaotic and eroded chaotic seismic facies. Subsidiary exposure events of more restricted impact are recognized in cores from Liuhua 11-1 field and from the distribution of concave seismic reflectors in the upper Zhujiang carbonate. These events interrupt what is thought to have been a continuously drowning sequence. Absolute timing of exposure events is difficult to assess due to the lack of marker biota within the carbonate interval.
5. Porosity of the upper carbonate was significantly reduced as a result of meteoric-phreatic cementation during periods of exposure.
6. The layering evident on high-resolution reprocessed seismic data in Liuhua 11-1 field is consistent with the regional sedimentological model for upper Zhujiang carbonates. Diagenetic textures in cores are in accordance with the exposure model. Integration of data demonstrates that models based solely on seismic data miss the details provided by geochemical and petrographic data, whereas those based only on rock data miss critical aspects of regional geometry.

ACKNOWLEDGMENTS

Acknowledgment is made to Amoco Production Company and China Offshore Oil Nanhai East Corporation for permission to publish this report. Funds to offset AAPG publication costs were graciously provided by Amoco Orient Production Company. The authors are grateful to E. R. Shaw and E. K. Chau for their superb seismic processing, and to R. D. Boutell for his valuable assistance with the interpretation of the seismic stratigraphy. Thanks are also due to W. Schlager, N. Turner, R. Erlich, G. Bell, and all members of the China Projects Team at Amoco Production Company for their insightful feedback at various points throughout the study. Editorial comments by B. Leclerc and constructive critiques by P. M. Harris, D. Wiggins, and one anonymous reviewer enhanced earlier versions of the manuscript. Finally, Rick Wall and Eva Moldovanyi extend special thanks to R. W. Dudley for his incessant support and mentorship.

REFERENCES CITED

Back, W., B.B. Hanshaw, J.S. Herman, and J.N. Van Driel, 1986, Differential dissolution of a Pleistocene reef in the ground-water mixing zone of coastal Yucatan, Mexico: Geology, v. 14, p. 137–140.

Erlich, R.N., S.F. Barrett, and B.J. Gou, 1990, Seismic and geologic characteristics of drowning events on carbonate platforms: AAPG Bulletin, v. 74, p. 1523–1537.

Fontaine, J.M., R. Cussey, J. Lacaz, R. Lanaud, and L. Yapaudjian, 1987, Seismic interpretation of carbonate depositional environments: AAPG Bulletin, v. 71, p. 281–297.

Guo, Zhenxuan, 1989, Source rock characteristics in offshore Cenozoic basins of China: China Earth Science, v. 1, p. 15–20.

Haq, B.U., J. Hardenbol, and P.R. Vail, 1988, Mesozoic and Cenozoic chronostratigraphy and eustatic cycles, *in* C.K. Wilgus, B.S. Hastings, C.G. St. C. Kendall, H.W. Posamentier, C.A. Ross, and J.C. Van Wagoner, eds., Sea-level changes: an integrated approach: SEPM Special Publication 42, p. 71–108.

James, N.P., and P.W. Choquette, 1987, Paleokarst: New York, Springer Verlag, 416 p.

Li, Desheng, 1984, Geologic evolution of petroliferous basins on continental shelf of China: AAPG Bulletin, v. 68, p. 993–1003.

May, J.A., and D.R. Eyles, 1985, Well log and seismic character of Tertiary Terumbu carbonate, South China Sea, Indonesia: AAPG Bulletin, v. 69, p. 1339–1358.

Mitchum, R.M., P.R. Vail, and J.B. Sangree, 1977, Seismic stratigraphy and global changes of sea level, Part 6: Stratigraphic interpretation of seismic reflection patterns in depositional sequences, *in* C.E. Payton, ed., Seismic stratigraphy: applications to hydrocarbon exploration: AAPG Memoir 16, p. 117–133.

Popenoe, P., F.A. Kohout, and F.T. Manheim, 1984, Seismic-reflection studies of sinkholes and limestone dissolution features on the northeastern Florida shelf: *in* B.F. Beck, ed., Sinkholes: their geology, engineering and environmental impact: Proceedings of the First Multidisciplinary Conference on Sinkholes, Orlando, Florida, p. 43–57.

Read, J.F., 1985, Carbonate platform facies models: AAPG Bulletin, v. 69, p, 1–21.

Roberts, D.G., 1988, Basin evolution and hydrocarbon exploration in the South China Sea, *in* H.C. Wagner, L.C. Wagner, and F.L. Wong, eds., Petroleum resources of China and related subjects: Circum-Pacific Council for Energy and Mineral Resources Earth Science Series 10, p. 157–177.

Ru, K., and J.D. Pigott, 1988, Episodic rifting and subsidence in the South China Sea: AAPG Bulletin, v. 70, p. 1136–1155.

Rudolph, K.W., and P.J. Lehmann, 1989, Platform Evolution and Sequence Stratigraphy of the Natuna Platform, South China Sea, *in* P.D. Crevello, J.L. Wilson, J.F. Sarg, and J.F. Read, eds., Controls on carbonate platform and basin development: SEPM Special Publication 44, p. 353–361.

Schlager, W., 1992, Sedimentology and sequence stratigraphy of reefs and carbonate platforms: AAPG Continuing Education Course Note Series 34, 71 p.

Stoessell, R.K., W.C. Ward, B.H. Ford, and J.D. Schuffert, 1989, Water chemistry and $CaCO_3$ dissolution in the saline part of an open-flow mixing zone, coastal Yucatan Peninsula, Mexico: Geological Society of America Bulletin, v. 101, p. 159–169.

Taylor, B., and D.E. Hayes, 1980, The tectonic evolution of the South China Basin, *in* D.E. Hayes, ed., The tectonic and geologic evolution of Southeast Asian seas and islands: American Geophysical Union Monograph Series 23, p. 89–104.

Turner, N., and Pingzhong Hu, 1991, The lower Miocene Liuhua carbonate reservoir, Pearl River Mouth Basin, offshore People's Republic of China: 23d Annual Offshore Technology Conference Transactions, p. 113–123.

Tyrrell, W.W., and H.E. Christian, 1992, Exploration history of Liuhua 11-1 field, Pearl River Mouth Basin, China: AAPG Bulletin, v. 76, p. 1209–1223.

Wagner, P.D., 1993, Reservoir compartmentalization caused by calcite cementation below a paleo-exposure and an oil-water contact, Liuhua, Offshore China (abs.), *in* A. Saller, D. Budd, R. Mitchell, and P.M. Harris, eds., Unconformities and porosity development in carbonate strata: recognition, controls, and predictive strategies: AAPG Hedberg Research Conference, Vail, Colorado.

Ward, W.C., A.E. Weidie, and W. Back, 1985, Geology and hydrogeology of the Yucatan and Quaternary geology of northeastern Yucatan Peninsula: New Orleans Geological Society, New Orleans, 160 p.

Watson, H.J., 1982, Casablanca field, offshore Spain, a paleogeomorphic trap, *in* M.T. Halbouty, ed., The deliberate search for the subtle trap: AAPG Memoir 32, p. 237–250.

Chapter 7

Porosity Development and Diagenesis in the Orfento Supersequence and Its Bounding Unconformities (Upper Cretaceous, Montagna Della Maiella, Italy)

M. Mutti
Swiss Federal Institute of Technology
Zurich, Switzerland

ABSTRACT

This paper discusses the development and evolution of porosity associated with different subaerially exposed unconformities of different hierarchical significance in the Orfento Formation (Campanian–Maastrichtian), integrating depositional facies, duration of exposure, and paleoclimate. The unit documents an early aggradational and a later progradational stage and is composed of rudist debris ranging in size from silt to rudite, associated with megabreccias. Two major unconformities, subaerially exposed on the shelf, bound the unit. The prograding sediment bodies consist of shingled, coarse-grained prograding units which contain several minor-order unconformities.

The strata of the Orfento Formation are characterized by high depositional porosity (average 20–30%), which reflects distribution of depositional facies. Primary porosity is increased by moldic porosity (average 15–20%), selectively on aragonitic grains. Early meteoric diagenesis was responsible for aragonite dissolution and precipitation of calcite cements. Calcite cementation is scarce and has a very heterogeneous distribution. The occurrence of moldic porosity and calcite cements is maximum in the progradational units and was controlled by stratigraphy, as related to minor-order unconformities, and by facies distribution. Cathodoluminescence patterns and stable isotopes suggest that precipitation of calcite cement occurred in localized freshwater systems, associated with minor-order erosional unconformities within the progradational unit.

The lower and upper supersequence boundaries were both associated with prolonged subaerial exposure (ca. 5–6 and 8–10 m.y., respectively), but responded differently with respect to porosity formation and preservation. Beneath the lower supersequence boundary, fabric-selective aragonite dissolution and extensive meteoric calcite cementation decreased the porosity. Cathodoluminescence patterns and stable isotopic compositions of the cements indicate precipitation in a stable freshwater system. The unconformity is capped by a calcrete soil typical of semi-arid conditions. Beneath the upper supersequence boundary, silica cements precipitated in lenses with an irregular distribution and do not significantly reduce the porosity. Cement textures and distribution suggest that silica precipitated either in a vadose/phreatic meteoric environment associated with a silcrete soil, typical of semi-arid conditions, or in the mixing freshwater/seawater zone.

Although not directly responsible for generating porosity, the unconformities bounding the Orfento Formation are significant, as they act as breaks on the regional distribution of porosity.

INTRODUCTION

Modern carbonate sediments are characterized by high primary porosity, which commonly decreases soon after deposition because of compaction and precipitation of cements (Enos and Sawatsky, 1981). The relative role of shallow (e.g., Halley and Harris, 1979; Longmann, 1980) versus deep burial processes (Bathurst, 1975; Scholle and Halley, 1985) in the modification of primary depositional porosity has generated much discussion. Subaerially exposed unconformities bring fresh water in contact with marine sediments, causing extensive diagenetic modifications of the depositional textures, such as dissolution and cementation. Factors controlling dissolution and cementation during subaerial exposure can be intrinsic, such as mineralogy and facies, or extrinsic, such as hydrology, time, and climate (Saller et al., 1994). However, it is difficult to predict the relative importance of specific processes because of the complex interplay among the different factors.

The Orfento Formation (Campanian–Maastrichtian) in the south-central Apennines illustrates the role that unconformities of different hierarchical significance play on the preservation or destruction of primary porosity. Within the Orfento Formation porosity is high, but has a heterogeneous distribution. The Orfento Formation is composed of rudist debris ranging in size from silt to breccias and is bounded by two unconformities, which have a regional distribution. These unconformities are referred to as major, as contrasted with local, minor unconformities which occur within the supersequence. The purpose of this paper is: (1) to test with a diagenetic study whether the major surfaces recognized as supersequence boundaries with field criteria also correspond to the major subaerial exposure events, and (2) to discuss the role that unconformities of different orders in the Orfento Formation play on the destruction or preservation of porosity.

The Upper Cretaceous strata of the Orfento Formation is an excellent case study to document porosity development and evolution in shingled, coarse-grained prograding bodies that underwent several episodes of subaerial exposure. The spectacular outcrops, the good stratigraphic control, and the peculiar sedimentological facies association make this setting ideal for an integrated study of the effects on porosity formation and modification by depositional facies, duration of exposure, and paleoclimate.

REGIONAL SETTING

The Maiella platform margin crops out in the Southern Apennines, in a frontal anticline of detached thrust sheet, trending north-south, perpendicular to the platform margin (Figure 1). Typically, Apenninic Mesozoic platform margins have been the site of tectonic decoupling during orogeny and are rarely preserved. The special orientation of the Maiella margin with respect to the thrust direction allowed the preservation of a nearly undeformed platform-basin transect approximately 20 km in length. This transect documents the exposed northwestern margin of a much larger platform, the Apulia platform, which was situated along the southern margin of the Mesozoic Tethys ocean (Bernoulli, 1972; Bernoulli and Jenkyns, 1974). The evolution of the northeastern margin has been described by Bosellini et al. (1993). In the Early Cretaceous, the peri-Adriatic Tethyan platforms were situated in the equatorial belt, between 10 and 30°N, and migrated during the Late Cretaceous across 30° latitude toward the north (Scotese et al., 1989).

PLATFORM EVOLUTION

The general stratigraphic framework of the Maiella platform has been well established (Bally, 1954; Crescenti et al., 1969), as well as the description of the major depositional geometries (Accarie, 1988; Vecsei, 1991) and seismic-scale depositional sequences (Vecsei, 1991; Eberli et al., 1993). The platform displays a two-stage evolution: aggradation from at least the Jurassic to late Cretaceous (Campanian), followed by progradation from the late Cretaceous (Maastrichtian) to Miocene (Eberli et al., 1993). During the Early Cretaceous, the Maiella was part of an isolated platform with a steep escarpment, which separated the shallow-water areas in the south from deep-water areas in the north (Figures 2 and 3) (Crescenti et al., 1969; Accarie,

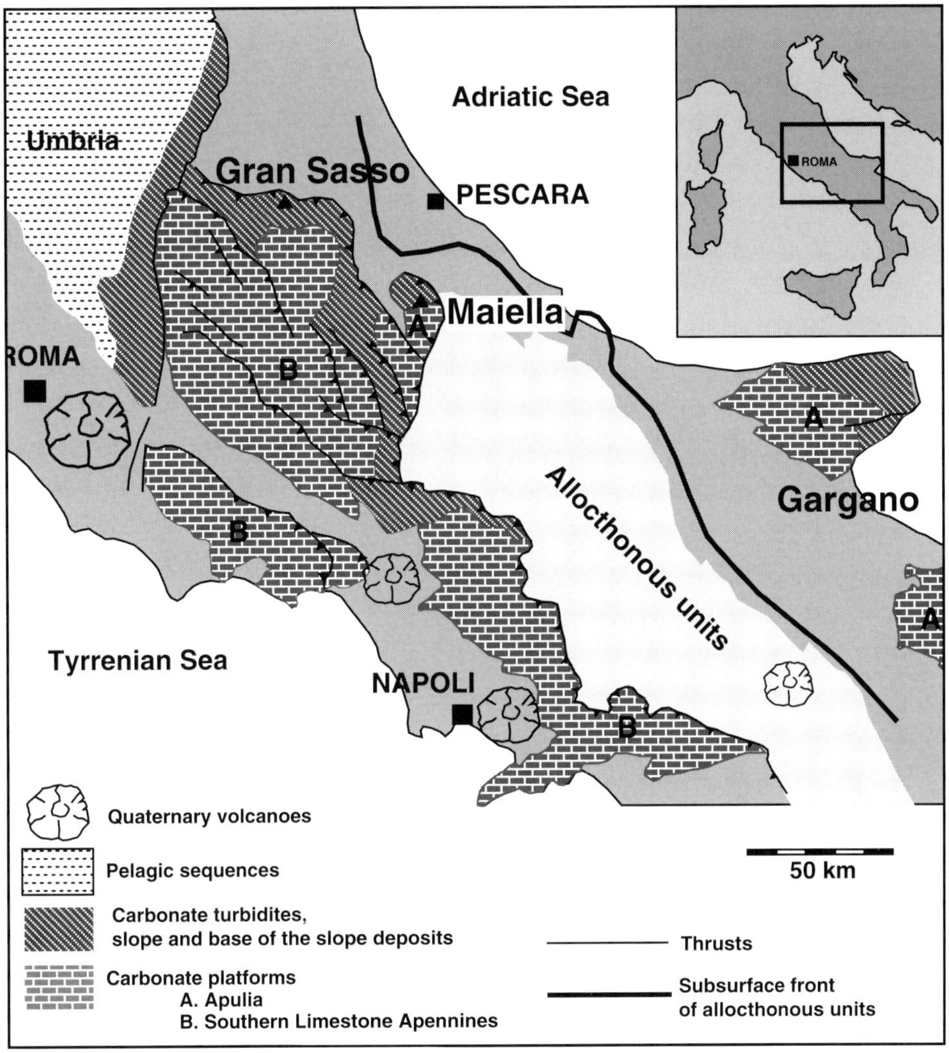

Figure 1. Location and paleogeographic position of the Maiella within the structural frame of the central-southern Apennines (modified from Eberli et al., 1993).

1988; Eberli et al., 1993). The escarpment was 1000 m high at the end of the Early Cretaceous and became progressively buried during the Late Cretaceous with the deposition of the Tre Grotte Formation (Bernoulli et al., 1992; Eberli et al., 1993). Several megabreccia beds and calcareous turbidites shed from the platform, together with pelagic sediments ("Scaglia" of Italian authors) progressively aggraded in the basin and onlapped onto the escarpment. Progressive burial of the escarpment allowed for a subsequent progradation of shallow-water sediments of the Orfento Formation over the former basinal strata of the Tre Grotte Formation. The result was a change from aggradation with a steep margin during deposition of the Cima delle Murelle Formation, to progradation on a distally steepened ramp of the Orfento Formation.

The Upper Cretaceous succession comprises on the shelf the Cima delle Murelle Formation (Cenomanian–Campanian) and the Orfento Formation (late Campanian–Maastrichtian), and in the basin the Tre Grotte Formation and the basinal part of the Orfento Formation (Figures 2 and 3). These can be grouped into two supersequences (*sensu* Haq et al., 1988; Vecsei, 1991; Bernoulli et al., 1992; Eberli et al., 1994). The first supersequence (lower Cenomanian–lower Campanian) on the platform consists in the inner platform of peritidal cycles that grade toward the margins into bioclastic grainstones and rudist biostromes (Eberli et al., 1993). The second supersequence, the Orfento supersequence (upper Campanian–upper Maastrichtian), on which this paper will focus, consists of rudist debris ranging in size from silt to coarse sand, interbedded with megabreccias (Vecsei, 1991). The Orfento supersequence was deposited during the change from aggradation to progradation on the Maiella platform margin (Eberli et al., 1993). The three-dimensional facies organization of this supersequence is very complex, resulting in strong variations in geometries, facies, and sediment composition (Mutti et al., 1994).

THE ORFENTO FORMATION

The Orfento Formation (Vecsei, 1991) comprises the lithostratigraphic unit termed "Calcari Cristallini" (Catenacci, 1965), Orfento member of the Acquaviva Formation (Crescenti et al., 1969) or "Barre Jaune" (Accarie, 1988), and coincides with the Orfento

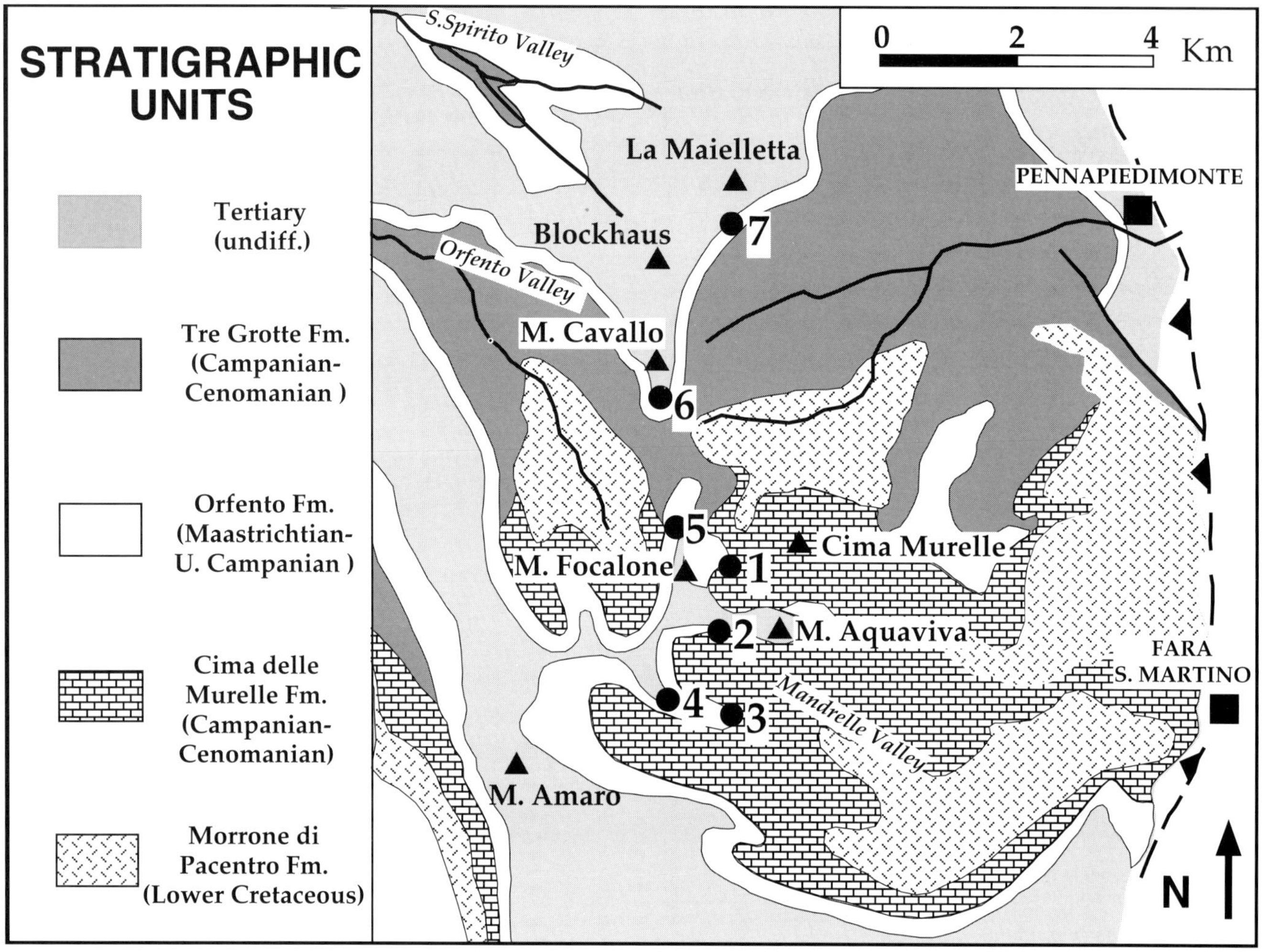

Figure 2. Simplified geological map of the central part of the Maiella showing the outcrops of the Orfento Formation (modified after Accarie, 1988; Carta Geologica d'Italia, 1970; Vecsei, 1991). The numbers indicate the location of the investigated sections.

supersequence as defined by Vecsei (1991). The Orfento Formation is characterized by poorly cemented, highly porous skeletal grainstones and rudstones, derived almost entirely from the breakdown of rudist shells in a high-energy environment. Rudist types include *Radiolites* and *Hippurites,* and their fragments can be recognized both in the rudstones and in the coarse grainstones. Subordinate components are rip-up clasts and cemented platform-derived lithoclasts. Carbonate mud is absent. Units with similar lithologies are common in this time interval in other areas of the Apulian platform margin, e.g., in the Gargano promontory (M. Acuto Formation., Bosellini et al., 1993; Borgomano and Philip, 1990). The deposition of the Orfento supersequence in the Maiella documents the change in platform morphology from aggradation to progradation, with up to 200 m of aggradation on the shelf and at least 4.5 km of basinward progradation over the former platform margin and the upper slope.

Sequence Boundaries

The Orfento supersequence is bounded on the shelf by two unconformities that can be traced into conformable strata in the deep basin (Crescenti et al., 1969; Vecsei, 1991). The two bounding surfaces have been identified as unconformities by (1) stratal geometries (erosional truncation), (2) lithological changes, and (3) local dating of the hiatus between strata below and above the surfaces (Vecsei, 1991). The supersequence can be internally subdivided into several third- and fourth-order depositional sequences, bounded by erosional surfaces (Figure 4) (Eberli et al., 1993; Mutti et al., 1994).

The lower supersequence boundary (LSSB) on the shelf is marked by an erosional surface that deeply incises the underlying strata. It is overlain by strata of the Orfento supersequence. Due to erosion, generally no soil surface is preserved, with the exception of the outer shelf/margin, where a calcrete crust caps the underlying platform facies and mantles an irregular relief (0–10 m at the scale of the outcrop). This surface will be described in detail in the diagenesis section. The hiatus on the platform spans from the early to the late Campanian (part of the G. Ventricosa Zone) (Vecsei, 1991), approximately corresponding to a maximum value of 6 m.y. (Haq et al., 1988). The hiatus

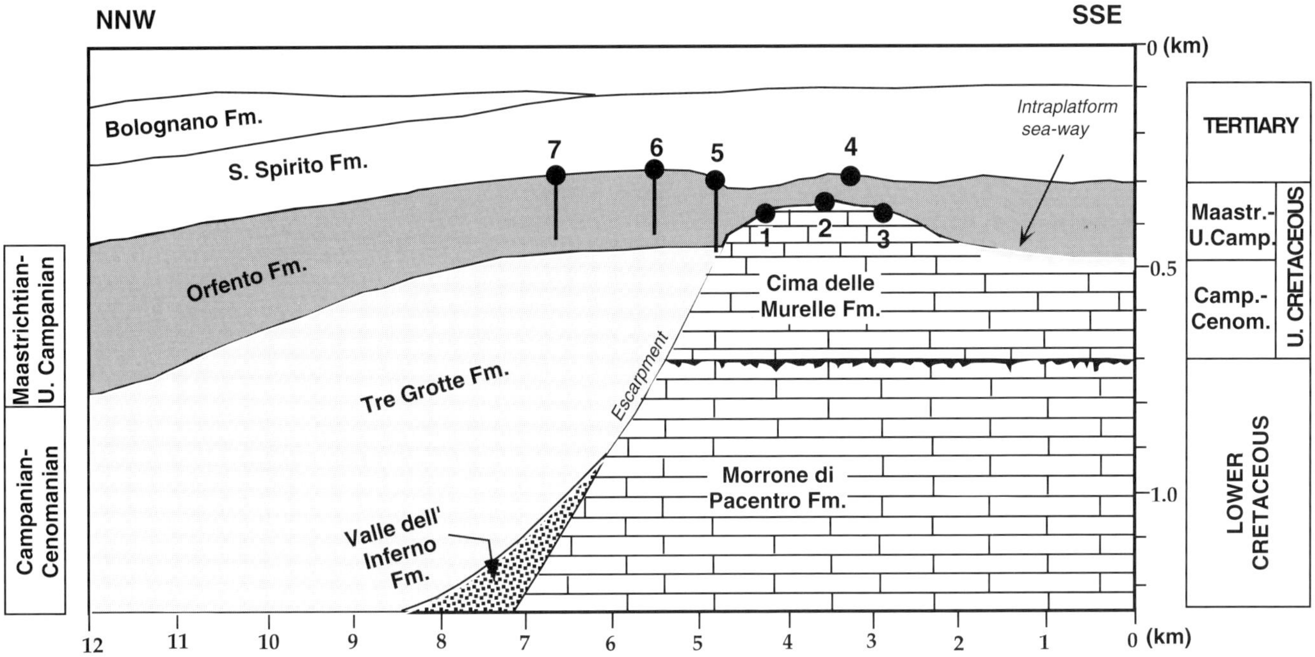

Figure 3. Schematic platform/basin cross section illustrating the stratigraphic and geometric relationships among the stratigraphic units in the Maiella. Numbers beside filled circles indicate locations of measured stratigraphic sections.

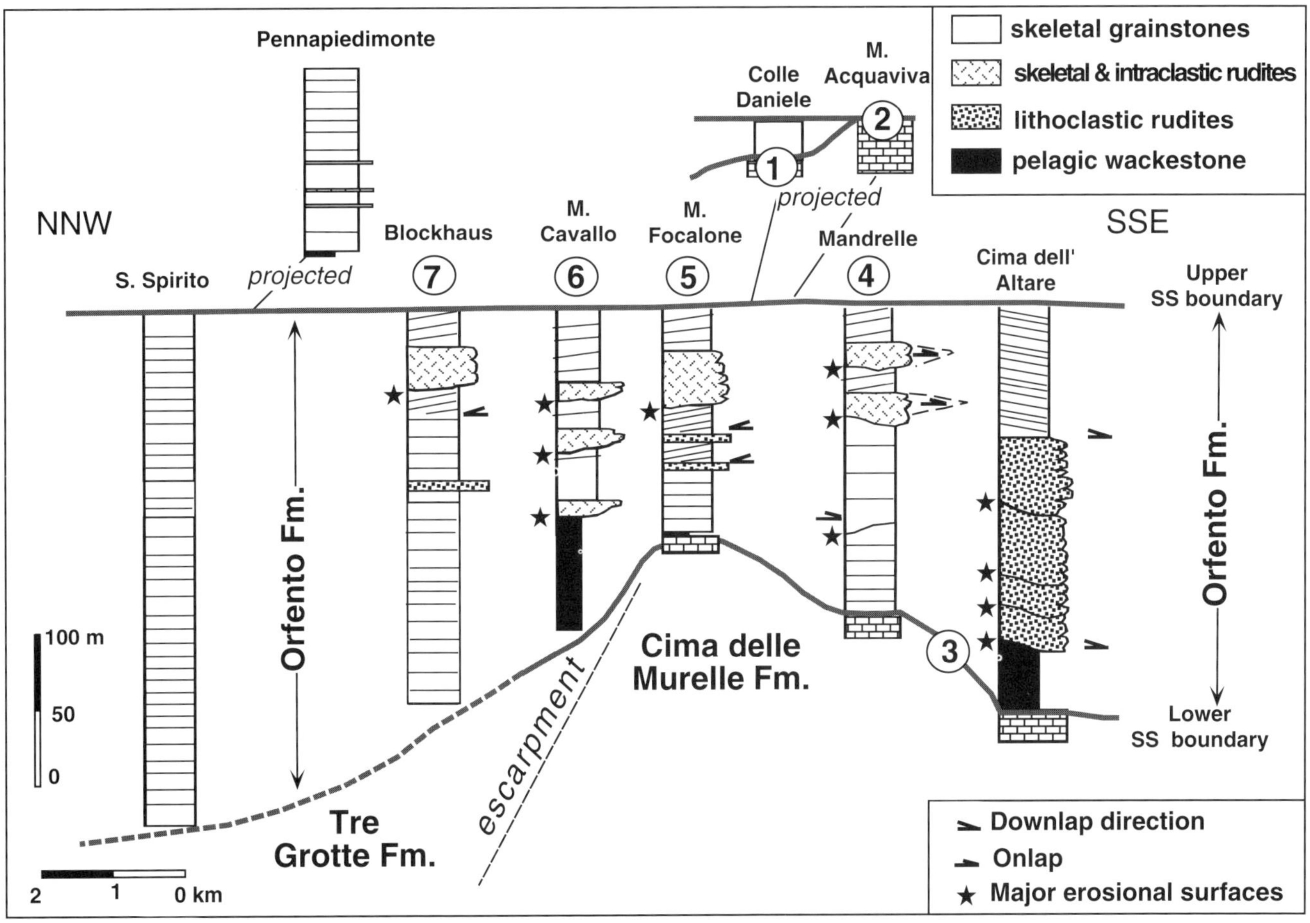

Figure 4. Internal stratigraphy of the Orfento supersequence. Locations of sections are marked in Figures 2 and 3. Horizontal bedding pattern indicates basinal, deeper-water sediments. Oblique bedding indicates progradational, shallower-water sediments.

could be, in fact, much less, if the eroded strata could be taken into account. On the escarpment, the boundary is expressed by the onlap of breccia and grainstone beds onto the escarpment. In the basin the boundary is marked by a bundle of breccia beds which erode into pelagic sediments (Calcarata Zone), but the hiatus can not be quantified biostratigraphically (Vecsei, 1991). Platform-derived breccia clasts commonly record vadose or phreatic meteoric diagenesis, suggesting that mechanical erosion and transport of the breccias was either synchronous or postdated the exposure on the platform. An intraplatform seaway, approximately oriented ENE-WSW and about 150–200 m deep, was generated on the shelf in association with the lower supersequence boundary (Figure 3). The origin of this relief is not clear, although abrupt changes in facies types and in thickness would indicate that at least one side of the seaway was fault bound. The presence of the intraplatform seaway strongly affected the patterns of sediment distribution within the unit, as will be discussed in the next section.

The upper supersequence boundary (USSB) is also associated with significant erosion on the platform/upper slope (Vecsei, 1991). The erosion associated with this surface is extensive, as it cuts down several tens of meters into the strata of the Orfento Formation. Locally, the Orfento Formation has been entirely removed by erosion. The associated hiatus spans from the late Maastrichtian to the early Thanetian (Vecsei, 1991), corresponding to approximately 8–10 m.y. (Haq et al., 1988). This surface coincides with the Cretaceous–Tertiary boundary which is a major regional unconformity in carbonate platforms of the Mediterranean area (Esteban, 1991).

Several erosional surfaces occur within the Orfento supersequence during the late stages of progradation and are interpreted as sequence boundaries of a minor hierarchical order (Mutti et al., 1994). These surfaces are characterized by pronounced erosional relief (ranging from a few to tens of meters) and affect the distribution of the sediment's grain size, resulting in a heterogeneous distribution of primary, interparticle porosity.

Internal Organization

The Orfento supersequence has a very complex stratigraphic organization and shows strong three-dimensional variations in depositional geometries as well as facies association (Figure 4). The presence of the intraplatform seaway strongly affected the patterns of sediment distribution within the unit. The overall internal organization of the Orfento supersequence records the infilling of the intraplatform seaway in two stages, with an initial accumulation of deep-water sediments and a later progradation of sediments in progressively shallower waters (Mutti et al., 1994). Sediments were initially confined to deep-water areas and occur in either channellized breccia bodies or in vertically stacked, bedded grainstones and breccias (Figure 4, Cima dell'Altare, sections 4 and 5). Breccias contain lithoclasts composed of cemented platform facies, derived from older platform facies, and partly lithified rip-up clasts of skeletal sands and rudist fragments, derived from intraformational erosion. Platform-derived clasts commonly record vadose or phreatic meteoric diagenesis, documenting exposure of part of the platform and its mechanical erosion. The grainstones are composed of fine to medium sized, sorted skeletal fragments and do not display any sedimentary structure, except where they are silicified. In this case, they are arranged in parallel beds with an erosive base, faintly graded, and capped by current ripples. Water escape structures have also been observed. Sediments aggraded in the intraplatform seaway or onlapped against the escarpment in the basin.

Progradation began when the relief associated with the seaway was partially filled and sediment was shed toward the ENE (Figures 4 and 5). The prograding sequence contains breccias, grainstones, and fine-grained sediments, and overall records a progressive shallowing of the depositional environment (Figure 4, sections 4 and 5; Figure 6). Initially, turbidites were deposited and are associated with water escape structures. Later, wave-reworking and hummocky cross-bedding indicate sedimentation near wave base. Breccia packages show low-angle cross-beds, and grade laterally downslope into breccia beds that are interbedded with grainstones, and finally into grainstones. Breccias are mainly composed of rudist fragments and intraclasts. Lithoclasts are very rare.

After the seaway was completely buried by sediments, progradation continued into the basin, toward the NNE, with shingled sigmoidal stratal geometries, composed of breccias and grainstones (Figure 4, sections 6 and 7; Figures 7 and 8). Each sigmoidal sequence is characterized by an erosive base, which cuts down into the underlying units up to several tens of meters and consists of a lower breccia package transitionally overlain by grainstones. The basal erosive surfaces can be traced downslope for a few hundreds of meters where they are not overlain by breccia beds but by skeletal grainstones, but they can no longer be recognized when they disappear in fine-grained sediments.

DIAGENESIS OF THE ORFENTO FORMATION

Primary depositional porosity was high in the Orfento Formation (average 20–30%) and reflected the distribution of depositional facies, grain size, and sediment sorting (Figures 6–8). Highest depositional porosity is found in breccias and coarse grainstones, in which inter- and intraparticle and shelter porosity dominate (Figures 6–8). Moldic porosity is also abundant in the Orfento Formation (average 15–20%). This type of porosity is facies-selective, occurring only in coarse lithologies, and mineral-selective, affecting only aragonitic grains. Aragonitic grains seem to be more abundant in coarse lithologies. Rudists shells show variable degrees of alteration and neomorphism.

Cementation by bladed and sparry calcite cement does not significantly affect porosity reduction and

Figure 5. View of the Orfento supersequence, M. Focalone (Figure 4, section 5), from north to south. Arrows indicate the lower and the upper bounding unconformities.

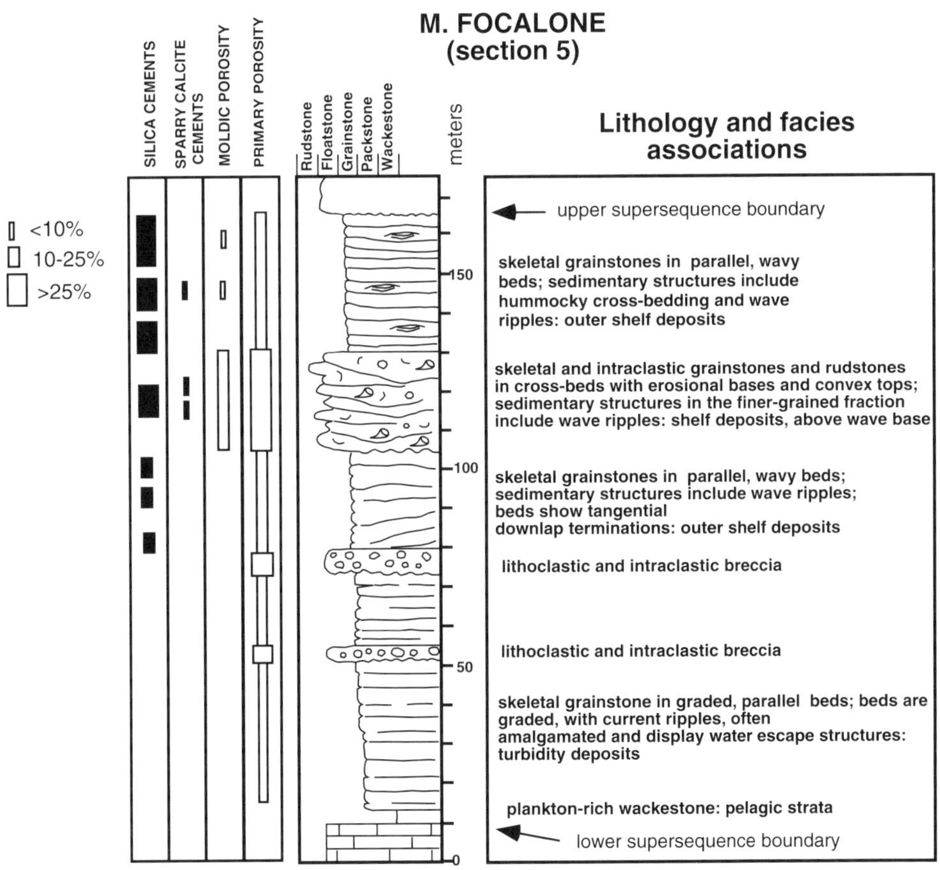

Figure 6. Monte Focalone. Stratigraphic section illustrating lithologies, facies associations, and distribution of diagenetic features. Erosional unconformities are marked with the undulated line.

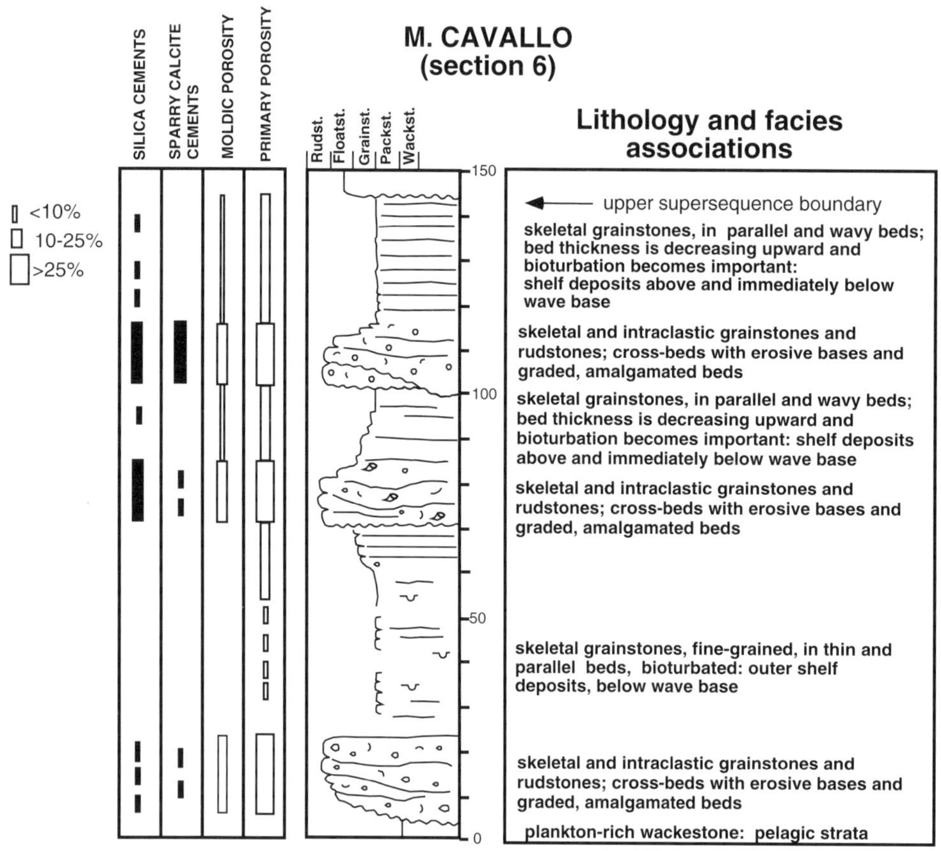

Figure 7. Monte Cavallo. Stratigraphic section illustrating lithologies, facies associations, and distribution of diagenetic features. Erosional unconformities are marked with the undulated line.

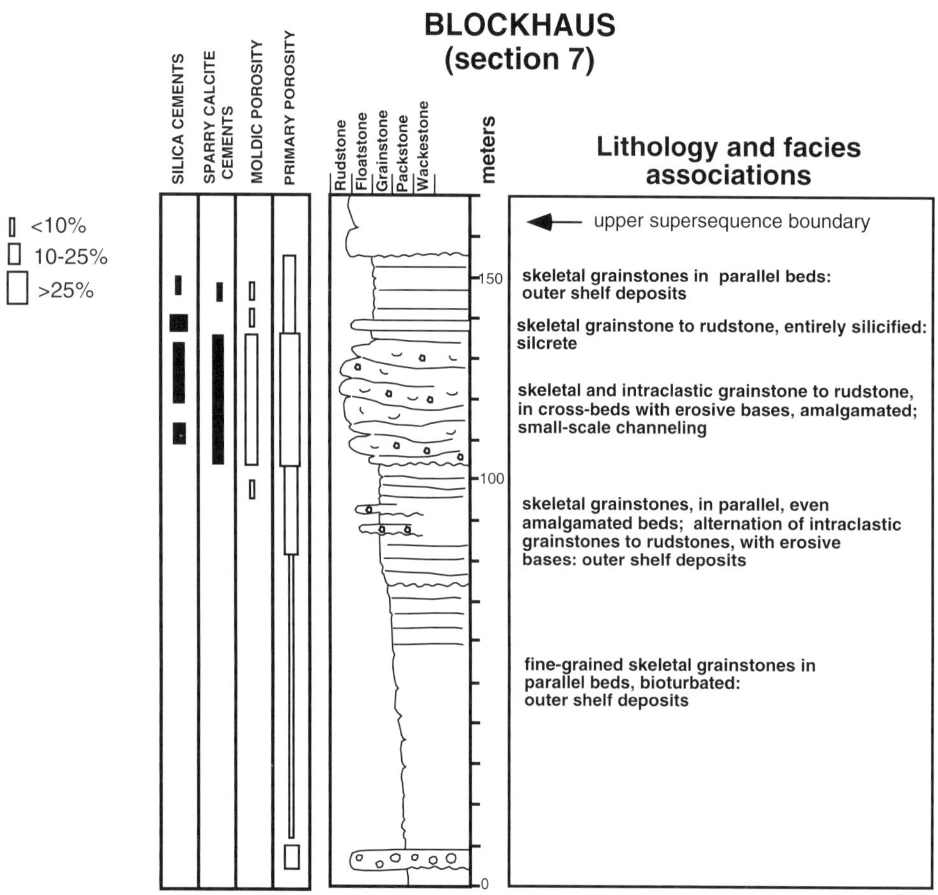

Figure 8. Blockhaus. Stratigraphic section illustrating lithologies, facies associations, and distribution of diagenetic features. Erosional unconformities are marked with the undulated line.

forms generally <5–10% of the rocks, although higher values occur in specific localities. Silica cementation is more abundant, although it has an irregular distribution. It will be discussed in detail later. Micritic rims are recrystallized to microspar.

Bladed cement consists of elongated crystals up to 50 μm long (average 10–25 μm) oriented tangentially with respect to the substrate. Crystals appear as poorly developed scalenohedra with rhombic terminations (bladed spar, James et al., 1976). The crystals grow selectively on rudist fragments as discontinuous grain linings in intergranular pores. This cement is not abundant. The precipitation of bladed cement predates moldic porosity, although some of the larger crystals postdate the formation of the molds. Crystals are characterized by a dull orange to patchy luminescence, indicating re-equilibration after precipitation. This cement is a common high-Mg precipitate in modern seawaters (James et al., 1976; Pierson and Shinn, 1980). Timing, texture, and cathodoluminescence characteristics would indicate that this cement precipitated from marine waters.

Sparry calcite cements are rare in the Orfento Formation. Two groups are distinguished based on size, luminescence characteristics, and distribution. The first group, sparry calcite I, consists of fine-crystalline, equant rim or mosaic cement characterized by 20 to 40 μm sized, euhedral, equant crystals that line pores (Figure 9A, B). Crystals are rich in dark, solid inclusions and often grow around bladed crystals. Sparry calcite is commonly nonluminescent, but in some places displays a thin orange rim in the outer borders of the crystal. Distribution of luminescence is localized, and bright zones can not be correlated from pore to pore in the same thin section. This cement commonly occurs as scattered crystals and, more rarely, as continuous pore lining. It is found in both intergranular and moldic pores, and grows in discontinuity over bladed cements.

The second group, sparry calcite II, consists of larger-sized (60 to 250 μm), anhedral to euhedral, equant crystals, confined to the upper part of the section in the Blockhaus–Maielletta area (Figure 4, section 7). This cement is most abundant in grainstones and breccias, as pore-fill in large primary and moldic pores, or as a replacement of former calcitic cements (Figure 9C, D). The coarser crystals under cathodoluminescence are bright orange. Sparry calcite I and II are postdated by silica cements.

Stable Isotopic Composition of Clasts and Early Cements

Microsamples of less than 1 mg were drilled from selected clasts and from single generations of early cements. Samples were reacted with 100% H_3PO_4 at 50°C and the CO_2 evolved was analyzed with a VG-903 mass spectrometer. The accuracy of duplication analyses is ±0.1‰ for $\delta^{13}C$ and ±0.2‰ for $\delta^{18}O$. Data are expressed relative to the PDB standard.

Thirty-eight samples have been measured. The samples include: rudist shells, fine-grained rip-up clasts, and sparry calcite cements (Figure 10). Fine-grained rip-up clasts show the most enriched composition, centered around $\delta^{18}O = -0.1‰$ and $\delta^{13}C = 2.2‰$. Rudist fragments are also centered around the same values, $\delta^{18}O = -0.1‰$ and $\delta^{13}C = 2.2‰$, but show a 2.5‰ range in oxygen. Sparry cements have an average value of $\delta^{18}O = -1.2‰$ and $\delta^{13}C = 2.0‰$, and a 5‰ range in oxygen. Rip-up clasts reflect mainly the marine composition, whereas sparry cements reflect precipitation from meteoric waters. Rudist shells show values transitional between marine and meteoric and reflect different degrees of alteration.

Sparry calcite cements from different areas show differences in their isotopic composition (Figure 11). Samples from the Blockhaus area are the closest to marine values, while samples from M. Cavallo and Mandrelle show more depleted $\delta^{18}O$ and $\delta^{13}C$ values. The depletion in $\delta^{18}O$ can be related to the position of the sampled stratigraphic section with respect to the slope, and is relatively increasing upslope.

DIAGENESIS ASSOCIATED WITH THE LOWER SUPERSEQUENCE BOUNDARY

Three localities have been investigated in detail to resolve the question whether subaerial exposure really occurred in association with the lower boundary of the Orfento supersequence and to document its effects on the underlying platform facies (see Figures 2 and 3: Colle Daniele indicated by 1, Acquaviva indicated by 2, and Mandrelle indicated by 3). These localities have been chosen on the basis of accessibility and the possibility of studying downslope variations in diagenesis. In all three sections, depositional textures of the Cima delle Murelle Formation have been extensively modified by dissolution and subsequent cementation (Figures 12 and 13). Individual grains, such as former aragonitic shell fragments, are commonly leached. Bioerosion of bioclasts is common in the Acquaviva area. Grains were wholly to partially micritized before being leached. Fracturing of the host rock at a macroscopic scale has been observed only in the outer margin (Cima Murelle, I. Stössel, 1993, personal communication). Calcite cements occur, and in paragenetic order they are: pore-lining microspar, rhombic cements, sparry calcite, and micritic cement.

Microspar cement occurs as fine, pore-lining, prismatic to equant, 2 to 10 μm crystals. The cement is substrate-selective for micritized grains, and it is generally isopachous. It is generally nonluminescent. It is most abundant near molds of leached grains.

Sparry calcite cement occurs as pore-lining and -filling mosaic cement and consists of equant, 10 to 150 μm sized crystals. Crystals increase in size toward pore centers. This cement commonly overlies the microspar; where it does, there is no clear petrographic boundary between the two fabrics. Fe oxides occur as a thin lining between microspar and sparry calcite cements in places. This cement shows slightly different cathodoluminescence characteristics in the three localities: sparry cements from Acquaviva (Figure 3,

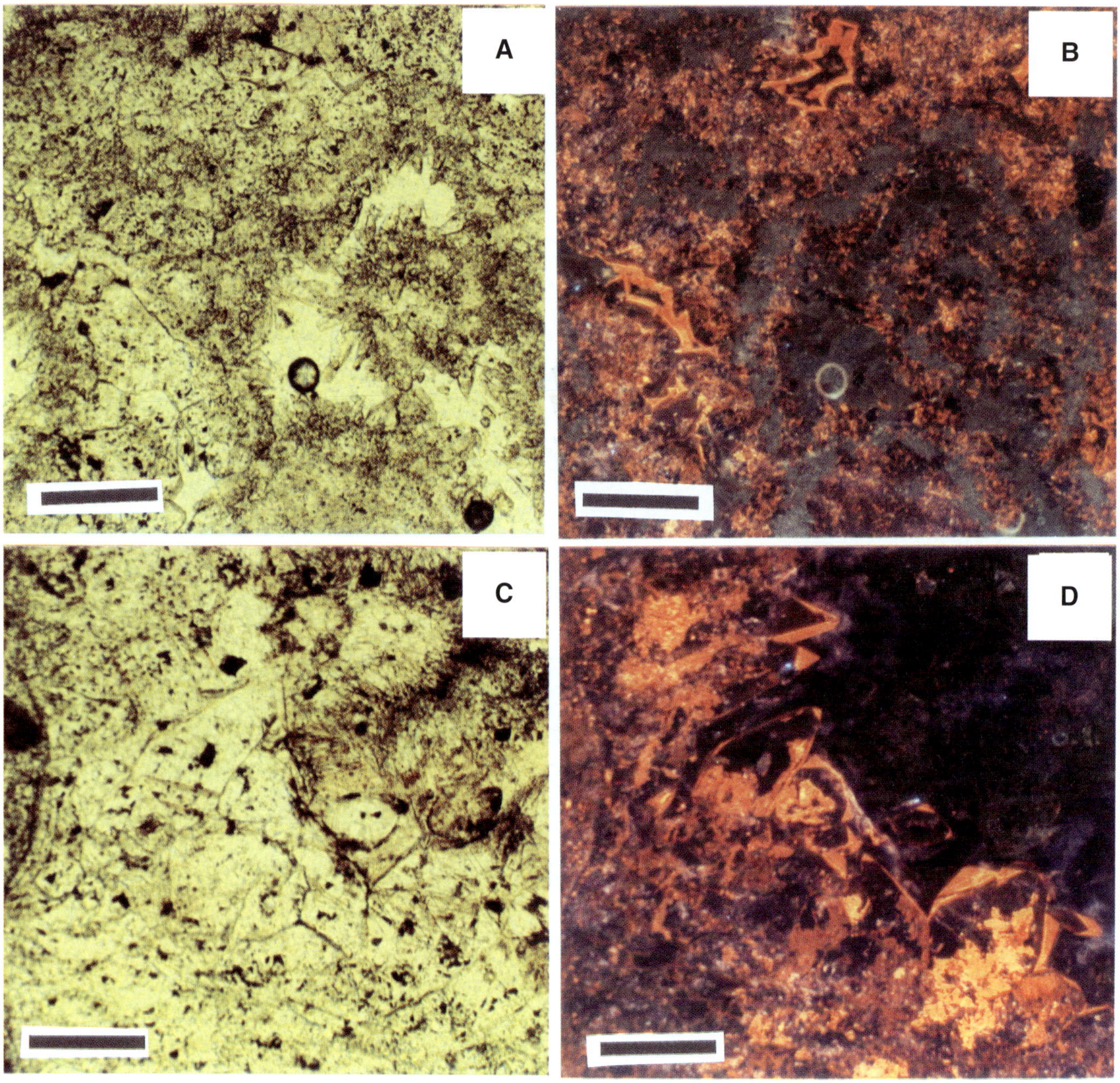

Figure 9. Photomicrographs of calcite cements in the Orfento Fm. (A) The photograph illustrates the relationships between micritized rims, bladed calcite, and sparry calcite cements. Scale bar is 200 μm. (B) Same as (A), in cathodoluminescence. Micritic rims and bladed calcite show a patchy luminescence, indicative of recrystallization after precipitation. Sparry calcite cements show three zones: first, a nonluminescent, growing on a patchy nucleus, followed by (second) a banded bright orange zone, and (third) followed by a dark, nonluminescent one. Note that the sparry calcite cement is less abundant than it appears in normal polarized light, part being patchy bladed calcite from Upper Mandrelle Valley (Figure 3, section 4). Scale bar is 200 μm. (C) Sparry calcite mosaic filling a rudist fragment mold. Note abundance of dark, solid inclusions. From Blockhaus (Figure 4, section 7). Scale bar is 200 μm. (D) Same image as (C) in cathodoluminescence. Crystals show a dark, nonluminescent nucleus followed by a thin bright orange zone. A third zone grows discontinuously over the former two and consists of bright orange-yellow color. Scale bar is 200 μm.

locality 2) and from Mandrelle (Figure 3, locality 3) are mainly nonluminescent or nonluminescent with two or three bright, thin zones toward the outer part of the crystals. The cements from Colle Daniele (Figure 3, locality 1) are nonluminescent with several bright zones in the outer parts that can be amalgamated into one thick, bright zone. Despite the differences in the thickness of the zones, the same pattern of nonluminescent to bright banded has been observed in the three localities.

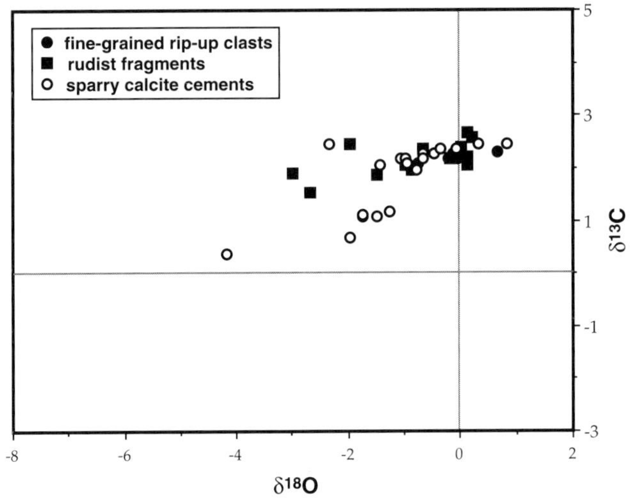

Figure 10. Stable isotopic composition of clasts and cements of the Orfento Formation.

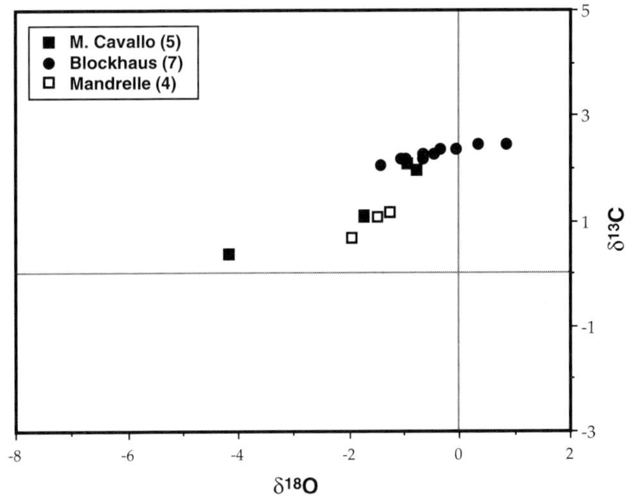

Figure 11. Stable isotopic composition of sparry calcite cements occurring in the Orfento Formation, distinguished according to the locality. Numbers in the legend refer to the localities mentioned in the text.

Scalenohedral cements are characterized by 20 to 100 µm long crystals. These crystals are rare and occur in large interparticle pores where they commonly overlie microspar cements (Figure 13A, B). They are generally nonluminescent but locally display bright orange rims, as described for the sparry calcite.

Micritic cements are characterized by dense micrite, generally 1 to 5 µm, with scattered Fe oxides. This cement postdates sparry calcite cement as the final pore-fill phase or, more rarely, occurs at grain contacts. Fe oxides are more abundant in the Mandrelle and Acquaviva areas.

The Calcrete at Colle Daniele

At Colle Daniele (Figure 3, location 3) a calcrete horizon caps the Cima delle Murelle Formation and is overlain by strata of the Orfento Formation (Figure 12A, B). Only the uppermost 30 cm of the interval affected by the calcrete could be sampled because the underlying cliff is inaccessible. The calcrete is replacive on the underlying lithologies, as suggested by the downsection gradation from a crystalline matrix into a skeletal-peloidal grainstone (Figure 12B). The grainstone is extensively cemented by sparry and rhombic calcite cements, occasionally with meniscus textures (Figure 13A, B). Single grains have extensive micritic coatings, are irregular in thickness, and display irregular protuberances which may form bridges between grains; these features are indicative of a microbial origin and typically occur in the vadose zone (Wright, 1986). This texture grades upsection into a crystalline matrix, with progressively more complete micritization of grains. The matrix consists of dense crystalline carbonate (micrite and microspar) with scattered rhombic crystals, 10 to 30 µm in size. The outer part of the crystals is commonly coated by micrite. Rhombic crystals are calcite or dolomite. The nature of the matrix is better revealed by cathodoluminescence, which shows a mosaic of irregularly shaped crystals with nonluminescent cores and brightly layered outer parts (Figure 13C, D). The multiple zones, separated from each other by dissolution surfaces, indicate that the crystals underwent multiple phases of dissolution and precipitation. Calcite-filled root molds are common in the matrix. The fabric is either laminated or mottled, with incipient crumbling fracturing, which is typical for alpha-type calcretes (Wright, 1990). This type of soil develops typically in semi-arid climates.

Stable Isotopic Composition of Early Cements

Early cements from 18 samples, as well as seven samples of sparry cements from megabreccia clasts, have been measured. The megabreccia clasts occur in the basin immediately above the supersequence boundary and have been measured to see whether they record the same signature as the sparry cements found in the platform. The early cement samples contain a mixture of microspar and rhombic and sparry calcite cement, since the single generations of cements are too small for separate analyses. The values $\delta^{18}O$ = –0.1‰ and $\delta^{13}C$ = 2.2‰, average of fine-grained rip-up marine clasts of the Orfento supersequence, are taken as the closest to the marine values. All measured samples are characterized by a different degree of depletion in $\delta^{18}O$ and $\delta^{13}C$ with respect to marine values (Figure 14). In general, $\delta^{18}O$ composition varies more than $\delta^{13}C$ composition. Samples show a strong covariance between oxygen and carbon. Samples from Acquaviva show the less depleted $\delta^{18}O$ and $\delta^{13}C$, progressively followed by Mandrelle and Colle Daniele.

The composition of these cements, characterized by a 7‰ range in oxygen and a 6‰ range in carbon, are similar to the known isotopic composition of carbonate meteoric environments (e.g., Allan and Matthews, 1982; Budd and Land, 1990). The depletion in oxygen

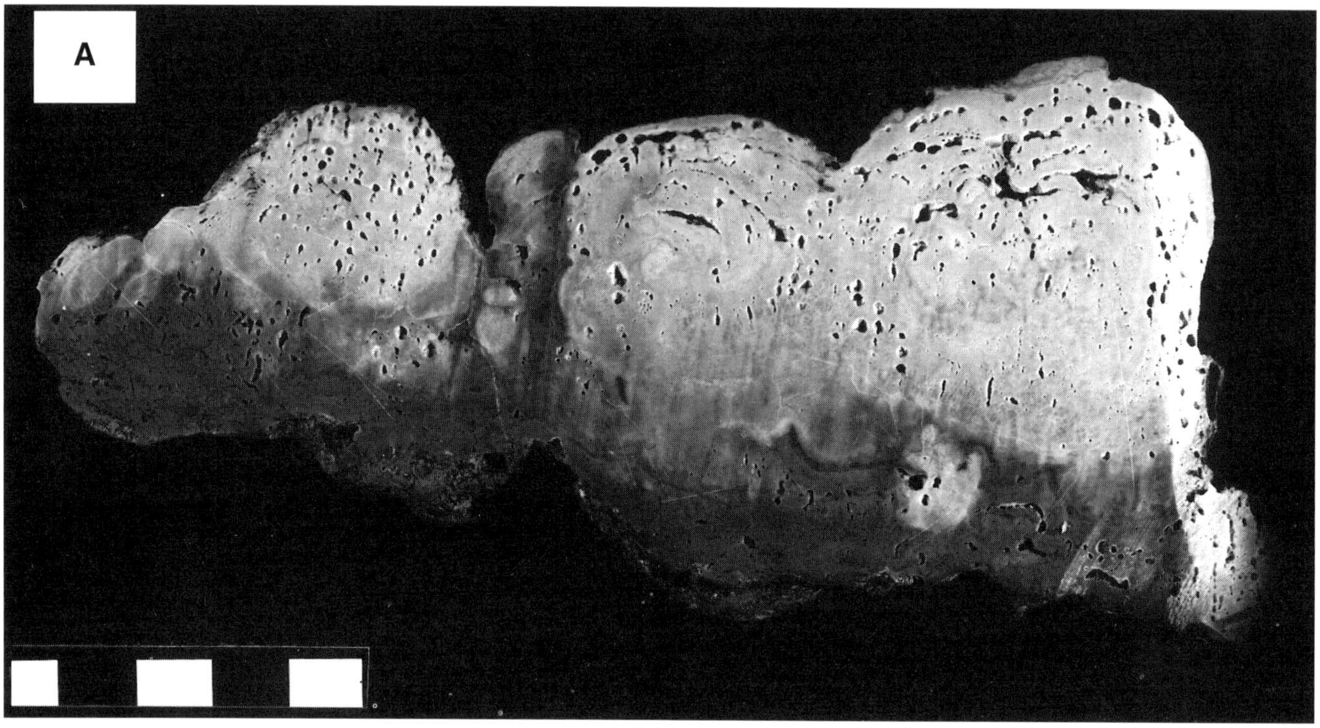

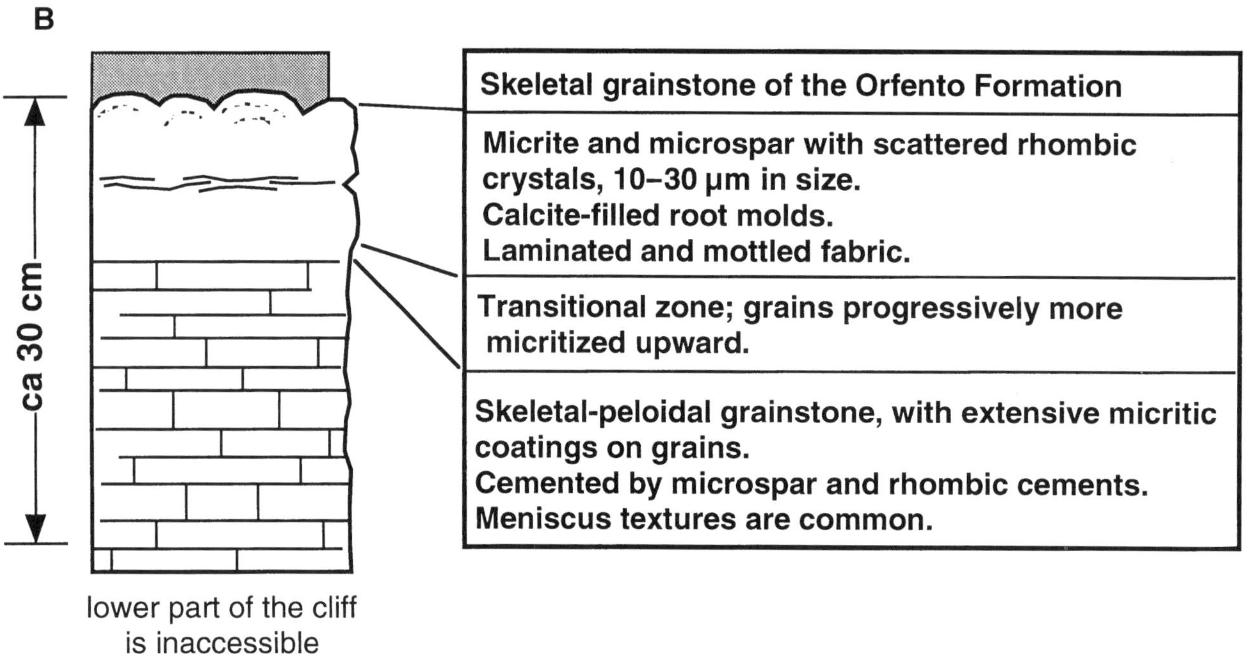

Figure 12. Calcrete at Colle Daniele. (A) Stratigraphic profile and petrographic characteristics of the different intervals recognized in the calcrete horizon. (B) Uppermost interval, completely micritized. The dark color results from a complex set of micrite-filled fractures and of Fe oxides. Note the root molds in the lighter-colored interval. Scale bar in centimeters.

with respect to marine values reflects precipitation from a coastal meteoric water at near-surface temperature. The depletion in $\delta^{13}C$ values requires input of light carbon (^{12}C) such as soil gas carbon or carbon liberated during the reduction of sulphates and the oxidation of organic matter. Depleted soil carbon could have been provided by unconformity-recharged meteoric fluids. Because of the heavily rock-dominated values of carbon, soil signatures rarely extend very far below caliche or soil horizons. The Colle Daniele $\delta^{13}C$ values are the only real negative values from the cements associated with the unconformity, and are the only data set that could be expected to give a soil carbon signature, the rest being erosional.

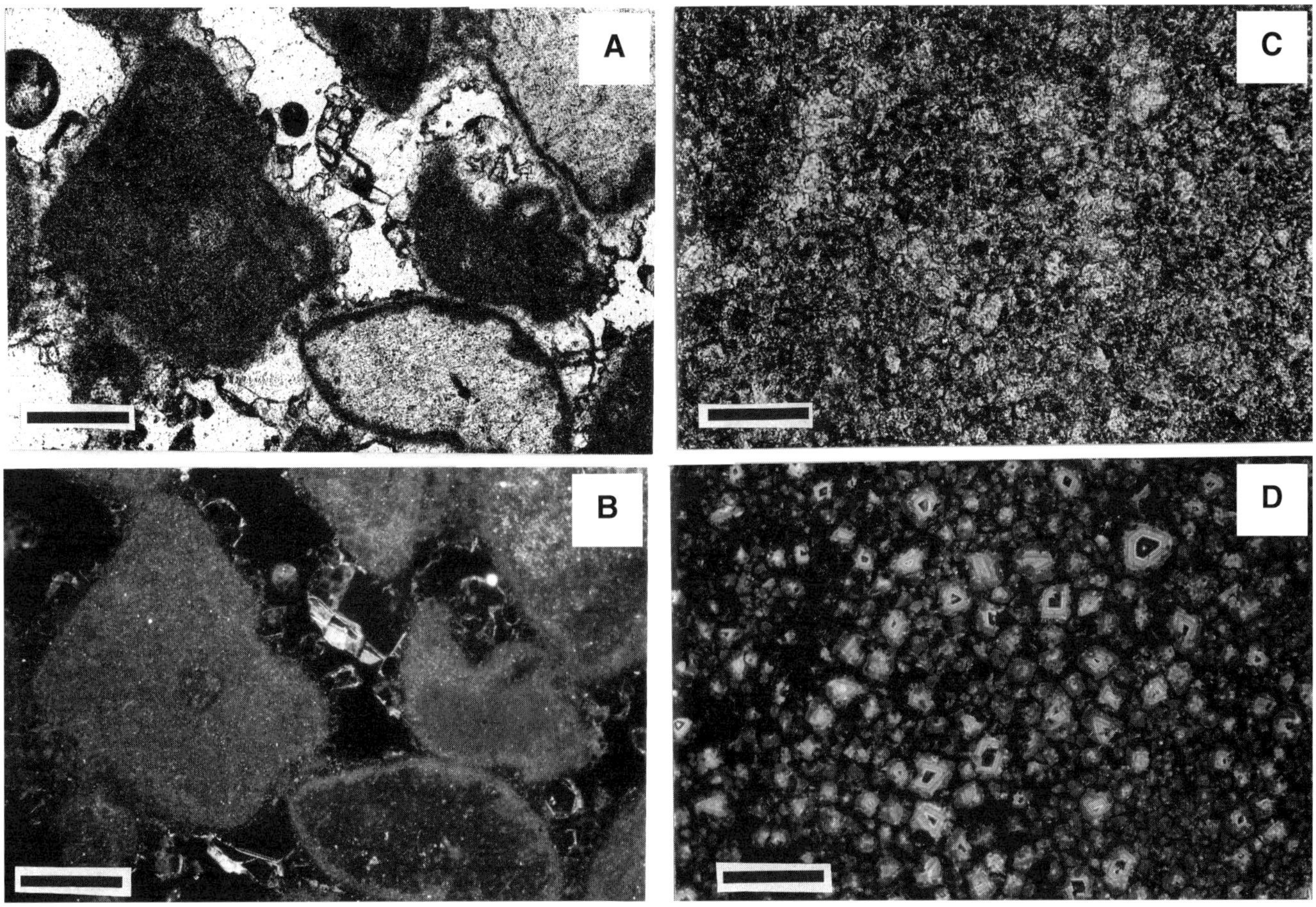

Figure 13. Photomicrographs of vadose features at Colle Daniele (Figures 2 and 3, locality 3). (A) Microspar and rhombic cements from the lower interval of Figure 12A. Note meniscus texture of the cements. Scale bar is 200 µm. (B) Same as (A) in cathodoluminescence. Grains are nonluminescent with brighter micritic rims. Cements are nonluminescent, with a few thin bright zones in the outer part. Scale bar is 200 µm. (C) Crystalline matrix, from the uppermost interval of Figure 12A. Scale bar is 200 µm. (D) Same as (C) in cathodoluminescence. The matrix is characterized by a mosaic of irregularly shaped crystals with nonluminescent cores and a bright, layered outer part. The rhombic crystals show multiple phases of dissolution and precipitation, as well as multiple zoning. This feature is typical of alpha-calcretes (Wright, 1986, 1990). Scale bar is 200 µm.

DIAGENESIS ASSOCIATED WITH THE UPPER SUPERSEQUENCE BOUNDARY

Subaerial exposure at this unconformity is inferred by the presence of significant erosion, the long hiatus, and the regional relationships. Although no soil surface is preserved, diagenesis of silica cements, occurring in the underlying sediments in the more proximal areas, is consistent with subaerial exposure.

Silica cements are abundant in the upper part of the Orfento Formation. Silica occurs in irregular, elongated patches (or lenses), which cross-cut lithological boundaries, typically 0.2–1 m thick and 0.5–5 m long. Only one crust, 0.5 m thick and several meters long, was found in the Blockhaus area (Figure 8). The most abundant silica fabrics include: microspherules, cryptocrystalline microquartz (chert), crystalline microquartz (<20 µm), and megaquartz (>20 µm).

Microspherules (average size 40–60 µm) consist of one or more concentric layers and are brown in plain polarized light due to the presence of bands of very small inclusions. Microspherules are now composed of microquartz. Opal typically precipitates as microspherules (Hesse, 1990), and the textural analogy suggests that the microquartz is possibly replacing a former opal precipitate. The microspherules occur either as isolated beads in intergranular pores, anchored to bladed calcite cements, or accumulated at the bottom of the pores with a geopetal orientation (Figure 15). Calcite cements underwent a phase of dissolution before precipitation of the silica cements (Figure 15D). Microspherules can also coalesce in larger sized lumps. Less commonly, microspherules occur as a selective replacement of fine-grained clasts. In this case, the primary textures are not well preserved.

Crystalline megaquartz with inclusions occurs as euhedral equant crystals (average size 40–200 µm), both replacing carbonates or as primary cements in intergranular pores (Figure 15A, B). The growth of crystals

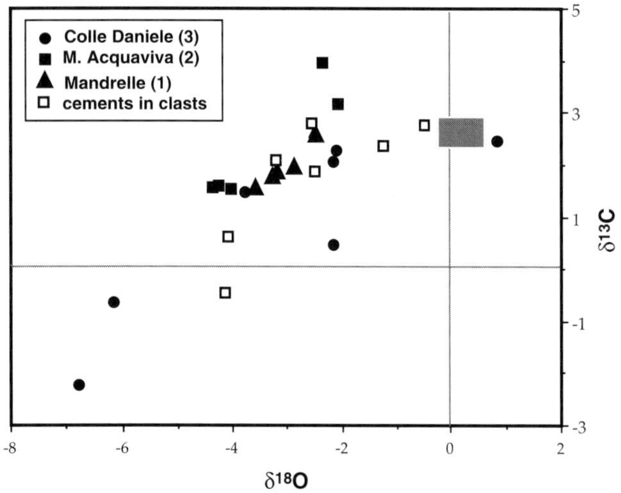

Figure 14. Stable isotopic composition of cements associated with the lower supersequence boundary of the Orfento supersequence. Numbers in the legend refer to the localities mentioned in the text. The field marked with the gray pattern indicates the average values of fine-grained rip-up marine clasts of the Orfento supersequence, taken as the closest approximation to the marine values.

is marked by thin rims of inclusions (<5 μm), brown in plain polarized light, spaced at regular intervals, and alternating with limpid quartz.

Crystalline micro- and megaquartz (<20 μm and >20 μm, respectively) are the most abundant silica types, and they typically occur both as replacement or as a pore-lining and pore-filling cement in intergranular and moldic porosity. Crystals are euhedral, equant, and range in size from 10 to 150 μ. This cement has a mosaic fabric similar to drusy calcite mosaics, with crystal size progressively increasing toward the center of the pore. Mosaics of crystals of the same size also have been observed. Where present, megaquartz is the last pore-filling silica phase.

Chert replaces fine-grained carbonate clasts, and, less commonly, rudist shell fragments. The fine-grained lithologies are replaced by chert only where they occur as clasts in high-porosity facies. The fine grain size makes the timing of chert formation with respect to the other silica phases unclear. Chert clasts have been observed isolated within calcitic breccias deposited during the late stages of progradation (e.g., Blockhaus), suggesting that at least some of them were silicified before transport. This timing would suggest that silicification had begun on the exposed, updip portions of the platforms, while sedimentation was still taking place in the downdip areas.

CONTROLS ON POROSITY DISTRIBUTION AND DEVELOPMENT

The Orfento Supersequence

Several cross-cutting relationships suggest that most moldic porosity is formed in association with the minor unconformities within the Orfento Formation: (1) breccias overlying the erosional unconformities commonly contain rip-up clasts with molds, whereas no molds occur in the breccia matrix; (2) moldic porosity is highest in the progradational stages, where several intra-Orfento unconformities occur, and (3) sparry calcite cements generally postdate the formation of moldic pores (but may be absent), as late silica cements commonly occur as the first mold-filling phase. Therefore, some of the molds may have formed after calcite cementation and before silicification. The occurrence of fabric- and mineral-selective moldic porosity indicates the flow of waters undersaturated with respect to aragonite, but supersaturated or in equilibrium with calcite.

Marine and meteoric cements are both rare in the Orfento Formation. Modern marine cements are abundant in areas of high currents, when seawater is pumped through the sediments (James et al., 1976; Aissaoui et al., 1986). However, marine grainstones are commonly not extensively cemented by marine cements, because a stabilized substratum is needed for the cements to precipitate (James and Choquette, 1990). The lack of a mechanically stable framework would provide an additional reason for the poor marine cementation in the Orfento Formation. In fact, the sedimentological characteristics of the Orfento Formation suggest deposition occurred in a high-energy environment, where grains were repeatedly reworked and redeposited.

The distribution of the meteoric cements is related to both facies and mineralogy, as well as to stratigraphic position. Breccias and coarse grainstones are better cemented than fine grainstones, as are facies that contained aragonitic fragments, now molds. Moldic porosity and sparry calcite cements occur mainly in the strata deposited during the progradational phases, when sedimentation was interrupted by several hiatuses. Calcite cements are generally more common below the minor-order unconformities, but can not be related precisely to any specific intra-Orfento exposure events. In fact, scarcity of calcite cements and other diagenetic phases makes it difficult to relate their occurrence to specific stratigraphic intervals. The meteoric cements in the Blockhaus area were precipitated during the last phase of progradation, and reflect higher rock/water interaction, with rock-buffered $\delta^{13}C$ values. Cathodoluminescence patterns, with zones noncorrelatable between pores of the same thin section, and stable isotope data are typical of cements precipitated in localized freshwater systems (Hammes, 1992).

The Lower Supersequence Boundary

Abundant leached and spar-filled molds indicate that initial meteoric waters were undersaturated with respect to aragonite, but were saturated with respect to calcite. The lack of macroscopic pores suggests that the aquifer never reached the stage of open flow and was confined to diffuse flow through the host rock. Cementation was associated with mineralogical stabilization, as suggested by the abundance of cemented samples that contain open or cement-filled molds of

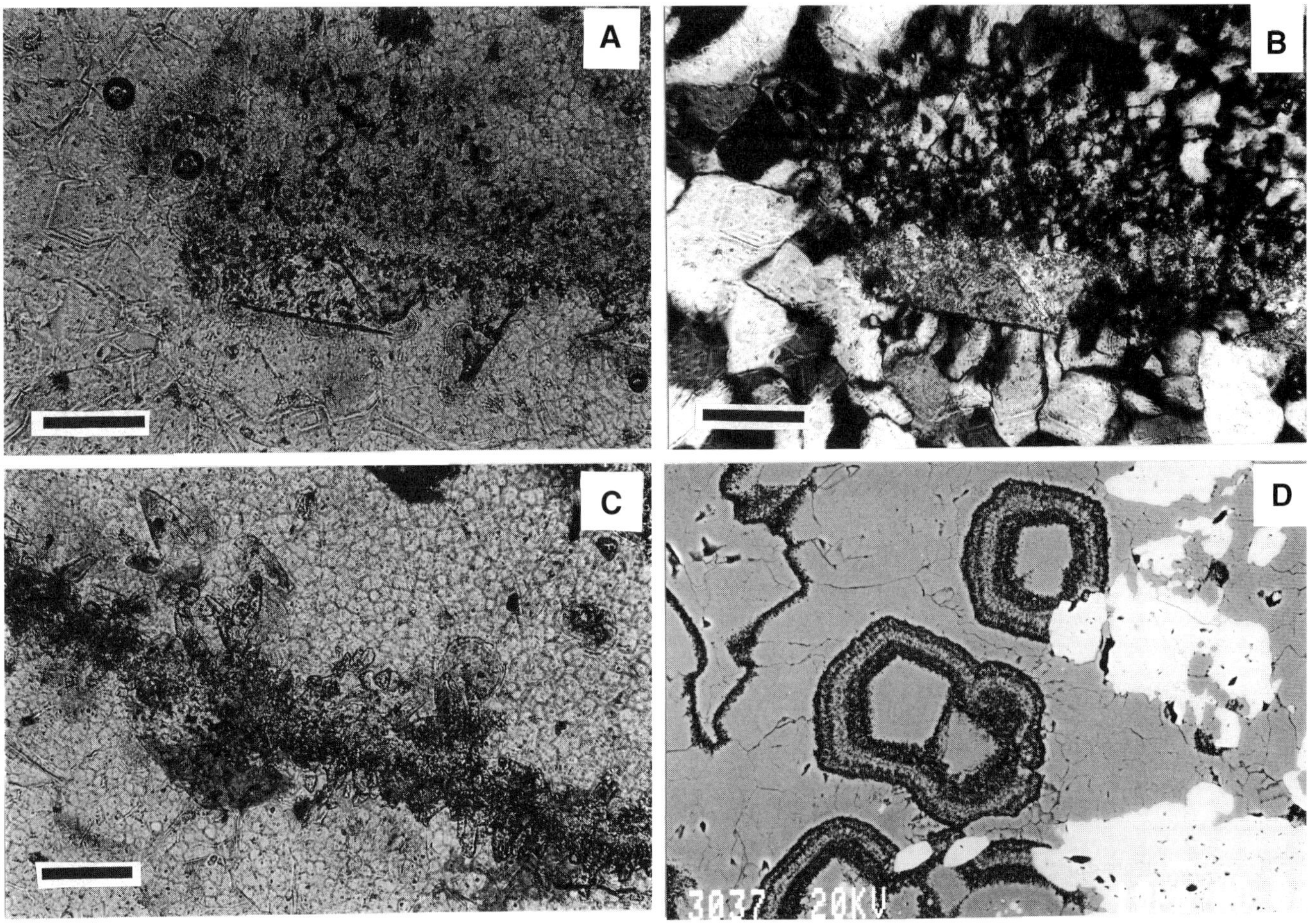

Figure 15. Photomicrograph of silica cements. (A) Microspherules occur as beads concentrated in a geopetal position. Some of the spherules are anchored to bladed calcite crystals. Inclusion-rich megaquartz grows as a later pore fill. Scale bar is 200 μm. M. Focalone (Figures 3 and 4, section 5) (B) Same as (A), in crossed nicols. Scale bar is 200 μm. (C) Enlarged view, detail, to illustrate the growing from a micritic rim, all that is left of a carbonate grain. Scale bar is 100 μm. (D) Back Scatter Electron Image: calcite is white, silica is gray. This image illustrates dissolution of calcite predating precipitation of silica and presence of inclusions within the microspherules.

former aragonitic bioclasts. Calcite cements were derived from localized sources through dissolution of aragonite and incongruent dissolution of high-Mg calcite. Cements, typically nonluminescent with minor bright zones, reflect the oxidizing nature of shallow meteoric waters (Niemann and Read, 1988). Abundance of Fe oxides in updip areas also indicates oxidizing conditions. Calcite cements show some consistent variations in cathodoluminescence patterns, which match possible downdip changes in Eh conditions and chemical gradients typical in carbonate acquifers (e.g., Edmunds and Walton, 1983). Carbon isotopic composition depleted up to 6‰ with respect to the marine values suggests medium to low rock-water interaction (Lohmann, 1988). Cathodoluminescence patterns correlatable across the three locations would indicate mineralogical stabilization and cementation in a stable freshwater system.

The characteristics of the calcrete soil at Colle Daniele suggest semi-arid climatic conditions during the development of the unconformity. Despite this indication, petrographic and isotopic data indicate that meteoric recharge was high enough to establish a stable freshwater lens. However, the long time interval involved in the formation of the unconformity (ca. 5 m.y.) would allow for the formation of the different features at different times, under changing climatic conditions.

Origin of Silica Cements

The timing of silica precipitation, based on the cross-cutting relationships with other features, and the distribution of silica lenses suggest that silicification is related to the upper unconformity (Figure 16). This timing is suggested by the facts that (1) silica distribution is highest on the shelf, down to 60 m below the USSB; (2) silica abundance decreases basinward as well as downsection; (3) silica cements have not been observed above the unconformity; and (4) clasts with silica cements are locally found reworked above the unconformity (Vecsei, 1992, personal communication).

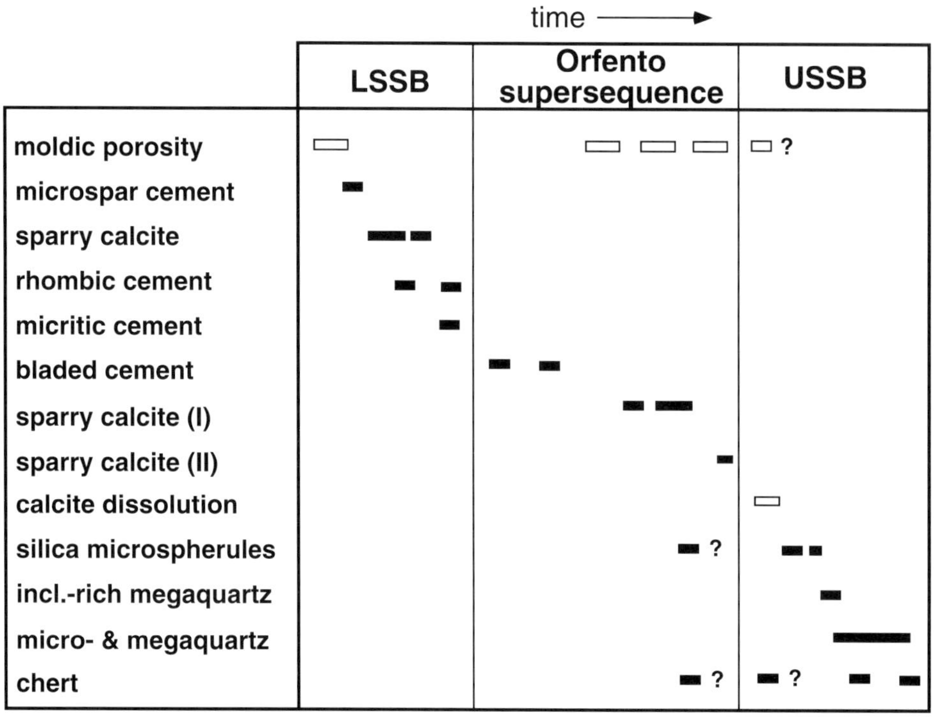

Figure 16. Summary of diagenetic events and their timing with respect to unconformities and depositional events. Black bars indicate cements; white bars indicate porosity.

However, silica cement distribution is irregular, and it cannot be excluded that silicification occurred in more than one episode, already during subaerial exposure events related to minor order sea-level drops within the progradational stage.

Silica cements associated with unconformities have been described in many carbonate platforms (e.g., Meyers, 1977). Petrographic observations indicate the following timing of events: (1) carbonate dissolution, (2) precipitation of silica microspherules, probably initially as opal, (3) precipitation/carbonate replacement by crystalline megaquartz with inclusions, (4) precipitation/carbonate replacement by crystalline micro- and megaquartz, and (5) chertification of fine-grained carbonate lithologies.

Direct quartz precipitation in a diagenetic environment affected by meteoric waters can be explained by the model of Knauth (1979). The model was conceived by analogy with the Dorag mixing-zone model for dolomitization, and the model suggests that silica is precipitated in the groundwater of mixed meteoric-marine coastal systems where dissolution of biogenic opal and mixing of marine and fresh waters can produce water that is highly supersaturated with respect to quartz and undersaturated with respect to calcite. The acidity of pore fluid is an additional important factor for silica precipitation (Lovering and Patten, 1962); in meteoric or mixed waters acidity is largely a function of the availability of carbonic acid or the partial pressure of CO_2. Knauth's model is consistent with the petrographic evidence observed in the Orfento Formation, such as direct precipitation of silica as opal, and silica associated with carbonate dissolution. However, there is no evidence to also apply this scenario to the precipitation of the other silica cements, such as micro- and megaquartz and chert.

Another mechanism of near-surface, inorganic silica precipitation is the formation of silcretes (Summerfield, 1983). Silcretes commonly form under semi-arid conditions with high evaporation rates and highly alkaline pore fluids that rise by capillary action. Precipitation of silica happens when mixing occurs with descending fluids of lower pH (Smale, 1973). Although silcretes most commonly occur at the surface, evidence has been presented that they may also form in the subsurface at depths of 20–30 m below an unconformity (Summerfield, 1983). Silicification of carbonates by groundwater (water-table silcretes) has been discussed by Thiry et al. (1988) who provide evidence for a hydrological control over the distribution of silicified lenses in carbonate strata. Abundance and thickness of quartzose lenses, typically at the meter scale, increase in areas of meteoric discharge. The depth of formation of silcretes below the unconformity is therefore a function of the position of the water table. The geometry of water-table silcretes is the one that best matches the distribution of silica lenses within the Orfento Formation.

In summary, the textures and the distribution of silica cements occurring in the Orfento Formation indicate that they precipitated either in the mixing zone, in the presence of meteoric water mixed with seawater, or in a vadose/phreatic meteoric system. As silcretes form in both humid (water table) and semi-arid (crusts) settings, they are not reliable paleoclimatic indicators by themselves, unless independent criteria exist. A primary problem, yet to be resolved, is posed by the nature of the silica source. In fact, no siliciclastic successions occur within the platform, and no significant amount of biogenic silica has been found in the strata in order to justify the present volume of silica cements. Wind-blown siliciclastic sand would provide a source of silica, but no evidence of such a material has been found in the strata.

CONCLUSIONS

The high porosity preserved in the Orfento supersequence results from a combination of high primary depositional porosity, increased by fabric-selective moldic porosity on aragonitic grains, and preserved due to a lack of extensive marine or meteoric cementation. Sedimentary facies (grain size, sorting, presence or lack of mud, and primary mineralogy of grains) control the distribution of primary porosity, and combined with stratigraphy (intra-Orfento unconformities), also control the distribution of moldic porosity and meteoric cements.

The surfaces with the most extensive record of meteoric diagenesis coincide with the most significant changes in depositional style. However, although both supersequence boundaries were subaerially exposed on the shelf, the diagenetic effects recorded in the strata are different. Meteoric diagenesis was important along the lower boundary, producing moldic porosity and calcite cementation. Silica cements, associated with the upper supersequence boundary, precipitated in lenses with an irregular distribution and reduce the porosity only locally.

The minor-order unconformities within the Orfento supersequence also left a record in the strata, although local. Moldic porosity and meteoric calcite cementation are most abundant in the progradational stage where several minor-order unconformities occur. Meteoric cementation is associated with aragonite dissolution, probably in localized freshwater systems.

The unconformities bounding the Orfento supersequence, although not efficient in creating porosity, are significant as they separate different lithologies and coincide with regional breaks in porosity distribution.

ACKNOWLEDGMENTS

This work is part of a project at E.T.H., coordinated by Daniel Bernoulli, and to which many people have contributed. I thank Daniel Bernoulli, Gregor Eberli, and Flavio Anselmetti for support and much discussion in the field. I am grateful to Gregor Eberli for fearlessly collecting the calcrete samples at Colle Daniele. The support from the Stable Isotope Laboratory at E.T.H. (Judy McKenzie and Stefano Bernasconi) is gratefully acknowledged. Daniel Bernoulli and Flavio Anselmetti read and commented on an early version of the manuscript. A. Vecsei and D. Sanders provided access to unpublished data. U. Gerber is thanked for his photographic work. AAPG reviewers P. D. Crevello, L. A. Melim, and P. M. Harris provided suggestions that helped improve the paper. Financial support for this project was provided by the Swiss National Science Foundation (Grants 20-35907.92 and 20-27457.89).

REFERENCES CITED

Accarie, H., 1988, Dynamique sédimentaire et structurale au passage platform/bassin. Les faciès Crétacés et Tertiaries: massif de la Maiella (Abruzzi, Italie): Ecole des Mines de Paris, Mémoires des Sciences de la Terre, n. 5, 158 p.

Aissaoui, D.M., Buiges, D., and Purser, B.H., 1986, Model of reef diagenesis: Mururoa Atoll, French Polynesia, *in* Schroeder, J.H., and Purser, B.H., eds., Reef Diagenesis: New York, Springer-Verlag, p. 27–52.

Allan, J.R., and Matthews, R.K., 1982, Isotopic signatures associated with early meteoric diagenesis: Sedimentology, v. 29, p. 797–817.

Bally, A., 1954, Geologisches Untersuchungen in den SE-Abruzzen, Ph.D. dissertation: University of Zürich, 289 p.

Bathurst, R.G.C., 1975, Carbonate sediments and their diagenesis: Elsevier, 658 p.

Bernoulli, D., 1972, North Atlantic and Mediterranean Mesozoic facies, a comparison, *in* C.D. Hollister, J.I. Ewing, et al., eds., Initial Reports Deep Sea Drilling Project, v. 11, p. 801–871.

Bernoulli, D., and Jenkyns, H.C., 1974, Alpine, Mediterranean and central Atlantic Mesozoic facies in relation to the early evolution of the Thetys, *in* R.H. Dott, Jr., and R.H. Shaver, eds., Modern and Ancient Geosynclinal Sedimentation: SEPM Special Publication 19, p. 129–160.

Bernoulli, D., Eberli, G.P., Pignatti, J.S., Sanders, D., and Vecsei, A., 1992, Sequence stratigraphy of Montagna della Maiella, Quinto Simposio di Ecologia e Paleoecologia delle comunità bentoniche, Paleobenthos, Roma: Libro-guida delle escursioni, p. 85–109.

Borgomano, J., and Philip, J., 1990, The rudist carbonate build-ups and the gravitary carbonates of the Gargano-Apulian margin (Southern Italy, Upper Senonian): Bollettino della Societa' Geologica Italiana, v. 40, p. 125–132.

Bosellini, A., Neri, C., and Luciani, V., 1993, Platform margin collapses and sequence stratigraphic organization of carbonate slopes: Cretaceous–Eocene, Gargano Promontory, southern Italy: Terra Nova, v. 5, p. 282–297.

Budd, D.A., and Land, L.S., 1990, Geochemical imprint of meteoric diagenesis in Holocene ooid sands, Schooner Cays, Bahamas: correlation of calcite cement geochemistry with extant groundwaters: Journal of Sedimentary Petrology, v. 60, p. 361–378.

Carta Geologica d'Italia, 1970, Foglio 147, Lanciano, 1:100.000; Roma, Servizio Geologico d'Italia.

Catenacci, E., 1965, Resoconto sommario delle osservazioni stratigrafiche compiute sulla Maiella (Appennino abruzzese): Bollettino del Servizio Geologico Italiano, v. 86, p. 17–25.

Crescenti, U., Crostella, A., Donzelli, G., and Raffi, G. 1969, Stratigrafia delle serie calcarea dal Lias al Miocene nella regione Marchigiano-Abruzzese (Parte II-Litostratigrafia, biostratigrafia, paleogeografia): Memorie della Societa' Geologica Italiana, v. 8, p. 343–420.

Eberli, G.P., Bernoulli, D., Sanders, D., and Vecsei, A., 1993, From aggradation to progradation: the Maiella platform (Abruzzi, Italy), *in* A.J. Simo, R. Scott, and J.-P Masse, eds., Atlas of Cretaceous Carbonate Platforms: AAPG Memoir 56, p. 213–232.

Edmunds, W.M., and Walton, N.R.G., 1983, The Lincolnshire Limestone-hydrochemical evolution over a ten-year period: Journal of Hydrology, v. 61, p. 201–211.

Enos, P., and Sawatsky, L.H., 1981, Pore networks in Holocene carbonate sediments: Journal of Sedimentary Petrology, v. 51, p. 961–985.

Esteban, M., 1991, Paleokarst: case histories, *in* Paleokarsts and Paleokarstic Reservoirs: Postgraduate Research Institute for Sedimentology, University of Reading PRIS Contribution no. 152, p. 120–158.

Halley, R.B., and Harris, P.M., 1979, Freshwater cementation of a 1,000 years old oolite: Journal of Sedimentary Petrology, v. 49, p. 969–988.

Hammes, U., 1992, Sedimentation patterns, sequence stratigraphy, cyclicity, and diagenesis of early Oligocene carbonate ramp deposits, Suwannee Formation, Southwest Florida, U.S.A.: Ph.D. dissertation, University of Colorado, Boulder, 344 p.

Haq, B.U., Hardenbol, J., and Vail, P.R., 1988, Mesozoic and Cenozoic chronostratigraphy and cycles of sea-level change, *in* Wilgus, C.K. et al., eds., Sea-level Changes—An Integrated Approach: SEPM Special Publication 42, p. 71–108.

Hesse, R., 1990, Silica diagenesis: origin of inorganic and replacement cherts, *in* McIlreath, I.A., and Morrow, D.W., eds., Diagenesis: Geoscience Canada, Reprint Series 4, p. 253–275.

James, N.P., and Choquette, P. W., 1990, Limestones—the sea-floor diagenetic environment, *in* McIlreath, I.A., and Morrow, D.W., eds., Diagenesis: Geoscience Canada Reprint Series 4, p. 13–34.

James, N.P., Ginsburg, R.N., Marszalek, D.S., and Choquette, P.W., 1976, Facies and fabric specificity of early sub-sea cements in shallow Belize (British Honduras) reefs: Journal of Sedimentary Petrology, v. 40, p. 457–462.

Knauth, L.P., 1979, A model for the origin of chert in limestones: Geology, v. 7, p. 274–277.

Lohmann, K.C., 1988, Geochemical patterns of meteoric diagenetic systems and their application to studies of paleokarst, *in* James, N.P., and Choquette, P.W., eds., Paleokarst: New York, Springer-Verlag, p. 58–80.

Longmann, M.W., 1980, Carbonate diagenetic textures from near surface diagenetic environments: AAPG Bulletin, v. 64, p. 461–487.

Lovering, T.G., and Patten, L.E., 1962, The effect of CO_2 at low temperature and pressure on solutions supersaturated with silica in the presence of limestone and dolomite: Geochimica et Cosmochimica Acta, v. 26, p. 253–284.

Meyers, W.J., 1977, Chertification in the Mississippian Lake Valley Formation, Sacramento Mountains, New Mexico: Sedimentology, v. 24, p. 75–105.

Mutti, M., Bernoulli, D., and Eberli, G.P., 1994, Progradation of Late Cretaceous sequences in the Maiella: the role of changes in sea-floor topography and sediment reworking in sequence organization (abs.): International Association of Sedimentologists 15th Regional Meeting, Ischia 1994

Niemann, J.C., and Read, J.F., 1988, Regional cementation from unconformity-recharged aquifer and burial fluids, Mississippian Newmann Limestone, Kentucky: Journal of Sedimentary Petrology, v. 58, p. 688–705.

Pierson, B.J., and Shinn, E.A., 1980, Cement distribution and carbonate mineral stabilization in Pleistocene limestones of Hosty reef, Bahamas, *in* Schneidmann, N., and Harris, P.M., eds., Carbonate Cements: SEPM Special Publication 36, p. 153–168.

Saller, A.H., Budd, D.A., and Harris, P.M., 1994, Unconformities and porosity development in carbonate strata: Ideas from a Hedberg Conference, AAPG Bulletin, v. 78, p. 857–872.

Scholle, P.A., and Halley, R.B., 1985, Burial diagenesis: out of sight, out of mind! *in* Schneidermann, N., and Harris, P.M., eds., Carbonate Cements: SEPM Special Publication 36, p. 309–334.

Scotese, C.R., Gahagan, L.M., and Larson, R.L., 1989, Plate tectonic reconstructions of the Cretaceous and Cenozoic ocean basins: Tectonophysics, v. 155, p. 27–48.

Smale, D., 1973, Silcretes and associated silica diagenesis in southern Africa and Australia: Journal of Sedimentary Petrology, v. 43, p. 1077–1089.

Summerfield, M.A., 1983, Petrography and diagenesis of silcretes from the Kalahari basin and Cape coastal zone, Southern Africa: Journal of Sedimentary Petrology, v. 53, p. 895–909.

Thiry, M., Bertrand Ayrault, M., Grisoni, J.-P., Menillet, F., and Schmitt, J.-M., 1988, Les Grès de Fontainbleau: silicification de nappes liées à l'evolution géomorphologique du bassin de Paris durant le Plio-Quaternaire: Bulletin de la Societé Géologique de France, v. 8, p. 419–430.

Vecsei, A., 1991, Aggradation und Progradation eines Karbonatplattform-Randes: Kreide bis Mittleres Tertiär der Montagna della Maiella, Abruzzen: Mitteilungen aus dem Geologischen Institut der Eidgenössischen Technischen Hochschule und der Universität Zürich, Neue Folge, n. 294, 171 p.

Wright, V.P., 1986, The role of fungal biomineralization in the formation of Early Carboniferous soil fabrics: Sedimentology, v. 33, p. 831–838.

Wright, V.P., 1990, A micromorphological classification of fossil and recent calcitic and petrocalcic microstructures, *in* Proceedings International Workshop Meeting Soil Micromorpholgy: San Antonio, 1988, Elsevier.

Chapter 8

Unconformity-Related Porosity Development in the Quintuco Formation (Lower Cretaceous), Neuquén Basin, Argentina

Neil F. Hurley
Marathon Oil Company
Littleton, Colorado, U.S.A.

Haydn C. Tanner
Marathon Oil Company
London, U.K.

Carlos Barcat
Marathon Oil Company
Buenos Aires, Argentina

ABSTRACT

Porous dolomites are present below a distinctive stratigraphic marker within the lower Quintuco Formation (Lower Cretaceous, Berriasian–lower Valanginian) in the eastern Neuquén basin, Argentina. Dolomitized packstones and wackestones with moldic and sucrosic porosity provide the main reservoir facies in Rio Neuquén field and perhaps other oil fields in the area.

Lower Quintuco carbonates are comprised of: (1) oolitic grainstones, (2) burrowed, dolomitized oolite-skeletal-peloid packstones/wackestones, (3) dolomudstones and bedded anhydrites, and (4) very fine-grained, superficially coated oolite grainstones. These sediments are commonly packaged into shoaling- and coarsening-upward parasequences.

Reservoir-quality porosity and permeability exist almost exclusively in burrowed, dolomitized packstones and wackestones. These strata are interpreted as off-bar facies deposited on the landward side of bar complexes, similar to modern facies analogs known in the Joulters Cay area of the Bahamas. In the lower Quintuco Formation, dolomite preferentially replaced carbonate mud. Below an inferred widespread paleo-exposure surface, ooid-skeletal-peloid grains were then dissolved to leave an open pore network with abundant moldic and intercrystalline porosity.

INTRODUCTION

The Neuquén basin, in west-central Argentina, is currently the country's most important hydrocarbon-producing region (Yrigoyen, 1993). The study area, which lies in the eastern part of the basin, is centered around Concession CNQ 17, the Sierras Blancas block (Figure 1). Rio Neuquén field, which lies to the south of the concession, and Loma de la Lata field, which lies to the west of the concession, are the most important nearby fields. Much of their production comes from carbonate rocks of the Quintuco–Loma Montosa Formation and from siliciclastic rocks of the Sierras Blancas Formation (all Upper Jurassic to Lower Cretaceous). Rio Neuquén field is estimated to have 9.2×10^6 m^3 (58 million bbl) of recoverable oil, and Loma de la Lata field is thought to have recoverable reserves of 245×10^9 m^3 (8.7 tcf) of gas and 33×10^6 m^3 (208 million bbl) of condensate (Hogg, 1993).

Sequence-stratigraphic studies in the Neuquén basin (Mitchum and Uliana, 1985, 1987; Legarreta, 1991; Eisner, 1991; Uliana and Legarreta, 1993) have suggested that the Quintuco–Loma Montosa–Vaca Muerta interval is a prograding, ramp-type deposit. Lithologies include oolitic grainstones and associated facies. Detailed microfacies descriptions of Quintuco–Loma Montosa samples are available in Carozzi (1989) and Carozzi et al. (1993).

This paper summarizes a study of cores, thin sections, wireline logs, and seismic data from the area in and around the Sierras Blancas block. Initial project goals were to determine the nature and origin of productive vs. nonproductive facies, relate porosity type and diagenetic history to facies, and derive a depositional model for oolitic carbonate sediments. As work progressed, unconformity recognition and the detection of unconformity-related porosity became major objectives. This study found that the best reservoir rocks occur in burrowed, sucrosic, dolomitized oolitic-skeletal packstones in the lowermost Quintuco Formation. Abundant moldic porosity in this facies is thought to have resulted from selective dissolution of nondolomitized allochems during a widespread subaerial paleo-exposure event.

GEOLOGIC SETTING

Stratigraphy

A stratigraphic column (Figure 2) shows the interval of interest. The most important hydrocarbon-producing units are carbonates of the Cretaceous Quintuco Formation (about 600 m thick; Figure 3) and sandstones of the Jurassic Sierras Blancas Formation (about 200 m thick).

Inconsistency exists in the stratigraphic nomenclature (Figure 2). Most authors believe that the Vaca Muerta Formation is the basinal equivalent of at least part of the Quintuco–Loma Montosa Formation (Mitchum and Uliana, 1985, 1987; Uliana and Legarreta, 1993; Carozzi et al., 1993). Carozzi et al. (1993, their Figure 3) show Loma Montosa as the shelf and Quintuco as the argillaceous shelf-to-basin transitional unit that lies between Loma Montosa and the Vaca Muerta shales. Uliana and Legarreta (1993, their Figure 8) show a cross section that includes the Quintuco–Loma Montosa–Vaca Muerta interval with the 126 Ma and 138.5 Ma sequence boundaries of Haq et al. (1987). Uliana and Legarreta (1993) equate Loma Montosa to the lower half of the carbonate section updip of the Vaca Muerta. The upper half is termed Quintuco. Because of these inconsistencies, the entire updip carbonate part of the Quintuco–Loma Montosa–Vaca Muerta interval in the eastern Neuquén basin is termed Quintuco Formation in this study (Figure 2).

Paleontologic age dating of the sequence is based on ammonites and palynomorphs (see references cited in Kugler, 1987). Unpublished reports by YPF (Yacimientos Petroliferos Fiscales) paleontologists determined that the Quintuco is Berriasian–early Valanginian in the study area. Carozzi et al. (1993) report similar ages in their study. If Uliana and Legarreta (1993) are correct, their 126 Ma and 138.5 Ma ages would place the Quintuco–Loma Montosa–Vaca Muerta in the Tithonian to early Valanginian.

Known source rocks occur in shales and dolomudstones of the Vaca Muerta Formation (Kugler, 1987). Geochemical analyses of a limited number of Vaca Muerta samples from a cored well in this study indicated total organic carbon (TOC) contents of 1 to 3%. This is consistent with Kugler's (1987) results of TOC = $2.11 \pm 1.62\%$ ($n = 85$). Pyrolysis studies suggest that oil-prone, type II amorphous kerogen is dominant. Today, the Vaca Muerta is within the liquid hydrocarbon-generation window in much of the Neuquén basin.

Structure

The Neuquén basin was once a back-arc basin that is now part of the sub-Andean foreland. The basin subsided actively from the Jurassic through Tertiary (Keeley et al., 1992; Hogg, 1993). Figure 1 shows major structural elements. The Agrio fold belt is present along the western margin, and the Huincul (Dorsal) arch is a major positive feature along the southern margin of the basin. The Sierras Blancas concession lies along the Añelo trough and the east-northeast flank of the basin.

Oil fields such as Rio Neuquén and Loma de la Lata (Figure 3) are thought to be combination structural-stratigraphic traps. Vaca Muerta source rocks in the area of Rio Neuquén field entered the oil window at the end of the Cretaceous (Kugler, 1987). Presumably, structures had to predate late Cretaceous oil migration if they were to hold large volumes of oil.

LITHOFACIES: QUINTUCO FORMATION

The Quintuco Formation is composed of variable quantities of limestone, dolomite, anhydrite, and

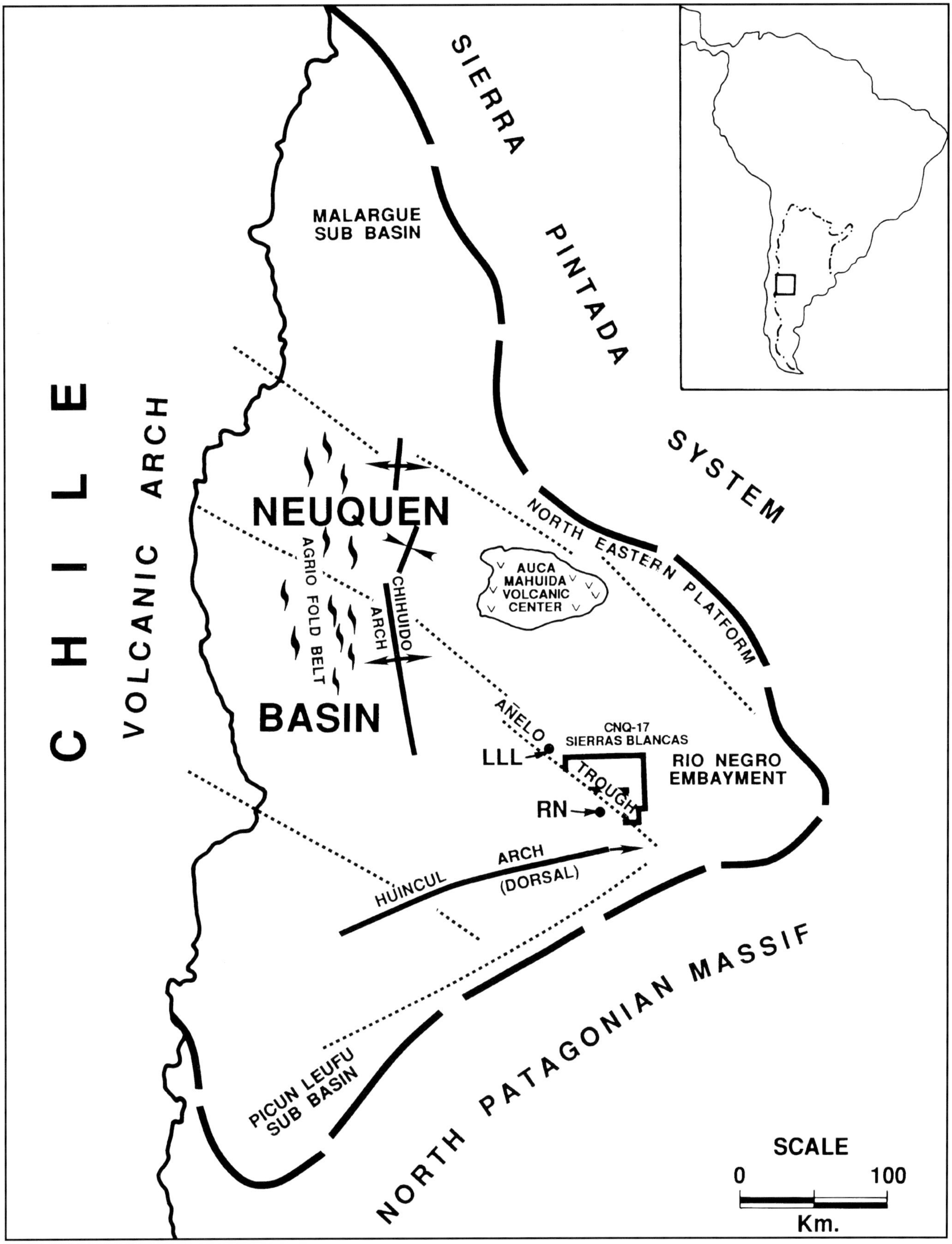

Figure 1. Index map showing Neuquén basin and major structural elements. Note the location of the Sierras Blancas concession, Rio Neuquén field (RN), and Loma de la Lata field (LLL).

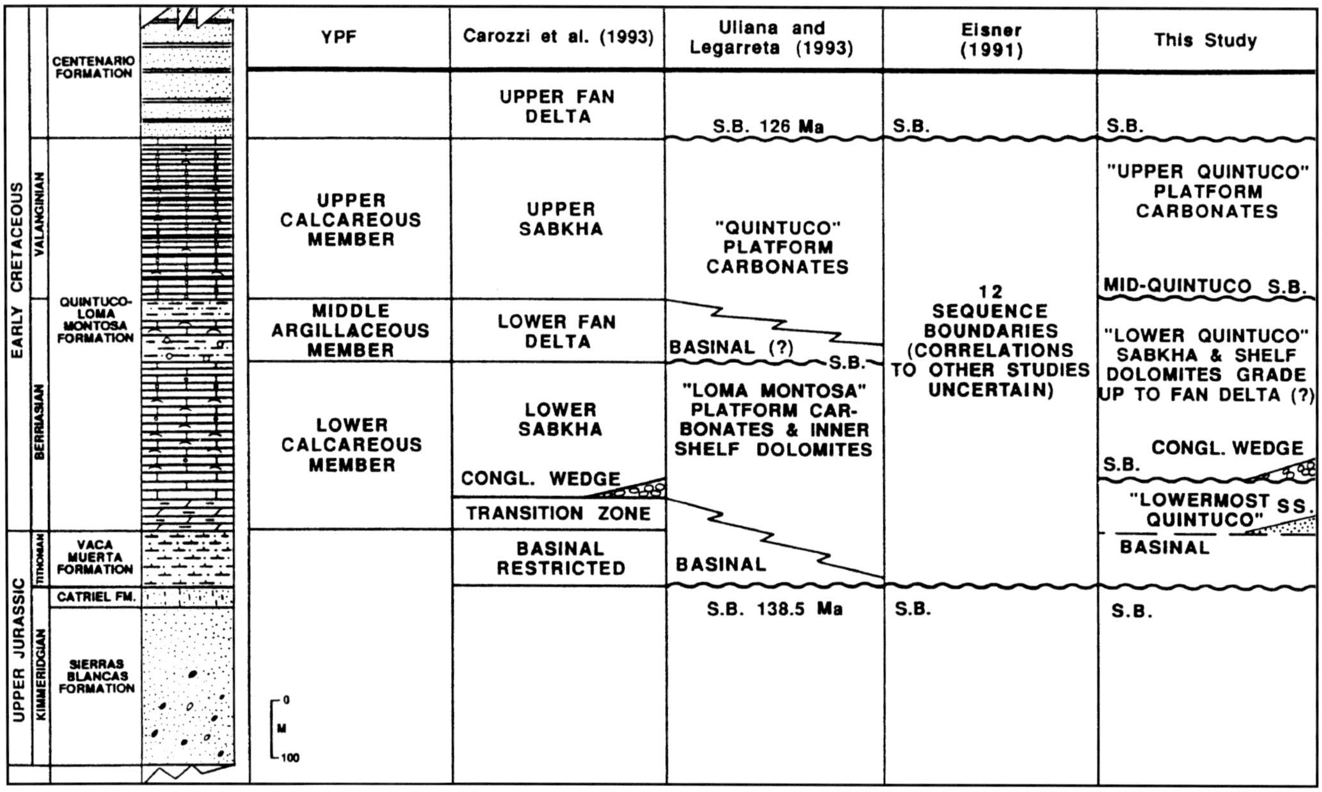

Figure 2. Stratigraphic column showing Cretaceous and Jurassic formations. This figure compares sequence boundaries and stratigraphic nomenclature used by various workers. Abbreviations: congl. = conglomerate; S.B. = sequence boundary; ss. = sandstone; Ma = million years ago.

quartz/volcaniclastic sand. Fine to medium quartz and volcaniclastic sand grains are mixed with all carbonate lithologies, and they commonly occur as the nuclei of Quintuco ooids. Discrete sandstone beds are relatively rare, although an exception occurs in the basal 20–30 m of the Quintuco where sandstone and siltstone layers do occur.

In the Quintuco–Vaca Muerta section, 33 cores from 13 wells and 183 thin sections have been examined (Table 1; Figure 3). In addition, Marathon and YPF cuttings descriptions were available for the AA.x-1, AB.x-1, ECr.x-1, and CS.x-1 wells. All cores used in this study were drilled by YPF as exploratory and development wells. Many wells had two, three, or more short (8 to 10 m) cores drilled at widely spaced intervals. Vast thicknesses of section were never cored. Because of this, it was important to be able to recognize facies from logs, and then make correlations to uncored intervals. All core descriptions, therefore, were plotted on montages that showed all available logs. Generally, SP-resistivity logs were available for each well. Many wells also had some combination of GR, sonic, density, and neutron logs. Note that the presence of clay-rich volcaniclastic particles throughout the section makes the GR log unreliable for lithology determinations and V_{shale} calculations. The following section discusses lithofacies and petrophysical characteristics of rocks within the Quintuco Formation.

Oolite Grainstone

This facies, which is common throughout the Quintuco, is characterized by cross-bedded, fine- to medium-grained ooids that are generally tightly cemented by rim and equant calcite cements (Figure 4). Broken fossil debris, such as crinoid fragments and mollusk shells, are abundant. Because of extensive cementation, oolite grainstones commonly have small SP deflections (15 to 45 mV to the left of the shale line) and relatively high resistivities (10 to 100 ohm m). In cases where significant SP deflections occur, thin-section examination shows that some interparticle porosity remains. Shallow, medium, and deep resistivity curves generally overlie each other in this facies (Figure 5). Because the resistivity of mud filtrate is higher than that of formation water, the lack of separation between deep and shallow resistivity readings suggests no invasion, therefore, no permeability (Asquith and Gibson, 1982).

Oolite-Skeletal-Peloid Packstone to Wackestone

This facies is very common throughout the lower third of the Quintuco Formation. In general, lithology ranges from dolomitic to thoroughly dolomitized. Anhydrite occurs as interlayered, resistive beds as minor pore-filling cement, and as nodules throughout the section.

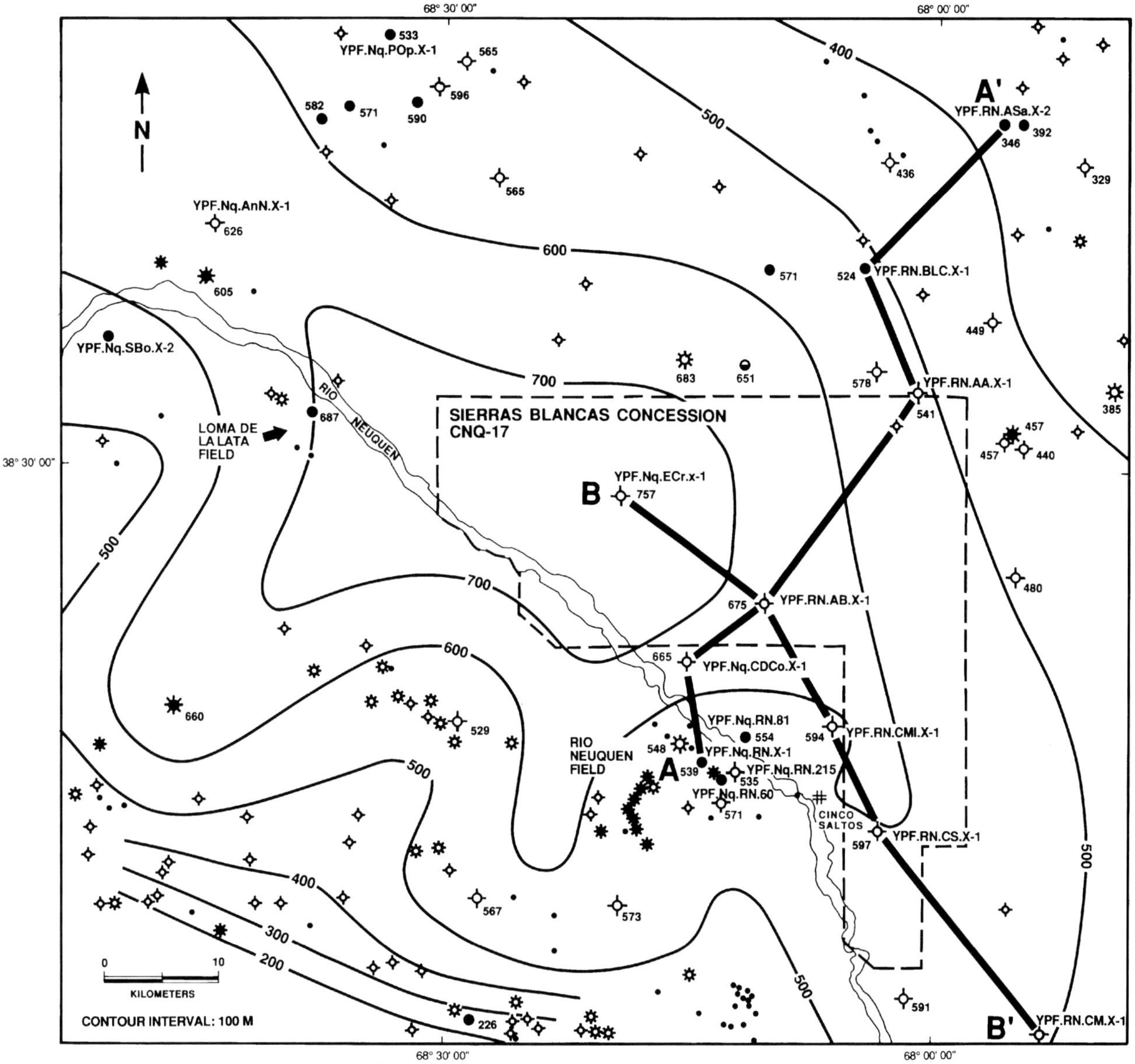

Figure 3. Isopach map of the Quintuco Formation showing location of cored wells and lines of cross section. The isopach map was constructed from wells shown with adjacent numbers. Cored wells and wells within cross sections A–A′ and B–B′ are labeled by name.

In core, sedimentary structures such as cross-bedding and laminations are rarely preserved. Instead, pervasive burrowing has churned the sediment. Burrows appear to be large (several centimeters across) tubular features somewhat analogous to *Thalassinoides* (Chamberlain, 1992). Y-shaped branches have not been observed. The net effect of burrowing has been homogenization of the sediment to packstone and wackestone textures, regardless of original texture. This is important because dolomite has preferentially replaced the mud. Such preferential dolomitization of carbonate mud has been documented by Murray and Lucia (1967) and others. A complete spectrum exists from partially dolomitized packstones (Figures 6A and 6B), to fully dolomitized packstones (Figure 6C), to fully dolomitized, moldic packstones (Figure 6D). In fully dolomitized rocks, intercrystalline porosity is very well developed (Figure 7). These rocks have porosities that can range from 15 to 30%, and permeabilities that can range from 10 to 230 md. Cores examined in this study suggest that moldic, sucrosic packstones are the main reservoir facies in Rio Neuquén field.

Dolomitized packstones and wackestones have distinctive log signatures (Figure 8). Deflections of the SP are commonly 60 to 90 mV to the left of the shale line. In Rio Neuquén field, SP deflections are suppressed by a hydrocarbon effect. Significant separation commonly exists between deep and shallow resistivity traces in water-saturated dolomitized packstones and wackestones. This separation of resistivity logs suggests mud-filtrate invasion and indicates a permeable

Table 1. Core samples and thin sections used in the Neuquén basin study, Quintuco–Loma Montosa–Vaca Muerta interval.

Well	Formation	Core Depth (m)	Thickness (m)	Thin Sections
YPF.RN.AA.x-1	Upper Quintuco	2156–2164	8	23
YPF.Nq.SBo.x-2	Upper Quintuco	2048–2056	8	1
YPF.Nq.AñN.x-1	Upper Quintuco	2614–2623	9	1
YPF.RN.AB.x-1	Upper Quintuco	2370–2378	8	6
YPF.RN.BLC.x-1	Upper Quintuco	2120–2128	8	15
YPF.RN.BLC.x-1	Upper Quintuco	2128–2135	7	14
YPF.RN.BLC.x-1	Upper Quintuco	2230–2248	18	19
YPF.RN.BLC.x-1	Upper Quintuco	2248–2254	6	6
YPF.Nq.RN.215	Lower Quintuco	2185–2204	19	0
YPF.Nq.RN.215	Lower Quintuco	2250–2260	10	0
YPF.Nq.RN.81	Lower Quintuco	2333–2342	9	0
YPF.Nq.RN.81	Lower Quintuco	2487–2494	7	2
YPF.Nq.RN.81	Lower Quintuco	2494–2496	2	0
YPF.Nq.RN.81	Lower Quintuco	2496–2504	8	2
YPF.Nq.RN.81	Lower Quintuco	2520–2528	8	1
YPF.Nq.RN.81	Lower Quintuco	2529–2530	1	0
YPF.Nq.RN.60	Lower Quintuco	2506–2508	2	0
YPF.Nq.RN.60	Lower Quintuco	2526–2528	2	0
YPF.Nq.RN.60	Lower Quintuco	2575–2578	3	1
YPF.RN.CS.x-1	Lower Quintuco	2694–2703	9	0
YPF.RN.ASa.x-2	Lower Quintuco	1893–1902	9	0
YPF.Nq.CDCo.x-1	Lower Quintuco	2666–2674	8	0
YPF.Nq.POp.x-1	Lower Quintuco	2610–2617	7	0
YPF.RN.AA.x-1	Lower Quintuco	2321–2324	3	14
YPF.RN.AA.x-1	Lower Quintuco	2414–2422	8	20
YPF.RN.AB.x-1	Lower Quintuco	2570–2578	8	8
YPF.RN.AB.x-1	Lower Quintuco	2643–2647	4	0
YPF.RN.AB.x-1	Lower Quintuco	2647–2655	8	8
YPF.RN.AB.x-1	Lower Quintuco	2757–2765	8	16
YPF.RN.BLC.x-1	Lower Quintuco	2438–2446	8	9
YPF.RN. BLC.x-1	Lower Quintuco	2491–2500	9	14
YPF.RN. CM.x-1	Vaca Muerta	2718–2721	3	1
YPF.RN. CM.x-1	Vaca Muerta	2736–2744	8	2
		Totals	243	183

Prefixes: YPF = company that drilled the wells; RN = Rio Negro Province; Nq = Neuquén Province.

formation (Asquith and Gibson, 1982). High SP deflection is also a permeability indicator because electrochemical currents need to flow along a permeable path. Crossplots of normalized SP vs. the ratio of shallow to deep resistivity have been used to estimate the thickness of net permeable rock in a given well. Net thickness of reservoir-quality rock ranges up to 30 m.

Dolomudstone and Layered Anhydrite

Dolomudstones and layered anhydrites are also common in the lower part of the Quintuco Formation, generally interbedded with and overlying the previous lithofacies. Porosity (0 to 3%) and permeability (few millidarcys) are low throughout this facies package. The log signature typically contains moderate to low (15 to 45 mV) SP deflections and numerous high-resistivity (≥50 ohm m) streaks that correspond to thin (≤1 m) anhydrite layers. Anhydrite is also common as nodules and as cements in intercrystalline porosity.

Very Fine, Superficially Coated Oolite Grainstones

This facies is present in cores examined from basinal settings (e.g., RN.81, CDCo.x-1, and AB.x-1). Rocks are typically a mixture of quartz/volcaniclastic silt and superficial ooids, and cross-beds are common. In many cases, siliciclastic grains occur as the nuclei of ooids. These rocks are very dense and have little or no porosity or permeability.

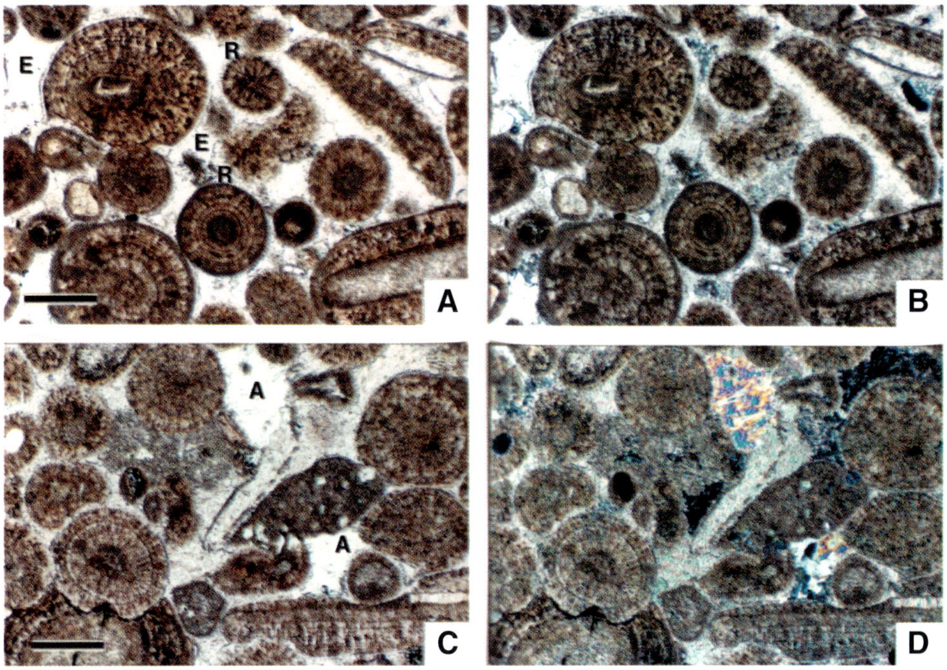

Figure 4. Photomicrographs, oolite grainstone facies, Quintuco Formation. (A) Nonporous oolite grainstone typical of this facies in the upper Quintuco Formation. Note excellent preservation of ooids. Cements include rim (R) and equant (E) calcites. BLC.x-1 well, 2130.5 m depth. Scale bar = 0.25 mm. (B) Same view as (A), but under cross-polarized light. (C) Nonporous oolite grainstone in the upper Quintuco Formation. Minor anhydrite (A) cement is present as well as rim and equant calcite. BLC.x-1 well, 2238.5 m depth. Scale bar = 0.25 mm. (D) Same view as (C), but under cross-polarized light.

SEQUENCE STRATIGRAPHY: QUINTUCO FORMATION

Cyclicity

Quintuco facies are commonly arranged in coarsening-upward, shoaling-upward parasequences that range from one to several meters in thickness. A common cycle consists of silty, very fine, superficially coated oolite grainstone that grades upward into fine- to medium-grained, cross-bedded oolite grainstone (Figure 5). Cyclicity in dolomudstone/bedded anhydrite intervals commonly occurs at a much smaller scale (≤1 m). Cycles have generally been homogenized in burrowed, dolomitized oolitic packstones/wackestones.

Different portions of the Quintuco Formation have distinctive groupings of lithofacies. For example, in well AA.x-1, the lower 100 m of the Quintuco is dominated by burrowed, dolomitized, moldic packstones/wackestones. This grades upward into interbedded dolomudstone and layered anhydrites. A silty facies about 50 m thick occurs in the middle Quintuco. The upper Quintuco (about 250 m thick) is composed mainly of nonporous, oolitic limestone with variable quantities of quartz/volcaniclastic sandstone and siltstone.

Cross Sections

To help derive a facies/porosity model in the study area, stratigraphic cross sections (Figures 9 and 10) were constructed from lithologic and wireline-log information. Lithologic information has been incorporated, where available, from the following sources of data: (1) core and thin-section descriptions, (2) cuttings descriptions for wells AA.x-1, AB.x-1, ECr.x-1, and CS.x-1, and (3) character recognition from SP, resistivity, and porosity logs.

The thickest oolite grainstones in the lower Quintuco occur in the vicinity of the AB.x-1 and ECr.x-1 wells (Figure 10). A broad zone of burrowed, dolomitized packstones apparently stretches from near the AB.x-1 well toward the AA.x-1 and BLC.x-1 wells to the northeast (Figure 9) and the CM.x-1 well to the east (Figure 10). Well ASa.x-2 (Figure 9), the farthest well to the northeast, appears to have age-equivalent intertidal-to-sabkha–type deposits in the lowermost Quintuco.

Overlying the lowermost Quintuco is an inferred paleo-exposure surface. This surface, which can be correlated among most wells in the study area, is shown as a wavy line and labeled "probable exposure surface" in Figures 9 and 10. A number of notable features occur in the vicinity of this surface: (1) drilling

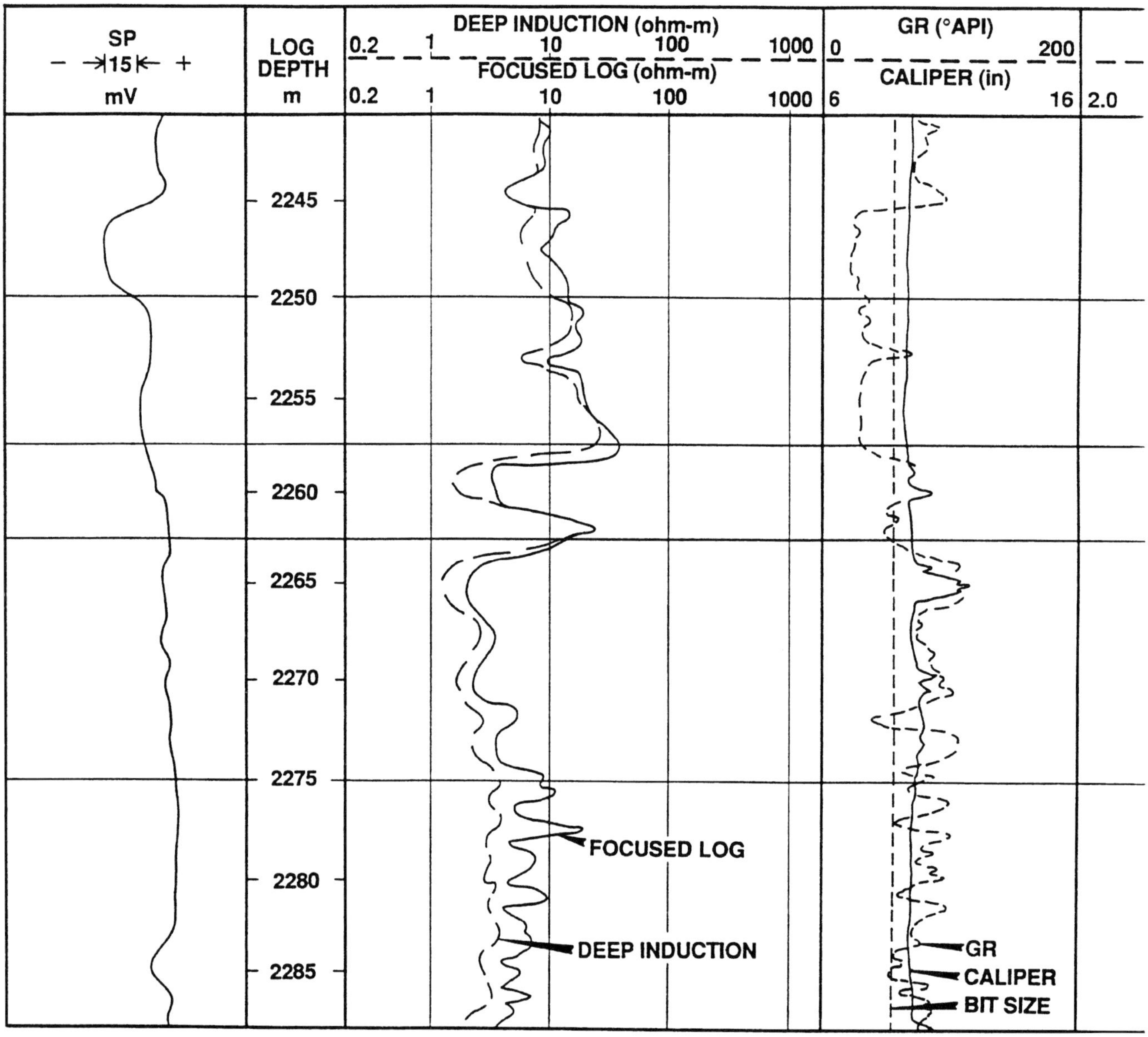

Figure 5. "Type log" of oolitic grainstone facies, upper Quintuco Formation. This log is from the BLC.x-1 well. Note the suppressed SP and porosity responses in most of this interval. High porosities in silty claystone layers (especially below 2258 m log depth) are unreliable because of hole washouts and corresponding high activity in the ΔRHO curve. In grainstone layers, resistivities tend to be blocky and high, with little separation

breaks, indicating a change from low to high porosity, occurred in most wells at this stratigraphic level, leading YPF to take many cores, (2) most of the perforations in Rio Neuquén field lie below this surface and wells have produced hundreds of BOPD from moldic, sucrosic dolomites, (3) two drill-stem tests in well AB.x-1 just below this surface produced a combined 900 BWPD from moldic, sucrosic dolomites, (4) core studies suggest that the facies above and below this surface are generally similar; however, layered anhydrite interbeds and dolomudstones are more common above the surface, (5) core studies show that moldic porosity is abundant only below this surface, and (6) a conglomeratic lag of volcaniclastic cobbles is seen just above the zone of moldic porosity in well BLC.x-1. This lag may have been introduced during a sea-level lowstand by a stream from the landmass to the north-northeast. Note that Carozzi et al. (1993, p. 438) described a conglomerate at essentially the same stratigraphic level in well 1 of their cross section V. This is the CM.x-1 well of cross section B-B′ in this study (Figure 10).

In certain wells, the "probable exposure surface" cannot be recognized. One example is the ECr.x-1 well, which was a relatively basinward well during deposition of the lowermost Quintuco (Figures 3 and 10). Sediments in this well may have been continuously subtidal while allochems in the vicinity of other wells

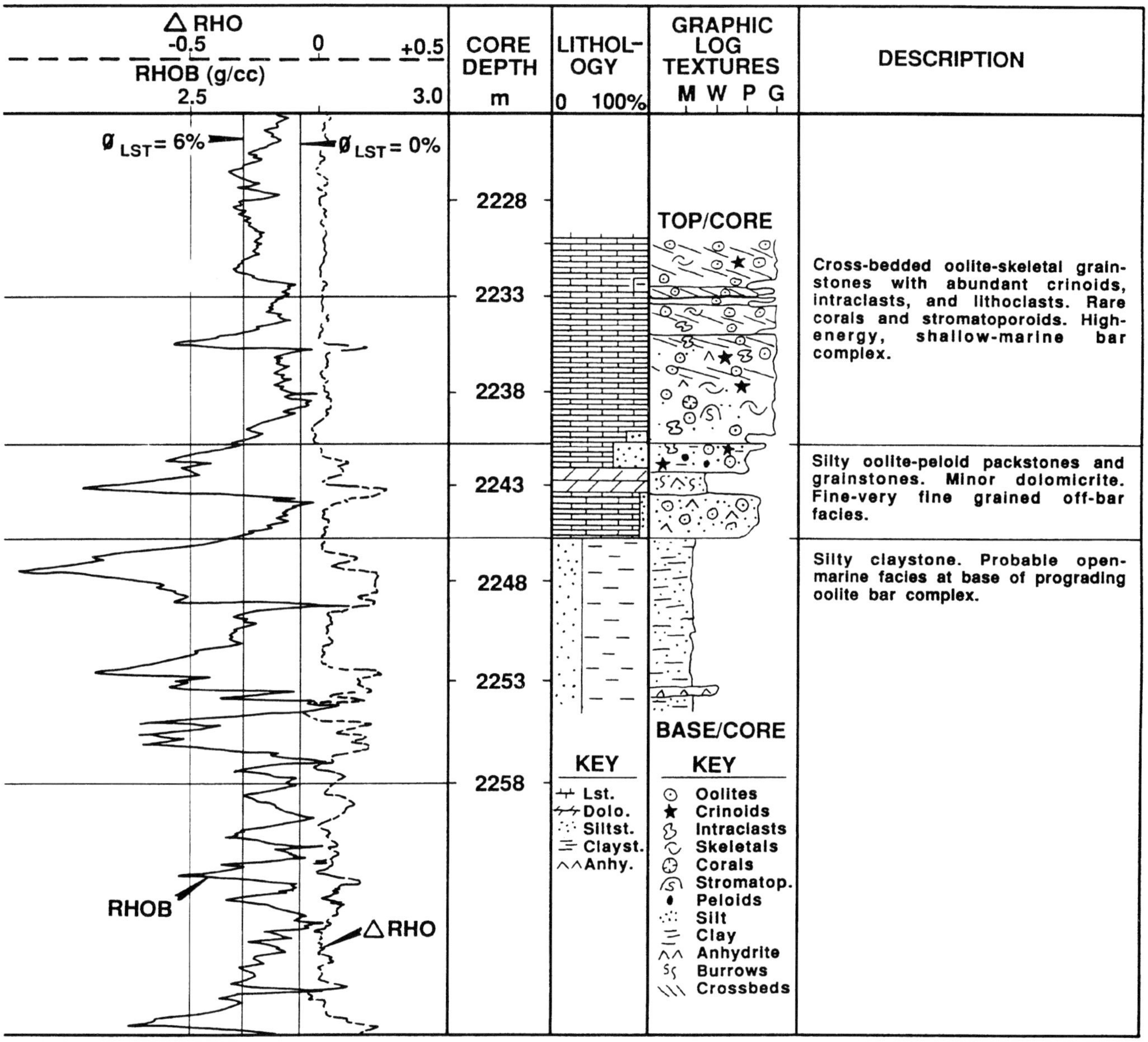

Figure 5 (*continued*).
between deep and shallow traces. The core description at the far right shows the main part of a coarsening-upward, shoaling-upward cycle common in Quintuco Formation parasequences. Note the position of the very fine grained oolites at the base of the oolitic part of the cycle.

were being dissolved. Another well in which the "probable exposure surface" cannot be recognized is well ASa.x-2 (Figures 3 and 9). This well, which is dominated by sabkha facies, has a thin Quintuco section and was relatively shelfward during deposition of the lowermost Quintuco.

Sequence Boundaries

Two major discontinuity surfaces have been identified from lithologic contrasts that occur within the Quintuco Formation. The lowermost surface is the "probable exposure surface" discussed in the previous section (Figures 9 and 10). Abundant oomoldic porosity occurs below this surface and excellent reservoir-quality rocks are present. A correlatable seismic marker also occurs at this level. This seismic marker probably would not have been mapped as a sequence boundary without indications of porosity from core, cuttings, and log studies.

Another surface higher in the section appears to be a major sequence boundary. This surface, which is termed the mid-Quintuco sequence boundary, is characterized by restricted-marine environment anhydrite-bearing rocks that are overlain by oolite bar complexes (Figures 9 and 10). A significant increase in accommodation space probably occurred above this surface. Unfortunately, large quantities of moldic porosity are

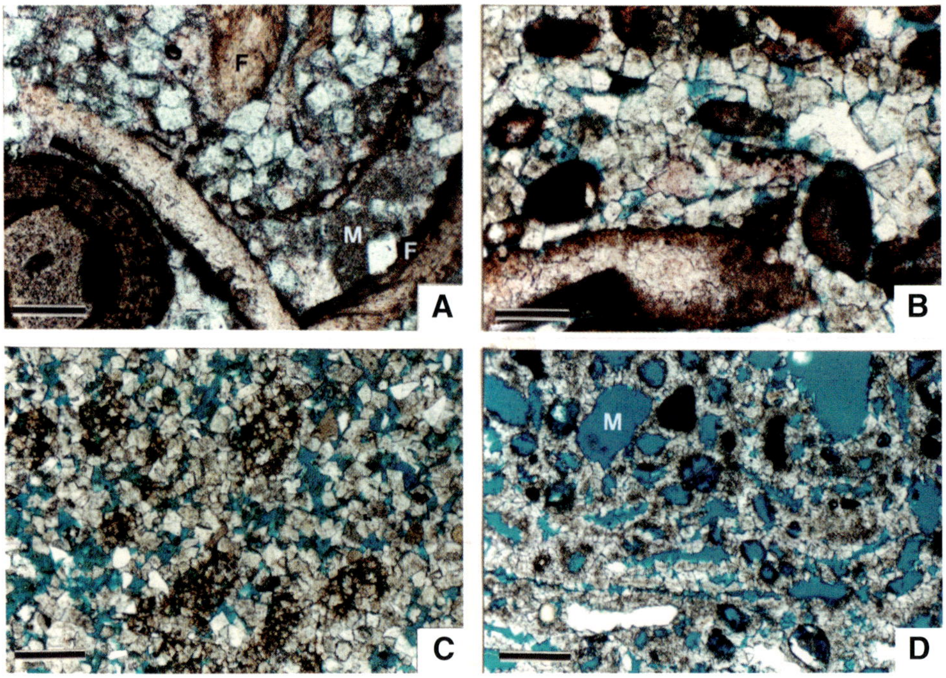

Figure 6. Photomicrographs of burrowed, dolomitized oolite packstone facies, Quintuco Formation. (A) Partially dolomitized oolite-skeletal packstone. Dolomite crystals, which are white-colored rhombs, preferentially replaced the lime-mud matrix (M). Mollusk fragments (F) and ooids are rarely replaced by dolomite. Upper Quintuco Formation, AA.x-1 well, 2159.0 m depth. Scale bar = 0.25 mm. (B) More fully dolomitized skeletal-peloidal-oolite packstone. Again, the white-colored dolomite rhombs have preferentially replaced the lime-mud matrix. Most grains are not dolomitized. Some intercrystalline porosity (blue) is present. Lower Quintuco Formation, AB.x-1 well, 2654.6 m depth. Scale bar = 0.25 mm. (C) Fully dolomitized oolite(?)-peloid(?) packstone with only relicts of grains (dark areas) preserved. This rock is from a perforated hydrocarbon-producing interval in Rio Neuquén field. No molds exist here, perhaps because grains were completely dolomitized before dissolution of calcitic allochems occurred. Lower Quintuco Formation, RN.60 well, 2576.6 m depth. Scale bar = 0.5 mm. (D) Fully dolomitized, then leached, oolite(?)-peloid(?)-skeletal packstone. Molds (M) represent precursor grains. This fabric preferentially occurs below an inferred paleo-exposure surface in the lower Quintuco Formation. AA.x-1 well, 2417.25 m depth. Scale bar = 0.5 mm.

not apparent below this surface, at least in the study area. Seismic work suggests that this sequence boundary is the most correlatable marker in the study area.

DEPOSITIONAL MODEL: QUINTUCO FORMATION

The isopach map in Figure 3 suggests that the thickest part of the Quintuco Formation occurred along the northwest-trending axis of the Añelo trough (Figure 1). An isopach map of the Vaca Muerta interval shows similar results. This evidence suggests that the Añelo trough probably subsided during Quintuco deposition.

Cross sections suggest that thick packages of oolite grainstone existed in the lower Quintuco in the vicinity of the AB.x-1 and ECr.x-1 wells. These rocks have poorer reservoir quality than moldic, sucrosic dolomites in the correlative interval in wells AA.x-1, BLC.x-1, and CS.x-1, for example (Figures 9 and 10). Core descriptions suggest that moldic, sucrosic dolomites occur as replacements of more shelfward oolite packstones and wackestones.

By analogy with modern carbonates, cross-bedded oolite grainstone bars probably occurred in a high-energy setting. Bar crests were probably emergent at mean low tide and under 1 to 2 m of water depth during high tide. Very fine grained, superficially coated oolite grainstones are also commonly cross-bedded. However, the grains were removed from the bar crest before they could receive further coatings. A subtidal depositional environment is inferred for this facies, with deposition most likely on the seaward side of bar

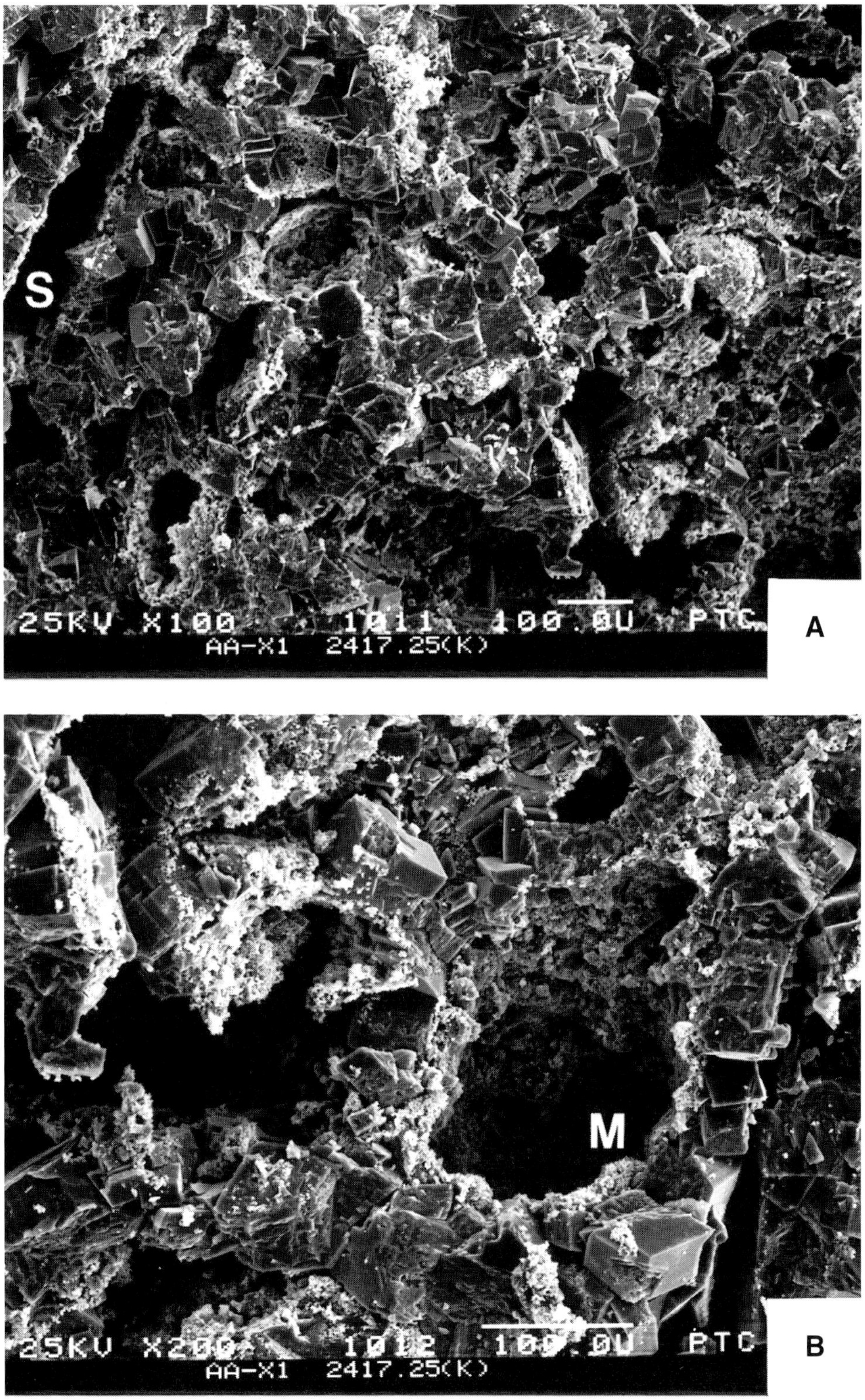

Figure 7. SEM images of burrowed, dolomitized oolite packstone facies, lower Quintuco Formation. (A) Sucrosic dolomite and moldic porosity, including a skeletal mold (S). AA.x-1 well, 2417.25 m depth. Scale bar = 100 µ. (B) Sucrosic dolomite and moldic porosity (M) at double the magnification of Figure 7A. Clays and/or hydrocarbons apparently line some pores. AA.x-1 well, 2417.25 m depth. Scale bar = 100 µ.

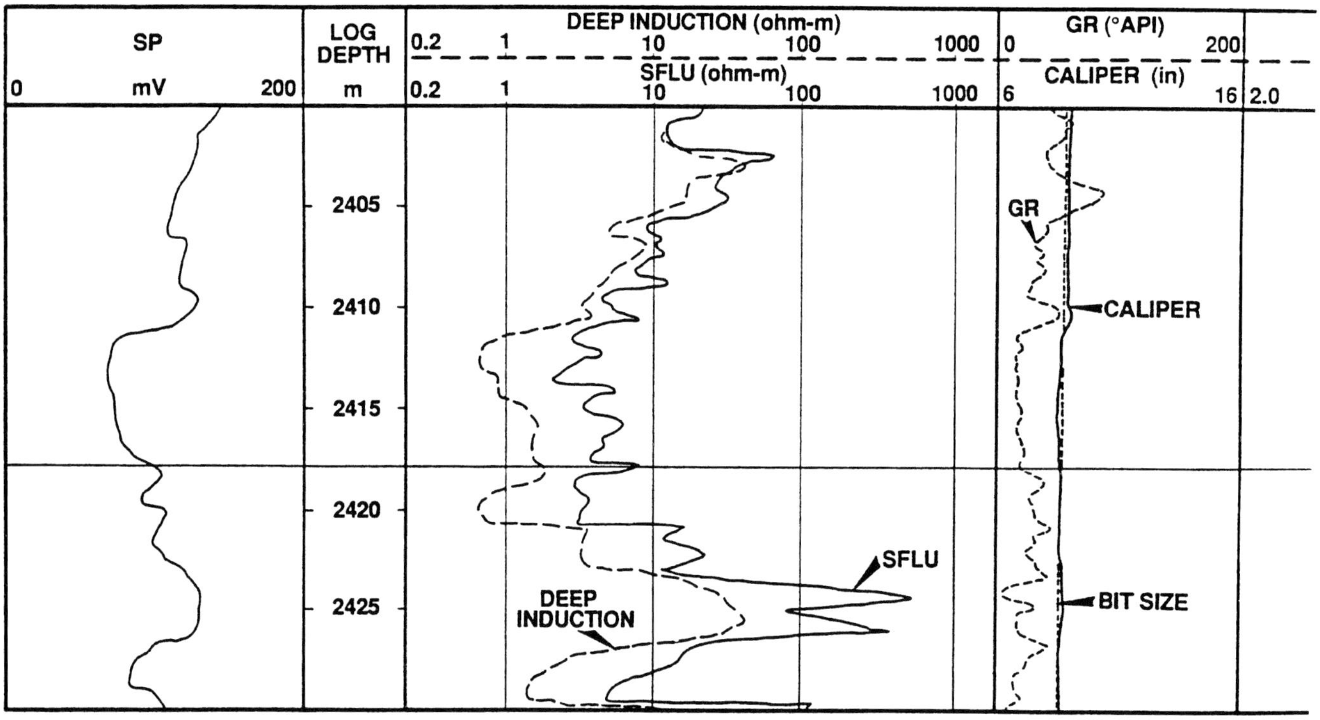

Figure 8. "Type log" of burrowed, dolomitized oolite packstone facies, lower Quintuco Formation. This log is from the AA.x-1 well in the vicinity of the inferred lower Quintuco paleo-exposure surface (log depth = 2412 m). Note the high SP and large separation between deep and shallow resistivity curves. Both responses

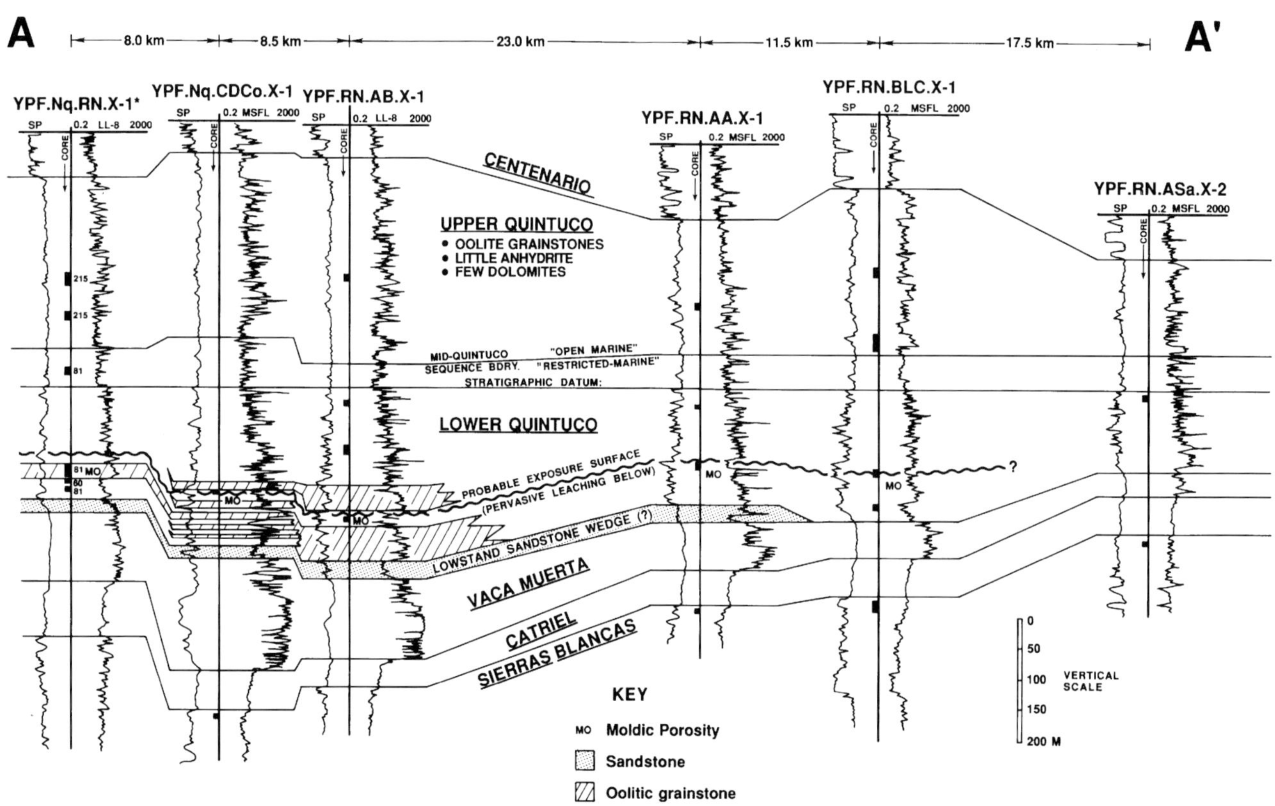

* Cores from Rio Neuquen field wells 60, 81, and 215 are projected into well x-1.

Figure 9. Stratigraphic cross section A–A'. See Figure 3 for location.

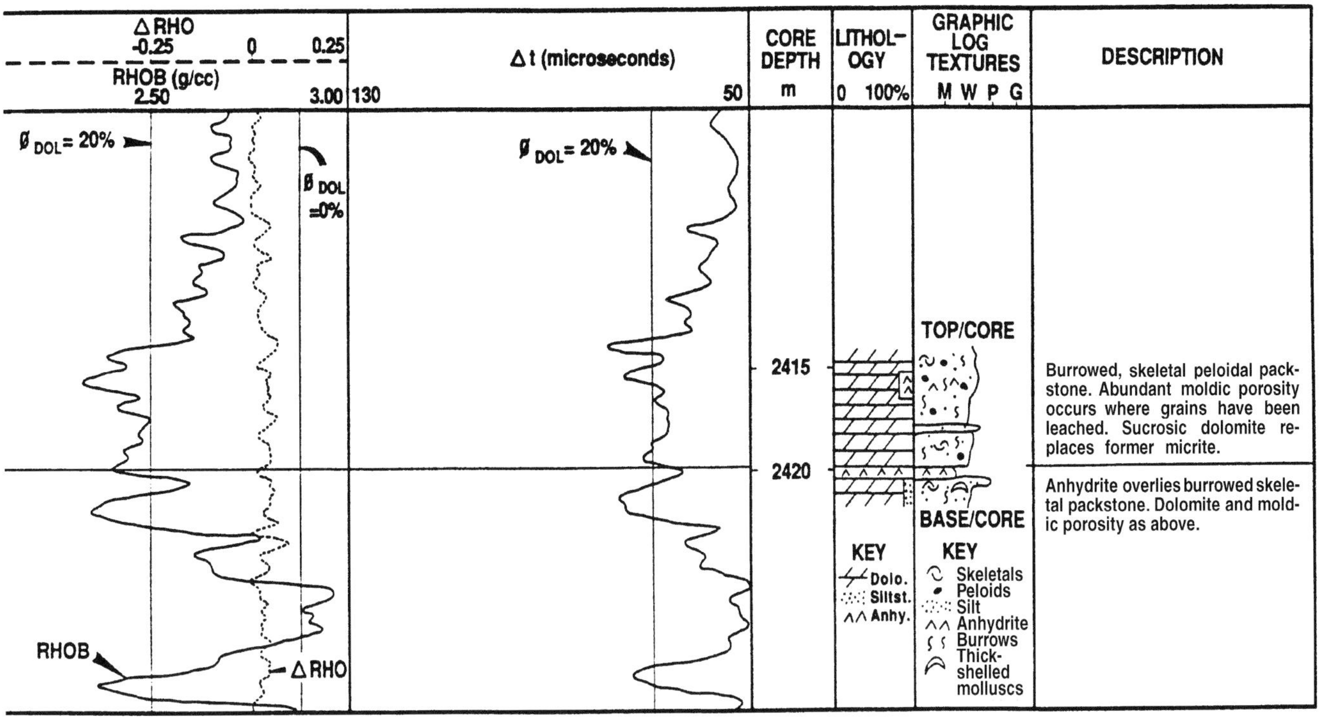

Figure 8 (*continued*).
suggest high permeability in a water-saturated formation. Also note very high porosities indicated by the density and sonic logs. The core description (far right) shows lithologies observed in this interval.

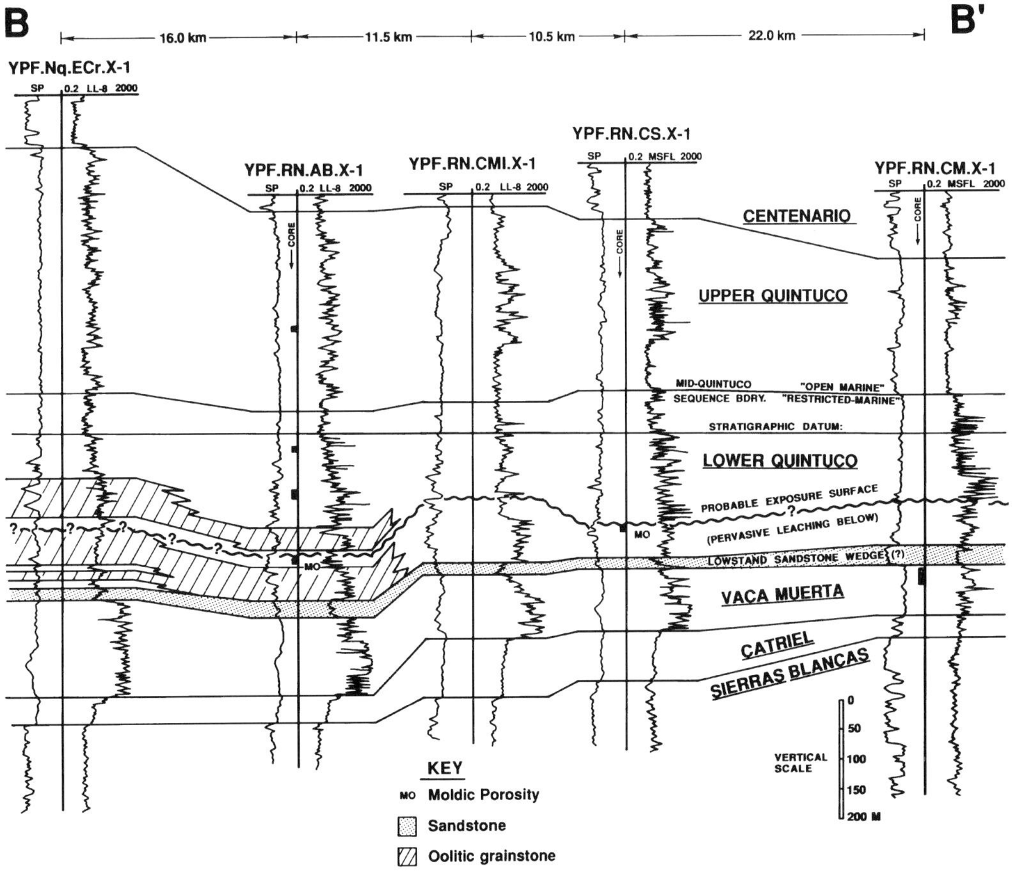

Figure 10. Stratigraphic cross section B–B′. See Figure 3 for location.

crests (Figure 11). Dolomitized oolite-skeletal-peloid packstones to wackestones are also an off-bar facies. These rocks probably formed in a broad subtidal lagoon on the landward side of oolite bars (Figure 11). A modern analogy exists in the 25 km wide burrowed packstone belt that lies shelfward of the Joulters Cay oolite bars in the Bahamas (Harris, 1979, 1984). Dolomudstones and layered anhydrites are probably intertidal to sabkha-type deposits that occurred only in the most landward areas. Wedges of siliciclastic debris could have come from the north-northeast (area marked LAND in Figure 11).

DISCUSSION

Unconformity-Related Origin of Sucrosic, Moldic Porosity

In the study area, the best reservoir-quality porosity and permeability occur as moldic porosity in dolomitized oolite-skeletal-peloid packstones/wackestones in the lowermost Quintuco Formation. This porosity occurs below a surface that can be widely correlated on wireline logs. Significant moldic porosity is not observed above this surface. In at least two wells, this surface is overlain by a conglomeratic lag of volcaniclastic pebbles. This surface can be traced on seismic and has been mapped as a sequence boundary. We infer that this is a paleo-exposure surface that postdated dolomitization of lime-mud matrix. Although exposure may not have been very long (a few thousand years?), it was long enough to generate extensive freshwater dissolution of calcitic allochems.

The proposed sequence of events is: (1) deposition of oolite bars and associated sediments, (2) contemporaneous burrowing of a broad area shelfward of oolite bars to form oolite-skeletal-peloid packstones and wackestones, (3) cementation of most oolite grainstones by shallow marine, vadose, and/or freshwater phreatic cements, (4) dolomitization of the lime-mud matrix of oolite-skeletal-peloid packstones and wackestones by reflux dolomitization from age-equivalent sabkha deposits to the north and east, (5) widespread exposure and dissolution of calcitic allochems (and

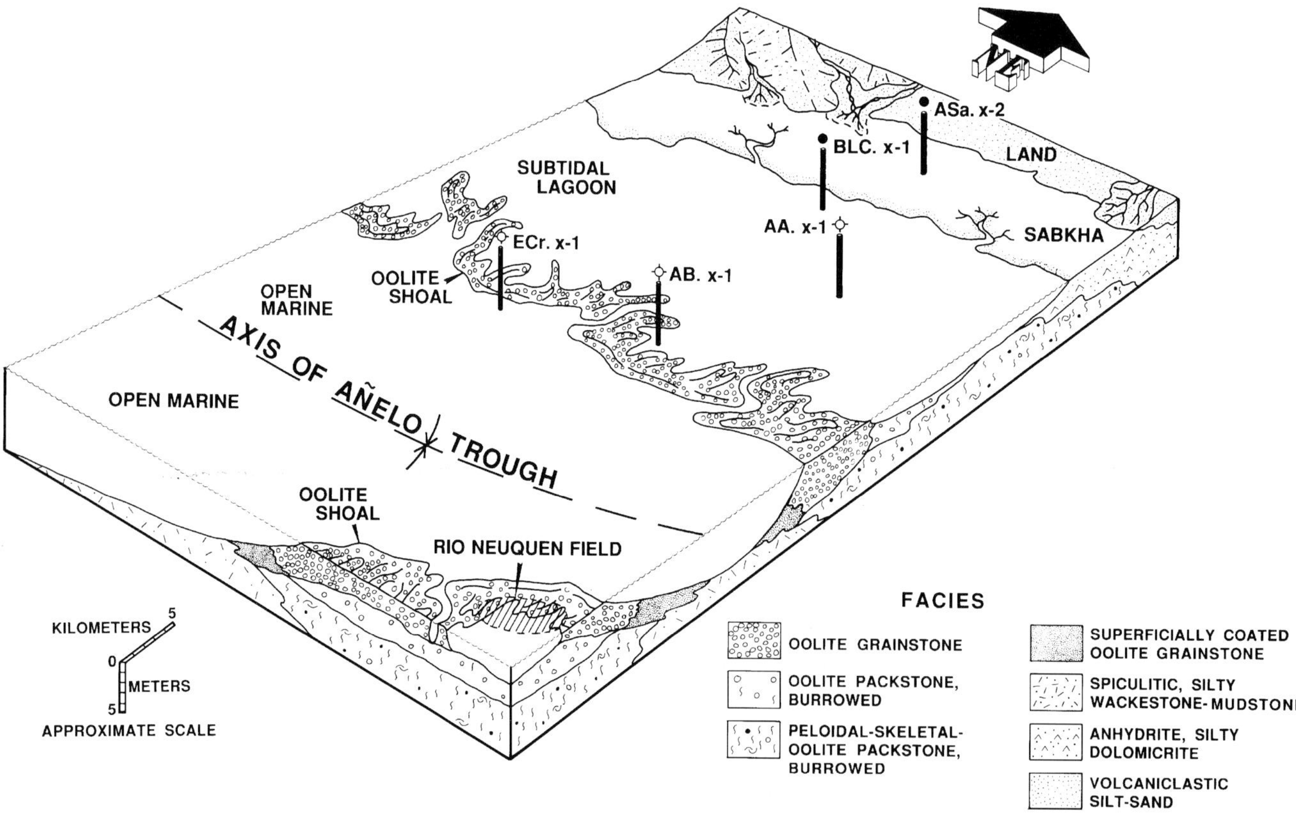

Figure 11. Depositional model for the lower Quintuco Formation in the vicinity of concession CNQ-17, Neuquén basin, Argentina. This diagram represents a "snapshot" in time just prior to the paleo-exposure that occurred on top of the lower Quintuco. Prior to exposure, micritic matrix was dolomitized in the broad expanse of oolite packstone and peloidal-skeletal-oolite packstone facies. During exposure, calcitic grains (oolites, peloids, skeletal fragments) were dissolved. Following exposure, in the BLC.x-1 and perhaps in the CM.x-1 well locations, transgressive lithoclast lags were deposited, perhaps from fan deltas that were present in the area shown as "LAND." Isopach thicknesses suggest that the Añelo trough existed during Quintuco deposition. The facies present at Rio Neuquén field appear to be most analogous to those in the vicinity of the AB.x-1 well on the northern side of the Añelo trough. Diagram is modified from Loucks and Anderson (1985).

perhaps anhydrite?) in dolomitic oolite-skeletal-peloid packstones and wackestones, (6) deposition of volcaniclastic conglomerate lags associated with the exposure surface, and (7) marine transgression followed by deposition of oolite bar, lagoon, and sabkha deposits in the remainder of the lower Quintuco. These overlying sabkha deposits could have re-introduced anhydrite cements and nodules into the underlying sediments.

Although we believe porosity in the lowermost Quintuco is unconformity related, at least three other possibilities exist: (1) ooids and other allochems below this surface may have been aragonitic (more soluble), whereas allochems above this surface were calcitic (less soluble), (2) ooids and other allochems were dissolved during the dolomitization process and moldic porosity was not related to exposure, and (3) moldic porosity is the result of burial diagenesis, as suggested by Carozzi et al. (1993).

Long-term secular variation in ooid mineralogy has been documented by Sandberg (1983), Wilkinson et al. (1985), and others. Short-term variation has been documented in the Upper Jurassic Smackover Formation by Moore et al. (1986) and Swirydczuk (1988). If Uliana and Legarreta (1993) are correct with their time estimate of 12.5 Ma for deposition of the approximately 600 m thick Quintuco–Loma Montosa–Vaca Muerta section, then the lowermost 100 m of the Quintuco may represent as much as 2.1 m.y. It is possible that a change from aragonitic to calcitic ooid mineralogy could have occurred during this time. However, preserved ooids from below this surface have radial fabric in Rio Neuquén field. Such fabric is typical of calcitic ooids, although it has also been observed in aragonitic ooids.

Hardie (1987) has stated that dolomitizing fluids commonly have the ability to dissolve limestones. This is a possible source of moldic porosity, but it does not explain the fact that although facies are generally similar and sediments are generally dolomitized above and below the inferred exposure surface, moldic porosity is abundant only below this surface.

In terms of burial diagenesis, it is possible that a porosity-forming process could have been confined to a particular stratigraphic unit within the lowermost Quintuco. This unit could have been a pathway for migrating basinal fluids, perhaps during hydrocarbon generation. However, although there is an increase in the number of bedded anhydrites, there is no obvious permeability barrier, such as a shale, above the inferred exposure surface. Calcitic allochems and partially dolomitized rock exist at other levels in the formation, but abundant moldic porosity has only been observed in the lowermost Quintuco.

Comparison to Previous Work

This work in the Quintuco Formation can be compared to other studies done in the same area that used different techniques. Carozzi (1989) and Carozzi et al. (1993) conducted detailed microfacies work in the Neuquén basin. Many of the wells used in their study were re-examined here. A number of seismic/sequence-stratigraphic studies have also been done (Mitchum and Uliana, 1985, 1987; Legarreta, 1991; Eisner, 1991). A recent article by Uliana and Legarreta (1993) contains a cross section (their Figure 8) that crosses the present study area. Figure 2 is a comparison of stratigraphic interpretations done by various authors. Because of the time-transgressive nature of some formation boundaries, this diagram is drawn in the area of the AA.x-1 and ECr.x-1 wells, common wells to all studies.

The first stratigraphic interpretation column in Figure 2 shows the YPF interpretation, a descriptive classification that appears in many final well reports. The Quintuco is informally subdivided into an upper calcareous member, a middle argillaceous member, and a lower calcareous member.

Carozzi et al. (1993) make no attempt to designate sequence boundaries. They feel that much of the Vaca Muerta predates Quintuco–Loma Montosa deposition (see their Figure 3). Their widespread "transition zone" overlies the Vaca Muerta. In well CM.x-1 (their well 1, cross section V), the transition zone is overlain by a volcaniclastic conglomerate. This transition zone appears to be correlative with the lowermost Quintuco of this study. It is interesting to see in cross sections (their Figures 10, 11, 13, and 14) how oolitic buildups abruptly appear on top of this transition zone. If their buildup interpretations are correct, this may provide further evidence that the top of the transition zone is a sequence boundary.

Seismic/sequence-stratigraphic studies of Mitchum and Uliana (1985, 1987) and Eisner (1991) apparently did not involve extensive subsurface sample examination. Eisner (1991) joined together a large number of un-reprocessed YPF seismic lines to do his regional interpretations. Mitchum and Uliana (1985, 1987) may have done the same. Eisner (1991) picked sequence boundaries based on reflection termination patterns such as onlap, downlap, and toplap. Eisner (1990, personal communication) also based his sequence boundaries partially on siliciclastic incursions as inferred from log interpretation. In our experience, volcaniclastic debris is mixed with otherwise clean carbonates, and lithologic interpretation from logs alone, especially GR logs, is unreliable.

Uliana and Legarreta (1993) presented an interpretation of sequence boundaries in the Quintuco–Loma Montosa–Vaca Muerta interval. Figure 2 shows that Uliana and Legarreta (1993, their Figure 8) recognized major sequence boundaries at the top, bottom, and middle of the Quintuco–Loma Montosa–Vaca Muerta section. Their middle sequence boundary apparently lies at the base of the "lower fan delta" complex of Carozzi et al. (1993). In this study, we place the mid-Quintuco sequence boundary at the top of the "lower fan delta" complex of Carozzi et al. (1993). This is based on correlations made using our grid of seismic lines. Eisner (1991) has shown a large number of sequence boundaries in the Quintuco–Loma Montosa–Vaca Muerta interval. None of these can be reliably correlated to sequence boundaries mapped in this study.

CONCLUSIONS

Carbonates in the Quintuco Formation are composed of: (1) oolitic grainstones, (2) burrowed, dolomitized oolite-skeletal-peloid packstones/wackestones, (3) dolomudstones and bedded anyhdrites, and (4) very fine grained, superficially coated oolite grainstones. These sediments are commonly packaged into shoaling- and coarsening-upward parasequences. Porosity in lower Quintuco oolitic grainstones is commonly occluded with pore-filling calcite cements. Reservoir-quality porosity and permeability are facies selective and are developed almost exclusively in burrowed, dolomitized packstones and wackestones. These rocks are sucrosic dolomites in which dolomite has preferentially replaced carbonate mud.

A widespread paleo-exposure surface has been inferred in the lower Quintuco Formation. Below this surface, ooid-skeletal-peloid grains were commonly dissolved to leave an open pore network with abundant moldic and intercrystalline porosity. Moldic, sucrosic dolomites are the main reservoir facies in Rio Neuquén field.

Well-log correlations and comparisons to known modern carbonate environments suggest that the burrowed, dolomitized facies probably extends laterally for tens of kilometers in a given oolitic parasequence. The burrowed, dolomitized facies is interpreted as an off-bar facies that occurs mainly on the landward side of bar complexes. Modern facies analogs are known in the Joulters Cay area of the Bahamas.

Different interpreters have picked different sequence boundaries in the same study area. No previously published study has related porosity development to sequence boundaries. It is unlikely that porosity related to the inferred paleo-exposure surface identified in this study would have been detected without core, cuttings, and wireline-log analysis.

ACKNOWLEDGMENTS

We are grateful to YPF (Yacimientos Petroliferos Fiscales, the Argentine State oil company) for allowing access to core materials and facilities in Buenos Aires and Plaza Huincul. The manuscript was improved by comments of reviewers M. W. Longman, W. D. Dawson, and A. H. Saller. Other Marathon personnel who contributed to this study are W. Head, A. Golovchenko, A. Gilmer, M. McBride, and J. Just. The authors thank Marathon Oil Company for permission to publish this paper.

REFERENCES CITED

Asquith, G. B., and C. R. Gibson, 1982, Basic well log analysis for geologists: AAPG Methods in Exploration Series, 216 p.

Carozzi, A. V., 1989, Carbonate rock depositional models: a microfacies approach: Prentice Hall, Englewood Cliffs, New Jersey, 604 p.

Carozzi, A. V., I. A. Orchuela, and M. L. Rodriguez Schelotto, 1993, Depositional models of the Lower Cretaceous Quintuco–Loma Montosa Formation, Neuquén basin, Argentina: Journal of Petroleum Geology, v. 16, p. 421–450.

Chamberlain, C. K., 1992, Trace fossils for petroleum geologists with emphasis on cores: RMAG Short Course Notes, 39 p.

Eisner, P. N., 1991, Tectonostratigraphic evolution of Neuquén basin, Argentina: Master's thesis, Rice University, Houston, Texas, 56 p.

Haq, B. U., J. Hardenbol, and P. R. Vail, 1987, Chronology of fluctuating sea level since the Triassic: Science, v. 235, p. 1156–1166.

Hardie, L. A., 1987, Perspectives: dolomitization: a critical view of some current views: Journal of Sedimentary Petrology, v. 57, p. 166–183.

Harris, P. M., 1979, Facies anatomy and diagenesis of a Bahamian ooid shoal: Sedimenta VII, University of Miami, Miami Beach, Florida, 163 p.

Harris, P. M., 1984, Cores from a modern carbonate sand body: the Joulters Cay ooid shoal, Great Bahama Bank, *in* P. M. Harris, ed., Carbonate Sands—A Core Workshop: SEPM Core Workshop 5, p. 429–464.

Hogg, S. L., 1993, Geology and hydrocarbon potential of the Neuquén basin: Journal of Petroleum Geology, v. 16, p. 383–396.

Keeley, M., M. Light, S. Hogg, and C. Urien, 1992, Argentina's exploration licensing round covers 145 tracts in 15 basins: Oil and Gas Journal, v. 90, no. 41, p. 85–88.

Kugler, R. L., 1987, Regional petrologic variation, Jurassic and Cretaceous sandstone and shale, Neuquén basin, west-central Argentina: Unpubl. Ph.D. dissertation, University of Texas, Austin, Texas, 523 p.

Legarreta, L., 1991, Evolution of a Callovian–Oxfordian carbonate margin in the Neuquén basin of west-central Argentina: Facies, architecture, depositional sequences, and global sea-level changes: Sedimentary Geology, v. 70, p. 209–240.

Loucks, R. G., and J. H. Anderson, 1985, Depositional facies, diagenetic terranes, and porosity development in Lower Ordovician Ellenburger dolomite, Puckett field, West Texas, *in* P. O. Roehl and P. W. Choquette, eds., Carbonate petroleum reservoirs: Berlin, Springer-Verlag, p. 19–38.

Mitchum, R. M., Jr., and M. A. Uliana, 1985, Seismic stratigraphy of carbonate depositional sequences, Upper Jurassic–Lower Cretaceous, Neuquén basin, Argentina, *in* D. R. Berg and D. G. Woolverton, eds., Seismic Stratigraphy II: An Integrated Approach to Hydrocarbon Exploration: AAPG Memoir 39, p. 255–274.

Mitchum, R. M., Jr., and M. A. Uliana, 1987, Regional seismic-stratigraphic analysis of Upper Jurassic–Lower Cretaceous carbonate depositional sequences, Neuquén basin, Argentina, *in* A. W. Bally, ed., Atlas of Seismic Stratigraphy, v. 2: AAPG Studies in Geology 27, p. 206–212.

Moore, C. H., A. Chowdhury, and E. Heydari, 1986, Variation of ooid mineralogy in Jurassic Smackover limestones as control of ultimate diagenetic potential (abs.): AAPG Bulletin, v. 70, p. 622–623.

Murray, R. C., and F. J. Lucia, 1967, Cause and control of dolomite distribution by rock selectivity: Geological Society of America Bulletin, v. 78, p. 21–36.

Sandberg, P. A., 1983, An oscillating trend in Phanerozoic nonskeletal carbonate mineralogy: Nature, v. 305, p. 19–22.

Swirydczuk, K., 1988, Mineralogical control on porosity type in Upper Jurassic Smackover ooid grainstones, southern Arkansas and northern Louisiana: Journal of Sedimentary Petrology, v. 58, p. 339–347.

Uliana, M. A., and L. Legarreta, 1993, Hydrocarbons habitat in a Triassic-to-Cretaceous Sub-Andean setting: Neuquén basin, Argentina: Journal of Petroleum Geology, v. 16, p. 397–420.

Wilkinson, B. H., R. M. Owen, and A. R. Carroll, 1985, Submarine weathering, global eustasy, and carbonate polymorphism in Phanerozoic marine oolites: Journal of Sedimentary Petrology, v. 55, p. 171–183.

Yrigoyen, M. R., 1993, The history of hydrocarbons exploration and production in Argentina: Journal of Petroleum Geology, v. 16, p. 371–382.

Chapter 9

Reservoir Degradation and Compartmentalization below Subaerial Unconformities: Limestone Examples from West Texas, China, and Oman

P. D. Wagner
Amoco Exploration and Production Technology
Tulsa, Oklahoma, U.S.A.

D. R. Tasker
G. P. Wahlman
Amoco Exploration and Production Co.
Houston, Texas, U.S.A.

ABSTRACT

This paper describes how meteoric cementation enhanced the hydrocarbon trapping and/or producing potential of three limestones. Petrophysical effects of meteoric diagenesis on carbonates vary between two perfect end members of pure *seal formation* and pure *reservoir enhancement*. Net porosity and permeability changes are inferred to be a simplistic function of water availability and the exposed terrane's chemical reactivity. *Meteoric tight zones* form under conditions of low water availability and high terrane reactivity (e.g., a semi-dry climate exposure of Mg-calcite sediment). *Solution-enhanced reservoirs* form under conditions of high water availability and low terrane reactivity (e.g., a rain forest exposure of stoichiometric dolomite).

Examples of meteoric tight zones are shown in cores from west Texas, offshore China, and central Oman. Petrographic and geochemical data were used to define the causes of reservoir degradation. From an exploration/exploitation standpoint, these intervals form potential top-seals for hydrocarbon trapping and/or intraformational permeability barriers that compartmentalize hydrocarbon production. More generally, meteoric tight zones may be a critical trapping factor in many similar hydrocarbon accumulations—both producing (but not recognized as such) and prospective. A more thorough investigation through the current inventory of fields might show meteoric seal formation is as economically important in trap formation as its much better studied "karsting" counterpart. Either end member should be easily recognized by its unusual petrographic and geochemical signature and overwhelming petrophysical effect on the rock.

Methodical searches for meteoric diagenetic traps should be most productive in areas with moderate drilling density (for rock control), relatively simple facies distributions (to minimize facies prediction problems), and accentuated paleotopography (where cross-formational meteoric tight zones can form appropriate trapping geometries). Drowning successions should also be more prospective because of a greater incidence of ephemeral exposure of unstabilized sediments during brief sea level drops.

INTRODUCTION

Petrophysical effects of meteoric diagenesis on carbonates vary between two hypothetical end members of seal formation and reservoir enhancement (Wagner, 1991). Whether an exposed carbonate becomes more "seal-like" or "reservoir-like" depends on a multitude of variables such as the water's chemical saturation state, water flux, chemical reactivity of exposed solid, terrane permeability, nucleation kinetics, time available for processes to operate, etc. But in a simplistic sense, reservoir modification can be estimated if only two general factors are known—*terrane reactivity* and *water availability* (Figure 1). As used here, *terrane reactivity* summarizes all rock factors that describe chemical susceptibility to meteoric diagenesis—mineralogy, crystal size, stoichiometry, etc. *Water availability* summarizes all fluid factors that describe water flux conditions in the rock volume undergoing meteoric diagenesis—amount of rainfall, evaporation intensity, sediment permeability, etc. Although the use of these composite-factor terms is simplifying in one sense, they can be just as difficult to predict "ahead of the drill bit" as more discrete factors. Some rock control in the area of interest is usually needed to make seal and reservoir predictions of meteorically affected intervals believable. Obviously, the assumption that sufficient time is available for porosity and permeability changes to occur is implicit in the simplified model.

Porosity and permeability loss occurs when a chemically reactive carbonate terrane is exposed to dry- or semi-dry climatic conditions (e.g., a low- to moderate-level rainfall exposure of aragonite and Mg-calcite–bearing sediments). This type of climatic effect on porosity changes was briefly addressed in James and Choquette (1990; Figure 2). Presumably, reservoir properties are diminished because a large source of dissolved carbonate is available for cementation, and water flux is low enough that dissolved material is not efficiently swept away. The net effect of high terrane reactivity and low water availability conditions is high cementation potential (Figure 3). *"Meteoric tight zones"* form under these conditions, commonly with some degree of stratigraphic thinning as water availability level rises (Figure 4).

Reservoir enhancement occurs when a chemically stable terrane is exposed to very wet climatic conditions (e.g., exposure of dolomite rock to monsoonal climatic conditions). Reservoir properties are enhanced because low chemical reactivity of the substrate helps keep water saturations low, and high fluid flux efficiently removes dissolved material. Highly leached karstified zones form in this way; stratigraphic thinning occurs as karsting enters its more mature stages of development (e.g., Esteban and Klappa, 1983).

Between end-member cases of reservoir degradation and reservoir enhancement, a complex combination of net leaching and cementing possibilities exists—including a hypothetical "steady-state" case where net pore *space* is conserved, but pore *type* changes from primary to secondary. Variability from one stratigraphic interval to another can be dramatic. In one case presented here, reservoir-enhanced intervals associated with separate paleo-exposures exist within hundreds of feet of a meteoric tight zone. Extreme variability in petrophysical effects with changing exposure conditions suggests that seal and reservoir predictions should be attempted only when supporting data from nearby equivalent rock intervals

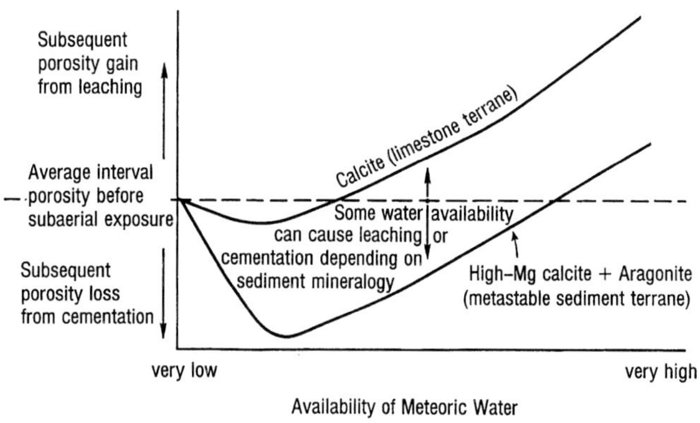

Figure 1. Porosity change of the exposed carbonate is shown schematically as a simple function of the availability of meteoric water and terrane reactivity (represented as "mineralogy," but other factors such as crystal size and stoichiometry also affect reactivity). Note that at low water availabilities, all calcium carbonate terranes lose porosity. At some intermediate levels of water availability, calcitic terranes gain porosity by leaching while equivalent metastable terranes lose porosity by cementation. At very high water availabilities, all carbonate terranes gain porosity through excessive leaching.

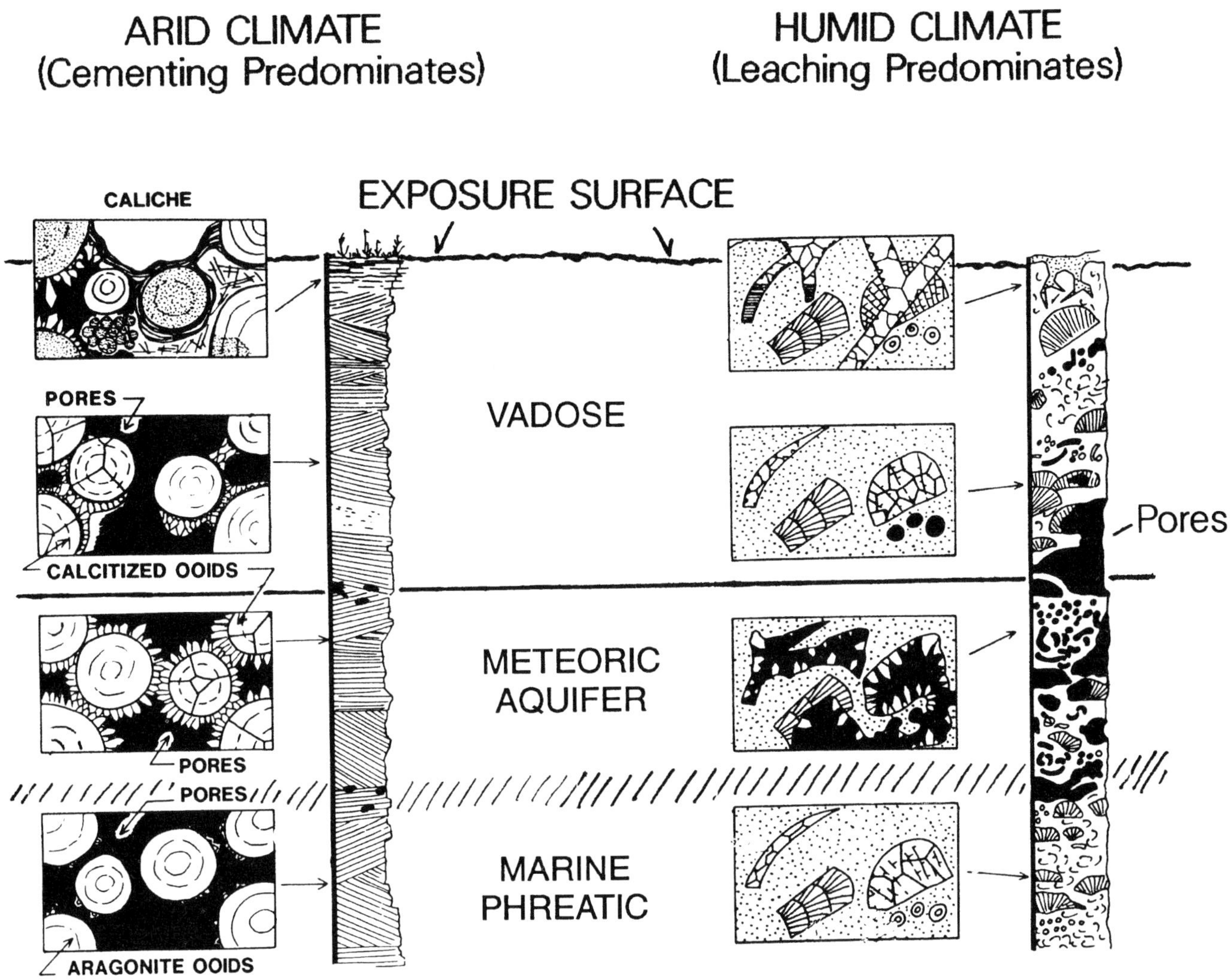

Figure 2. Some of the petrographic differences are shown that are expected to develop between carbonates exposed to meteoric diagenesis under "dry" and "wet" climates (modified from James and Choquette, 1990). Note that under drier conditions, carbonate cementation predominates, allowing for the possible formation of intensely cemented meteoric intervals. Increasingly wetter conditions cause progressively more secondary porosity to form because dissolved carbonate is removed more efficiently by flushing.

are available—and then preferably only if exposure conditions were near one of the two end members.

Petrophysical examples of three meteoric tight zones are shown. Particular examples are shown in part because reservoir-enhanced and reservoir-degraded rock co-occur in some closely spaced cored intervals, demonstrating the changeability of petrophysical-controlling factors over relatively short distances and/or brief times. Prefacing the three field examples are short case summaries.

HYDROCARBON TRAPPING RELATED TO METEORIC DIAGENESIS

Reservoir Enhancement

The importance of leaching in the formation and/or enhancement of carbonate reservoir rocks is well documented (e.g., Bathurst, 1975; Roehl and Choquette, 1985; James and Choquette, 1988). A positive aspect of meteoric leaching is formation of an irregular surface geometry that is mappable with seismic data. If correlation of reservoir location with a seismic signature can be confirmed, then seismic data can serve as an excellent proxy to define geographic limits of reservoir occurrence. In many cases, however, either no angularity exists for seismic recognition or data quality is insufficient for stratigraphic resolution. In those cases, other data sets would be required to identify and geometrically characterize the meteorically altered intervals (discussed later).

A negative aspect of meteoric leaching is that it does not form a hydrocarbon trap. Karstification may produce a world-class reservoir rock, but the lack of an overlying seal would render it useless for those purposes. For a trap to form, karsting must be followed by deposition of an overlying seal, or formation of a seal through later surface- or burial-related cementation

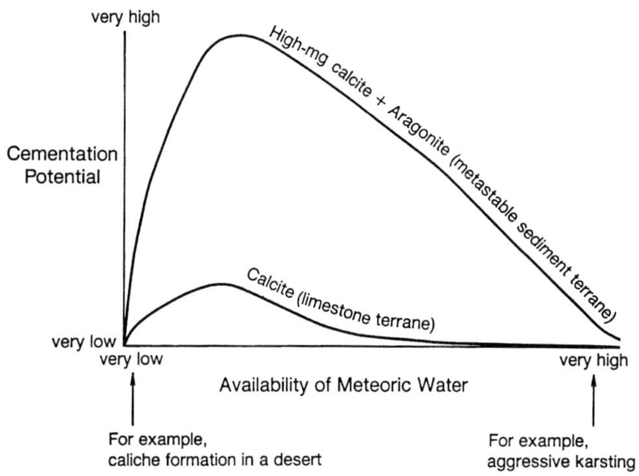

Figure 3. Cementation potential is shown as a function of meteoric water availability and terrane mineralogy. Note that cementation potential is highest with relatively low water availability and a highly reactive solid (Mg-calcite and aragonite). "Meteoric tight zone" formation occurs readily under these conditions, but can also occur with stable calcitic terranes if water availability is low enough.

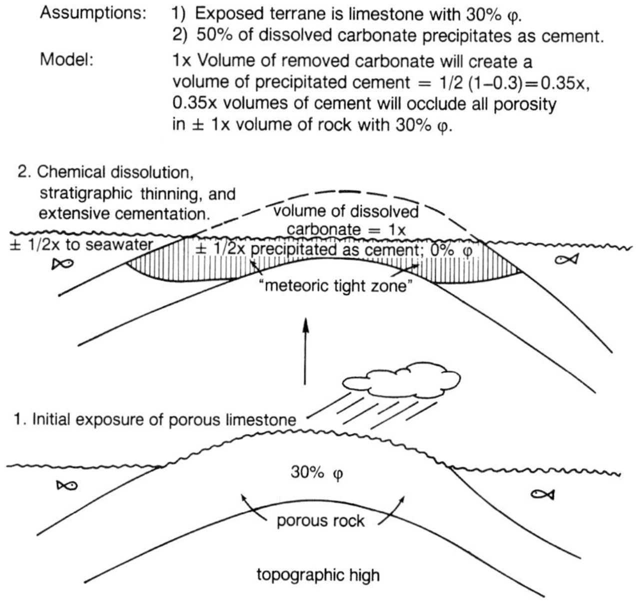

Figure 4. A model of meteoric tight zone formation under low to moderate rainfall is shown. Note that under favorable water availability and terrane reactivity conditions, stratigraphic thinning can be accompanied by formation of a highly cemented interval of great lateral extent.

(e.g., Yates field, west Texas; Craig, 1988). If a rock of high permeability is deposited over the karsted interval, any trapping potential is greatly diminished. An example would be the karstified wedge edge of the Devonian Wabamun Formation in Alberta; it is overlain by permeable Cretaceous sands that bled off any significant volume of hydrocarbons migrating through the underlying carbonate.

Porosity and Permeability Loss

Even though the importance of meteoric cementation as a trap-forming mechanism is poorly documented in the literature, it is easy to see from the preceding discussion why it is potentially important: Meteoric cementation represents a contributory process for forming trap geometries in the typical surficial limestone section (i.e., one with high porosities and permeabilities). Meteoric cementation may form geometries of low porosity and permeability limestones that could be a first-order control on later distributions of seals. Meteorically affected rock also carries with it all the potential of recognition (of karst) by seismic, although particular care must be taken to correctly interpret the position of "off structure" reservoir rocks when selecting a drilling location.

A disadvantage of meteoric cementation to form effective, long-lived seals is a tendency toward rock brittleness at shallow to intermediate burial depths. The considerable compaction and related section shortening during early stages of burial can fracture the early cemented interval. For this reason, meteoric cementation cannot always be depended on to form an effective seal for large columns of hydrocarbons unless burial cementation also occurs. In two of the three examples described here, burial fracturing did occur, and burial cementation healed the damage in both cases.

DATA TYPES FOR DEFINING GEOMETRY OF METEORIC PROCESSES

Petrography is a traditional means for identifying rock intervals that have undergone meteoric diagenesis. Esteban and Klappa (1983) described petrographic features associated with subaerial exposure of carbonates. James and Choquette (1990), Longman (1980), and Bathurst (1975) also presented good summaries of petrographic features that commonly form in meteoric phreatic intervals.

The major advantage of petrography as a meteoric diagenesis discriminator is the diagnostic value of certain cement types, paleosols, or other features. The major disadvantage of petrography is that many sections do not contain diagnostic features even though they have undergone meteoric diagenesis. Especially if the section is mud-rich, few large pore spaces are available as hosts for specific cement types. Mechanical erosion can also remove diagnostic petrographic features such as paleosols, leaving no readily identifiable visual record behind.

Interpretation of geochemical data in stratigraphic context is also an established method for identifying rock intervals exposed to meteoric diagenesis (Allan and Matthews, 1977, 1982; Wagner, 1983, 1994; Beier, 1987; Saller and Moore, 1991). These past studies have shown that geochemical values (carbon isotopes, oxygen isotopes, ppm strontium, ppm magnesium, etc.) on either bulk or component samples can be used, with proper supporting evidence, to identify diagenetic boundaries such as paleo-exposures and tops of paleo-aquifers (Figure 5). In the meteoric diagenetic environment, stratigraphic patterns of depletions and enrichments are controlled by water flow direction and chemistry changes during progressive water-rock interaction (Figure 6).

The major advantage of geochemical data as a meteoric discriminator is its stratigraphic thickness of signal. Even if the uppermost section of meteorically affected rock is removed by mechanical erosion, the remaining interval will still record a meteoric imprint. The major disadvantage is that any stratigraphic pattern of relative enrichments and depletions in a single geochemical parameter can be caused by several factors. Collection of multiple types of geochemical data is recommended to reduce ambiguity of interpretation. In most applications of this type, geochemical data are best used as a secondary line of evidence in support of petrography.

WEST TEXAS EXAMPLE

Case Summary

The best of three examples of porosity and permeability loss due to meteoric diagenesis presented here comes from the Lower Permian (Wolfcampian) Hueco Limestone along the margin of the Central Basin platform in the Permian basin of west Texas. This same stratigraphic section provides a good example of meteoric leaching in a different horizon.

The Hueco Limestone shelf-margin facies complex is divided into a lower bank unit and an upper shoaling unit (Figure 7). The lower bank unit is a shelf-margin biohermal buildup that consists of phylloid algal bafflestones and *Tubiphytes* boundstones, and interstitial and flanking *Tubiphytes*-fusilinid-algal packstone-grainstones. Wahlman (1988) discussed and illustrated many of the biohermal microfacies from the same cored stratigraphic sections analyzed for this study. The biohermal facies commonly have very low porosity because of syndepositional submarine cementation and later diagenetic burial cementation. Porosity is better developed in the grain-rich flank-bed facies of the lower bank unit. The overlying upper shoaling unit, which onlaps and buries pre-existing topography, consists of a series of shallowing-upward cycles capped by oolitic and bioclastic grainstones. Porosity in this upper unit is best developed in the capping grainstone shoal facies of the cycles.

An integrated petrographic and isotope-stratigraphic data set from subsurface conventional cores was used to define a minimum of two paleo–subaerial-exposure surfaces in the shelf-margin complex. One paleo-exposure caps the lower bank unit, and the other caps the upper shoaling unit. Petrophysical effects of exposure were completely different for the two intervals. The bank unit was extensively cemented shortly after deposition, with cementation being most intense immediately below the paleo-exposure surface. This meteoric tight zone is interpreted to be the

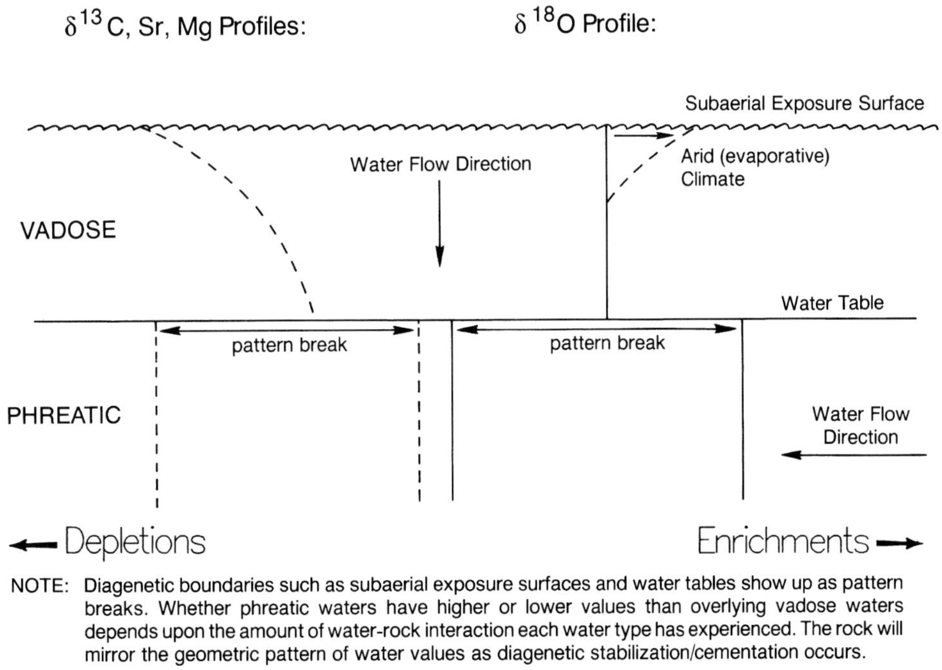

Figure 5. Expected depth-related isotopic and cationic patterns are shown that should develop in a carbonate undergoing meteoric diagenesis (modified from Wagner, 1983). Patterns are controlled by water flow directions and chemistry changes during water-rock interaction. Intense evaporation also causes a pattern break in oxygen isotopic data. Note that the subaerial exposure surface and water table are two diagenetic boundaries that may be well defined by geochemical data.

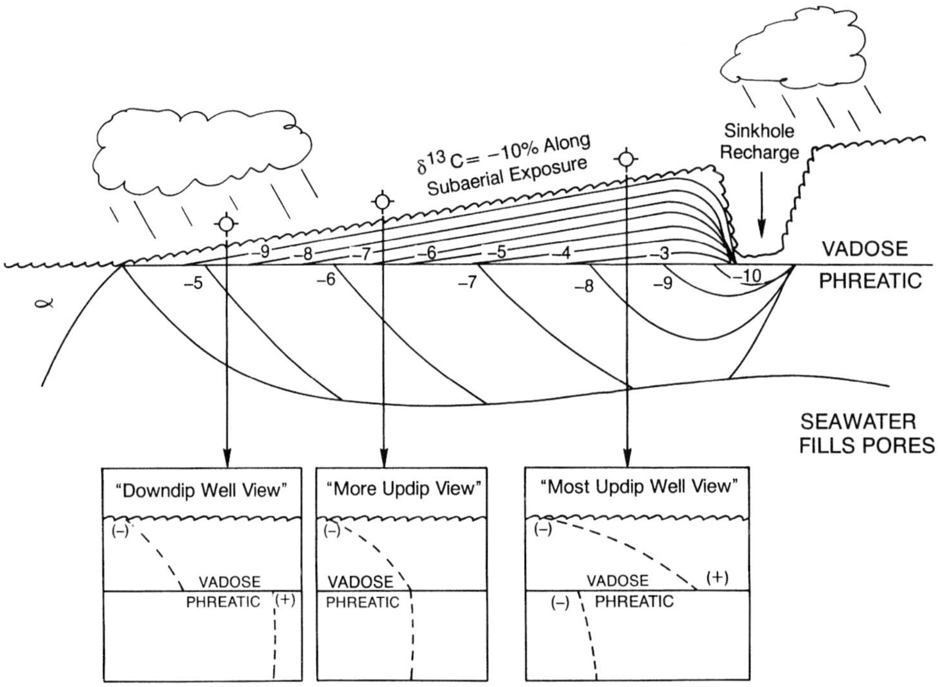

Figure 6. A hypothetical two-dimensional model is shown for the carbon isotopic compositions of meteoric waters; carbonates stabilized in these waters will record a similar spatial pattern of values. Source of very light carbon isotopic values is decaying organic matter in the soil zone. Note that meteoric phreatic waters have been contaminated by a large input of isotopically light vadose water from a sinkhole. A well drilled nearby would encounter very light carbon isotopic values in both the uppermost vadose and phreatic intervals. Wells drilled down-aquifer would encounter increasingly heavier carbon isotopic values in the meteoric phreatic interval from ongoing effects of water-rock interaction. Water lines reflect water flow direction, degree of water-rock interaction, and water mixing. Rock values mirror the geometric pattern of water values as diagenetic stabilization/cementation occurs.

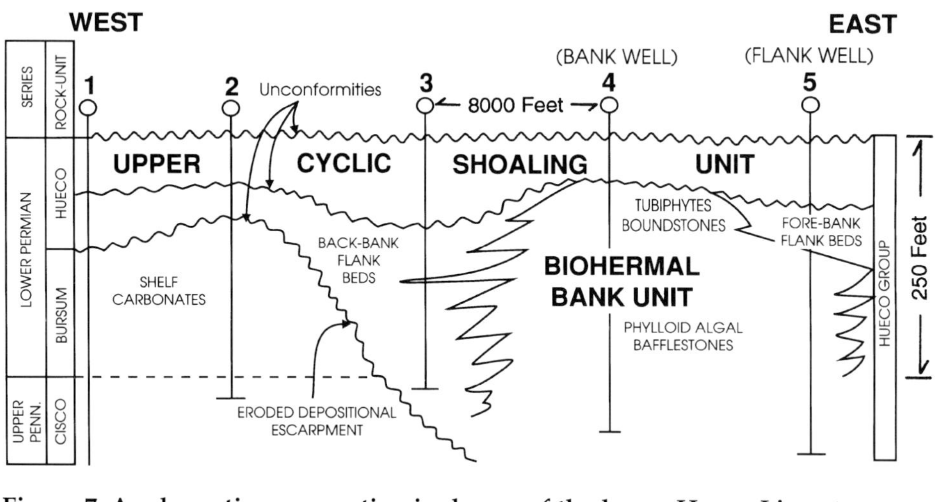

Figure 7. A schematic cross section is shown of the lower Hueco Limestone on the margin of the Central Basin platform, west Texas. The lower bank unit is capped by a paleo–subaerial-exposure surface, immediately below which lies the meteoric tight zone. Topography at the top of the bank unit probably formed from both depositional and erosional processes. The upper shoaling unit buried the bank unit under a series of cycles that shallow upward into grainstone shoals. The upper shoaling unit is also capped with a paleo–subaerial-exposure surface.

top seal for the underlying bank reservoir facies. After (and intermittently during?) deposition of the upper shoaling unit, extensive meteoric leaching produced abundant solution-enhanced primary porosity and moldic porosity in more grain-rich limestone facies. Some meteoric cementation also occurred, but its effects are judged visually to be subordinate to reservoir-enhancing effects of leaching in grainstone intervals. Later burial and associated cementation degraded porosity and permeability to varying degrees in porous and permeable sections.

Meteoric Tight Zone Capping Bank Unit (West Texas)

The meteoric tight zone in the uppermost interval of the bank unit has very low porosity and permeability across the field, even though it is predominantly a grainstone. Hydrocarbon production pressures in grainstone pay intervals above and below the uppermost bank tight interval are quite different, demonstrating that the meteoric tight zone has a sealing nature over at least a time scale of hydrocarbon exploitation efforts. Petrographic examination of the interval shows the cause of this sealing nature to be extensive calcite cementation. Along the bank crest, the low-porosity character of the tight zone continues stratigraphically downward into the biohermal facies, which is heavily cemented by syndepositional isopachous and radial fibrous submarine cements (see Wahlman, 1988), and to a lesser degree by later diagenetic ferroan calcite cements. However, off the bank crest, porous and permeable grain-rich flank-bed reservoirs are overlain and sealed by this capping meteoric tight zone (Figure 8).

The uppermost interval of the bank unit has a number of unusual petrographic and isotopic features that distinguish it from all other carbonate intervals in the shelf-margin facies complex. The interval contains a caliche paleosol and pendant cements (Figure 9). Porosity-occluding calcite cements generally have an equant morphology, but they resemble meniscus cements near some grain contacts. Carbon and oxygen isotopic profiles show values that are lighter (more negative) toward the top of the bank (Figures 10 and 11).

These data are interpreted to indicate the presence of a meteorically affected interval at the top of the lower bank unit. The uppermost bounding surface is interpreted to represent a paleo–subaerial-exposure developed on a terrane with moderate topographic relief, and the capping highly cemented interval is inferred to be a meteoric tight zone.

Of particular interest is the concentration of the most intense cementation at, and immediately below, the inferred paleo-exposure surface. Intense cementation of this grainstone/boundstone appears to have mimicked paleotopography, and did not extend

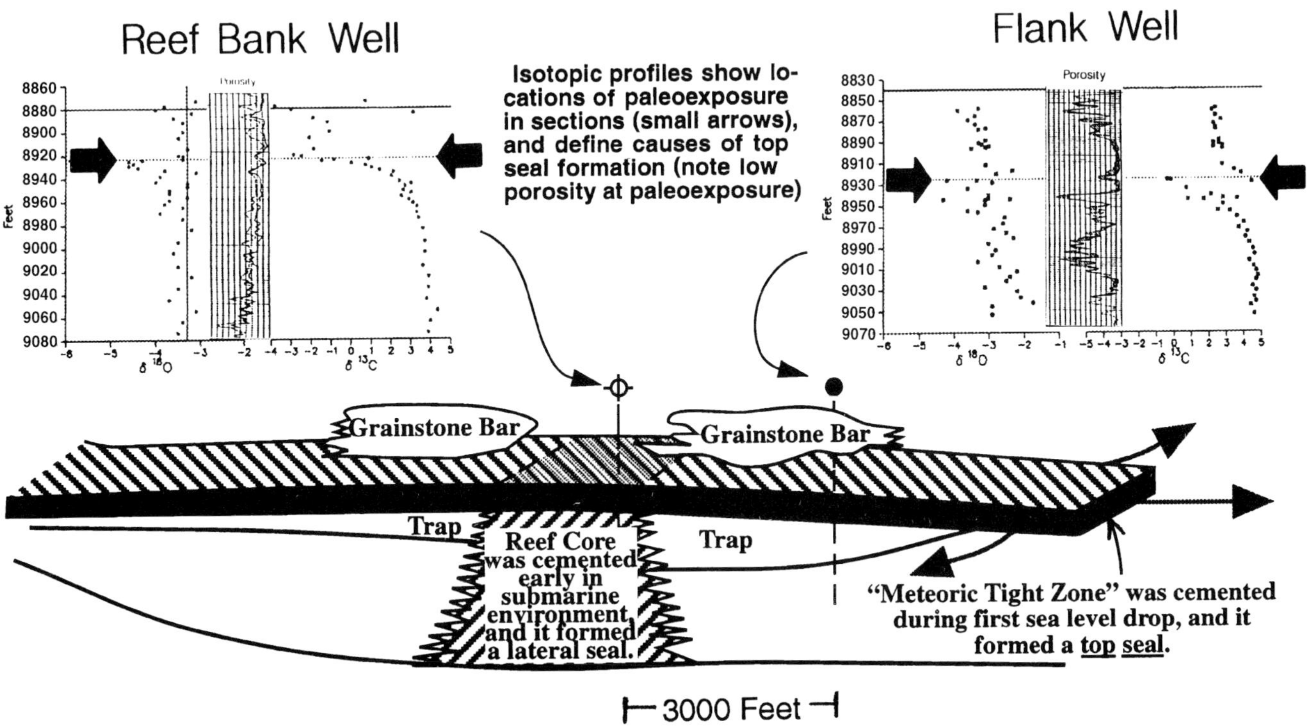

Figure 8. A model of a cementation trap is shown from the Lower Permian Hueco Limestone of west Texas. Porous carbonate sands were sealed at the top by unconformity-related cements (note porosity log for "flank well"), and sealed laterally by marine-related cements ("reef bank well"). Paleo-exposures were identified with an integrated isotopic and petrographic data set. Note the well-developed "vadose flags" in carbon isotopic profiles associated with porosity loss below the paleo-exposures (small arrows). Details of both wells are given in later figures.

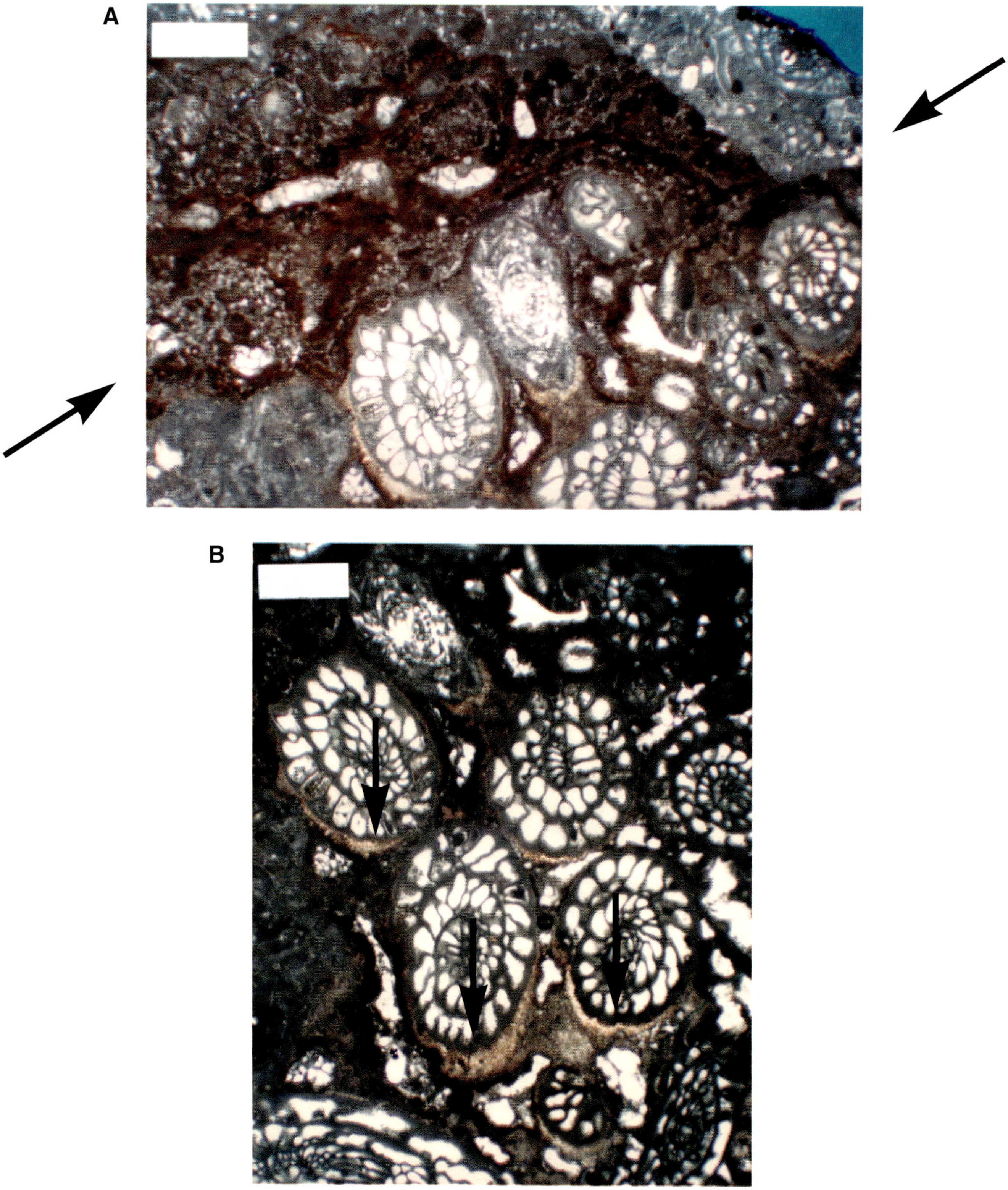

Figure 9. Dark brown caliche (A, arrows) and pendant cements (B, arrows) in the uppermost bank unit's meteoric tight zone, "reef bank well," Hueco Limestone, west Texas. Scale in both photographs is 1 mm.

downward into underlying coarse-grained intervals. A strong geometrical tie of the cementation process to the paleo-exposure surface is indicated. Both the morphology of cements and the overall geometry of porosity plugging suggest that porosity occlusion occurred primarily in the uppermost paleo-vadose interval. The position of the paleo–subaerial-exposure surface at the top of the overall shallowing-upward shelf-margin

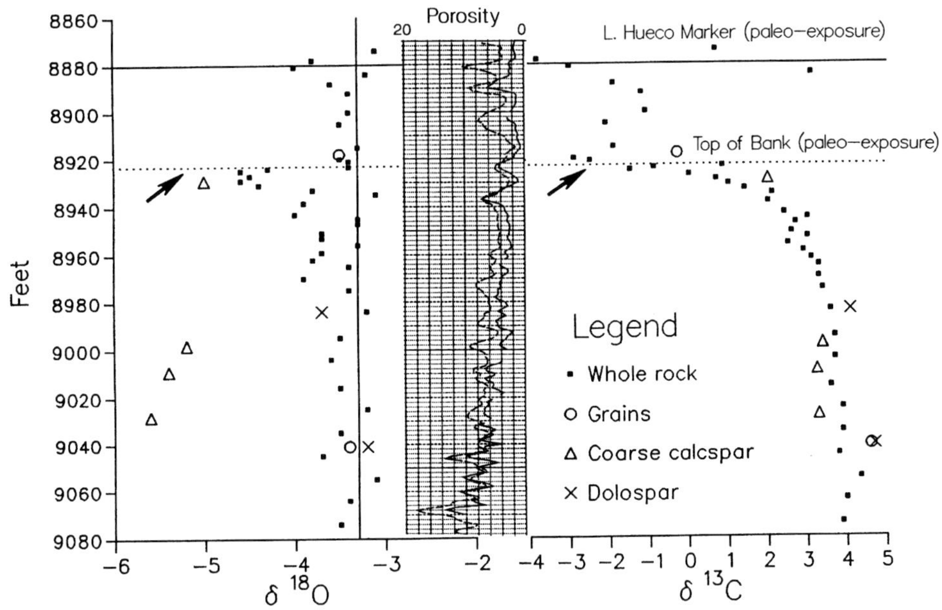

Figure 10. Isotopic depth profiles are shown for a well at the crest of the bank unit in the Hueco Limestone, west Texas. A "vadose flag" in carbon isotopic values (Allan and Matthews, 1977) helps define the top of the bank unit as a paleo-exposure surface. Lighter oxygen isotopic values also help define the top of the bank as a diagenetic boundary. The caliche and pendant cements of Figure 9 come from the top of this interval. The middle strip log indicates low porosity throughout the interval of interest (about 8924–8950 ft); low porosity is from meteoric cementation immediately below the paleo-exposure and from syndepositional marine and much later burial cementation below the meteoric tight zone (8950+ ft).

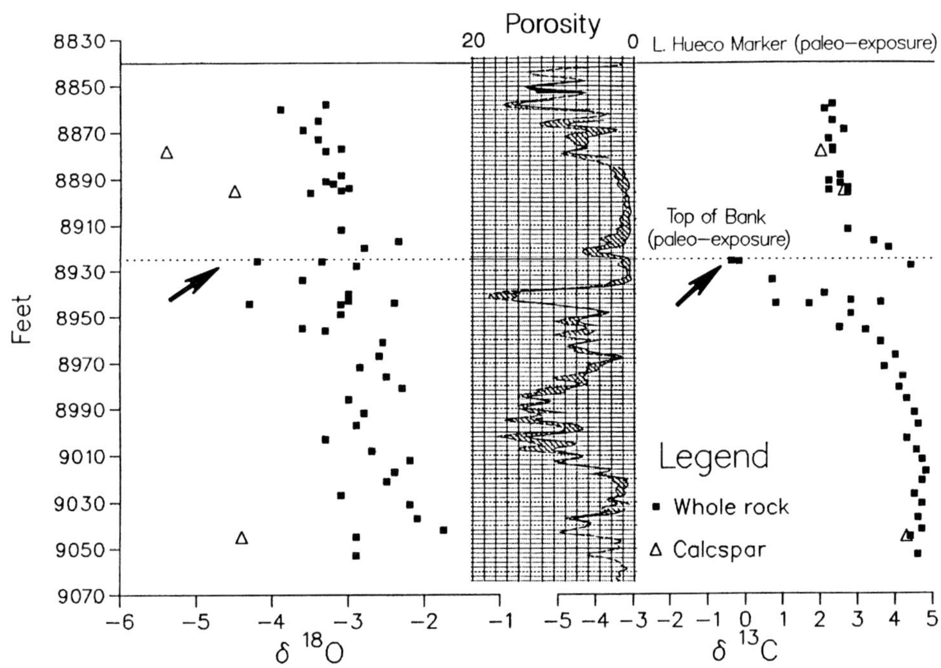

Figure 11. Isotopic depth profiles are shown for a well on the flank of the shelf-margin bank in the Hueco Limestone, west Texas. A "vadose flag" in carbon isotopic values helps define the top of the bank as a paleo-exposure in this flank well. The highly cemented grainstone from about 8925–8938 ft is interpreted to be a meteoric tight zone. Production data show that oil pays in the grainstones below the meteoric tight zone are not in pressure continuity with shallower pays—demonstrating the sealing nature of the inferred meteoric tight zone on a minimum time scale of field exploitation.

bank stratigraphic package further suggests that the exposed terrane was composed of newly deposited, chemically reactive carbonate sediments. Cementation under arid to semi-arid climatic conditions is inferred from the presence of a caliche paleosol (e.g., James and Choquette, 1990).

The economic effect of meteoric exposure at the top of the bank unit was the formation of an internal seal within the Hueco Limestone where none would have been expected using only traditional depositional models. Position of the intraformational seal impacts how exploration for new hydrocarbon pools is conducted and helps define how existing pools can be more efficiently exploited (e.g., water flooding).

Meteoric Leaching in the Upper Shoaling Unit (West Texas)

The upper shoaling unit is composed of a series of shallowing-upward successions that also have a number of unusual petrographic and isotopic features. The interval contains meniscus cements in grain-rich intervals and abundant early formed moldic porosity throughout. Carbon and oxygen isotopic profiles through the unit show values become lighter (more negative) toward the top of the interval (Figure 11). No paleosols were recognized.

These data are interpreted to indicate the presence of a thick, meteorically affected interval that makes up most or all of the upper depositional unit. The top of the unit is interpreted to be a paleo–subaerial-exposure surface, but multiple exposures might have occurred during cyclic deposition. Net petrophysical effect of exposure is difficult to access on a whole-interval basis, but was porosity generating on the scale of individual grainstone beds. Dissolved calcareous material may have been reprecipitated in muddier intervals in the lower parts of the cycles.

OFFSHORE CHINA EXAMPLE

Case Summary

The second example of petrophysical effects of meteoric diagenesis comes from the Miocene Zhujiang Limestone of offshore China (Figure 12). The Zhujiang complex is composed of a series of back-stepping platforms on a large horst block (Christian and Tyrrell, 1991; Erlich et al., 1991; Turner and Hu, 1991). The cores and cuttings for this example come from Amoco's Liuhua discovery area. Figure 13 shows the local stratigraphy. Rocks of the upper shelf interval in the field area are comprised primarily of grainstones

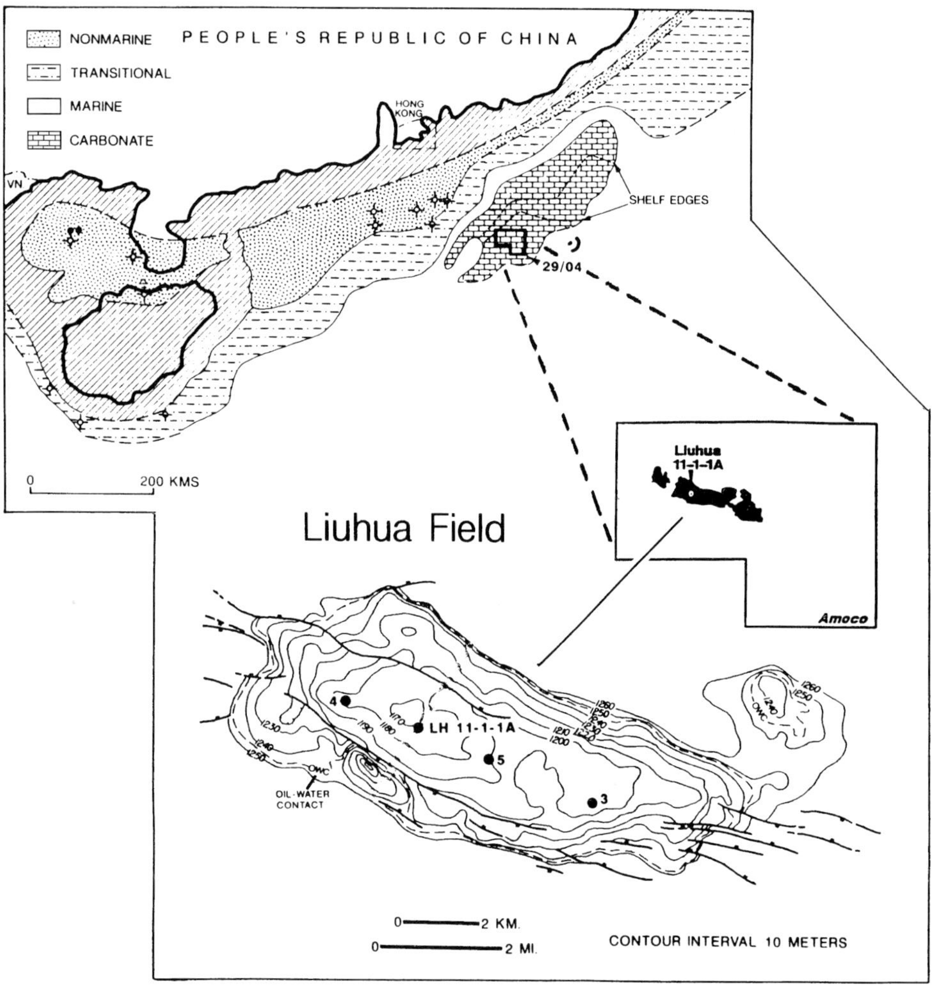

Figure 12. The location of Amoco's Liuhua Block 29/04 lease is shown in the Pearl River Mouth Basin of offshore China (modified from Christian and Tyrrell, 1991, and Turner and Hu, 1991). The block is located over the large basement horst feature known as the Dongsha Massif. Well locations are indicated on the structure contour map.

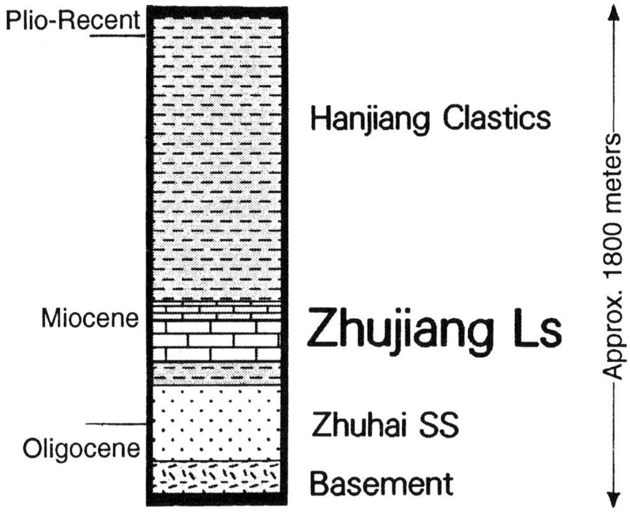

Figure 13. Local stratigraphy is shown for the Liuhua area. The Zhujiang Limestone forms a series of back-stepping platforms over the Dongsha Massif. Hanjiang clastics form the top seal for the Liuhua accumulation. The Zhuhai Sandstone was a hydrocarbon migration conduit in portions of the area. Both the Zhujiang Limestone and Zhuhai Sandstone are good stratigraphic conduits for migrating fluids.

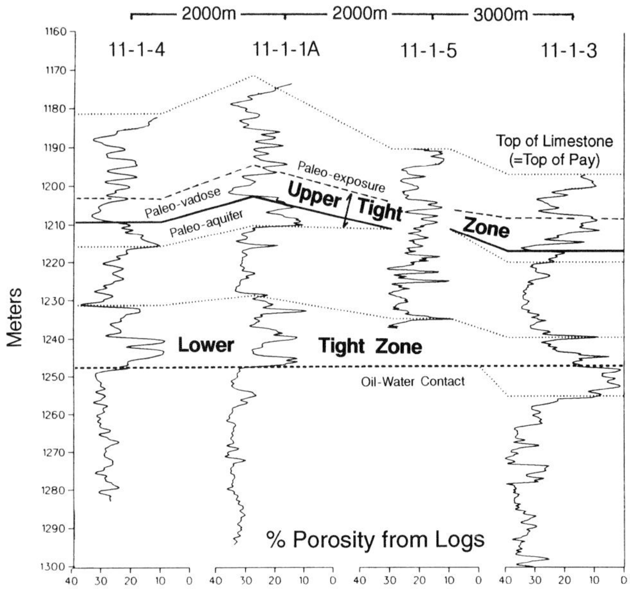

Figure 14. A porosity cross section of wells from the Liuhua discovery is shown modified from Turner and Hu (1991). Locations of wells are shown in Figure 12. The so-called "upper tight zone" and "lower tight zone" are indicated. The datum of the section is the oil-water contact for the field. The thick unbroken line within the upper tight zone is the presumed top of the paleo-phreatic interval. The dashed line immediately above the paleo-phreatic is the inferred paleo-exposure surface based on isotopic, petrographic, and GR log information. Note that porosity is generally decreased by meteoric cementation in the inferred paleo-phreatic interval, but may also be locally reduced near the paleo-exposure. A comparatively brief exposure is inferred. Petrographic examination indicates that the marked decrease in porosity near the paleo-exposure surface in the 11-1-3 well was caused by localized chemical compaction during burial, and not by early meteoric cementation. The "lower tight zone" formed by a combination of burial processes unrelated to the earlier exposure event.

and packstones. Figure 14 shows a cross section of porosity logs from field wells. Note in particular that two intervals with low porosity and permeability characteristics exist within the overall pay section. The shallower interval is informally termed the "upper tight zone"; the deeper interval is termed the "lower tight zone."

A single paleo–subaerial-exposure surface is interpreted to exist in the section. Meteoric cementation below this surface is interpreted to have formed the "upper tight zone." Petrographic and isotopic data indicate that effects of meteoric diagenesis were of limited areal and stratigraphic extent. Subaerial exposure by a brief glacio-eustatic sea level drop is inferred. Porosity log data indicate that net pore-space losses occurred in both the paleo-vadose and paleo-phreatic intervals, although significantly more loss occurred in the paleo-phreatic interval due to meteoric cementation. Given the timing and limited areal extent of exposure, dissolved material for cementation is inferred to have been derived from the overlying paleo-vadose interval. A downward shift is indicated in most-intense cementation from the uppermost vadose interval in the west Texas example to the meteoric phreatic interval in this case.

Meteoric Seal Formation during an Ephemeral Island-Type Exposure (China)

The informally termed "upper tight zone" has anomalously low porosity across most of the field (Figure 13). This interval has a number of unusual petrographic and isotopic features that distinguish it from all others in the complex. The interval contains meniscus cements and abundant early formed moldic porosity that is partially filled and/or lined with equant calcspar. Porosity and permeability are diminished in the upper tight zone due to extensive calcite cementation, although the interval is as grain rich in depositional texture as underlying pay-zone rocks. Carbon and oxygen isotopic profiles show values become lighter (more negative) toward the top of the interval; isotopic values of overlying and underlying limestones are much heavier (Figure 15).

These data are interpreted to indicate the presence of a meteorically affected interval near the top of the limestone platform complex. The uppermost bounding surface is interpreted to represent a paleo–subaerial-exposure, and the highly cemented interval is inferred to be a partially developed meteoric tight zone. The lighter (more negative) oxygen and carbon isotopic characters near the top of the carbonate shelf section are interpreted to indicate the limited

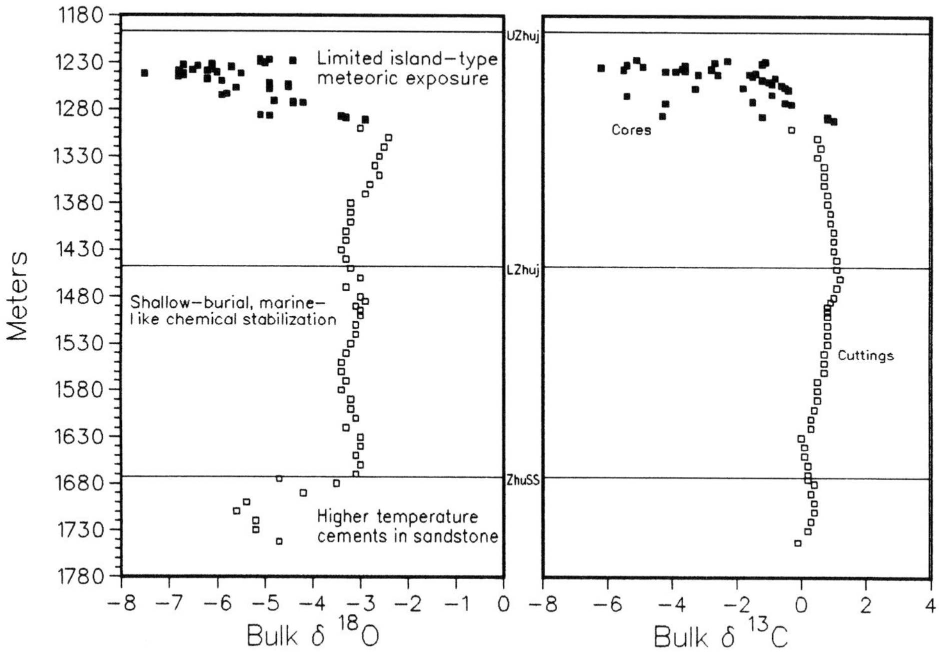

Figure 15. Stable isotopic stratigraphy of the Liuhua discovery well is shown; values are from bulk rock and are relative to PDB. The shallowest two units are termed the upper and Lower Zhujiang Limestone; the deepest unit is the Zhuhai Sandstone. Note that pronounced isotopic depletions exist in bulk rock limestone values in the upper part of the Zhujiang platform at the Liuhua location. Isotopic values from limestones in other wells above and below the isotopically depleted interval are much heavier and are nearly constant in value; these other limestones are interpreted to have been chemically stabilized in the shallow subsurface in a fluid with seawater-like character. Lighter oxygen isotopic values from the Zhuhai Sandstone are from ferroan dolospars that are interpreted to have precipitated at elevated temperatures in the subsurface.

stratigraphic extent of early meteoric alteration. Adjacent intervals with much heavier, more uniform isotopic values are interpreted to have chemically stabilized under shallow-burial conditions in a fluid type very similar to seawater. Lighter oxygen and carbon isotopic values from the deepest part of the well come from dolomitic cements in sandstone, and are interpreted to be due to burial cementation processes. No other meteorically affected intervals are interpreted to exist in the limestone platform at the Liuhua location.

A more detailed examination of the meteorically affected interval and comparison with other well-constrained meteoric systems shows that the most-intense meteoric cementation apparently occurred in the Liuhua's paleo-phreatic interval. Figure 16 shows details of two cored intervals interpreted to have undergone meteoric diagenesis. Diagenetic boundaries are indicated. Boundaries were inferred from (1) isotopic stratigraphic comparisons with known meteoric systems (Figure 5), (2) petrographic observations of unusual cement types, and (3) GR logs (Turner and Hu, 1991). Intervals with lowest porosity in the meteorically affected interval generally contain more meteoric cements. Petrographic examination also indicated that some porosity loss in the paleo-vadose interval of well 11-1-3 occurred by much later burial-induced chemical compaction.

Comparison of Liuhua data patterns with geochemical profiles from a limestone undergoing meteoric diagenesis on Barbados, West Indies, were particularly helpful. Figure 17 shows depth profiles of mineralogy, oxygen isotopes, and carbon isotopes from borehole 33, on the south coast of the island (Wagner, 1983). The "vadose flag" in borehole 33 is poorly developed because of fairly arid climatic conditions and attendant retarded chemical reaction of vadose sediments in the local area (e.g., Harrison and Steinen, 1978). The underlying meteoric phreatic interval is much further along in the chemical stabilization process because of greater (lateral) water availability, and has a well-developed lighter (more negative) isotopic signature from stabilization in meteoric water. Bulk porosity in the present meteoric phreatic interval is visually estimated to have been decreased by chemical stabilization and associated calcite cementation. It is inferred that an interruption of the meteoric diagenetic alteration of this Barbadoan interval, and subsequent stabilization of remaining carbonate in shallow-burial, nonmeteoric water, would produce identical geochemical stratigraphies and porosity-reduction patterns to those observed in the Liuhua cores.

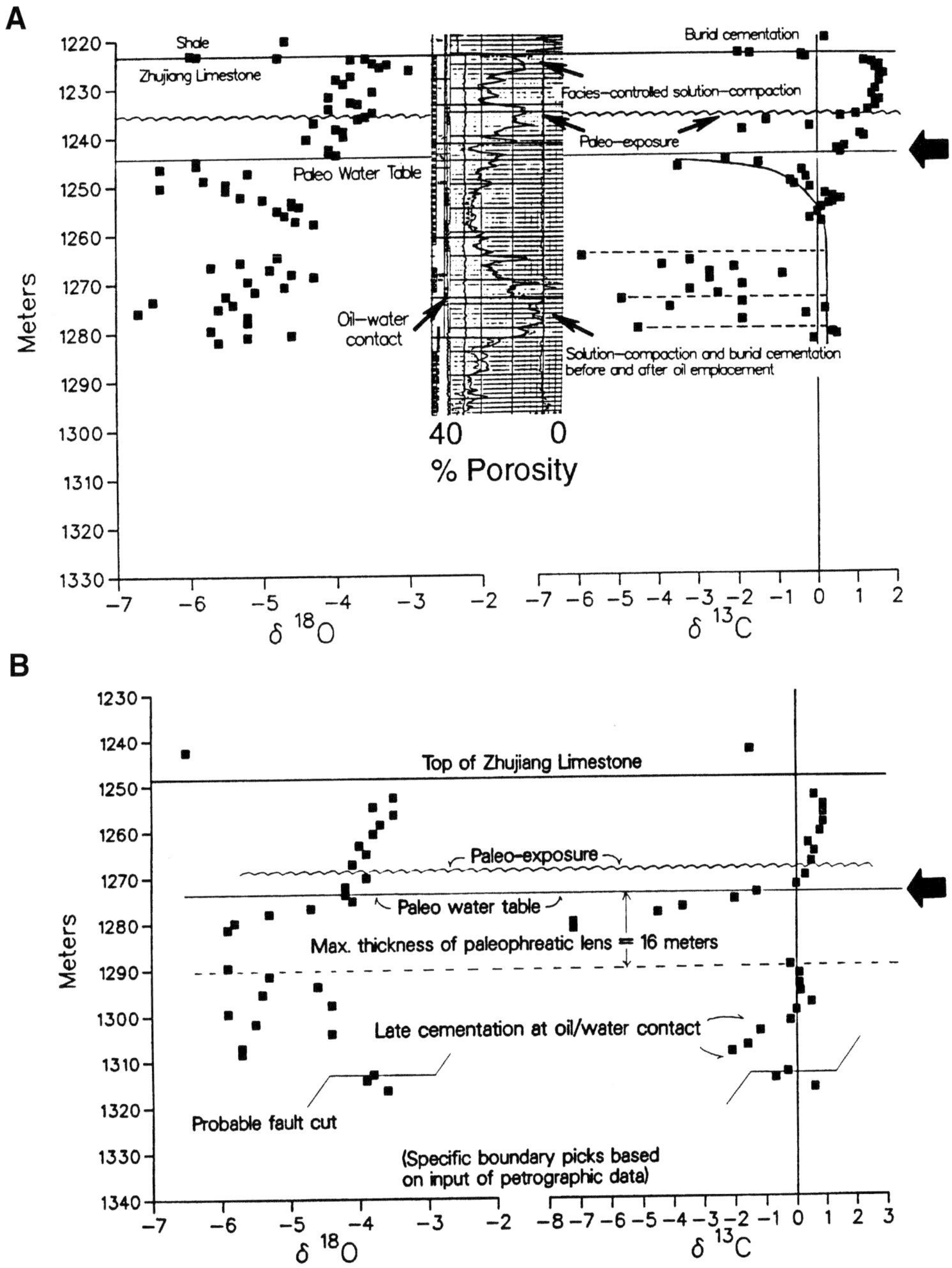

Figure 16. Isotopic stratigraphies from cored intervals in two exploration wells are shown. (A) Amoco Liuhua 11-1-3. (B) Amoco Liuhua 11-1-4. Diagenetic boundaries are inferred from isotopic, petrographic, and logging information. The "upper tight zone" extends from the inferred paleo–subaerial-exposure in each well down to a somewhat indistinct deep-phreatic layer estimated from isotopic data. A relatively complicated relationship of isotopic stratigraphy to diagenetic environments is interpreted to be a consequence of comparatively brief exposure (see Figure 17 for comparison). The "lower tight zone" formed much later from burial processes of chemical compaction and burial cementation. Note that without petrographic control, it is impossible to discriminate similar isotopic stratigraphies generated by different diagenetic processes.

The "upper tight zone" within the Liuhua pay section is interpreted to have been produced by a brief exposure caused by a temporary relative sea level drop that punctuated the overall rising sea level conditions prevalent at that time (Erlich et al., 1991). The ephemeral nature of the subaerial exposure is inferred from two observations. First, the amount of diagenetic alteration of carbonate in the paleo-vadose interval is limited. Only the very uppermost portion of the vadose shows development of a "vadose flag" in carbon isotopic values. Second, the overall drowning nature of the platform system suggests a limited

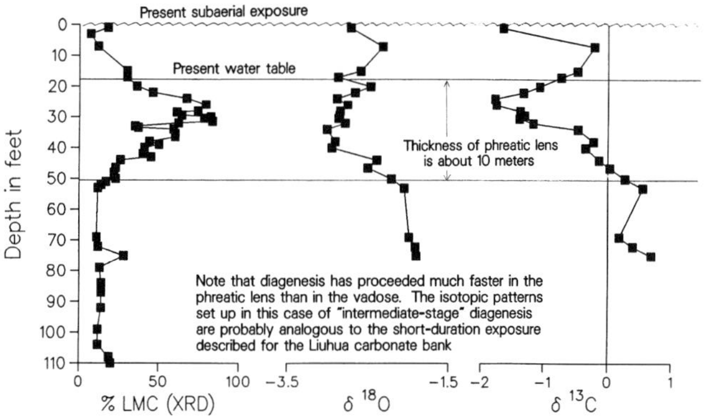

Figure 17. Isotopic stratigraphies from a cored interval on the relatively dry south coast of Barbados, West Indies, are shown; modified from Wagner (1983). The present exposure and phreatic (water table) surfaces are indicated. X-ray diffraction data show that the phreatic interval has undergone the most extensive chemical stabilization from metastable sediments to stable limestone; visual estimates indicate net porosity has decreased in the phreatic. Note also that only minor chemical stabilization has occurred immediately below the exposure surface, but that it has imparted a perceptible isotopic signature. It is inferred that a similar type of incomplete meteoric stabilization occurred in the Liuhua section; rapid sea level rise abruptly terminated meteoric diagenesis. Remaining metastable mineralogies are inferred to have stabilized under shallow-burial, marine-water–like conditions that produced a heavier, stratigraphically constant isotopic signal.

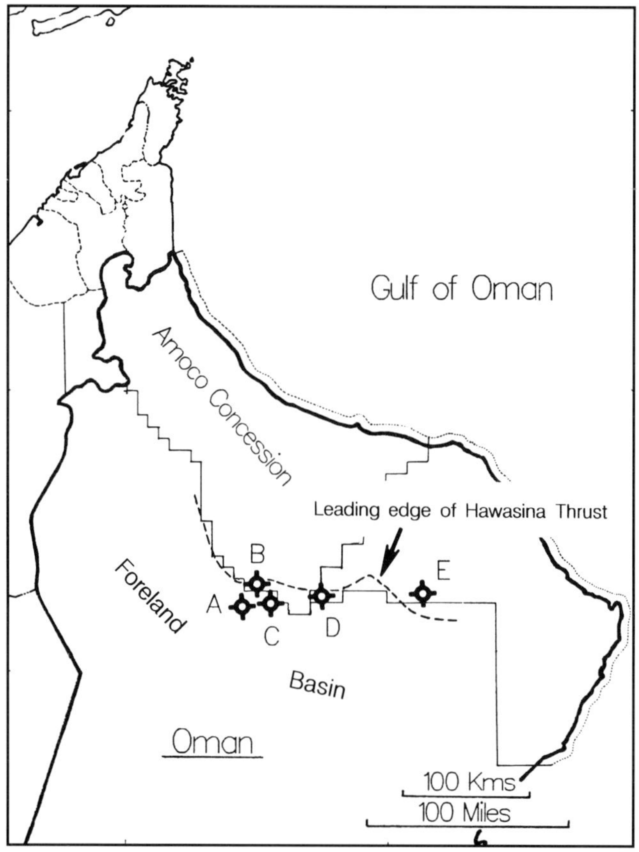

Figure 18. A map of central Oman and the northern United Arab Emirates (U.A.E.) is shown modified from Wagner (1990). Wells A through E form a 100 mile transect near the leading edge of the Hawasina thrust nappe (dashed line). Well D is on the western flank of the salt-cored Shabiyah high.

opportunity to develop and to sustain meteoric diagenesis. Exposure of metastable carbonate sediments is inferred.

The economic effect of meteoric cementation in this example is profound. Ultimate development of the Liuhua discovery depends on retarding upward water incursion from the underlying active aquifer. Current plans are to drill horizontal development wells just above the intraformational meteoric tight zone to guard against this possibility.

CENTRAL OMAN EXAMPLE

Case Summary

The third example that demonstrates different petrophysical effects of meteoric exposure is from Cretaceous limestones in central Oman (Figure 18). The Omani depositional setting was a very broad platform with large intrashelf subbasins and scattered salt-induced swells (e.g., the Shabiyah and Lekhwair highs). Northwestern portions of the limestone platform were intensely structured and/or overridden during emplacement of the Hawasina thrust nappe during the Late Cretaceous. Carbonate porosity below the thrust sheet was generally destroyed by burial-induced stylolitization and associated cementation (Wagner, 1990). Sections of the Cretaceous shelf located outboard of thrusting, in the foreland basin areas, are composed dominantly of very porous wackestones and fine-grained packstones. Figure 19 shows the local stratigraphy. The primary exploration targets are Natih and Shuaiba limestones.

As in the preceding west Texas example, at least paleo–subaerial-exposure surfaces are inferred to exist in the limestone section of interest. The Omani case differs from the west Texas case, however, in the predominance of muddy carbonates. Very few large interparticle pore spaces existed to accommodate the unusual cement morphologies characteristic of meteoric diagenesis. As a result of this difference in depositional textures, petrography played only a minor role in defining intervals exposed to meteoric processes. Isotopic data and stratigraphic information formed the basis of most final interpretations.

Two boundaries inferred to be paleo–subaerial-exposure surfaces are described. The first, and stratigraphically youngest, is the top of the Cretaceous shelf sequence in the area (i.e., top of Natih Limestone); a net increase in porosity and permeability characteristics is tentatively inferred to have occurred during this exposure event. The second boundary is interpreted to be a

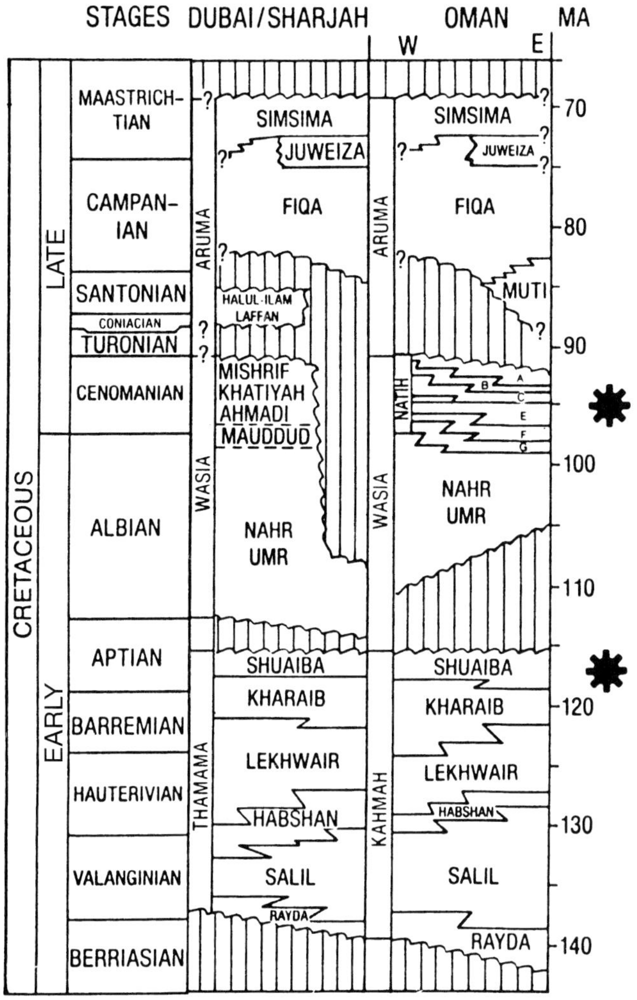

Figure 19. The Cretaceous stratigraphy of central Oman and the U.A.E. is shown modified from Scott (1990). The Natih and Shuaiba limestones are primary exploration targets in the area. The Natih Formation represents the last development of limestone platforms in the area. Clastic units immediately above the Natih and Shuaiba limestones act as regional seals.

localized "island-type" paleo-exposure within the Natih Limestone; a net decrease of porosity and permeability is inferred to have occurred during exposure.

Meteoric Leaching at Top of Cretaceous Shelf (Top of Natih Limestone, Oman)

Stratigraphic and paleontologic studies suggest the Natih Formation was truncated by a regional paleo-exposure (e.g., Glennie et al., 1973; Scott, 1990). Geochemical data support this inference (Wagner, 1990). Figures 20 and 21 show the carbon and oxygen isotopic stratigraphies from five exploration wells across a 100 mile transect in central Oman. Carbon and oxygen isotopic profiles show that values become lighter (more negative) toward the top of the limestone. Petrographic evidence of meteoric exposure is ambiguous in the upper Natih Limestone.

These limited data support the presence of an upper-bounding paleo-exposure surface for the Natih Limestone. Lighter (more negative) carbon isotopic values are interpreted to represent "vadose flags." Lighter (more negative) oxygen isotopic values are interpreted to discriminate between the meteoric-diagenetic character of this section and the underlying shallow-burial–diagenetic character (e.g., see Figure 15 of the preceding China example).

The effect of exposure on the uppermost Natih Limestone was somewhat variable. Many wells outboard of the Hawasina nappe's leading edge show a distinct porosity enhancement toward the paleo-exposure surface in an otherwise monotonous wackestone section. Other foreland basin wells show little or no increase in porosity values near the paleo-exposure. The cause of this difference is unknown, but it could be related to any one of a number of factors, including variations in terrane permeability during exposure and differential porosity loss during burial.

Local Meteoric Tight Zone Formation on Flanks of Paleotopographic High (in Natih Limestone, Oman)

The uppermost part of the Natih E member has anomalously low porosity in well D (see Figure 22). Nearby wells show a highly porous character in the equivalent interval. The upper E member in well D has a number of unusual petrographic and isotopic features that diagenetically distinguish it from stratigraphically adjoining intervals and equivalent intervals in nearby wells. Petrographically, the muddy texture of the uppermost E member has been neomorphosed to a coarse microspar texture, unlike adjoining muddy intervals in the same well, or the equivalent interval of nearby wells. Geochemically, carbon and oxygen isotopic profiles show that values become lighter (more negative) toward the interval's top; bulk strontium values are also anomalously depleted toward the interval's top (Figure 22). These geochemical patterns do not appear in equivalent sections in nearby wells, or in adjoining limestone members.

These geochemical and petrographic data are interpreted to indicate the presence of a meteorically affected interval within the uppermost E member of the Natih Limestone. The uppermost bounding surface is interpreted to represent a paleo–subaerial-exposure, and the highly cemented underlying interval is inferred to be a meteoric tight zone. Extensive fracturing in portions of the highly cemented interval appear to have been completely healed by later calcspar cementation.

Available rock and stratigraphic information indicates the inferred meteoric tight zone in the E member was locally developed on the flanks of the Shabiyah high at well D. An updip well shows that the entire Natih Limestone is absent from the crest of the Shabiyah high. Geometries of seal and reservoir rocks suggest the presence of a hydrocarbon trap (Figure 23).

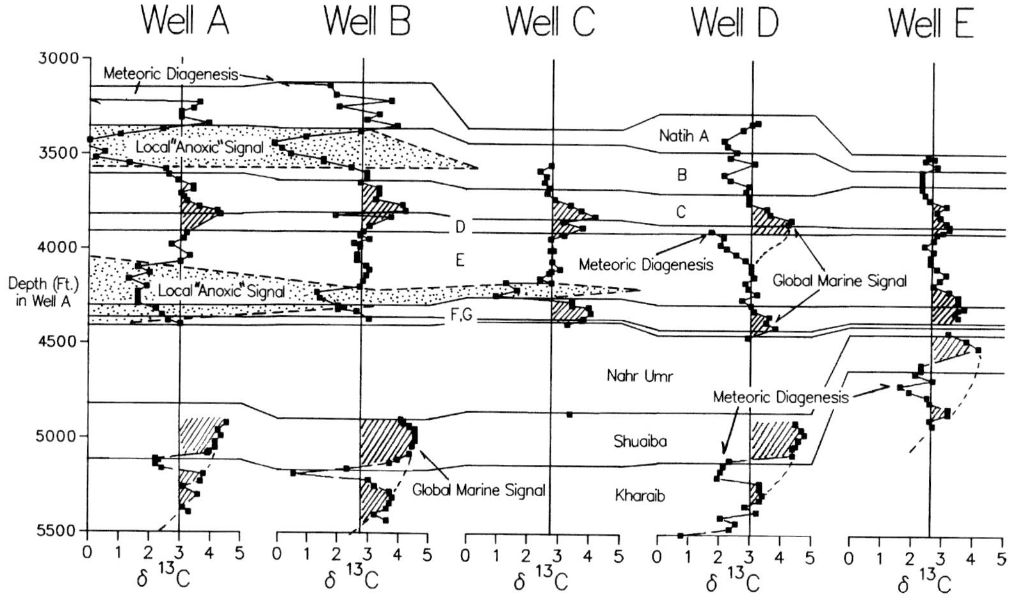

Figure 20. A portion of the Cretaceous carbon isotopic stratigraphy of central Oman is shown; modified from Wagner (1990). The top of the Natih Limestone is the upper bounding level; the base of the Kharaib Limestone is the lower bounding level. The carbon isotopic stratigraphy is interpreted to have formed from a combination of (1) the global marine signal imparted during deposition, (2) diagenetic overprinting by meteoric diagenesis at paleo-exposures, and (3) burial cementation in and around organic-rich intervals (local "anoxic" signal). Note the dramatic stratigraphic pattern shift in Natih E member values at well D compared to surrounding wells.

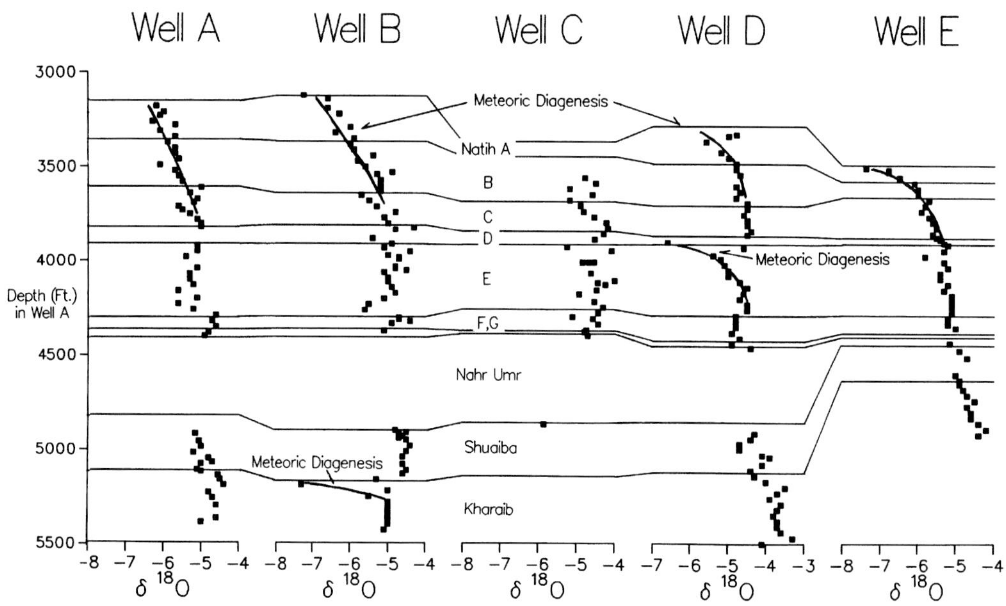

Figure 21. A portion of Cretaceous oxygen isotopic stratigraphy of central Oman is shown; modified from Wagner (1990). The bounding surfaces are the same as in Figure 20. The oxygen isotopic stratigraphy is interpreted to be a simple two-component mixture of meteoric and shallow-burial stabilization signatures, similar to those observed in the offshore China example of Figure 15. Note again the anomalous isotopic deflections toward lighter values at the top of the Natih E member of well D.

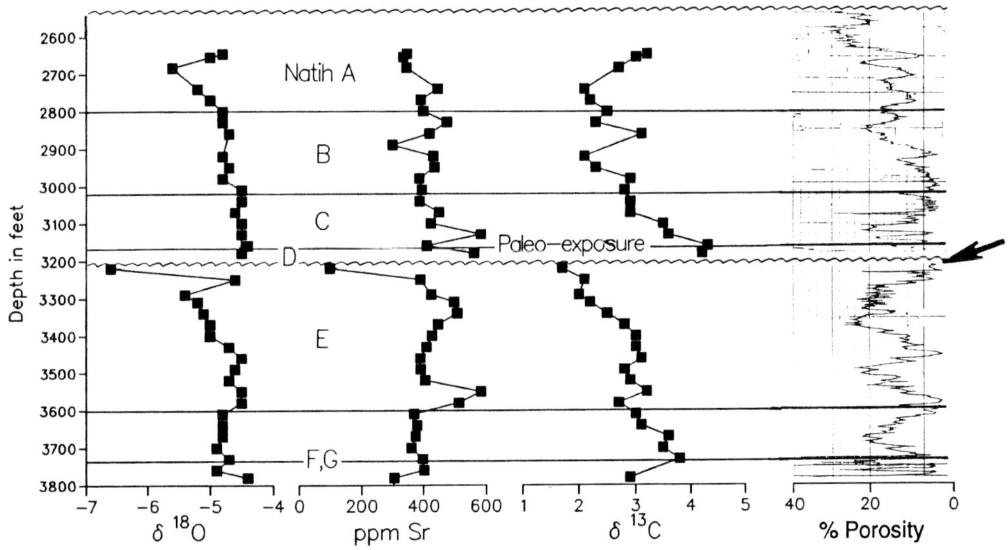

Figure 22. A detailed look at geochemical stratigraphies of the Natih Limestone in well D is shown along with a porosity log. Note that isotopic and cationic depletions occur in a stratigraphic trend toward the top of the E member and are consistent with patterns expected for meteoric diagenesis. Porosity values in the uppermost Natih E member are very low, unlike the porous character of the E member in surrounding wells. Limestone in this tighter interval exhibits an aggrading neomorphism texture that is petrographically unusual for the section. The presence of a meteoric tight zone is inferred. Also note, for comparative purposes, that the greatest porosity values in the carbonate interval are associated with the paleo–subaerial-exposure at the top of the A member.

PREDICTION OF DIAGENETIC TRAPS FORMED BY METEORIC TIGHT ZONES

Inferring the presence of a meteoric tight zone in a stratigraphic section is generally easier than predicting the location of one. Engineering studies might not require any prediction, but exploration efforts certainly do. Based on observations and inferences from the preceding three examples of meteorically induced reservoir degradation and compartmentalization, some guidelines for predicting these types of diagenetic features are indicated. Methodical searches for meteoric diagenetic traps should be most productive in areas with the following characteristics:

- Drilling density should be moderate.
- Facies distributions should not be overly complex.
- At least moderate paleotopography is desirable.
- An overall drowning sea level history punctuated by brief sea level drops is desirable.

Moderate drilling density is desirable because it provides sufficient rock control for stratigraphic high-grading purposes, but enough breathing room to allow for sizable hydrocarbon accumulations. Relatively simple facies distributions allow the explorationist to concentrate on diagenetic patterns without running the risk of complete randomization of reservoir distribution from a variable facies mosaic. Accentuated paleotopography is desirable because it allows cross-formational meteoric tight zones to form suitable trapping geometries. And finally, a depositional history involving carbonate drowning would be desirable because it maximizes the possibilities of ephemeral exposure of highly reactive lime sediments during short-lived sea level drops.

The data set of choice for predicting diagenetic traps in carbonates is petrography. Integration of geochemical data is desirable as an independent check on petrographic interpretations. Geochemical data should be used as the lead interpretive tool only where visual rock information is lacking. Integration of seismic data is desirable, especially as mapping needs increase with a decrease in well control, and as the quality of seismic porosity prediction increases with decreasing depth of burial.

OTHER CONCLUSIONS

Meteoric tight zones are important trapping or reservoir-compartmentalizing components in some fields. The "deliberate search for the subtle trap" should include petrographic and geochemical data sets collected in stratigraphic context to address trapping potential in limestones previously exposed to meteoric diagenesis. More work needs to be done on the existing inventory of fields to determine the

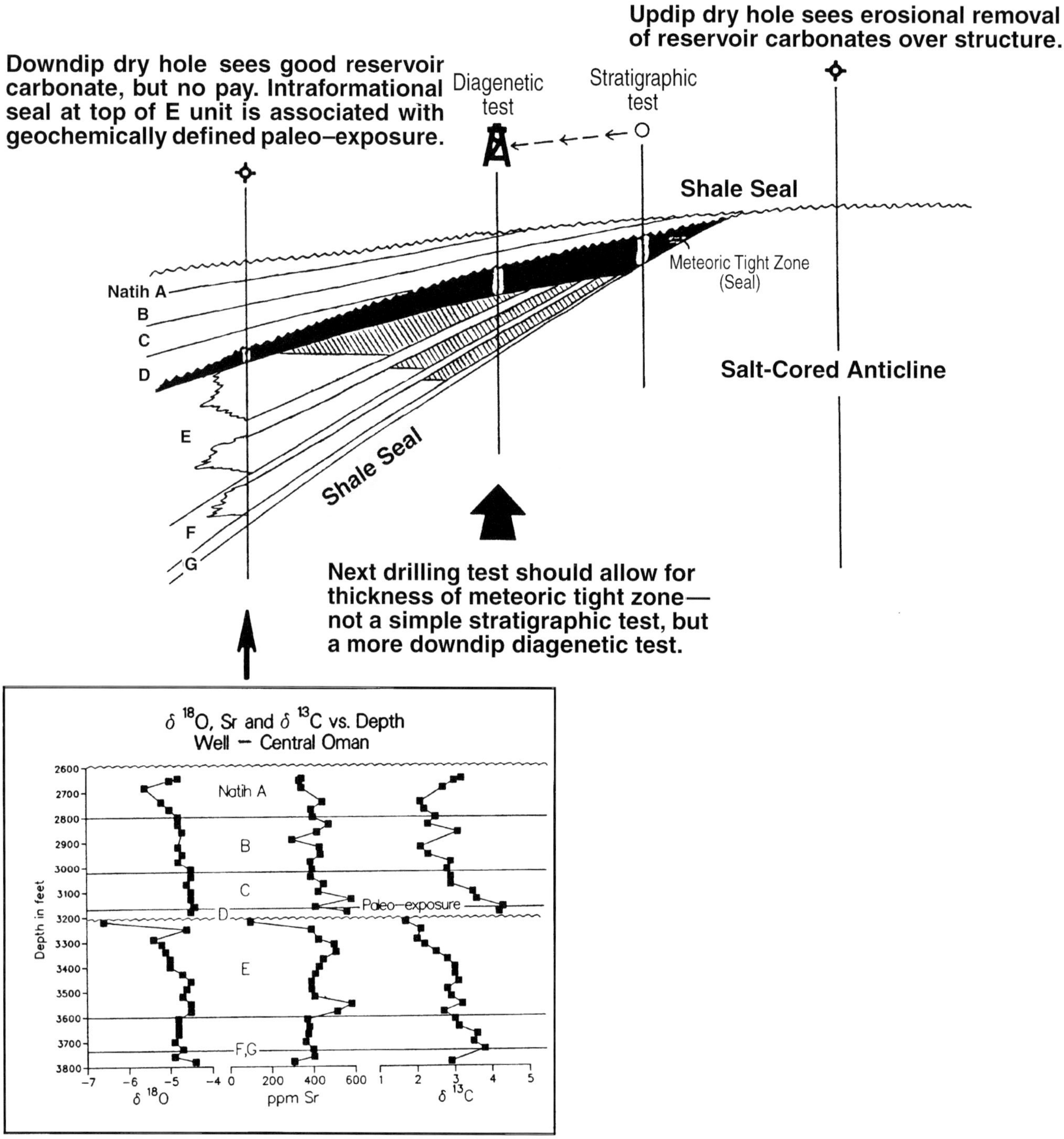

Figure 23. An example of a possible diagenetic cementation trap in central Oman is shown. The overall reservoir section is truncated over a salt swell. Isotopic, cationic, and petrographic data indicate the presence of a meteoric tight zone (black) below an intraformational paleo-exposure in the flank well (well D). Updip drilling demonstrated the erosional removal of target limestones from the crest of the salt swell. Further drilling should be located downdip from a stratigraphic test after allowing for the thickness of meteorically cemented carbonate (large arrow).

relative abundance of meteoric tight zones. Such work might increase reservoir reserve estimates and provide more exploration models.

Predicting lateral extents of flow barriers is important in reservoir production modeling. Data from examples in this study suggest that meteoric tight zones can be areally extensive. In field studies, emphasis should be placed on determining how a flow barrier formed. If a flow barrier formed by meteoric cementation, then the potential for good compartmentalization and resistance to water invasion is enhanced.

Field data suggest that a stratigraphic change in intensity of meteoric diagenesis occurs as water availability (climate, etc.) conditions change. Under very arid conditions, cementation seems to be focused in the uppermost vadose interval. As water availability increases marginally, most-intense cementation seems to shift stratigraphically downward into the underlying meteoric phreatic interval. Under very high water availability conditions (wet climate, high rock permeability), dissolved material is generally swept from the system to produce abundant secondary porosity and improved permeability.

REFERENCES CITED

Allan, J. R., and Matthews, R. K., 1977, Carbon and oxygen isotopes as diagenetic and stratigraphic tools—data from the surface and subsurface of Barbados: Geology, v. 5, p. 16–20.

Allan, J. R., and Matthews, R. K., 1982, Isotopic signatures associated with early meteoric diagenesis: Sedimentology, v. 29, p. 797–817.

Bathurst, R. G. C., 1975, Carbonate sediments and their diagenesis: Springer Verlag, 658 p.

Beier, J. A., 1987, Petrographic and geochemical analysis of caliche profiles in a Bahamian Pleistocene dune: Sedimentology, v. 34, p. 991–998.

Christian, H. E., and Tyrrell, W. W., 1991, Exploration history of the Liuhua 11-1-1A discovery, Pearl River Mouth Basin, China, *in* Proceedings of 23rd Annual Offshore Technology Conference, May 6–9, Houston, Texas, p. 93–100.

Craig, D. H., 1988, Caves and other features of Permian karst in San Andres Dolomite, Yates field reservoir, west Texas, *in* N. P. James and P. W. Choquette, eds., Paleokarst: Springer-Verlag p. 342–363.

Erlich, R. N., Barrett, S. F., and Guo, B. J., 1991, Drowning events on carbonate platforms: a key to hydrocarbon entrapment?, *in* Proceedings of 23rd Annual Offshore Technology Conference, May 6–9, Houston, Texas, p. 101–112.

Esteban, M., and Klappa, C. F., 1983, Subaerial exposure, *in* P. A. Scholle, D. G. Bebout, and C. H. Moore, eds., Environments of Carbonate Deposition: AAPG Memoir 33, p. 1–54.

Glennie, K. W., Boef, M. G. A., Hughes Clark, M. W., Moody-Stuart, M., Pilaar, W. F. H., and Reinhardt, B. M., 1973, Late Cretaceous nappes in the Oman Mountains and their geologic evolution: AAPG Bulletin, v. 57, p. 5–27.

Harrison, R. S., and Steinen, R. P., 1978, Subaerial crusts, caliche profiles, and breccia horizons: Comparison of some Holocene and Mississippian exposure surfaces, Barbados and Kentucky: Geological Society of America Bulletin, v. 89, p. 385–396.

James, N. P., and Choquette, P. W., eds., 1988, Paleokarst: Springer-Verlag, 416 p.

James, N. P., and Choquette, P. W., 1990, Limestones—the meteoric diagenetic environment, *in* I. A. McIlreath and D. W. Morrow, eds., Diagenesis: Geoscience Canada Reprint Series 4, p. 35–73.

Longman, M. W., 1980, Carbonate diagenetic textures from near-surface diagenetic environments: AAPG Bulletin, v. 63, p. 461–487.

Roehl, P. O., and Choquette, P. W., 1985, Carbonate petroleum reservoirs: Springer-Verlag, 622 p.

Saller, A. H., and Moore, C. H., 1991, Geochemistry of meteoric calcite cements in some Pleistocene limestones: Sedimentology, v. 38, p. 601–621.

Scott, R. W., 1990, Chronostratigraphy of the Cretaceous carbonate shelf, southeastern Arabia, *in* A. H. F. Robertson, M. P. Searle, and A. C. Ries, eds., The Geology and Tectonics of the Oman region: Geological Society Special Publication 49, London, p. 89–108.

Turner, N. L., and Hu, P. Z., 1991, The lower Miocene Liuhua carbonate reservoir, Pearl River Mouth Basin, offshore People's Republic of China, *in* Proceedings of 23rd Annual Offshore Technology Conference, May 6–9, Houston, Texas, p. 113–123.

Wagner, P. D., 1983, Geochemical characterization of meteoric diagenesis in limestone: development and applications: Ph.D. dissertation, Brown University, Providence, Rhode Island, 386 p.

Wagner, P. D., 1990, Geochemical stratigraphy and porosity controls in Cretaceous carbonates near the Oman Mountains, *in* A. H. F. Robertson, M. P. Searle, and A. C. Ries, eds., The Geology and Tectonics of the Oman Region: Geological Society Special Publication 49, p. 127–137.

Wagner, P. D., 1991, Identifying paleo-exposures in carbonates, and using them for reservoir and seal prediction, *in* J. C. Dolson, ed., Unconformity Related Hydrocarbon Exploitation and Accumulation in Clastic and Carbonate Settings: Proceedings of Rocky Mountain Association of Geologists Meeting, November 12–13, p. 23–27.

Wagner, P. D., 1994, Defining paleo-exposures and paleo-aquifers in carbonates with isotopic and cationic data: a brief review, *in* J. C. Dolson, M. L. Hendricks, and W. A. Wescott, eds., Unconformity-Related Hydrocarbons in Sedimentary Sequences: Rocky Mountain Association of Geologists, p. 59–68.

Wahlman, G. P., 1988, Subsurface Wolfcampian (Lower Permian) shelf-margin reefs in the Permian Basin of west Texas and southeastern New Mexico, *in* W. A. Morgan and J. A. Babcock, eds., Permian Rocks of the Midcontinent: Society of Economic Paleontologists and Mineralogists, Midcontinent Section, Special Publication 1, p. 177–204.

Chapter 10

The Post-Rotliegend Reservoirs of Auk Field, British North Sea: Subaerial Exposure and Reservoir Creation

Volker C. Vahrenkamp*
Shell Research
Rijswijk ZH, The Netherlands

ABSTRACT

The Auk field is situated in U.K. Block 30/16 in the Central North Sea. It was discovered in 1971 and has been producing since 1975 under a strong waterdrive from Lower Permian (Rotliegend) to Lower Cretaceous reservoirs. The post-Rotliegend to Lower Cretaceous stratigraphy of the Auk Horst is extremely heterogeneous. Laterally, stratigraphic sections change over very short distances across faults with stratigraphic heterogeneity resulting from the interplay between uplift, erosion, and deposition related to Central Graben tectonism. Significant hiatuses exist. Subaerial exposure has been instrumental in creating porous and permeable successions. Three effects of subaerial exposure on reservoir creation and character are highlighted:

1. The Zechstein reservoir of the Auk Horst is a 10 m thick well-defined sabkha dolomite layer with porosity mainly confined to molds of evaporite minerals. The timing of evaporite leaching and the composition of the aggressive fluids, however, are poorly constrained. The absence of any gypsum/anhydrite on the structurally elevated Auk Horst and the occurrence of massive evaporite layers in the nearby Zechstein Basin suggest that dissolution is most likely a by-product of subaerial exposure and circulation of meteoric waters during the Triassic, Jurassic, and/or Early Cretaceous. Pervasive fracturing, caused by mechanical instability of the rock and Central Graben tectonism, connected isolated vuggy pores to form a highly permeable reservoir.

2. In a tectonically active terrain, subaerial exposure, erosion, and deposition may create reservoirs in structurally low areas. A mixed-mineralogy clastic breccia, which was previously interpreted to be of a solution-collapse origin, was a product of such interplay between subaerial exposure and erosion on parts of the Auk Horst during the Early Cretaceous. The distribution of this reservoir is areally restricted and structurally controlled.

*Present address: Shell Sarawak Berhad, Sarawak, Malaysia.

3. Highly permeable vuggy and fracture porosities are probably common in exposure-related reservoirs. They introduce nonstandard petrophysical characteristics of the pore network which may lead to a significant underestimation of hydrocarbon saturations. In the Zechstein reservoir of the Auk field, values for cementation factor and saturation exponent are significantly below 2. In addition, the invasion of drilling solids into the pore system affected core and probably log-derived petrophysical properties: porosity and permeability values are underestimated, whereas the saturation exponent and the cementation factor are overestimated.

INTRODUCTION

The Auk field is situated in U.K. Block 30/16 in the Central North Sea (Figure 1). Auk was discovered in 1971 and has produced 89 million stock tank barrels (MMSTB) of oil under a strong water drive, mainly from Upper Permian Zechstein carbonates (73 MMSTB) and from Lower Permian Rotliegend sandstones (16 MMSTB). After initially high flow rates of oil from many wells, the Zechstein reservoir currently produces from only three wells with watercuts of 86% or more. Rotliegend sandstone reservoirs with lower permeability and gross production per well presently provide the main reservoir target. Heward (1991) has described these reservoirs in detail.

A complex interplay between regional tectonism, deposition, and erosion at the southwestern flank of the North Sea Central Graben at Auk and the nearby Argyll field (Pennington, 1975; Bifani and Smith, 1985) has created stratigraphically complex reservoirs in super-Rotliegend successions (Figures 1 and 2). Hydrocarbon storage and production in these fields are profoundly influenced by the effects of subaerial exposure during episodes in the Triassic, Jurassic, and Early Cretaceous. Previous workers (Brennand and Van Veen, 1975; Buchanan and Hoogteyling, 1979) recognized the influence of subaerial exposure on reservoir development and production behavior in the early stages of field development.

This paper summarizes available information on the super-Rotliegend reservoirs of the Auk field with emphasis on the role of subaerial exposure in creating porosity, permeability, and reservoir units. Special attention is given to the character of the pore system and its petrophysical properties.

DATA

In addition to previously available data (Brennand and Van Veen, 1975; Buchanan and Hoogteyling, 1979), 400 m of cores through the Upper Permian to Upper Cretaceous section of the Auk field were described and correlated with GR and porosity logs. Thin sections were analyzed with standard petrographic techniques.

REGIONAL GEOLOGY AND STRATIGRAPHY OF THE AUK HORST

Structural Setting

The Auk field is located in the western part of a horst along the western flank of the Central Graben in a structurally high and complex area dominated by a NNW–SSE-trending tectonic grain (Figure 1; Glennie, 1986). Erosionally truncated Rotliegend, Zechstein, and Lower Cretaceous reservoir-quality rocks are capped by Triassic shales and Upper Cretaceous chalks (Figures 1 and 2). The present structure of the Auk Horst is a result of multiple periods of faulting and uplift that continued from the Permian through the mid-Cretaceous. These structural episodes had a profound influence on the present distribution of strata and reservoirs.

Stratigraphy

The stratigraphy of the Auk field is very complex and is summarized in Figure 2 with emphasis on the interval from the Upper Permian to the Upper Cretaceous. Recognizing its complexity provides a key in understanding the depositional and structural history of the field area and, ultimately, the distribution and behavior of its reservoirs. Descriptions of the respective lithologies and their depositional environments are given below.

Sub-Upper Permian (Devonian–Rotliegend)

A thick succession of Devonian limestones and sandstones overlies lower Paleozoic basement and, in turn, is unconformably overlain by 300 m of Lower Permian Rotliegend sandstones (see Heward, 1991, for a comprehensive review of the Rotliegend). The majority of the Rotliegend sandstones are eolian in origin. However, evidence near the top of the section indicates reworking and deposition in a subaqueous environment. These waterlain deposits drape onto local dune highs and indicate a low relief at the end of the Early Permian (Heward, 1991).

Upper Permian (Zechstein)

The Zechstein succession along the western flank of the Central Graben is incomplete compared to the classical Zechstein stratigraphic sequence described by

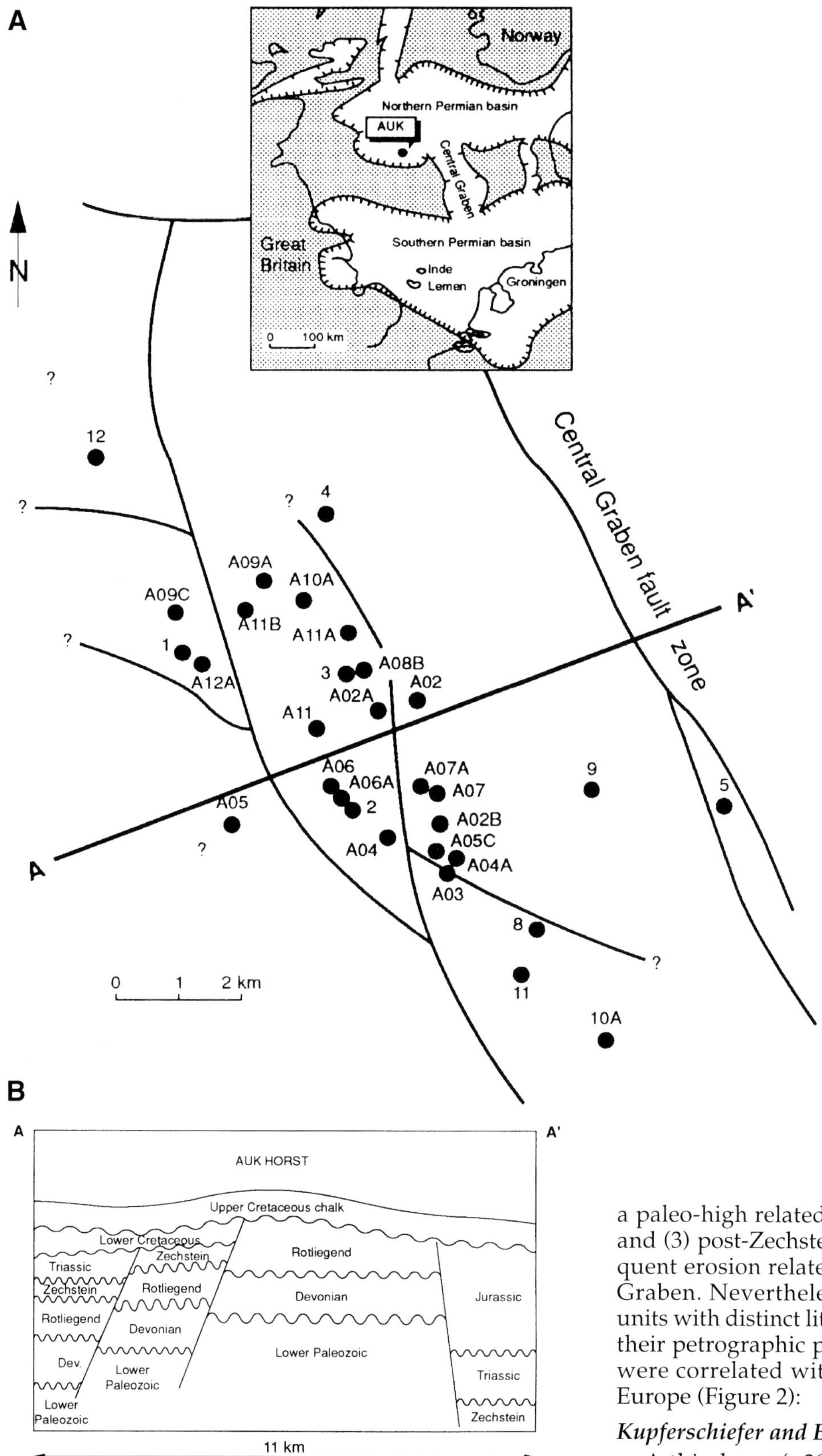

Figure 1. Auk field location map and schematic cross section. (A) Location map of the Auk field with block definition and well locations. The location of some block boundaries is rather tentative due to lack of seismic control (note question marks). (B) Schematic cross section of the Auk field (modified from Brennand and Van Veen, 1975).

Clark (1980). This may reflect the lack of deposition or the erosional removal of some of the Zechstein cycles in this region and may be controlled by (1) proximity to the Mid North Sea High, (2) the possible existence of a paleo-high related to Rotliegend dune topography, and (3) post-Zechstein tectonic movement and subsequent erosion related to the formation of the Central Graben. Nevertheless, several distinct stratigraphic units with distinct lithofacies were identified, based on their petrographic properties and log signatures, and were correlated with Zechstein units from northern Europe (Figure 2):

Kupferschiefer and Basal Dolomite

A thin layer (<30 cm) of euxinic black shale conformably overlies Rotliegend sandstones on structurally low blocks in the Auk field. The shale unit reflects an anoxic event at the beginning of the Late Permian and is interpreted to be equivalent to the Kupferschiefer, which is found throughout the

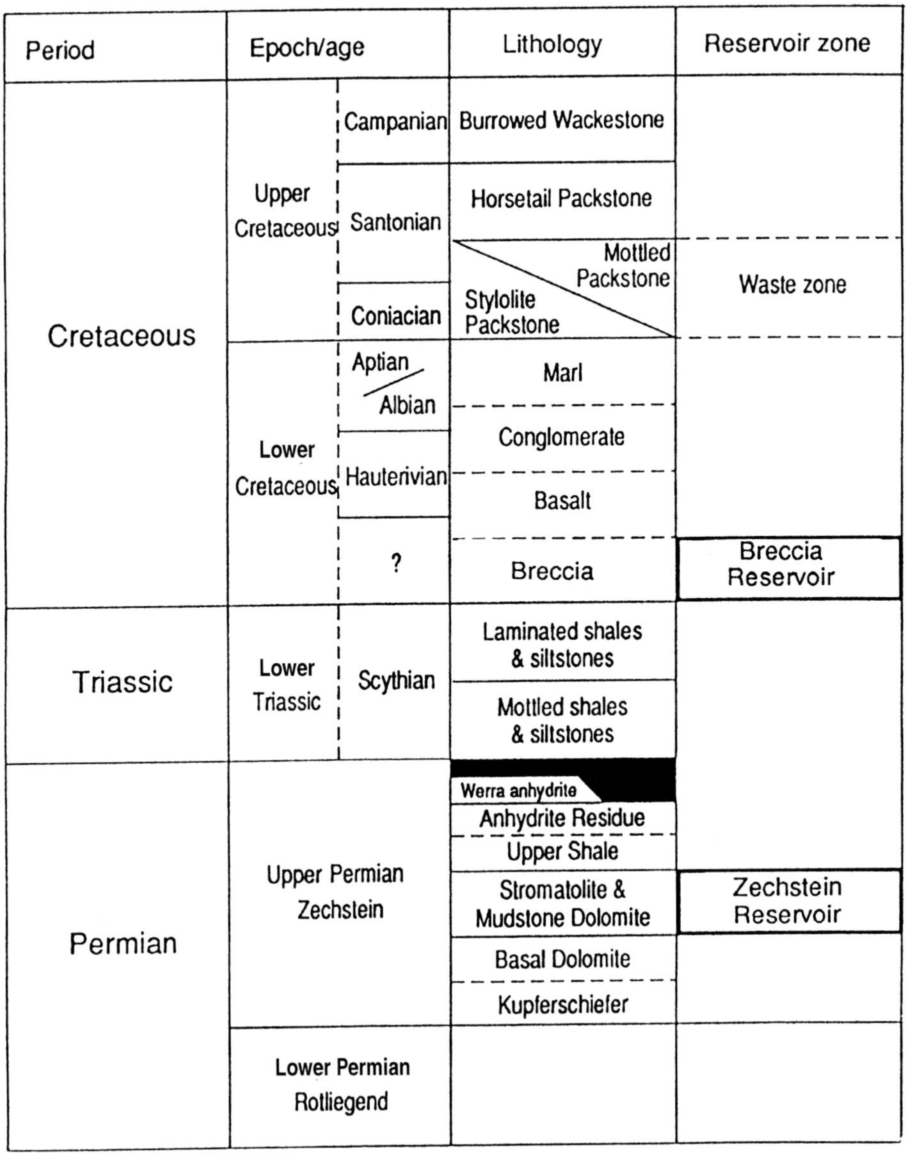

Figure 2. Permian to Cretaceous stratigraphy of the Auk area.

Permian basin. The Kupferschiefer was probably deposited below wave base (Taylor, 1984). These shales grade upward into a dense, up to 1 m thick, gray dolomite, rich in both organic and siliciclastic material and capped by a subaerial exposure surface (Figure 3A). The occurrence of current ripples suggests a shallow-water environment of deposition for this basal dolomite. Collectively, the Kupferschiefer and basal dolomite correspond to a prominent GR peak that provides an excellent fieldwide and regional correlation marker (Figure 4).

Stromatolite and Dolomudstone Lithofacies

A 10 m thick dolomite overlies the subaerial exposure surface (Figure 3). The dolomite consists mainly of a featureless to laminated mudstone intercalated with multiple layers of algal lamination (stromatolites) (Figure 5). Features such as planar to domal algal lamination and ubiquitous traces of evaporite minerals (Figures 5 and 6) suggest that deposition occurred in an inter- to supratidal, highly saline environment similar to the modern sabkhas of the Persian Gulf (e.g., Butler et al., 1982). The stromatolite and dolomudstone lithofacies forms the main reservoir zone in the Auk field (see below).

Upper Shale and Anhydrite Residue

A distinct, organic-rich shale unit overlies the sabkha dolomites. This unit is 2–3 m thick and consists of thin beds of laminated black shale with calcite, dolomite, and silica replacing wrinkled and contorted thin gypsum bands (Figure 3B), which were originally a major component of the facies. The deposits of this unit are rich in siliciclastics, with insoluble residues up to 60%. The preservation of fine, regular millimeter-scale lamination at the top of the unit suggests that the replacement of gypsum did not greatly disturb the host deposits. Furthermore, evidence of large-scale

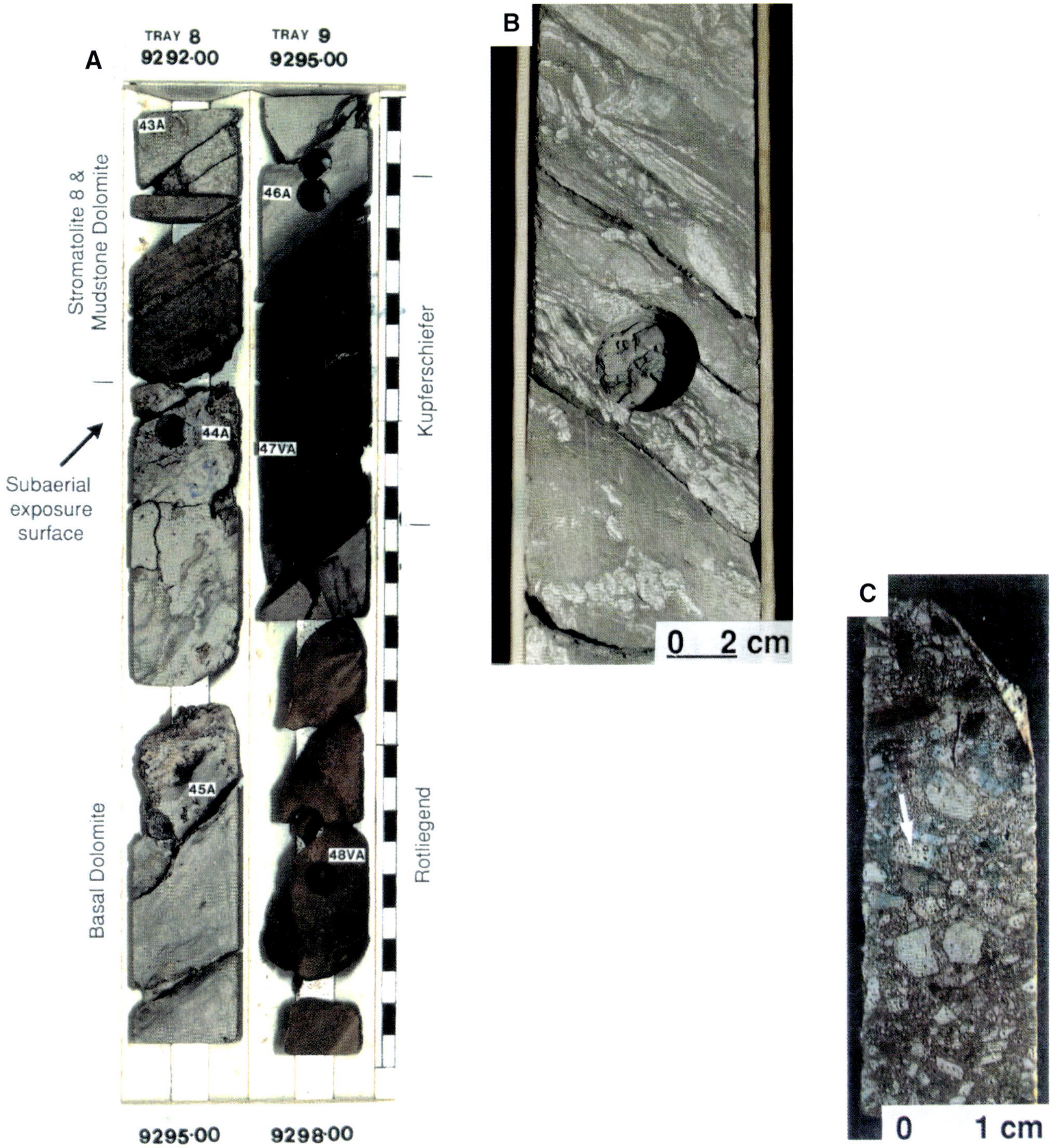

Figure 3. (A) Contacts between Rotliegend sandstones, Zechstein Kupferschiefer, and basal dolomite lithofacies. Note the brecciation in the upper part of the basal dolomite indicating subaerial exposure prior to the deposition of the overlying stromatolite and dolomudstone lithofacies. Core is from the highly deviated A06 well; depths shown on photograph are in feet. Core interval 2832.2–2834 m (9292–9298 ft). (B) Core photograph of upper shale and anhydrite residue. Note the chaotic light banding of now calcitized, dolomitized and/or silicified former evaporite layers. Core is from the highly deviated A11A well at 2999 m (9839 ft). (C) Small triangular sidewall core (MCT core) from well 30/16-1. Calcitic components are stained red with Alizarin-S. Note the angularity of the clasts and the large oval vugs (arrow) in the unstained dolomitic clasts, typical of the Zechstein stromatolite and dolomudstone lithofacies.

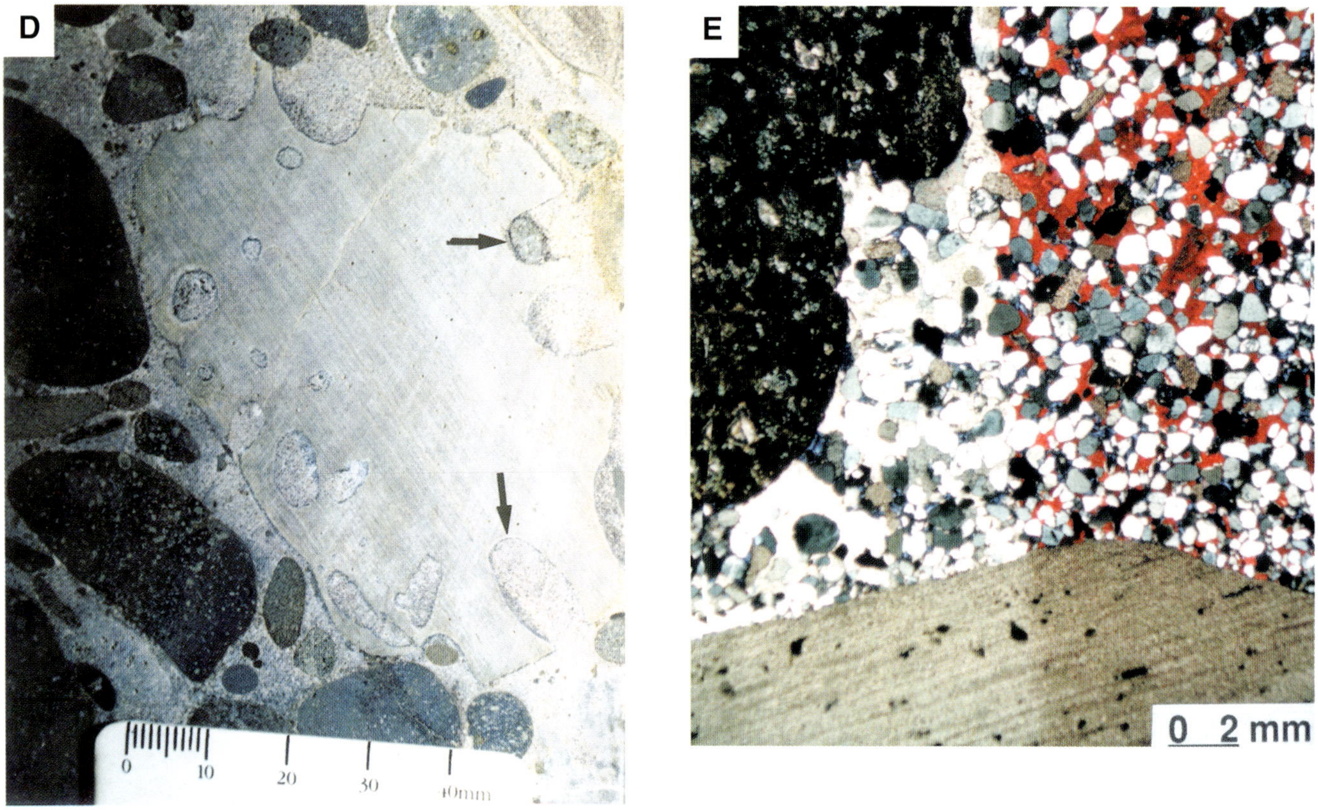

Figure 3 (*continued*). (D) Conglomerate lithofacies: core photograph of a large Zechstein cobble and a number of smaller greenish basalt pebbles in a calcite-cemented sandstone matrix. The Zechstein cobble is riddled with clam borings (arrows), indicating a shallow-marine environment of deposition, probably along a rocky coast line. Borings exhibit geopetal sand fills with two different orientations, indicating periods of quiescence with syndepositional cementation followed by re-activation and redeposition. Sample from well A11B at 3517 m (11,538.5 ft). (E) Thin-section photomicrograph of the conglomerate lithofacies. A laminated Zechstein dolomite clast (bottom) and Lower Cretaceous basalt pebble are embedded in a sandstone matrix consisting mainly of rounded quartz grains, probably of eolian (Rotliegend) origin. The matrix is cemented by calcite (see red Alizarin-S stain on the right half). Cross-polarized light. Sample from well A11B at 3519 m (11,545.5 ft).

collapse after dissolution of massive gypsum or anhydrite layers has not been observed. Deposition probably occurred in an evaporitic playa environment, evidently with high rates of siliciclastic influx. Similar to the Kupferschiefer unit, the upper shale is characterized by a high GR-log signature, making it ideal for fieldwide and regional correlations (Figure 4).

Massive Anhydrite

A thick anhydrite unit overlies the upper shale in the southwestern part of the field (well A05) but is absent elsewhere (Figures 1 and 4). It continues into the Forth Approaches Basin (well 29/25-1), where it is overlain by volcanic rocks and thick salt deposits (Figure 4). It is interpreted to be stratigraphically equivalent to the Werra or Z1 anhydrite of the northern European Zechstein (Clark, 1980).

Other Upper Permian Deposits

Latest Permian is not represented in the Auk field area. A section previously interpreted as the collapsed residue of an upper Zechstein carbonate cycle (Brennand and Van Veen, 1975) is now considered to be of Early Cretaceous age (see below).

Triassic

The Triassic section is comprised of red-brown to gray-green silty claystones and has been dated as Early Triassic (Scythian) on the basis on pollen assemblages. The Triassic shales and claystones are locally more than 160 m thick (well A05). Color and sedimentological features such as laminations, mottling, burrows, fluid escape structures, etc., suggest a nonmarine, probably fluviatile origin (S. Flint, 1988, personal communication). These deposits are undisturbed apart from some synsedimentary folds, some fluid-escape structures, and fractures. No evidence for large-scale collapse caused by the solution of underlying Zechstein evaporite units has been observed. The Triassic deposits are probably part of the Smith Bank formation of the Central North Sea area (Fisher, 1984).

Jurassic

Jurassic sediments are not present on the Auk Horst. It is not known whether this is due to erosion or nondeposition.

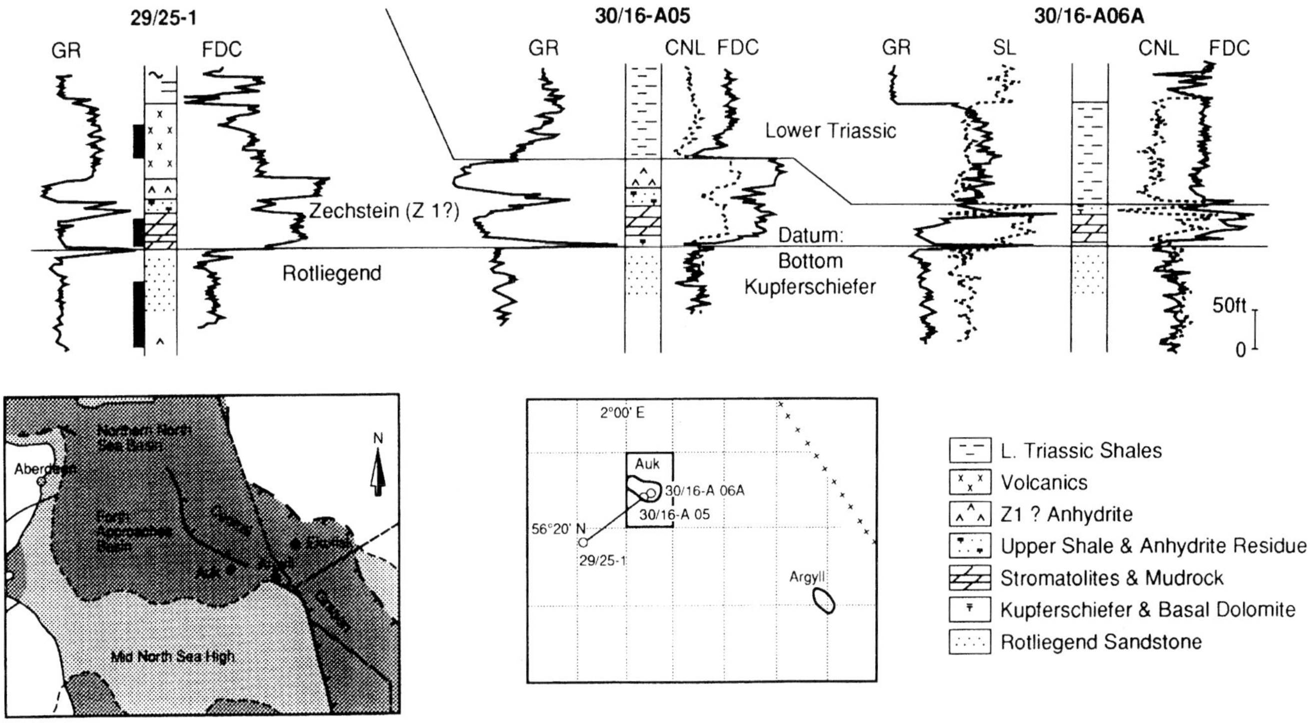

Figure 4. Cross section showing the onlap of Zechstein anhydrite (Werra anhydrite?) onto the Auk Horst. This anhydrite unit can be correlated from the Forth Approaches Zechstein Basin to the southwestern extension of the Auk Horst (well A05). There, it is overlain by Lower Triassic (Scythian) shales and siltstones. No anhydrite occurs closer to the center of the Auk Horst (wells A06, A06A, and 30/16-2), indicating either nondeposition or erosion prior to the deposition of Triassic sediments in this part of the field. The distance between wells 29/25-1, 30/16-05, and 30/16A is 17 km and 2 km, respectively. The size of block 30/16 is about 11 × 16 km. Modified from Brennand and Van Veen (1975).

Lower Cretaceous

Four Lower Cretaceous units are identified. In ascending order, they include: (1) a breccia, (2) a Hauterivian basalt flow, (3) an Aptian–Albian conglomerate, and (4) a marl.

Breccia

The breccia consists of Zechstein dolomite fragments and some shale clasts of Triassic(?) origin (Figure 3C). The intervening matrix includes well-rounded (Rotliegend) sand grains, calcite, pyrite, and some clay minerals, and contains an open-marine fauna, probably Neocomian in age.

This unit comprises a second post-Rotliegend reservoir interval. It was previously interpreted to be Late Permian in age (post upper shale and anhydrite residue). Brecciation was previously interpreted to be due to collapse caused by the dissolution of evaporite deposits, and the Neocomian fauna was thought to have been washed in during Cretaceous exposure (Brennand and Van Veen, 1975). However, some observations contradict this interpretation:

- the breccia occurs only where Triassic sediments are absent.
- the breccia is nowhere overlain by Triassic sediments.

Unless there were preferential pre-Triassic erosion of the breccia in areas where Triassic sediments are found today, this stratigraphic configuration suggests that the breccia is post-Triassic in age rather than pre-Triassic. In addition, there is no direct evidence for the presence of a former thick evaporite layer on the Auk Horst, as would be necessary to explain collapse brecciation of a second Zechstein cycle on a large scale. Only to the west and southwest, further into the Zechstein Basin, has the onlap of thick evaporites onto the Auk Horst been documented (Figure 4). Furthermore, it is easier to explain the Neocomian fauna in the matrix as a depositional admixture rather than an infiltration into a pre-existing collapse frame. All of this suggests that this reservoir zone is of Early Cretaceous age and a by-product of subaerial exposure caused by block faulting during tectonic movement of the Central Graben (Figure 1B). Deposition is interpreted to have occurred in structurally low blocks with sediments shed from erosion of lower Zechstein and Rotliegend at the crest of the Auk Horst. Thus, the extent of the breccia corresponds to the disposition of structurally low fault blocks.

The Basalt Flow

A thick basalt flow occurs in the NE corner of the field (Figure 7, wells A11A and A10A). Radiometric dating suggests a Hauterivian age (Heward, 1991).

Figure 5. Core and thin-section photomicrographs of the stromatolite and dolomudstone lithofacies. The facies is characterized by three porosity types: pinhead vugs (A and B), oval vugs (C and D), and molds of gypsum rosettes (D and E). (A, B) Pinhead vugs. In hand sample (A) the rock is riddled with pinhead vugs (0.25 to 2 mm) giving it a sponge-like appearance. Sample from well A11A at 3003 m (9852 ft); coin diameter = 1.5 cm. In thin section (B), these pinhead vugs were identified as anhydrite crystal molds. These vugs are partially invaded along fractures by blue plastic impregnation, thereby illustrating the originally isolated character of the pores (white dots) and the importance of fractures in providing permeability. Sample from well A11B at 3545 m (11631 ft). (C, D) Oval vugs. Millimeter-size oval vugs are concentrated between stromatolitic algal laminae and resulted from the dissolution of anhydrite that had replaced the dolomitized stromatolites. The pervasive impregnation of the thin section (upper half of [D]) with blue plastic indicates the good horizontal permeability. Sample (C) from well A06A at 2827 m (9274 ft); coin diameter = 1.5 cm, deviated well; sample (D) from A11B at 3543.6 m (11,626 ft). (D, E) Gypsum rosettes. Large (up to mm-size) gypsum rosettes and disks grew within dolomudstone and were subsequently leached. Pervasive fracturing causes good connectivity of the pore system (lower half of [D]), which is filled with blue plastic and drilling solids. Sample (E) coin diameter = 1.5 cm. Sample (D) from A11B at 3543.6 m (11,626 ft).

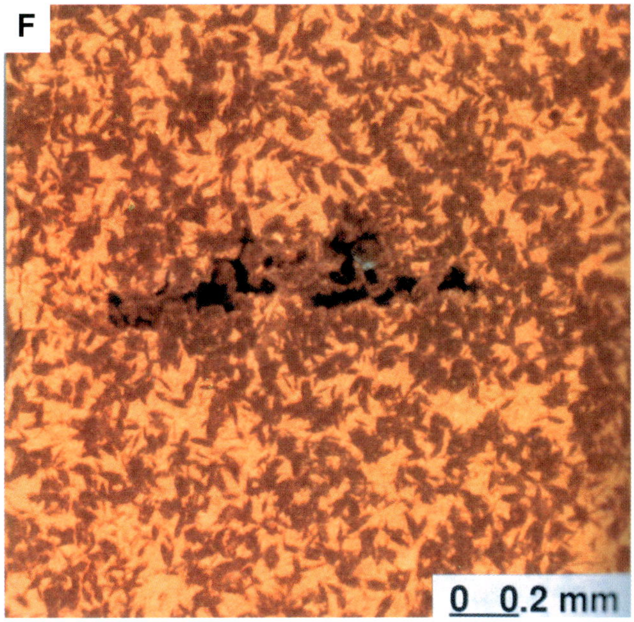

Figure 5 (*continued*). (F) Cathodoluminescence photomicrograph of a dolomudstone with vugs after dissolved gypsum (black). The freckled fabric shows that the dull luminescent dolomite is a replacement of tiny gypsum crystals (<50 µ), indicating that the original sediment was probably composed of a carbonate/gypsum mush. Sample from well A11B at 3542.6 m (11,622.8 ft).

Figure 6. Pore casts of the stromatolite and dolomudstone lithofacies (SEM photomicrographs). Dolomite was dissolved with HCl, leaving behind a plastic cast of the pore space. (A) Note the gypsum rosettes as well as the many sheet-like fractures that connect individual vugs. Sample from well A11B at 3544 m (11,626 ft); 1 bar = 1 mm. (B) Close-up of (A). Note sheet fractures (arrows) and spongy appearance of the pore casts. "Spongy pores" in pore casts are after soluble drilling solids which also dissolved during acid leaching.

The source of this basalt flow has not been determined. However, the extent of a conglomerate fringe (see below) at the edges of the basalt flow suggests that the basalt may have extruded along a local fault, or, alternatively, further east, within the SW Central Graben fault zone itself.

The Conglomerate

A conglomerate with rounded to well-rounded basalt and Zechstein pebbles (Figure 3D and 3E) fringes the basalt flow (e.g., Figure 7B, wells 30/16-3 and A11B). The calcite-cemented matrix consists mainly of rounded quartz and feldspar grains (probably of Rotliegend origin) (Figure 3E) and some reworked shell fragments. Larger Zechstein clasts contain clam borings (Figure 3D). Shell fragments are dated as Aptian–Albian and reflect an open-marine environment. Deposition is interpreted to have occurred during a marine transgression in the Aptian–Albian.

The Marl

A Lower Cretaceous marl is penetrated in wells 30/16-1 and 30/16-2 where it overlies either the breccia or older Triassic sediments. This marl is the only Lower Cretaceous deposit that overlies Triassic rocks on the Auk Horst. The marl on the Auk Horst is interpreted to have been deposited in a relatively deep-water marine setting, away from the coast where the conglomerates and breccias were deposited.

Summary

The four Lower Cetaceous lithologies are partly laterally equivalent and hence partly time-equivalent (Figures 2 and 7). The breccia and the basalt flow probably reflect a period of subaerial exposure and faulting at the beginning of the Cretaceous. However, the breccia does not contain basalt clasts suggesting a pre-Hauterivian origin. The conglomerate and marl may have formed during a subsequent marine transgression in the Aptian–Albian. With the exception of the marl, these Cretaceous clastic and volcanic units do not overlie Triassic deposits. Instead, they overlie older Zechstein strata. The widespread lack of Triassic deposits indicates either removal by erosion or nondeposition and suggests that the Auk Horst was probably a topographic high during the Early Cretaceous.

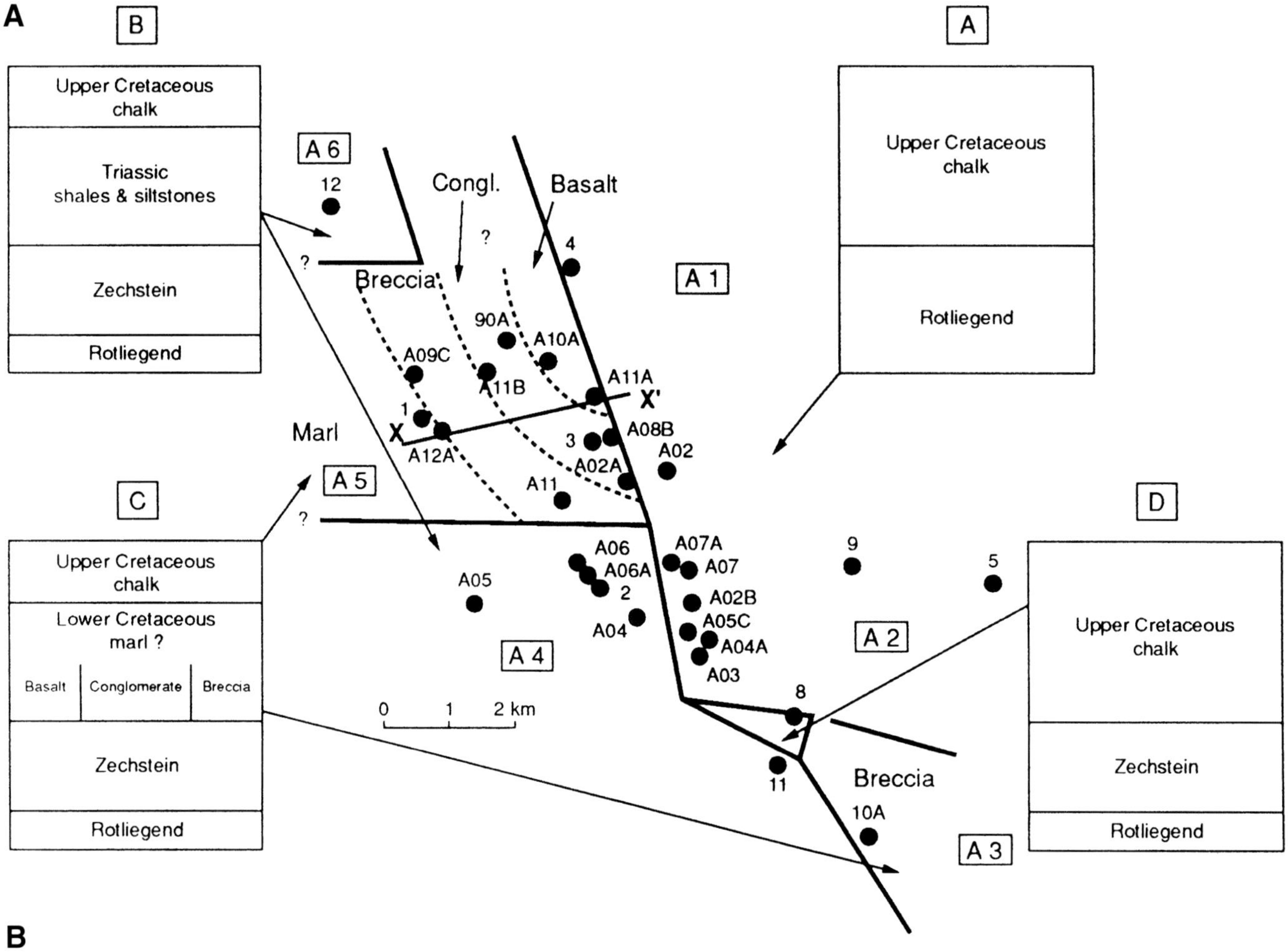

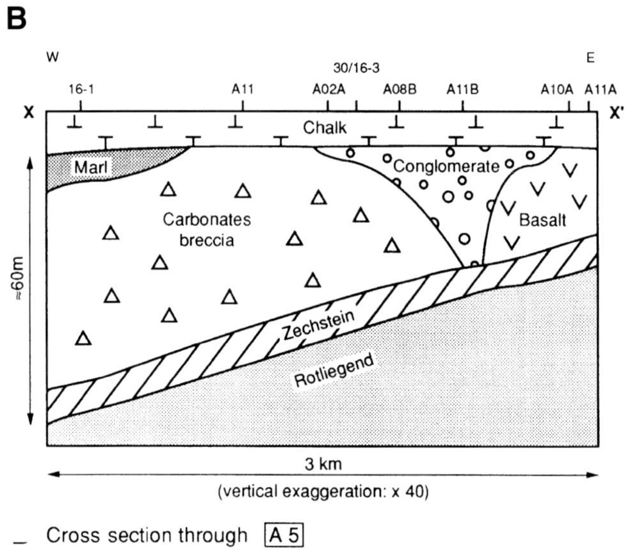

Figure 7. Stratigraphic units of the Auk field and their areal distribution. (A) Four different Permian to Upper Cretaceous stratigraphic units are recognized in the Auk area, reflecting various degrees of erosion/nondeposition in different parts of the field. The Upper Cretaceous chalk subcrop map is subdivided into six areas defined by one of the stratigraphic section types. (B) Cross section along X-X′. Note all boundaries are unconformable with the exception of the boundary between the Rotliegend–Zechstein succession.

Upper Cretaceous

The Upper Cretaceous sequence in the Auk area consists of chalks ranging from foraminiferal packstones of Coniacian and Santonian age, to overlying foraminiferal wackestones of Campanian age. The occurrence of burrows and large, relatively well preserved *Inoceramus* shell fragments indicates an autochthonous origin. These chalks unconformably overlie Lower Cretaceous, Triassic, Zechstein, and Rotliegend deposits and collectively, with the overlying Tertiary successions, provide an effective cap rock for the Auk oil accumulation. The lower part of the chalks is hydrocarbon bearing. However, economic production rates have not been achieved, and it thus constitutes a waste zone.

Areal Distribution of the Upper Permian to Upper Cretaceous Lithofacies

Tectonic movements both on a regional and a local scale were the dominant factors creating the complex stratigraphy of the Auk field. The lack of good seismic reflection data makes it difficult to predict the stratigraphy outside the area of well control. The lithofacies distribution is shown in Figure 7 and can be summarized as follows:

- The Upper Permian to Lower Cretaceous section is missing in the NE/E part of the field (Figure

7A; area A1—stratigraphic type A). Facies patterns of Rotliegend sandstones (Heward, 1991) and the uniform thickness and facies distribution of the Zechstein succession in adjacent areas suggest that the absence of Zechstein deposits in this area is the result of Jurassic–Early Cretaceous erosion rather than of nondeposition. Furthermore, the Lower Cretaceous breccia unit is largely composed of clasts derived by erosion from the Zechstein and Rotliegend successions from area A1 (Figure 7A).

- Zechstein strata are present in areas A2–A5 (stratigraphic types B–D). Kupferschiefer/basal dolomite and upper shale/anhydrite residue form excellent markers on a regional scale sandwiching Zechstein dolomites with a rather uniform thickness of 8–10 m (Figure 4).
- A Zechstein anhydrite unit (Werra anhydrite?) is found in the southwestern part of the Auk field (Figure 4; well A05, area A4). On the Auk Horst itself, this unit is missing, and Triassic deposits overlie older Zechstein deposits without any indication of the dissolution of underlying massive evaporites (Figure 7A, and areas A4 and A6, stratigraphic type B). Hence, unless the anhydrite was completely eroded before deposition of the Lower Triassic sediments, an onlap of the anhydrite onto the Auk Horst is indicated.
- In areas A2, A3, and A5, Zechstein strata are present but Triassic deposits have been removed by erosion (Figure 7A, stratigraphic types C and D).
- Four laterally equivalent Lower Cretaceous units (breccia, marl, basalt flow and conglomerate) occur in areas A3 and A5 (Figure 7A and 7B). They are a by-product of Early Cretaceous tectonic activity and erosion and a subsequent marine transgression. The mixed carbonate/siliciclastic and volcanic deposits dip and thicken toward the southwest. The source and eastward extent of the basalt flow is not known.
- Lower Cretaceous and older strata are unconformably overlain by an Upper Cretaceous chalk of Coniacian to Campanian age. The chalk and overlying Tertiary deposits form the reservoir seal.

RESERVOIR UNITS

Three reservoir units occur in the Auk field in addition to the Rotliegend reservoirs described by Heward (1991). The Zechstein and the Cretaceous breccia reservoirs are both related to subaerial exposure; the chalk reservoir forms a waste zone.

Zechstein Reservoir

The Zechstein reservoir is a pervasively fractured dolomite with secondary moldic porosity (Figures 5 and 6). By the end of 1991, the Zechstein reservoir had contributed more than 80% of the oil produced from the Auk field. The production rate at that time was about 5 MSTB/d from the three remaining Zechstein wells (A10A, A08B, A12A), or about 50% of the total daily field production.

A major NNW–SSE trending fault cuts the field into two parts. In the eastern part (area A1 of Figure 7A) the Zechstein is missing, probably as a result of Jurassic to Early Cretaceous erosion. In the other areas the Zechstein reservoir has an almost uniform thickness of about 8–10 m and consists only of the stromatolite and dolomudstone lithofacies.

Diagenesis

Synsedimentary dolomitization of the primary evaporite/carbonate mush (Figure 5F) produced a tight dolomite mudstone with a significant evaporite content. Repetitive episodes of dolomite/evaporite mineral precipitation/dissolution led to the replacement of the smaller evaporite minerals with dolomite and the growth of larger (>0.5 mm) euhedral gypsum crystals in the dolomudstones and anhedral gypsum ovoids between the algal laminae. Subsequent infusion of waters undersaturated with respect to gypsum/anhydrite resulted in the complete dissolution of evaporite minerals, leaving the rock riddled with vuggy secondary pores (Figures 5 and 6).

Stylolites commonly overprint organic-rich algal laminae (Figure 8). They nestle between horizontal rows of vugs and appear not to be influenced by the porosity created during the leaching of evaporite minerals (Figure 8). These features indicate that pressure-solution seams developed before leaching of the evaporite minerals, probably during a first episode of burial diagenesis concurrent with deposition of the Triassic Smith Bank formation. Dissolution is interpreted to have occurred from meteoric waters from post-Triassic to pre-Early Cretaceous. Porosity generation related to local post-Triassic uplift and subaerial exposure is further supported by the absence of leaching of evaporite minerals in similar Zechstein rocks away from the Auk Horst farther into the Forth Approaches Basin (i.e., well A05).

Pervasive fracturing with apparent random orientation occurred during reburial and/or as a result of tectonic movements related to Central Graben faulting. The culmination of multiple phases of diagenesis is a pervasively fractured, moldic dolomite reservoir.

Drilling Solids Invasion

A significant part of the vuggy and fracture porosity in the Zechstein cores is filled with fine-grained, unconsolidated material (Figures 5 and 9). It is composed mainly of anhedral, abraded dolomite fragments as well as traces of barite, mica, and other silica minerals. Whereas some of it may be of natural origin (i.e., insoluble residue of dissolved evaporite minerals or material washed in during exposure), a significant part has been introduced into the pore system during coring. This interpretation is supported by the infilling of late, possibly drilling-induced fractures, the size and shape of the particles, the occurrence of mica flakes which likely originate from the drilling mud, and the scarcity of sedimentary features indicating natural processes (i.e., lamination, geopetals) (Figure 9). The pervasive invasion of core porosity suggests that some of the pore space adjacent to the well bore was also invaded by drilling solids. Thus, the invasion

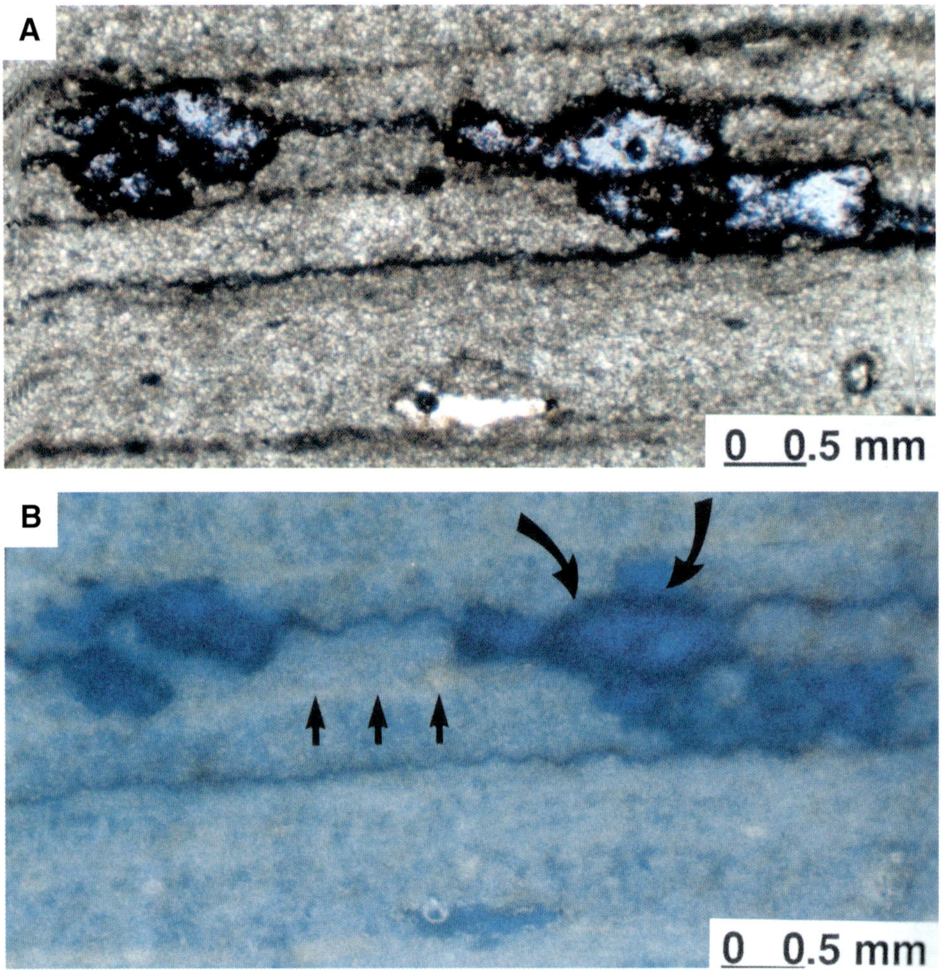

Figure 8. Plain light (A) and UV light (B) thin-section photomicrographs of stylolites overprinting stromatolite laminae in the stromatolite and dolomudstone lithofacies. A stylolite crosses a vug formerly occupied by anhydrite (curved arrows in [B]). Pressure solution must have taken place prior to evaporite dissolution since an insoluble residue seam could not form in the middle of a pore space. Straight arrows point to a stromatolite lamina rich in organic material (light color). Sample from well A11B at 3544 m (11,626 ft).

of drilling solids may significantly alter both core- and log-derived petrophysical properties. In fact, drilling-solids invasion in exposure-related reservoirs with vuggy and fracture porosity may be an often overlooked phenomenon.

Porosity (1.8–26%; Arithmetic Average, 13%)

Primary porosity of the original muddy sediment was completely destroyed either prior to or during early dolomitization. All porosity is secondary and resulted from the dissolution of evaporite minerals, likely during subaerial exposure, and the subsequent mechanical collapse of the rock (Figures 5 and 6). Because of the variable origin of the pores, pore sizes are heterogeneous and range from less than 100 µ to centimeter size. Correspondingly, pore throats and pore geometry are also extremely heterogeneous (Figure 6). Image analyses suggest that approximately 25–50% of the original porosity is occupied by silt-sized drilling solids. Thus, average formation porosity prior to invasion was probably 18%.

Permeability (0.02–620 md; Geometric Average, 53 md)

Permeabilities measured from core analysis range from less than one millidarcy to almost one darcy. These values are probably conservative due to (1) biased plug sampling in more competent rock with lower porosities and permeabilities, (2) inferred drilling-solids invasion, and (3) the fact that larger vugs and high-permeability fractures cannot be sampled adequately on a plug scale. The significantly higher permeabilities (tens of darcys) measured during production and buildup tests (Buchanan and Hoogteyling, 1979) reflect the influence of pervasive fracturing.

Cementation Factor (m) and Saturation Exponent (n)

The cementation factor and the saturation exponent were measured at $m_{ave} = 1.94$ (range: 1.81–2.25) and $n = 1.73$ (7 and 1 samples, respectively). The deviation from an "Archie" average of 2 toward lower values for both parameters is in agreement with observations in other fractured vuggy carbonate reservoirs (Rasmus, 1983; Focke and Munn, 1987). The invasion of drilling solids into plugs used for the measurements diminished the connectiveness between pores (Figure 9). Thus, it is likely that actual m and n values for the uninvaded reservoir are even lower yet.

Reservoir Model and Its Implications

The Zechstein reservoir in Auk is a sheet deposit of almost uniform 8–10 m thickness, cut by faults having various orientation and throw. Porosity is exclusively

Figure 9. Core photographs and photomicrographs showing evidence for artificial-solids invasion into the Zechstein reservoir rock. (A) Core photograph. Large vugs and fractures are occupied by a light, unconsolidated material (arrows). Sample from well A11A at 3005 m (9860 ft); coin diameter = 1.5 cm. (B) Close-up of broken core segment. Note that unconsolidated material has pervasively invaded all porosity over the full width of the core. Sample from well A11A at 3005 m (9860 ft). (C, D) Side and frontal views of a plug taken from (A). Despite cleaning of the plugs for petrophysical measurements, pores and fractures are still filled with the densely packed, yet unconsolidated, material (arrows). Porosity: 16.5%; permeability: 1600 md; cementation factor $m = 1.82$; saturation exponent $n = 1.73$; plug diameter = 2.54 cm (1 in.); sample from well A11A at 3005 m (9860 ft). (E) Thin-section photomicrograph of solids invasion in a fractured stromatolite. Note the progression of invasion into horizontal layers away from the fracture (arrows). Porosity impregnated with blue plastic. Sample from well A11 at 2359.2 m (7740 ft).

secondary, consisting of both fractures and vugs. The great variation in pore size and pore distribution (micron to centimeter range) and the variable fracture density cause a significant heterogeneity of the reservoir with respect to porosity and permeability on a millimeter to meter scale. Measured rock permeabilities are significantly lower than field-measured permeabilities (Buchanan and Hoogteyling, 1979), indicating that small-diameter plugs and even whole-core samples do not adequately represent reservoir characteristics. The highly permeable fracture system may also cause early water breakthrough and high water cuts, whereas oil is drained slowly from the less well-connected matrix pore system. This may partly explain the continuous but low rate of oil production from the Zechstein, as well as the high water cuts. However, because of the prominent occurrence of fractures and faults, the Kupferschiefer and basal dolomite units are not sufficiently thick to provide a seal for the underlying Rotliegend reservoirs. Thus oil and water may also leak upward into the Zechstein reservoir.

Breccia Reservoir

The Lower Cretaceous breccia reservoir consists mainly of dolomitic clasts in a sandy to marly matrix (Figure 3C). This unit is found in areas A3 and A5 (Figure 7) of the field and is absent wherever Triassic sediments are present. It has been partially cored in only three wells (30/16-1, 30/16-3, and 30/16-10A); hence, it is not as well understood as the Zechstein reservoir. Approximately 7 MMSTB of oil have been produced, predominantly from two wells (A09C and A12A) in area 5 and an uncored sandstone in well 30/16-10A in the southeast area of Auk. By the end of 1991, only well A12A was still producing at a rate of 350 BOPD with a 96% water cut. It is unclear whether communication exists with the underlying Zechstein reservoir and what portion of the production from these wells should be attributed to the Zechstein (Buchanan and Hoogteyling, 1979).

The breccia is interpreted to be a by-product of subaerial exposure and erosion; thus, its origin is entirely related to the development of the local unconformity that stripped away Triassic lithologies and accommodated deposition of Zechstein and Rotliegend clasts eroded from other areas (Figure 7, area A1).

Porosity (0.5–25%; Geometric Average, 10%; n = 53)

In addition to the porosity within breccia clasts (i.e., vuggy porosity in Zechstein clasts) interparticle porosity is partially preserved due to incomplete calcite cementation. Local variations in clast and matrix composition and packing cause large variability in porosity.

Permeability (0.01–100 md; Geometric Average, 2.5 md; n = 16)

Only a few permeability determinations could be made because of the friable nature of the reservoir rock and the limited core material. Measurements are most likely biased toward the poorer reservoir rock with low porosities and low permeabilities. Permeability due to fracturing was not assessed.

Chalk Reservoir

A third reservoir unit occurs in the lowermost part of the Upper Cretaceous chalk of Coniacian to Santonian age. Production tests were conducted on these chalks, but economic production rates were not achieved. This unit, therefore, constitutes a waste zone.

CONCLUSIONS

1. The super-Rotliegend to Lower Cretaceous stratigraphy of the Auk Horst is extremely heterogeneous. Significant hiatuses exist. Laterally, stratigraphic sections change over very short distances across faults related to Central Graben tectonism. Such stratigraphic heterogeneity is indicative of the interplay between uplift, erosion, and deposition in a tectonically active terrain.
2. Leaching, erosion, and deposition occurred on the Auk Horst during Triassic to Early Cretaceous intervals of subaerial exposure and were instrumental in creating reservoirs.
3. The Zechstein reservoir of the Auk Horst is a well-defined sabkha dolomite layer with porosity mainly related to the leaching of evaporites, most likely during Early Cretaceous subaerial exposure. Pervasive fracturing related to mechanical instability of the rock and Central Graben tectonism connected isolated vuggy pores to form a highly permeable reservoir rock with values for cementation factor and saturation exponent significantly below 2.
4. Invasion of drilling solids into the well-connected vuggy pore system of the Zechstein reservoir affects core- and, probably, log-derived petrophysical properties. Measurements of porosity and permeability are underestimated, whereas the saturation exponent and the cementation factor are overestimated. Drilling-solids invasion may be a common feature in some exposure-related reservoirs with their typical vuggy porosity and high permeabilities.
5. A Lower Cretaceous, mixed-mineralogy clastic breccia forms another reservoir. It is a product of subaerial exposure and erosion on parts of the Auk Horst during the Early Cretaceous. The distribution of this reservoir is areally restricted and structurally controlled. Previously, this reservoir was interpreted to be a solution-collapse breccia.

ACKNOWLEDGMENTS

Many colleagues with Shell have contributed to the knowledge on the Auk field through time. I especially thank H. Bolz, R. Buchanan, E. v.d. Graaff, and A. Heward for their technical expertise. The manuscript benefited greatly from detailed and constructive reviews by Art Saller, Emily Stoudt, and Jack Wendte. Permission from Shell Research B.V. and Shell Expro to publish this paper is gratefully acknowledged.

REFERENCES CITED

Bifani, R., and C.A. Smith, 1985, The Argyll field after a decade of production: Society of Petroleum Engineers Paper 13987.

Brennand, T.P. and F.R. Van Veen, 1975, The Auk oil field, *in* A.W. Woodland, ed., Petroleum geology and the continental shelf of NW Europe: Applied Science, Barking Publisher, v. 1, p. 275–283.

Buchanan, R., and L. Hoogteyling, 1979, Auk field development: a case history illustrating the need for a flexible plan: Journal of Petroleum Technology, v. 31, p. 1305–1312.

Butler, G.P., C.G.St.C. Kendall, and P.M. Harris, 1982, Recent evaporites from the Abu Dhabi coastal flats, *in* C.R. Handford, R.E. Loucks, and G.R. Davies, eds., Depositional and diagenetic spectra of evaporites: SEPM Core Workshop 3, p. 33–64.

Clark, D.N., 1980, The diagenesis of Zechstein carbonate sediments, *in* H. Füchtbauer and T.M. Peryt, eds., The Zechstein basin with emphasis on carbonate sequences, Contributions to Sedimentology: Stuttgart, Schweizerbart'sche Verlagsbuchhandlung, v. 9, p. 167–203.

Fisher, M.J., 1984, Triassic, *in* K.W. Glennie, ed., Introduction to the petroleum geology of the North Sea: Blackwell, p. 113–132.

Focke, J.W., and D. Munn, 1987, Cementation exponents in Middle Eastern carbonate resevoirs: Society of Petroleum Engineers, Formation Evaluation, p. 155–167.

Glennie, K.W., 1986, The structural framework and the pre-Permian history of the North Sea area, *in* K.W. Glennie, ed., Introduction to the petroleum geology of the North Sea (2nd edition): Oxford, Blackwell.

Heward, A.P., 1991, Inside Auk—the anatomy of an eolian oil reservoir, *in* A.D. Miall and N. Tyler, eds., The three dimensional facies architecture of terrigeneous clastic sediments and its implications for hydrocarbon discovery and recovery: SEPM Concepts in Sedimentology and Paleontology, v. 3, p. 44–56.

Pennington, J.J., 1975, The geology of the Argyll field, *in* A.W. Woodland, ed., Petroleum geology and the continental shelf of NW Europe: Applied Science, Barking Publisher, v. 1, p. 275–281.

Rasmus, J.C., 1983, A variable cementation exponent, m, for fractured carbonates: The Log Analyst, v. 11–12, p. 13–23.

Taylor, J.C.M., 1984, Late Permian—Zechstein, *in* K.W. Glennie, ed., Introduction to the petroleum geology of the North Sea: Blackwell, p. 87–111.

Chapter 11

Multiple Karst Events Related to Stratigraphic Cyclicity: San Andres Formation, Yates Field, West Texas

S. W. Tinker
Marathon Oil Company
Petroleum Technology Center
Littleton, Colorado, U.S.A.

J. R. Ehrets
Consulting Geologist
Boulder, Colorado, U.S.A.

M. D. Brondos
Marathon Oil Company
Petroleum Technology Center
Littleton, Colorado, U.S.A.

ABSTRACT

Open caves and solution-enhanced joints influence porosity distribution and fluid flow in Yates field. Therefore, applying an accurate model for cave formation, describing the distribution of cave and karst features, and quantifying the contribution of caves to total pore volume is important in order to characterize the reservoir. Prior work showed that karst and caves in Yates field were formed by meteoric processes acting on subaerially exposed islands following San Andres deposition, and predicted that the number of open caves should decrease with depth below the top of the San Andres Formation. The current work addresses three related issues: the possibility of multiple karst events in the San Andres of Yates field and the relationship of these events to stratigraphic cyclicity, the 3-D distribution of caves in the reservoir and the areas where future well deepenings might encounter caves in the field, and the effect of subaerial exposure on porosity and permeability.

There is a relationship between cave distribution and sequence stratigraphy in Yates field. Sequence-stratigraphic interpretation indicates that four major cycles aggraded and prograded from west to east. Each major cycle has a clinoformal geometry, shoals upward overall, and was capped by a subaerially exposed island complex. On these exposed islands separate cave lenses formed as a result of meteoric processes. Because the clinoforms and their associated island complexes aggraded and prograded from west to east, deeper drilling in Yates field should encounter additional caves to the west of known shallower caves in the east.

An important economic issue in carbonate reservoirs is the effect of subaerial exposure on porosity and permeability. Cavernous porosity formed beneath major unconformities in Yates field, and when examined by stratigraphic cycle, the data show that cycles with abundant caves have lower average matrix porosity and lower total porosity than the cycles with few caves. This results from a combination of matrix porosity reduction due to the cave-forming process and a lower pre-cave matrix porosity. Because Yates field is relatively shallow, most caves have not undergone mechanical collapse. Therefore, in contrast to the reduction in porosity observed in zones with many caves, the extensive network of open vertical caves and solution-enhanced fractures increases total vertical permeability of the system.

INTRODUCTION

Yates field was discovered in 1926, and has produced over 1.2 billion bbl of oil from over 4 billion bbl of original oil in place, in large part from the Permian (Guadalupian) San Andres carbonate reservoir. Very early in the field's history, cavernous dolomite within the San Andres Formation was known to contribute to remarkable flow rates, and numerous bit drops encountered during early field development provided evidence for an extensive network of caves within the San Andres Formation.

Three cave-forming events occurred in the Permian San Andres Formation in Yates field, located in west Texas between the Midland and Delaware basins on the southern tip of the Central Basin platform. Although many mechanisms for cave formation are possible for Yates field, such as "sulfuric acid oil-field karst" (Hill, 1992), or a regional meteoric aquifer model (Moore, 1989), the data support the island hydrologic model for karst and cave formation proposed by Craig (1988, 1990). According to this model caves formed as a result of mixing-zone dissolution related to subaerial exposure following San Andres deposition. Cave distribution was controlled by the presence of near-vertical joints, and by proximity to the meteoric water table and paleoshoreline. The present work recognizes three such meteoric surface karst and cave-forming events, and ties the position of these events into a sequence-stratigraphic framework in order to understand and predict cave distribution in three dimensions.

Yates field is characterized by an excellent log database from nearly 1800 wells, generally drilled on 10 acre (4 ha) spacing, along with a core database totaling approximately 23,000 aggregate ft (8500 m) from 118 wells. This extensive database provides an opportunity to describe in detail the features and distribution of karst within Yates field. Modern log suites document the presence of open caves, and highly altered sedimentary fabrics, infiltrated sediments, cements, and open vugs seen in cores provide supporting evidence. Evidence of surface karst in cores includes solution breccias and altered fabrics associated with the Permian unconformity at the top of the San Andres. Because Yates field is relatively shallow (1500–2000 ft; 457–610 m), many caves have not undergone mechanical collapse and remain open in the reservoir. The lack of collapse of early-formed caves is also evidence that Yates field was never buried significantly deeper than it is today.

STRUCTURAL AND STRATIGRAPHIC SETTING

Yates field is located in eastern Pecos County, Texas, approximately 90 miles (144 km) south of Midland (Figure 1). The field is situated on the southeastern tip of the Central Basin platform, which hosts numerous oil fields that produce from the same Permian formations encountered at Yates. The field is bounded to the east by the Midland basin and to the south by the Sheffield channel, and represents the highest present-day structural position of San Andres and equivalent-age strata on the Central Basin platform (Galloway et al., 1983). Northeastern and southeastern flanks of the structure display maximum dips of 3 to 5° into the Midland basin and the Sheffield channel, respectively. The western flank of the structure dips more gently at 1/2° toward the axis of the Central Basin platform (Figure 2).

Yates field produces from four Guadalupian-age formations, including the San Andres, Grayburg, Queen, and Seven Rivers (Figure 3). The San Andres Formation, comprised of interbedded, dolomitic wackestone and argillaceous mudstone in the west, and dolomitic fusulinid packstone and grainstone in the east, is the most prolific producing formation in the field, characterized by an initial gross pay interval approximately 450 ft (165 m) thick in the structurally highest portion of the field. The Grayburg Formation is characterized by interbedded siltstone and dolomite with variable porosity, ranges from <10 to >100 ft (3.5 to 35 m) in thickness, and produces from relatively tight siltstone and sandstone across the field and from

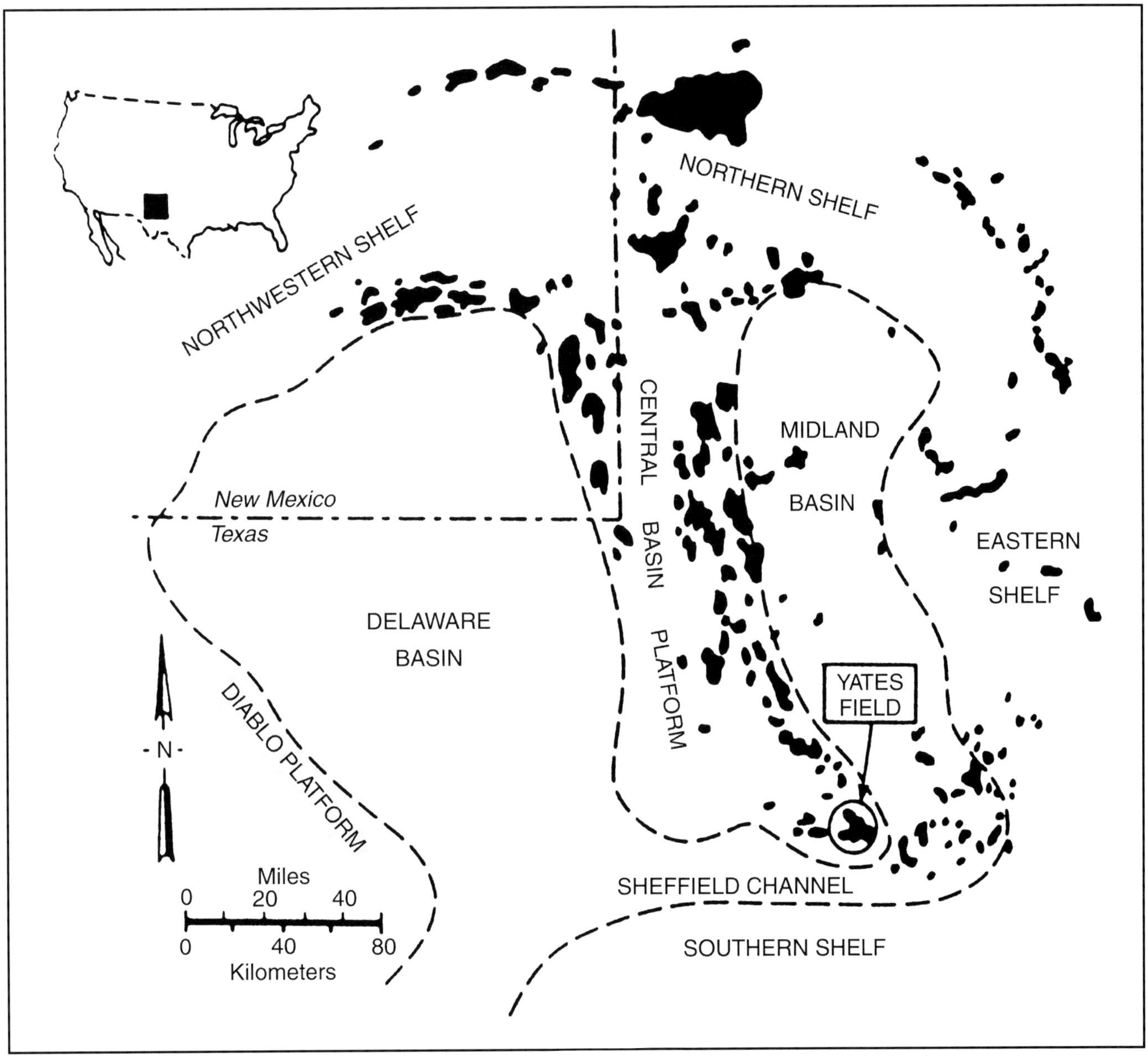

Figure 1. Regional paleogeographic elements of the Permian basin during the late Permian, with the distribution of oil and gas fields that produce from San Andres and Grayburg reservoirs and their age-equivalents shown in black. Yates field is located on the southeastern tip of the Central Basin platform (from Craig, 1990). Reprinted by permission of the Geologic Society of London.

dolomitic packstone and grainstone in the eastern portion of the field. The Grayburg is overlain by thin-bedded sandstone, siltstone, and dense dolomite of the Queen Formation, which ranges from 20 to 50 ft (7.5 to 18.5 m) thick in the field area. The Queen Formation, which produces locally from relatively tight sandstone and siltstone, is overlain by the evaporite-dominated Seven Rivers Formation, which serves as the principal fluid seal to the underlying reservoir sequence. Porous sandstone facies in the lower portion of the Seven Rivers are productive along the northeastern and southeastern flanks of the field. Thin, widespread siltstone marker beds within the lower Seven Rivers have been used as correlation markers for paleotopographic reconstruction in previous field studies (Craig, 1988). These markers are believed to have been deposited as near-horizontal time surfaces within the salina and sabkha facies of the Seven Rivers. One of these siltstone markers, labeled Seven Rivers 'M,' is used as a datum in this study to illustrate internal San Andres stratigraphic relationships.

A west-to-east stratigraphic section, from a 3-D geologic model, illustrates the general stratigraphic framework of Yates field (Figure 4A). The upper San Andres Formation is comprised of a ramp overlain by four major cycles, labeled Cycle 1, Cycle 2, Cycle 3, and Cycle 4, which aggrade and prograde to the east. Each major cycle shoals upward overall and is comprised of three to five minor cycles. A typical west-side minor cycle began with a dolomitic shale with a sharp

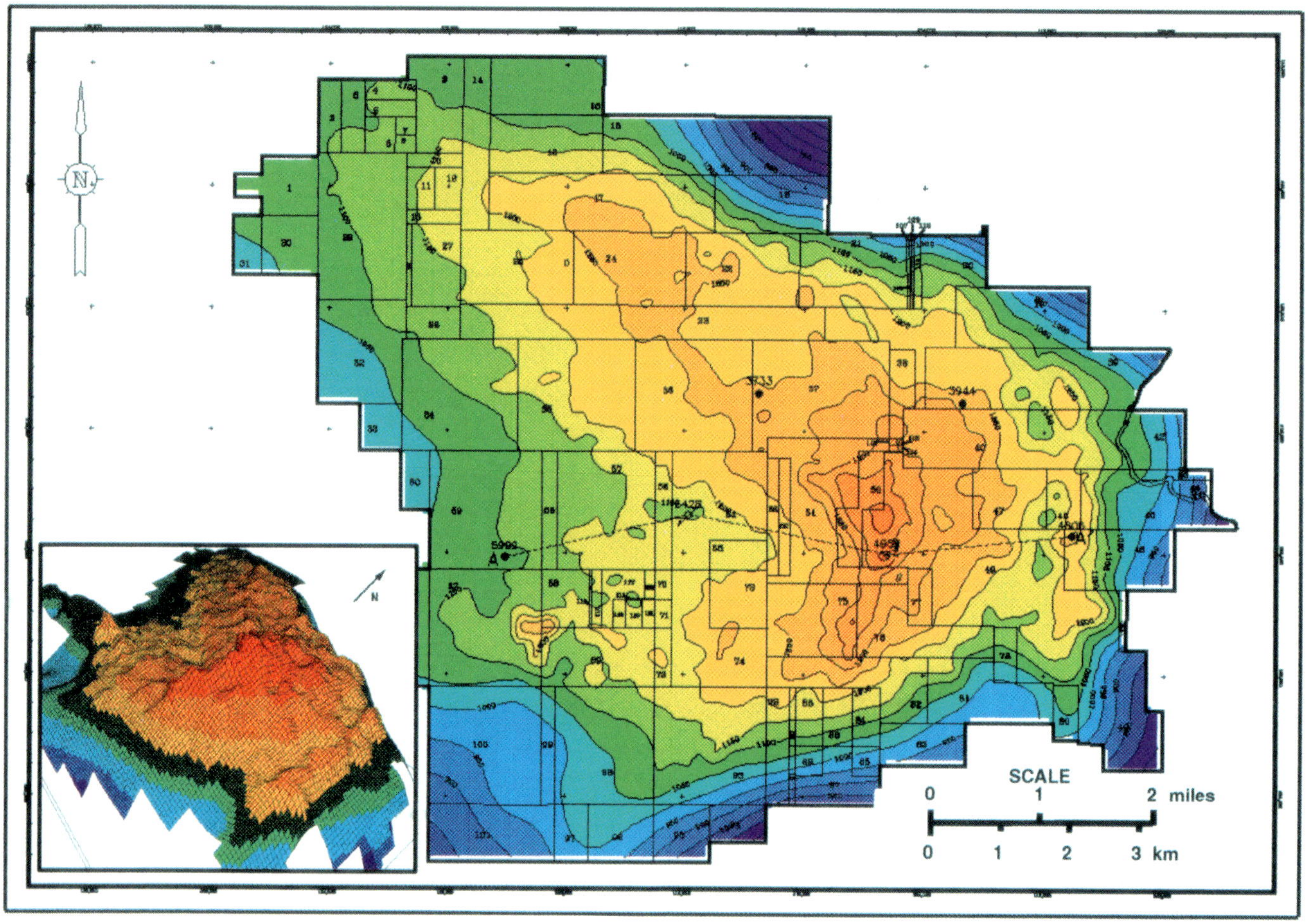

Figure 2. Structure on top of the San Andres Formation within the Yates field unit. Colors represent approximate present-day fluid levels, with blue the water zone, green the oil-water transition zone and oil column, and orange and red the secondary gas cap. Note the gentle structural dips to the west and the steeper dips to the south, east, and northeast. The inset shows the same surface in 3-D.

basal contact. Clays in the shales are predominantly illite (90%) and chlorite (10%), and were probably trapped in lagoons behind islands during times of relative sea level fall. Clays were later reworked during subsequent relative sea level rise and grade upward into argillaceous dolomitic wackestone.

Major cycle boundaries are projected from shales that could be correlated in the west, into stacked fusulinid shoals and inner shoals that could not be correlated to the east. The interpreted geometry of the major cycles is based, in part, on San Andres analogs exposed along the Algerita escarpment (Kerans et al., 1991) of the Guadalupe Mountains. A major erosional unconformity, marked by significant karst-related features, defines the top of the San Andres. Very little of Cycle 4 is preserved in the field area, owing to erosion following San Andres deposition, and therefore Cycles 3 and 4 are collectively referred to as Cycle 3 for the remainder of the paper. The San Andres unconformity is onlapped by sediments of the Grayburg Formation, which fill most karst-related topographic irregularities. Some caves are formed in the Grayburg Formation, probably as a result of mechanical collapse into caves of the underlying San Andres Formation, and such collapse caves are not discussed further in this work.

The San Andres Formation is comprised of three principal lithofacies: (1) west-side lagoonal shales, dolomitic mudstone, and wackestone that display high GR and low porosity log signatures; (2) higher porosity, low GR (except in areas of uranium) dolomitic, fusulinid packstone and grainstone shoal complexes found dominantly in the lower west side and upper east side of the field; and (3) relatively low GR, low-porosity dolomitic wackestone of the eastern slope and ramp. A west–east stratigraphic section of porosity illustrates the position of the three major San Andres lithofacies and their relationship to the major cycles (Figure 4B).

SAN ANDRES KARST AND CAVES

The extensive log and core databases in Yates field allow for a good understanding of the spatial distribution of surface karst beneath the unconformity at the top of the San Andres Formation and of caves within the San Andres Formation. Kerans and Parsley (1986)

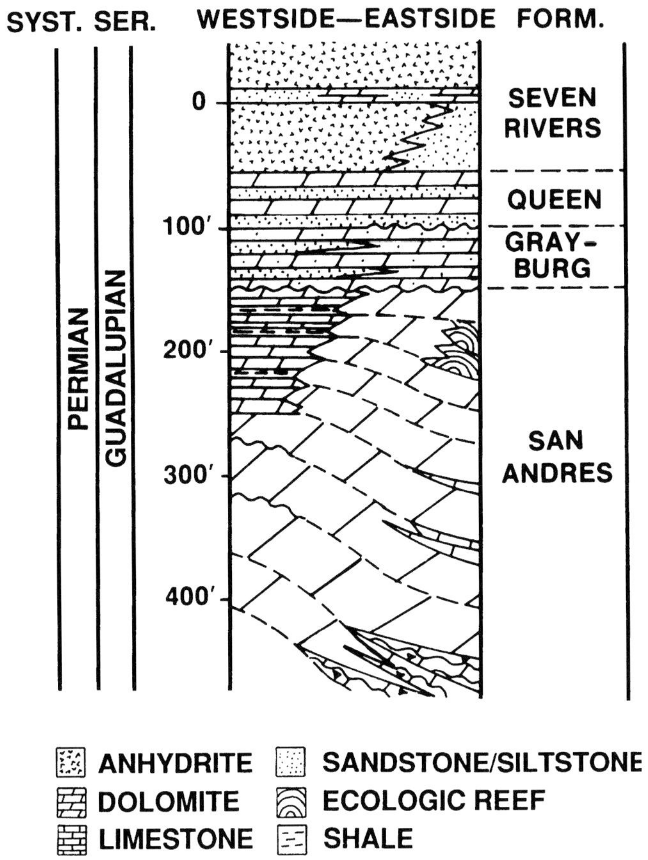

Figure 3. Stratigraphic nomenclature for Yates field, illustrating the position of the major unconformity at the top of the San Andres Formation. Depth scale shown is measured from the Seven Rivers 'M' marker datum (after Craig, 1990). Reprinted by permission of the Geologic Society of London.

described similar San Andres surface karst features in the Taylor Link West San Andres reservoir located 20 mi (32 km) west of Yates field, and minor paleokarst is reported at the top of the San Andres Formation in the Guadalupe Mountains in Last Chance Canyon (Sonnenfeld, 1991) and along the Dog Canyon escarpment (C. Kerans, 1993, personal communication). These occurrences suggest the possibility of interregional paleokarst, as defined by Choquette and James (1988).

San Andres karst is believed to have formed in porous and permeable limestone, as shown by extensive, early dissolution of limestone along joints. Some karst may have formed in conjunction with early dolomitization, but subsequent dolomitization has extensively altered many early karst fabrics.

Production Evidence of Open Caves

Very early in the history of Yates field cavernous dolomite within the San Andres Formation was known to contribute to remarkable flow rates at relatively shallow depths (Gester and Hawley, 1929; Adams, 1930). By 1929 over 200 wells had tested at rates in excess of 10,000 BOPD (1590 m^3/day), ranging upward to 200,000 BOPD (31,800 m^3/day) (Hennen and Metcalf, 1929; Shirley, 1987). These astounding production rates, along with numerous bit drops encountered during early field development, provide evidence of an extensive network of caves within the San Andres Formation.

Caves Observed on Logs

Caves were identified from a combination of caliper, GR, and bulk-density logs. A cave was identified if GR was less than 40 API (non-shale; anomalously high GR values in the uranium-rich areas on the east side of the field were ignored), bulk density was less than 2 g/cc (anomalously high porosity), and the caliper showed some excursion from a baseline (Figure 5). Intervals with interpreted washouts or bad logs were coded in the database and eliminated from the statistical analysis. The minimum cave height accepted was one foot (0.3 m), resulting in a database of 1050 caves (4477 cave ft; 1365 cave m). A cave foot is one vertical foot of cave; for example, a 5 ft high cave equates to 5 cave ft (1.5 cave m).

Many of the caves were identified in wells where the logs were run after artificial acid stimulation treatments. The two populations—"true caves" (before acid stimulation) and "post-acid caves"—are similar in terms of cave height in 95% of the data, indicating that only rarely did acid stimulation significantly enhance the vertical component of the cave. Therefore, the "post-acid caves" are considered valid and are included in all statistical analyses.

Cave and Karst Features Observed in Cores

The 3-D distribution of open caves in Yates field can be thoroughly evaluated only through the analysis of bit drops and log anomalies. Many cave-related features have been observed in cores, however, which supports the presence of open caves in the field (Figure 6). Infiltrated sediment, which variably fills solution-enlarged vugs and channels, is typically dark in color and is dominantly dolomitic silt and mud (>90%) with a minor component of quartz silt and clay (<10%). Remnant voids not filled by sediment are commonly partially to completely filled by multiple generations of dolomite cement. These cements in some places display dense to porous, travertine-like fabrics.

Vertical solution fabrics are perhaps the most common cave-related features. These fabrics range from highly altered, variably dense dolomite with solution-enlarged, sediment-floored, and cement-filled vugs to solution-enlarged fractures and channels completely devoid of cement or sediment. Dense, horizontal, dolomite-cemented fabrics that fill vugs might represent cave-fill deposition, especially in porous grainstone and packstone textures where these banded dolomites were definitely not primary deposits. Solution and collapse breccias are also observed in association with caves, but rapidly diminish in abundance with depth below the top of the San Andres. A significant number of cores recovered rubble from intervals 1 to 5 ft (0.3 to 1.5 m) thick, which could be the only rock record recovered after coring an open cave. Extremely vuggy zones may also be indicators of nearby open caves in the reservoir.

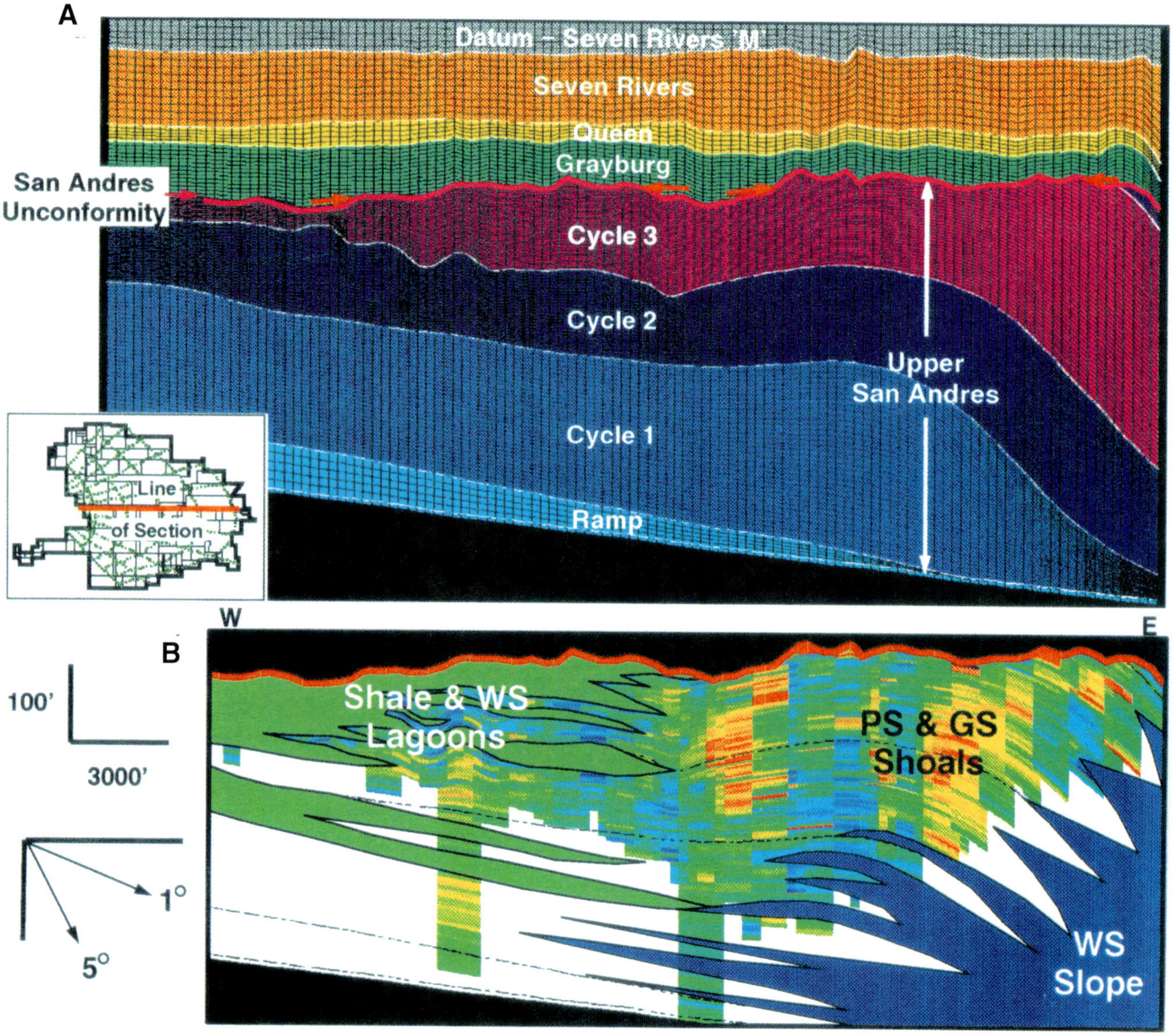

Figure 4. (A) Yates field stratigraphic cross section cut from a 3-D model; datum is on Seven Rivers 'M.' Grid cells indicate the layering scheme in the model. The irregular surface at the top of Cycle 3 and Cycle 4 represents erosional truncation along the San Andres unconformity, which is onlapped (red arrows) and filled by Grayburg silts and carbonate mudstone. Location of cross section, shown on inset map, is the same for all figures. (B) Porosity distribution, with hot colors (yellows and reds) representing porosity up to 35%, and cold colors (blues) representing porosity <10%. The three dominant San Andres lithofacies: (1) west-side lagoonal shales, dolomitic mudstone (MS), and wackestone (WS) (green), (2) dolomitic, fusulinid packstone (PS) and grainstone (GS) shoal complexes, and (3) wackestone (WS) of the eastern slope and deep ramp (blue), are illustrated.

Surface karst observed in cores (Figure 7) is primarily related to the unconformity at the top of the San Andres, although evidence of minor karst from earlier subaerial exposure events exists within the San Andres Formation. Solution breccias, ranging from slightly disrupted clasts to completely chaotic fabrics, are found just below the top of the San Andres. These near-surface breccias are commonly a few feet thick and are usually abruptly overlain by siltstones of the Grayburg Formation. Crude vertical fissures are filled with San Andres lithoclasts and quartz silt, which probably represent initial influx of Grayburg sediments. Geopetal features are commonly present, particularly in larger pores connected by solution-enhanced fractures and channels. Solution breccias represent one end member of a continuum which ranges to well-developed collapse breccias, solution channels, and small caves. Collapse breccias often contain clasts which are distinctly different from the adjacent host rock and can be traced to strata several feet above the collapsed zone. Irregular voids and shelters preserved within some collapse breccias are filled with a variety of material, including clastic sediment derived from the Grayburg Formation, dolomitic silt and mud, cave pearls, and multiple generations of dolomite cement.

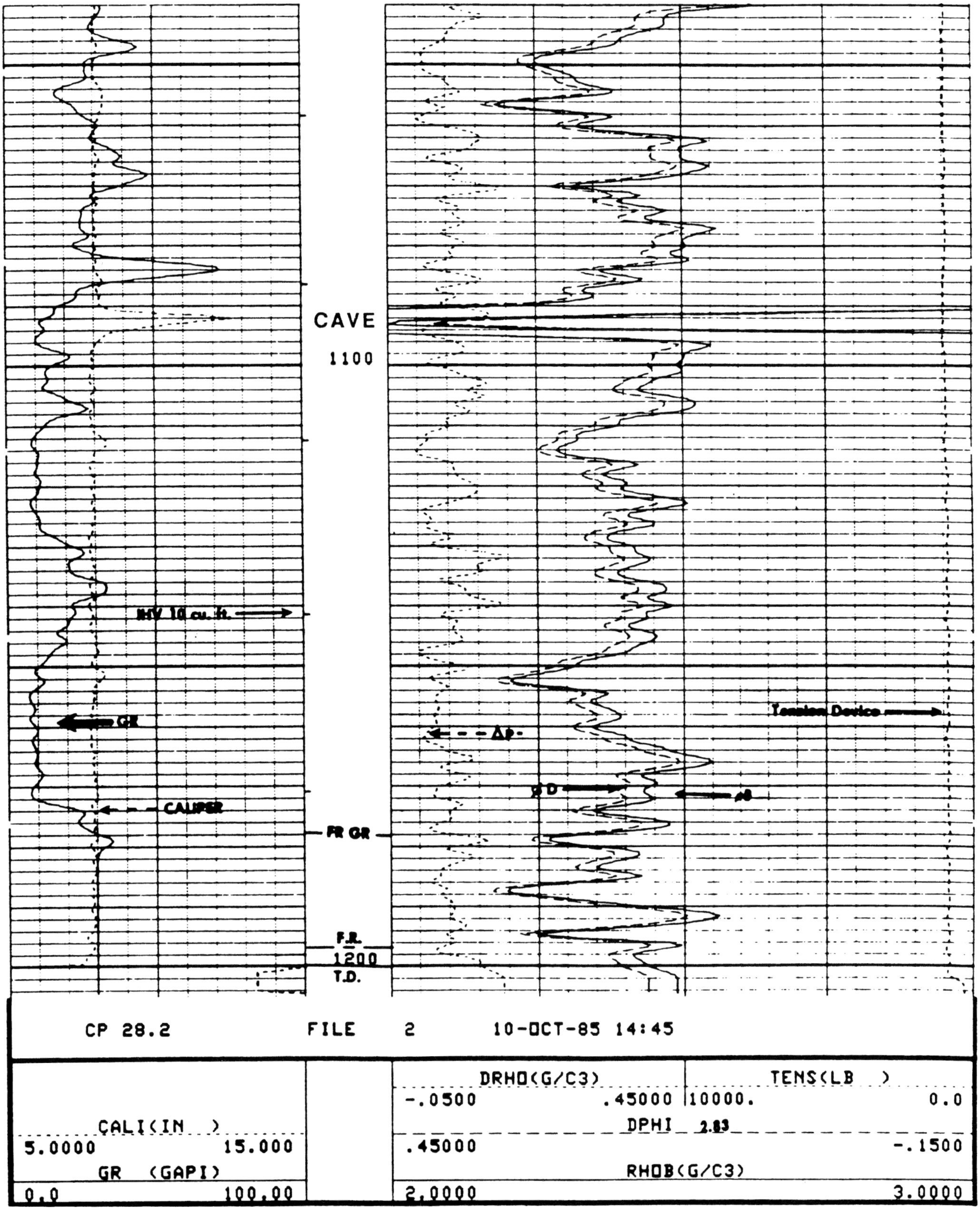

Figure 5. FDC log with GR illustrating the parameters used for cave selection. At a depth of approximately 1090 ft, note the (1) clean GR, (2) off-scale Rhob, and (3) caliper excursion of 4 inches (10 cm). This is a typical signature for a cave 4–5 ft (1.2 –1.5 m) high.

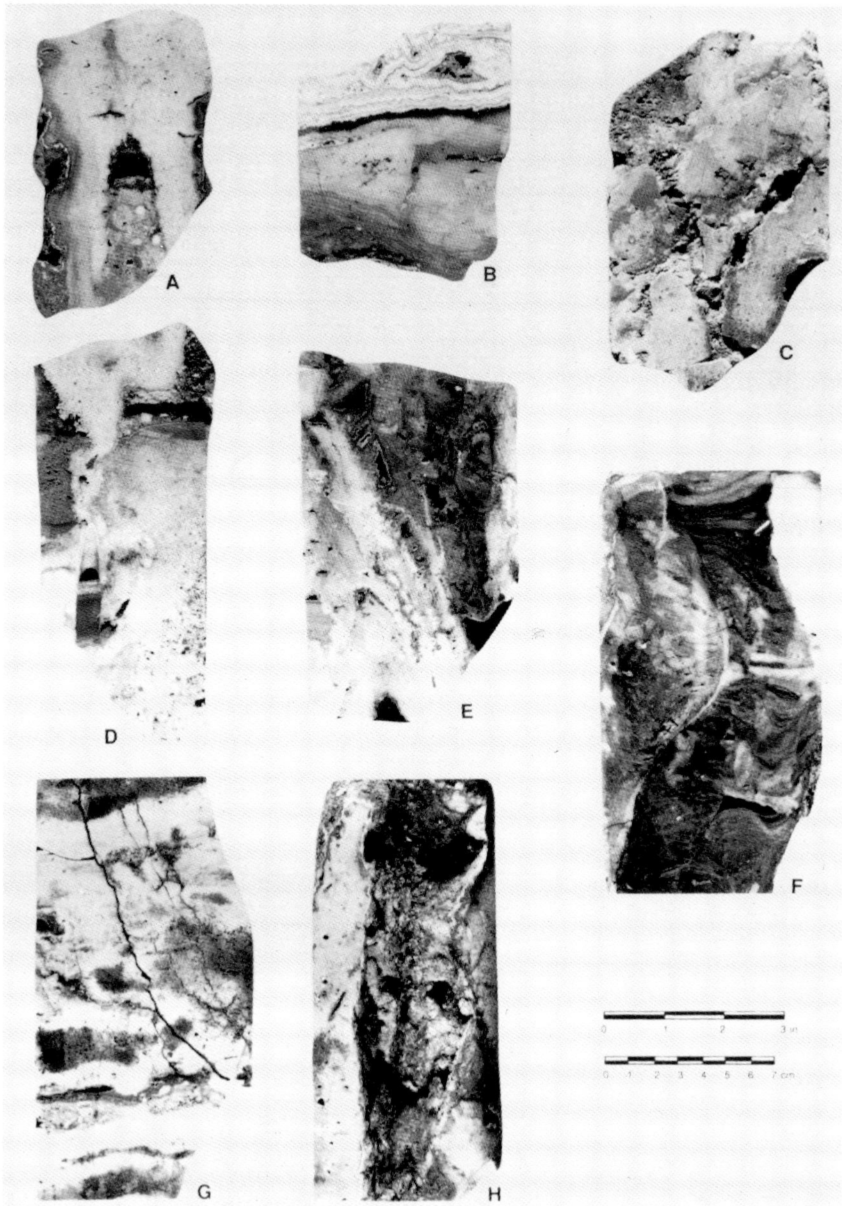

Figure 6. Core surface photographs illustrating lithified sediment, cements, and breccias associated with caves developed within the San Andres. All samples are from YU4959 (see Figures 11 and 12 for location). Textural terms reflect original depositional texture of the host rock. (A) 1267 ft, dolomite; peloidal wackestone–packstone. Irregular vugs are partly filled with carbonate sediment, then lined with dolomite cement. Sample is 63 ft (20 m) below the San Andres unconformity. (B) 1269 ft, dolomite; peloid–fusulinid grainstone. Travertine fabric reflected by banded cement fills a remnant solution void. Sample is 65 ft (21 m) below the San Andres unconformity. (C) 1281 ft, gypsiferous dolomite; peloid-fusulinid-skeletal packstone–grainstone. Irregular vugs within collapse breccia are partly filled by sediment and by patches of white gypsum cement. Sample is 77 ft (25 m) below the San Andres unconformity. (D) 1284 ft, dolomite; peloid-intraclast-fusulinid packstone–grainstone. Dark cave sediment forms the floor of irregular vugs and partly fills vertical, tubular void (burrow or mold?), the latter of which is then partly filled and lined with cement. Original rock fabric is visible in the lower right. Sample is 80 ft (26 m) below the San Andres unconformity. (E) 1291 ft, dolomite; peloid-skeletal packstone. Chaotic fabric results from irregular distribution of cave sediment and cement which fill a network of channels and vugs. Sample is 87 ft (28 m) below the San Andres unconformity. (F) 1292 ft, dolomite; peloid-skeletal packstone. Multiple generations of cave sediment and cement fill an irregular, near-vertical solution channel. Host rock is visible in upper right. Sample is 88 ft (28 m) below the San Andres unconformity. (G) 1308 ft, dolomite; peloid-skeletal-fusulinid wackestone–packstone. Light-colored areas of dolomite cement define an irregular horizontal and vertical pattern of alteration. Sample is 104 ft (33 m) below the San Andres unconformity. (H) 1359 ft, dolomite; peloid-skeletal-fusulinid packstone. A vertical solution channel remains essentially free of cave sediment and cement. Host rock is visible along the left margin of the slab. Sample is 155 ft (49 m) below the San Andres unconformity.

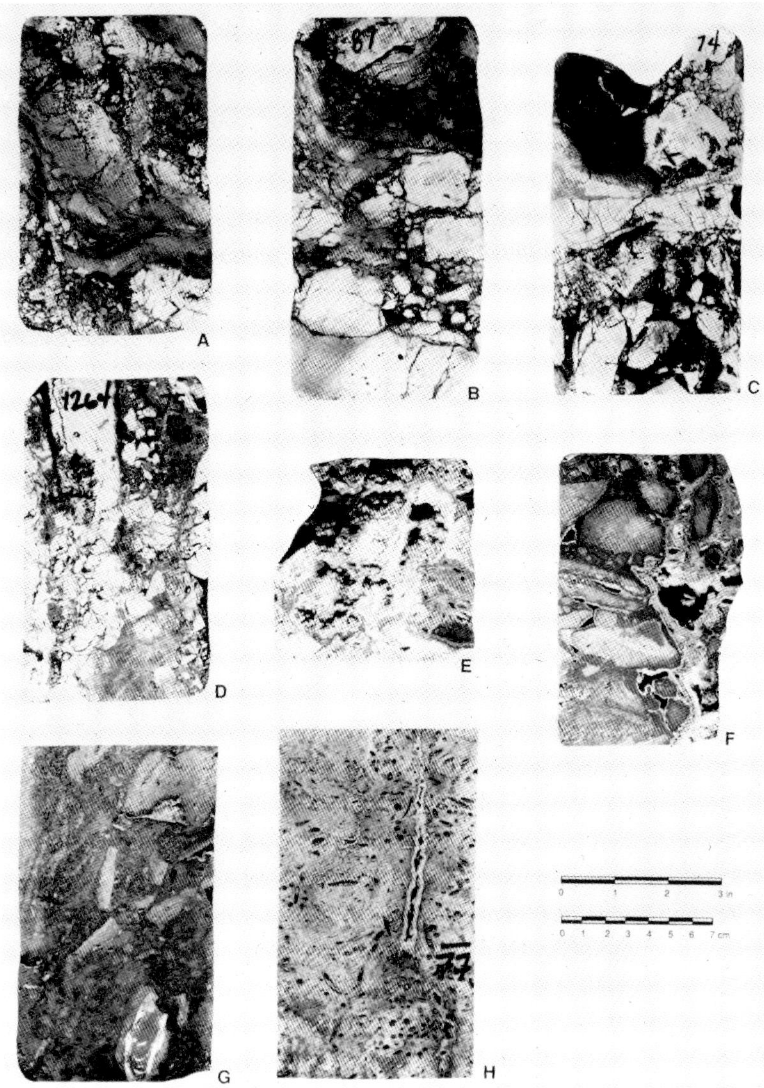

Figure 7. Core surface photographs illustrating fabrics associated with karst breccias and near-surface caves at the top of the San Andres Formation. Textural terms reflect original depositional texture of the host rock. See Figures 11 and 12 for core locations. (A) YU5909; 1507.5 ft, dolomite; peloid-skeletal wackestone. Infiltrated carbonate sediment with some quartz silt is contained within deformed and brecciated fabric of upper San Andres. Brecciated zone is approximately 5 ft (1.5 m) thick. Sample is 1 ft (0.3 m) below the San Andres unconformity. (B) YU5909; 1511.0 ft, dolomite; skeletal packstone. Subangular clasts within basal portion of brecciated interval, approximately 5 ft (1.5 m) below the San Andres unconformity. Skeletal molds contained within unaltered rock fabric are visible near the base of the slab. Dark areas represent heavily oil stained, porous dolomite matrix. (C) YU3733; 1263.5 ft, dolomite; peloid-lithoclast-skeletal wackestone–packstone. Brecciated fabric developed approximately 3 ft (0.9 m) below the San Andres unconformity. Brecciated interval is approximately 4 ft (1.2 m) thick, and is underlain by a 20 ft (6 m) interval of dense and shaly dolomite. (D) YU3733; 1264.0 ft, dolomite; peloid-lithoclast-skeletal wackestone–packstone. Pattern of breccia clasts suggests crude vertical solution fabric. Sample is 3.5 ft (1 m) below the San Andres unconformity. (E) YU3944; 1268.6 ft, slightly bituminous dolomite; peloid-skeletal wackestone–packstone. Large vugs developed within matrix are lined by dolomite cement and are partly filled with small spheroidal grains which represent cave pearls. Sample is 30 ft (9 m) below the San Andres unconformity. (F) YU3944; 1268.9 ft, slightly bituminous dolomite; peloid-fusulinid wackestone–packstone. Cement-lined vugs are irregularly developed within a collapse breccia. Portions of the original vug network are filled by infiltrated carbonate sediment. Sample is 30.5 ft (9.1 m) below the San Andres unconformity. (G) YU3944; 1269.5 ft, sample from just below F. Slightly pyritic dolomite; peloid-skeletal-fusulinid wackestone–packstone and lithoclast floatstone. Collapse breccia shown by sample F gives way below to a fabric which contains redeposited lithoclasts within a carbonate sand matrix. Sample is 31 ft (9.3 m) beneath the San Andres unconformity. (H) YU3944; 1276.7 ft, dolomite; peloid-fusulinid wackestone–packstone. Skeletal molds are visible throughout the dense dolomite matrix at the base of the zone of alteration depicted by E, F, and G. Vertical fractures and irregular vugs are lined by multiple generations of dolomite cement. Sample is 38 ft (11.4 m) below the San Andres unconformity.

Present-Day Distribution of Caves

Further evidence that cave formation was related to subaerial exposure, and not to later burial processes, comes from examining the distribution of San Andres logged feet and cave feet relative to present-day subsurface elevation (Figure 8A). A "logged foot" is one vertical foot of wellbore with modern log data; a "cave foot" is one vertical foot of cave. "Total logged feet" represent the database from which cave feet were picked. The barplot of logged feet from +1470 to +1090 (+448 to +332 m) mimics present-day San Andres structure, and logged feet fall off rather sharply below the water-oil contact because most wells were not drilled below that elevation. The modal class for cave feet is at +1170 (+357 m), 80 ft (24 m) higher than that for logged feet. The distribution of normalized cave feet (cave feet per 1000 logged feet) shows the greatest count at the top of the San Andres, and generally decreases overall with elevation (Figure 8B).

This present-day cave distribution would be unlikely if caves were formed by the "sulfuric acid oil-field karst" model of Hill (1992), because aggressive fluids migrating from a deep basin upward would form caves in the deep reservoir first. In addition, the distribution of surface karst features associated with the San Andres unconformity is difficult to explain by sulfuric acid karst, which might have no relationship to a depositional surface. However, it is possible that caves formed early by a different process were later enhanced in size by basin-derived sulfuric acid. According to Hill (1992), another significant criterion for identification of sulfuric acid oil-field karst is "a lack of linear or branching passages related genetically to identifiable recharge and discharge points." The distribution of caves in Yates field, in contrast, is linear and branching, and does appear to be related to meteoric processes at recharge and discharge points. This relationship is clarified by plotting the data relative to a stratigraphic datum.

Paleodistribution of Caves

The relationship between cave formation and subaerial exposure within the San Andres is apparent when cave feet and logged feet are examined relative to the Seven Rivers 'M' datum. Logged feet have a skewed distribution with a mode at 230 to 250 ft (70 to 76 m) below the Seven Rivers 'M' (Figure 9), and decrease with depth below 250 ft (76 m). Cave feet also show a skewed distribution around a mode at 190 ft (58 m) below the Seven Rivers 'M,' with a smaller peak at 290 ft (88 m) below 'M.' A crossplot of logged feet against cave feet (Figure 9; inset) shows the number of cave feet relative to logged feet is anomalously high from 150 to 210 ft (46 to 64 m; points labeled 1 through 4). If points 1 to 4, which are related to the unconformity at the top of the San Andres, are treated as anomalous outliers, a least-squares regression line can be fit to the remaining data with $r^2 = 0.95$. Assuming no geographic bias in well location, this indicates that the number of cave feet from 230 to 690 ft (70 to 210 m) below the Seven Rivers 'M' is purely a function of the number of logged feet, and therefore all data must be normalized.

The normalized distribution shows the cave feet count increasing from a depth of 150 ft to 190 ft (46 to 58 m) below 'M,' and then decreasing to 290 ft (88 m)

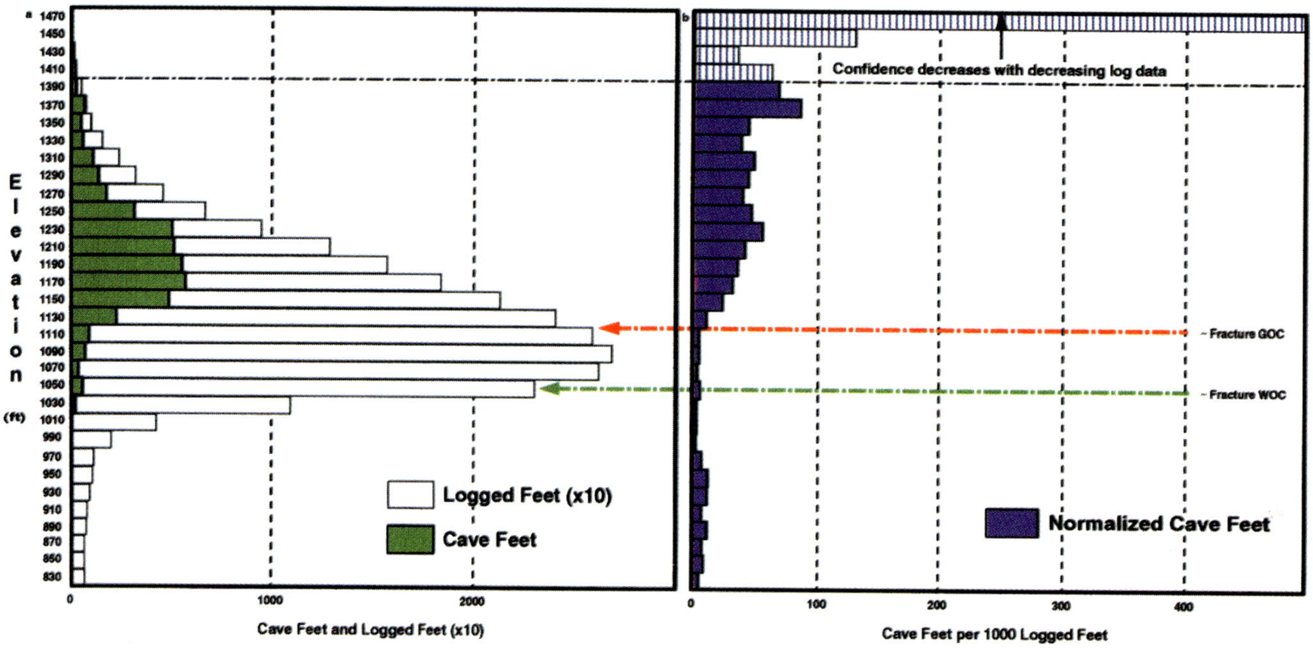

Figure 8. Bar plots of (A) San Andres cave feet (green) and logged feet (white), and (B) normalized cave feet (cave feet/1000 ft) as a function of present-day subsurface elevation. Vertical scale represents the midpoints of the bars. Confidence in the normalization is poor above +1400 ft. The approximate gas-oil contact (GOC) and water-oil contact (WOC) in fractures are shown.

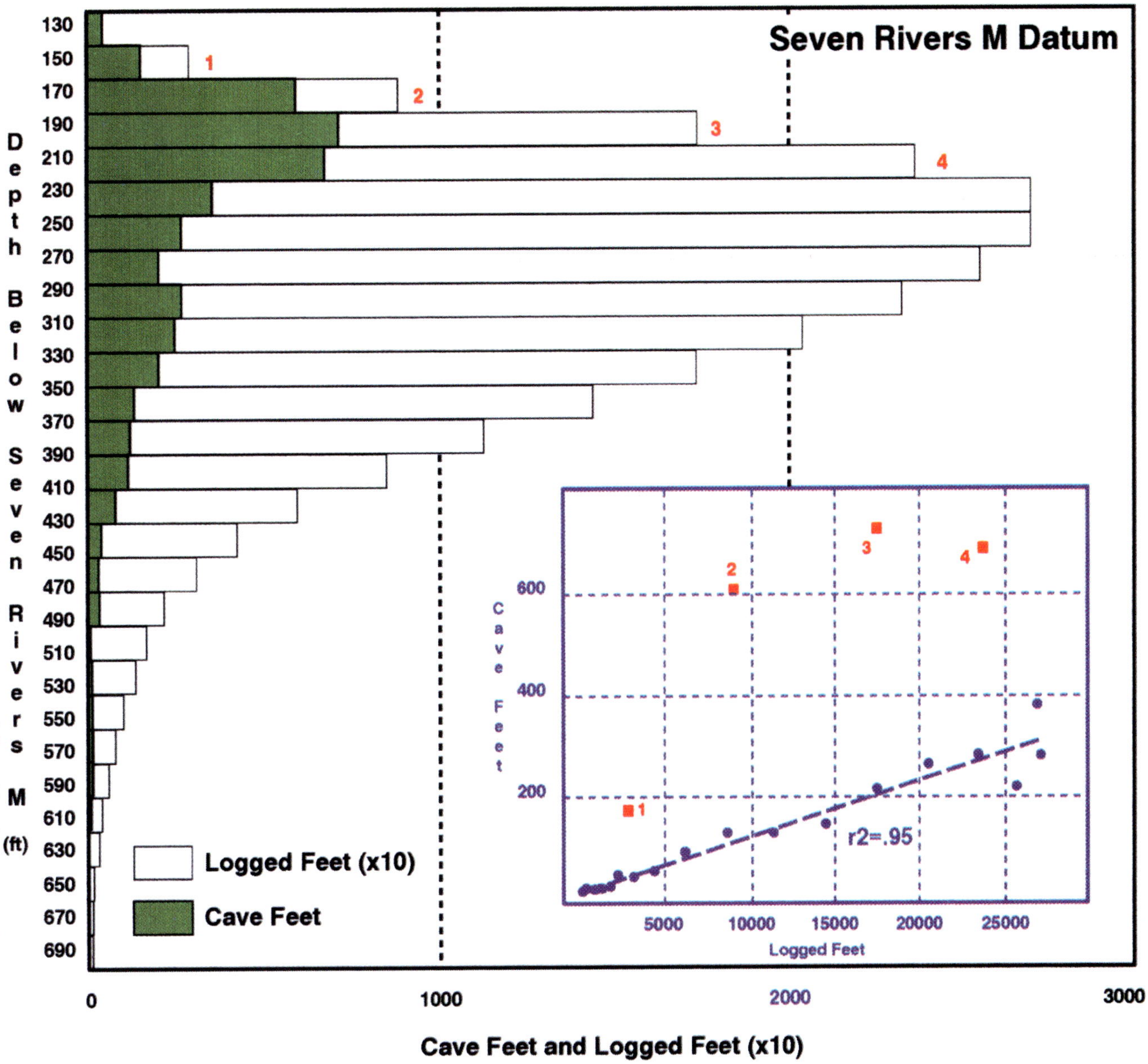

Figure 9. Bar plot of San Andres cave feet (green) and logged feet (white) as a function of depth below Seven Rivers 'M.' Vertical scale represents the midpoints of the bars. Inset is a cross plot of logged feet against cave feet for each 20 ft slice. A least-squares linear regression line is posted, which ignores points 1–4 labeled on the cross plot and bar plot.

below 'M' (Figure 10, Cycle 3). This pattern is repeated in Cycle 2 from 290 ft to 450 ft (88 to 137 m), and again in Cycle 1 from 470 ft to 530 ft (143 to 162 m). Below 530 ft (162 m), confidence in the normalization decreases because of the small number of logged feet.

When compared directly to mean cave height as a function of depth below 'M,' the same three cycles are repeated. The probable range of relative sea level fall in feet (vertical feet of island exposure) is indicated for each of the three cycles. Note that total cycle thickness decreases, but the range of relative sea level fall increases from Cycle 1 through Cycle 3 (Figure 10). Cave formation, which is related to the dynamic position of the island shoreline and the freshwater table (Craig, 1990), should be more intense at the top of Cycle 3 where relative sea level fell the most, and this is supported by a larger number of normalized cave feet and greatest cave heights. Cave formation was less intense at the top of Cycle 2, and least intense at the top of Cycle 1, where relative sea level fell the least.

An isochore map from the Seven Rivers 'M' datum to the top of the San Andres unconformity highlights the interval thinner than 205 ft (62 m) as land, and the interval thicker than 205 ft (62 m) as water (Figure 11). This represents one static shoreline position of emergent San Andres islands when relative sea level fell at the end of San Andres deposition. Cave feet and logged feet from boreholes that penetrate Cycle 3 are

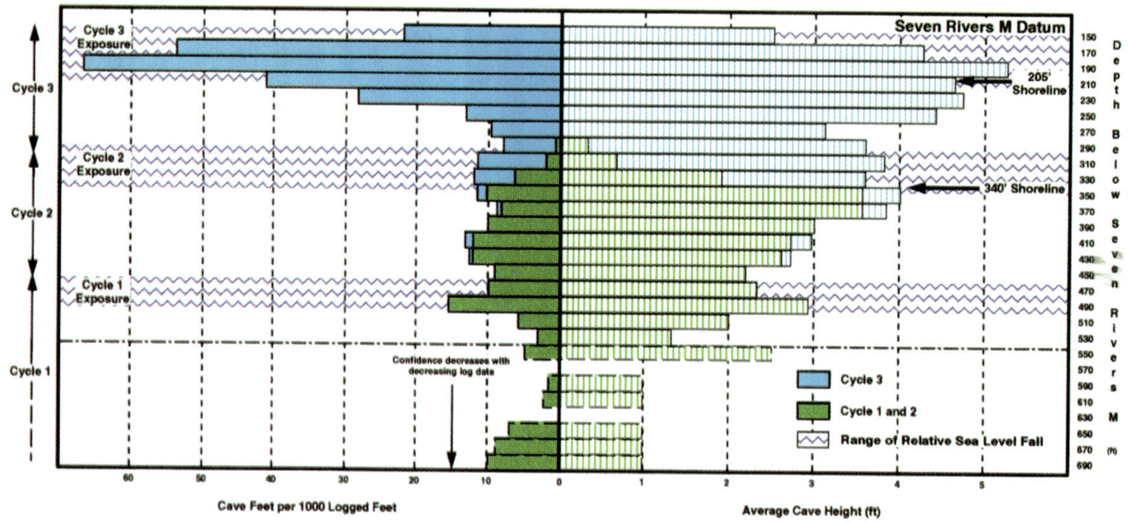

Figure 10. Bar plot of normalized cave feet and cave height as a function of depth below Seven Rivers 'M.' Cycles 1 and 2 are shown in green and Cycle 3 is shown in blue. The range of relative sea level fall for each cycle (maximum subaerial exposure) is illustrated. Vertical scale represents the midpoints of the bars. Shorelines for Cycle 3 at 205 ft and Cycle 2 at 340 ft are marked, which correspond to the shorelines illustrated in Figures 11 and 12.

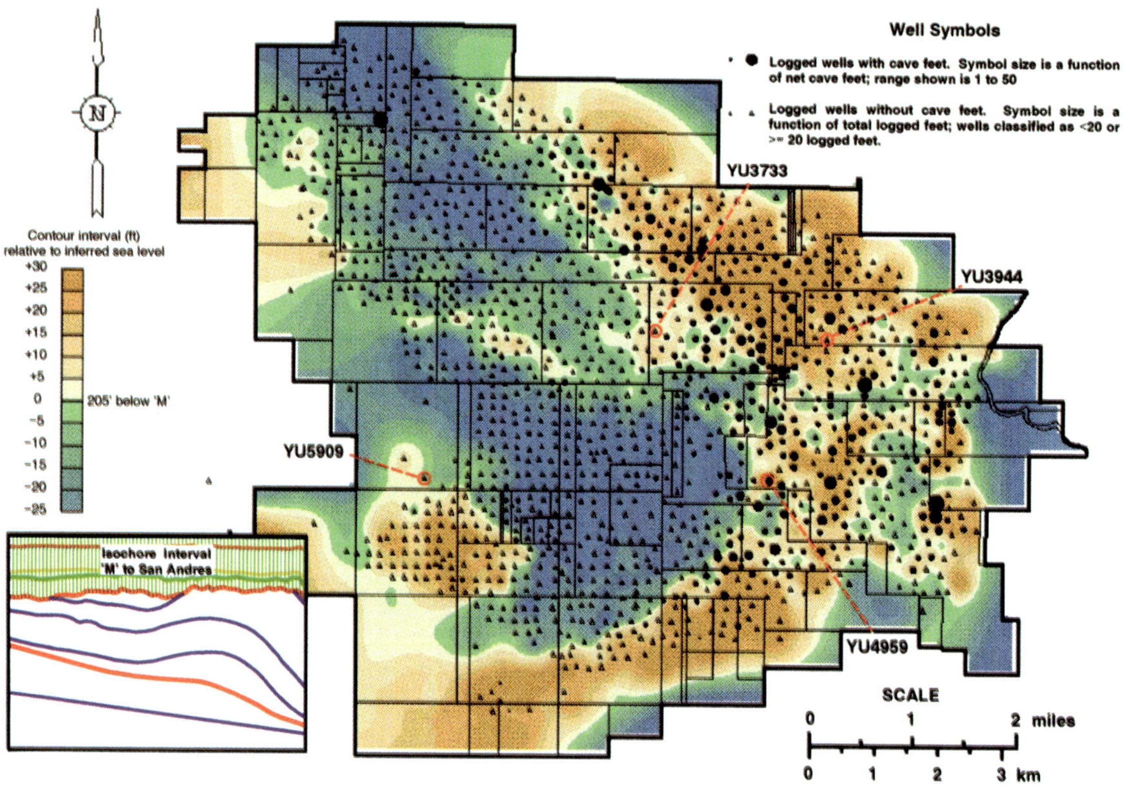

Figure 11. Late San Andres (Cycle 3) cave and log feet superimposed on the island complex. Islands are determined by thickness trends of the stratigraphic interval from the Seven Rivers 'M' marker to the top of the San Andres (inset). The static shoreline is represented at 205 ft (62 m) below 'M,' and the color bar is related to this shoreline. Cave feet are posted as solid circles, with the size of the symbol proportional to the total cave feet. Triangles have been posted where the interval has been logged and no caves are present. Small triangles are wells with <20 ft (6 m) of log data, and large triangles are wells with ≥20 ft (6 m) logged. The locations of cores included in Figures 6 and 7 are denoted by Yates field unit well numbers.

superimposed on this map to illustrate how well the distribution of caves can be explained by the exposed island complex. The small, blue areas on the island surface, approximately 1000 ft (305 m) in diameter, are interpreted as sinkholes, and the isolated tan areas (>30 ft above paleo-sea level) are interpreted as karst towers. The effect of early jointing is not readily apparent on this map. However, if individual 20 ft slices are mapped (i.e., a cave feet map of each bar of data from the Figure 10 bar plot), the caves line up along N50°W and N40°E trends, which is interpreted to be the orientation of the dominant conjugate joint set during early cave formation.

An isochore map from the Seven Rivers 'M' to the top of Cycle 2 highlights the interval thinner than 340 ft (107 m) as land, and the interval thicker than 340 ft (107 m) as water (Figure 12). Cave feet and logged feet from Cycle 1 and from Cycle 2 are superimposed on this map to illustrate the distribution of caves relative to the islands, and the limited number of borehole penetrations in these cycles. Although the prediction is not bad, it is apparent that these caves do not follow the island shoreline as well as caves followed the Cycle 3 shoreline. This is because the top of Cycle 2 on the east side of the field is purely a projected surface, and many of the wells only penetrate the upper few feet of Cycle 2. Even with these limitations, the most likely locations for additional Cycle 2 caves are island areas where log penetrations are limited (small triangles).

Cave Distribution Relative to the San Andres Unconformity

In other oil fields it is not always possible to use logs, core data, or seismic data to find a flat, time-line datum such as Seven Rivers 'M' that allows for reconstruction of the shape of a sequence boundary or exposure surface. It is therefore necessary to use the unconformity surface as the datum and to examine cave distribution relative to that surface. The resulting distribution of data from Yates field looks quite different from the distribution based on the Seven Rivers 'M' datum. The data show that caves are the most abundant just below the San Andres unconformity and decrease down to 170 ft (52 m) below the San Andres (Figure 13). Mean cave height mimics this trend. The surficial layer (of karst and caves) that follows topography and shows great variation relative to a horizontal datum is called *epikarst* (Mylroie and Carew, 1993). Data flattened on the top of Cycle 2 show that cave feet and cave height increase for 20 to 40 ft (6 to 12 m) before they decrease, and a second level of caves (Cycle 1) is apparent at 150 ft (46 m) below the top of Cycle 2 (Figure 14).

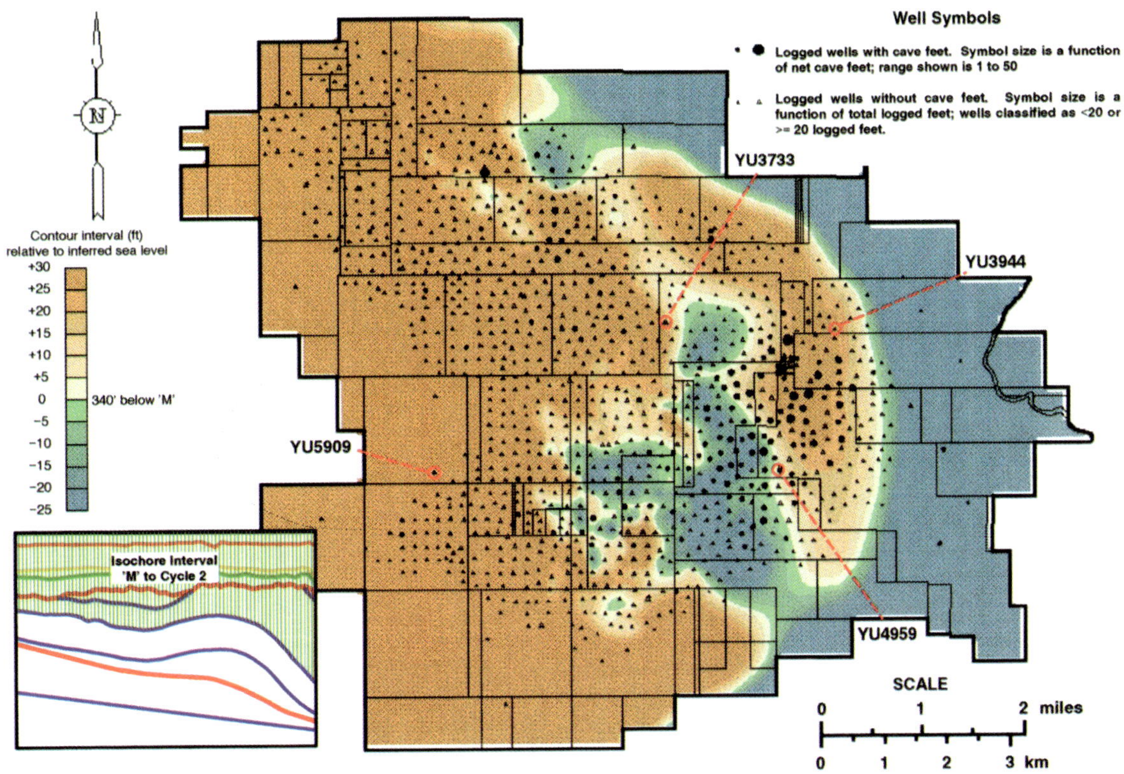

Figure 12. Cycle 1 and Cycle 2 cave and log feet superimposed on the island complex. Islands are determined by thickness trends of the stratigraphic interval from the Seven Rivers 'M' marker to the top of Cycle 2 (inset). The static shoreline is represented at 340 ft (104 m) below 'M,' and the color bar is related to this shoreline. Cave feet and log feet symbols are the same as Figure 11.

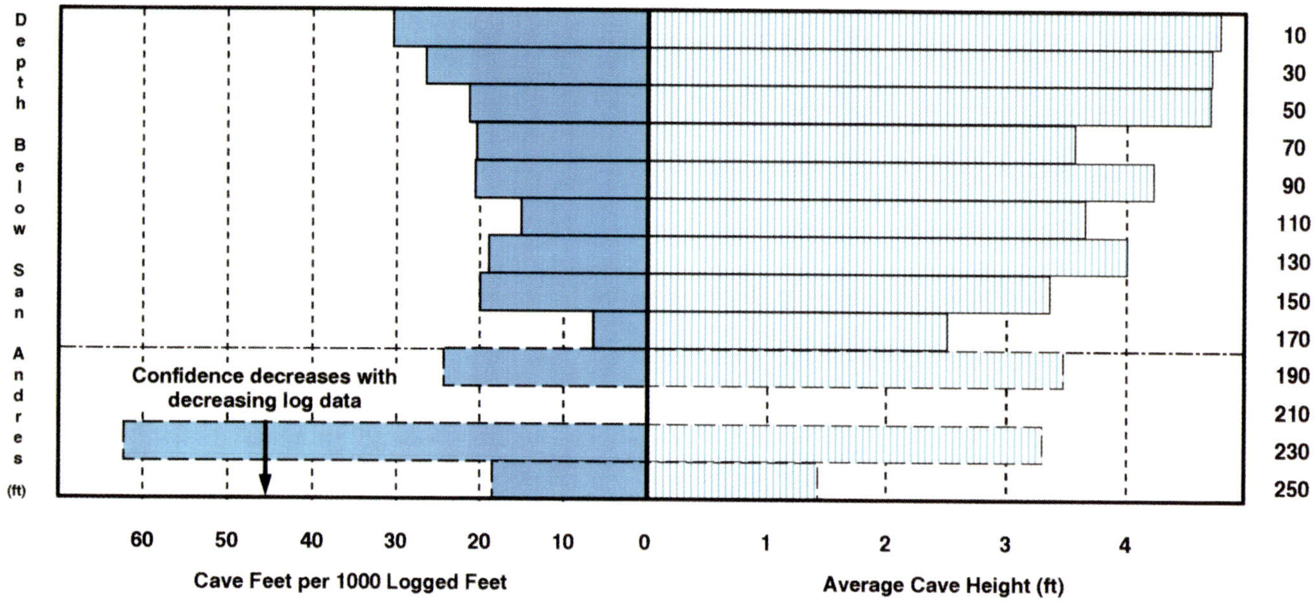

Figure 13. Bar plot of Cycle 3 normalized cave feet and cave height as a function of depth below the top of the San Andres. Vertical scale represents the midpoints of the bars.

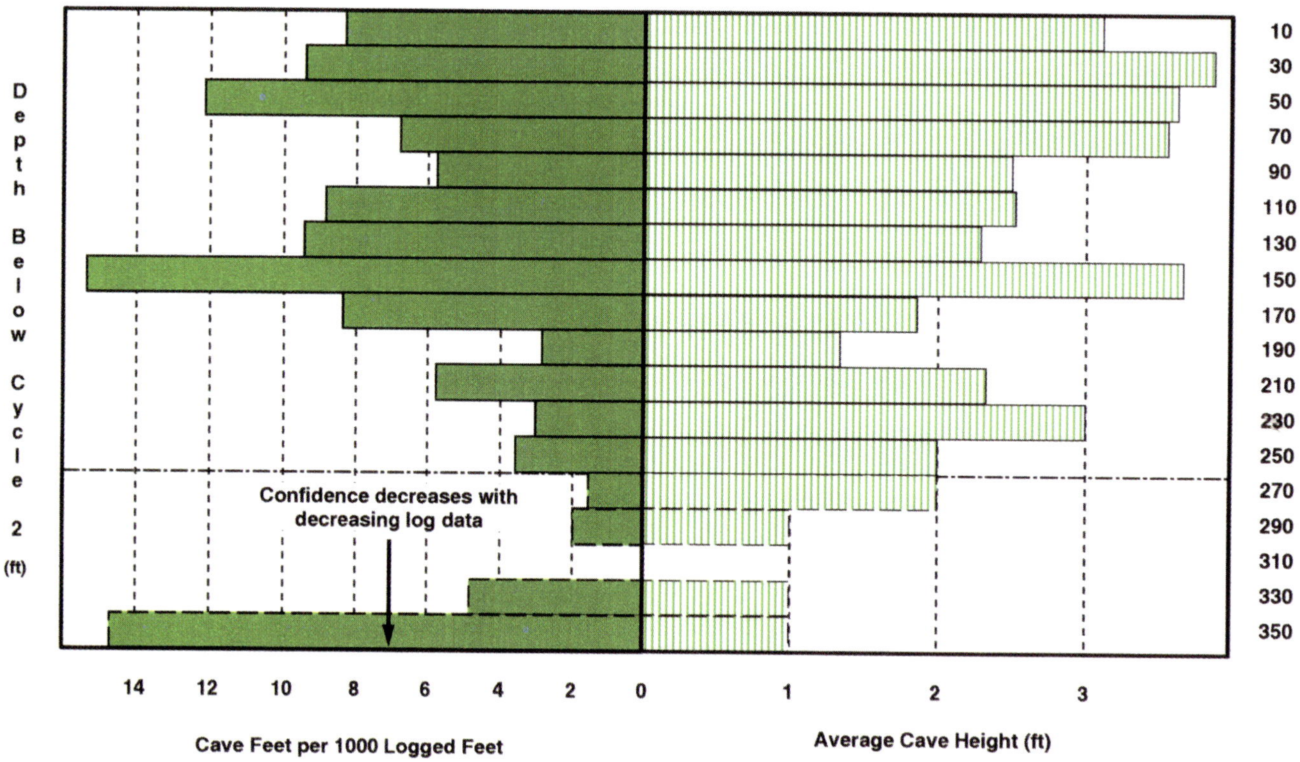

Figure 14. Bar plot of Cycle 1 and 2 normalized cave feet and cave height as a function of depth below Cycle 2. Vertical scale represents the midpoints of the bars.

Cave Porosity

In carbonate reservoirs, it is important to understand the distribution of porosity and permeability beneath an unconformity surface. The analysis requires a comparison of matrix and cave properties for a given volume of rock. In Yates field the Seven Rivers 'M' datum was used in order to analyze cave and matrix properties in terms of the paleogeographic distribution of caves, and each 20 ft slice below 'M' was treated as one fieldwide volume of rock. For the purpose of definition (fieldwide average in second parentheses):

- cave log porosity ($Ø_{cave\ log}$) = porosity measured by density logs in a cave (40.8%);

- matrix log porosity ($\O_{matrix\ log}$) = porosity measured by density logs in matrix (17.3%);
- cave porosity ($\O_{cave}$) = % of a volume that is cave pore space (0.7%);
- matrix porosity ($\O_{matrix}$) = % of a volume that is matrix pore space (17.0%);
- total porosity ($\O_{total}$) = sum of cave porosity and matrix porosity;
- pre-cave porosity ($\O_{pre\text{-}cave}$) = total porosity prior to cave formation (unknown).

The first two terms, $\O_{cave\ log}$ and $\O_{matrix\ log}$, are measurements of present-day properties, and the next two terms, $\O_{cave}$ and $\O_{matrix}$, are calculations made from the measured values. The stepwise procedure used to calculate $\O_{cave}$ and $\O_{matrix}$ follows, along with examples from two theoretical zones, one with abundant caves and one with few caves. In order to demonstrate the quantified impact on $\O_{total}$, $\O_{cave}$, and $\O_{matrix}$ the same numbers are used for $\O_{cave\ log}$, $\O_{matrix\ log}$, and log ft.

1. For each 20 ft stratigraphic slice, determine the:

	Abundant	Few
a. number of logged feet (*log ft*);	10,000	10,000
b. number of cave feet (*cave ft*);	1000	100
c. number of matrix feet (*matrix ft = log ft – cave ft*);	9000	9900
d. normalized cave feet (*cave ft*/1000 ft);	100	10
e. average $\O_{cave\ log}$; and	40%	40%
f. average $\O_{matrix\ log}$.	17%	17%

2. The following calculations assume uniform sampling and are essentially a weighted average of porosity, with weights being assigned as the ratio of cave feet or matrix feet to total logged feet in each slice:

	High	Low
a. $\O_{matrix\ log} \times matrix\ ft$	1530	1683
b. $\O_{cave\ log} \times cave\ ft$	400	40
c. $\O_{matrix} = 2a/log\ ft$	15.3%	16.8%
d. $\O_{cave} = 2b/log\ ft$	4.0%	0.4%
e. $\O_{total} = (2a + 2b)/log\ ft$	19.3%	17.2%

In the intervals with abundant caves $\O_{total}$ is 12% greater and $\O_{matrix}$ is 10% less than in intervals with few caves. In addition, $\O_{matrix}$ is 11% lower than $\O_{matrix\ log}$. In other words, given a fixed $\O_{matrix\ log}$, $\O_{matrix}$ is less and $\O_{total}$ is greater in cavernous intervals than in noncavernous intervals. The stability of this relationship can be tested with actual data from Yates field.

Density logs provide accurate porosity readings for caves and matrix. When the data are displayed relative to depth below 'M' (Figure 15), it is not obvious whether the measured values for average $\O_{matrix\ log}$ and average $\O_{cave\ log}$ are significantly influenced by variations in cave ft/1000 ft. This can be addressed by examining the calculated values for $\O_{matrix}$, $\O_{cave}$, and $\O_{total}$ for each major cycle (Table 1).

When the data are displayed relative to depth below 'M' (Figure 16), the ratio of $\O_{cave}/\O_{matrix}$ makes the zones with many caves at the top of Cycles 3 and 1 obvious. Curve shapes for $\O_{matrix}$ and $\O_{total}$ are remarkably sinusoidal through all three major cycles (Figure 17), and cycle boundaries occur at low to high porosity inflection

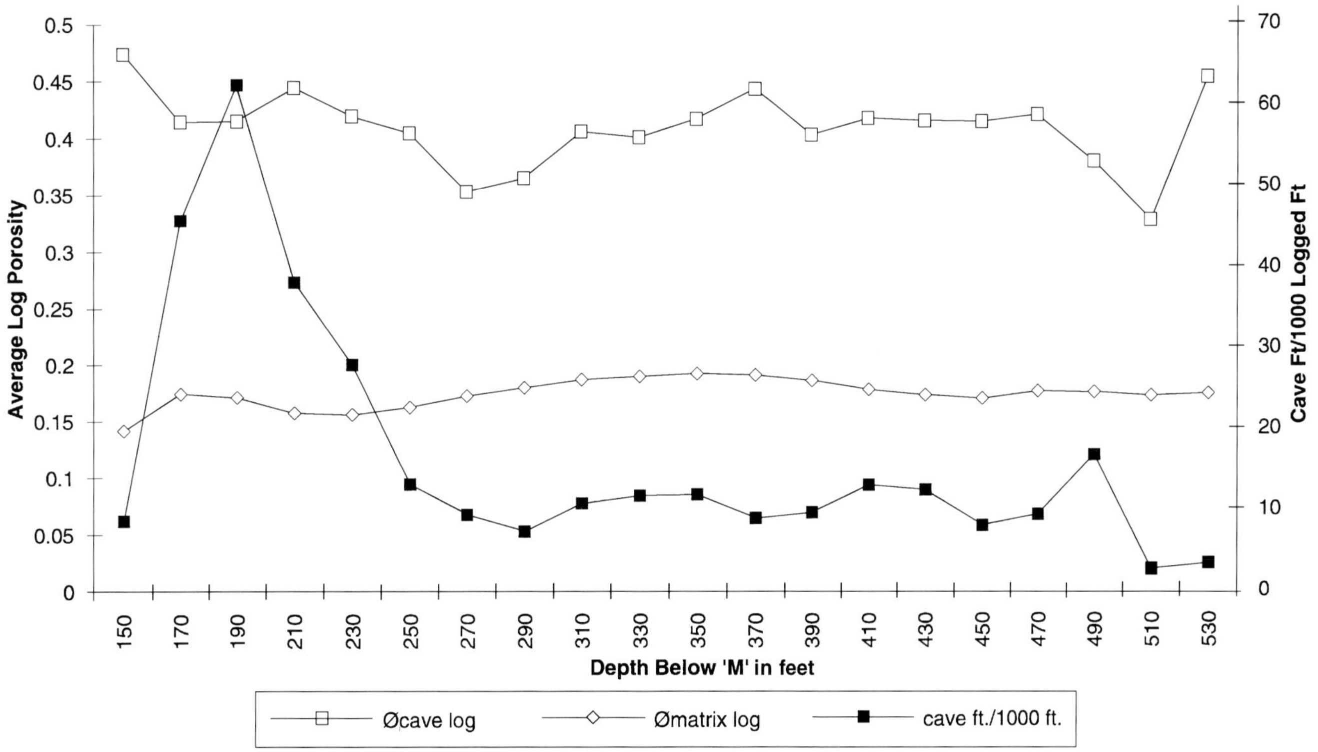

Figure 15. Average log porosity for caves ($\O_{cave\ log}$), average log porosity for matrix ($\O_{matrix\ log}$), and normalized cave feet (cave ft/1000 ft) as a function of depth below Seven Rivers 'M.' Note the separate Y-axis scale bars.

Table 1

Average	Full-Field	Cycle 1	Cycle 2	Cycle 3
cave ft/1000 ft	16.5	7.9	10.8	26.6
$Ø_{matrix\ log}$	17.3%	17.3%	18.2%	16.4%
$Ø_{cave\ log}$	40.8%	39.4%	41.4%	41.1%
$Ø_{matrix}$	17.0%	17.2%	18.0%	16.0%
$Ø_{cave}$	0.7%	0.3%	0.4%	1.1%
$Ø_{total}$	17.7%	17.5%	18.5%	17.1%

points on the curves. Low $Ø_{matrix}$ values represent poorer reservoir quality wackestone and packstone in the lower part of the cycle, which grade into better reservoir quality packstone and grainstone in the upper part of the cycle. Greater curve separation at cycle tops relative to cycle bases, especially in Cycle 1 and Cycle 3, indicates the position of cavernous porosity within each cycle.

The fieldwide average $Ø_{cave\ log}$ of 40.8% indicates that not all caves are totally open or tall enough for the logging tool to register 100% porosity. The average $Ø_{cave\ log}$ in each cycle does not vary significantly from the fieldwide average. This is not the case for $Ø_{matrix\ log}$. Unlike the theoretical example, where $Ø_{matrix\ log}$ was fixed at 17%, the Yates field data indicate that in the highly cavernous Cycle 3, measured $Ø_{matrix\ log}$ and calculated $Ø_{matrix}$ are lower than in the less cavernous Cycles 1 and 2. If $Ø_{pre\text{-}cave}$ represents porosity in the interval prior to cave formation, then the lower $Ø_{matrix}$ in Cycle 3 indicates that $Ø_{pre\text{-}cave}$ was reduced to the observed $Ø_{matrix}$ by the cave-forming process, or $Ø_{pre\text{-}cave}$ was lower, and therefore $Ø_{matrix}$ remains lower following cave formation, or both.

Although a precise $Ø_{pre\text{-}cave}$ cannot be determined, the rock types in each cycle are very similar fieldwide, and $Ø_{pre\text{-}cave}$ should also have been similar. If $Ø_{pre\text{-}cave}$ was the same for each cycle, then the decline in $Ø_{matrix}$ and $Ø_{total}$ in Cycle 3 relative to Cycles 1 and 2 would have been caused by the process of cave formation. At first glance this would seem to contradict the theoretical example, where $Ø_{total}$ increased in the zone of abundant caves. However, in that example, $Ø_{matrix\ log}$, not $Ø_{pre\ cave}$, was fixed at 17%, indicating that in zones with many caves, $Ø_{pre\text{-}cave}$ must have been greater, and was later reduced to 17% by the cave-forming process. Both examples illustrate that, given a similar $Ø_{pre\text{-}cave}$, the cave-forming process will reduce $Ø_{matrix}$ and $Ø_{total}$.

Total porosity should provide a close estimate of $Ø_{pre\text{-}cave}$ in zones with few caves, such as Cycles 1 and 2, because porosity will not have been altered significantly by the cave-forming process. Since $Ø_{total}$ in Cycle 3 is lower than in Cycles 1 and 2, $Ø_{pre\text{-}cave}$ of Cycle 3 was probably lower to begin with. However, the percent reduction in Cycle 3 $Ø_{total}$ compared to Cycles 1 and 2 (2% and 8%, respectively) is less than the percent reduction in Cycle 3 $Ø_{matrix}$ compared to Cycles 1 and 2 (7% and 11%, respectively). This indicates that the decrease in Cycle 3 $Ø_{matrix}$ and $Ø_{total}$ relative to Cycles 1 and 2 is not caused solely by the cave-forming process, but results from a combination of $Ø_{matrix}$ reduction due to the cave-forming process and a lower $Ø_{pre\text{-}cave}$.

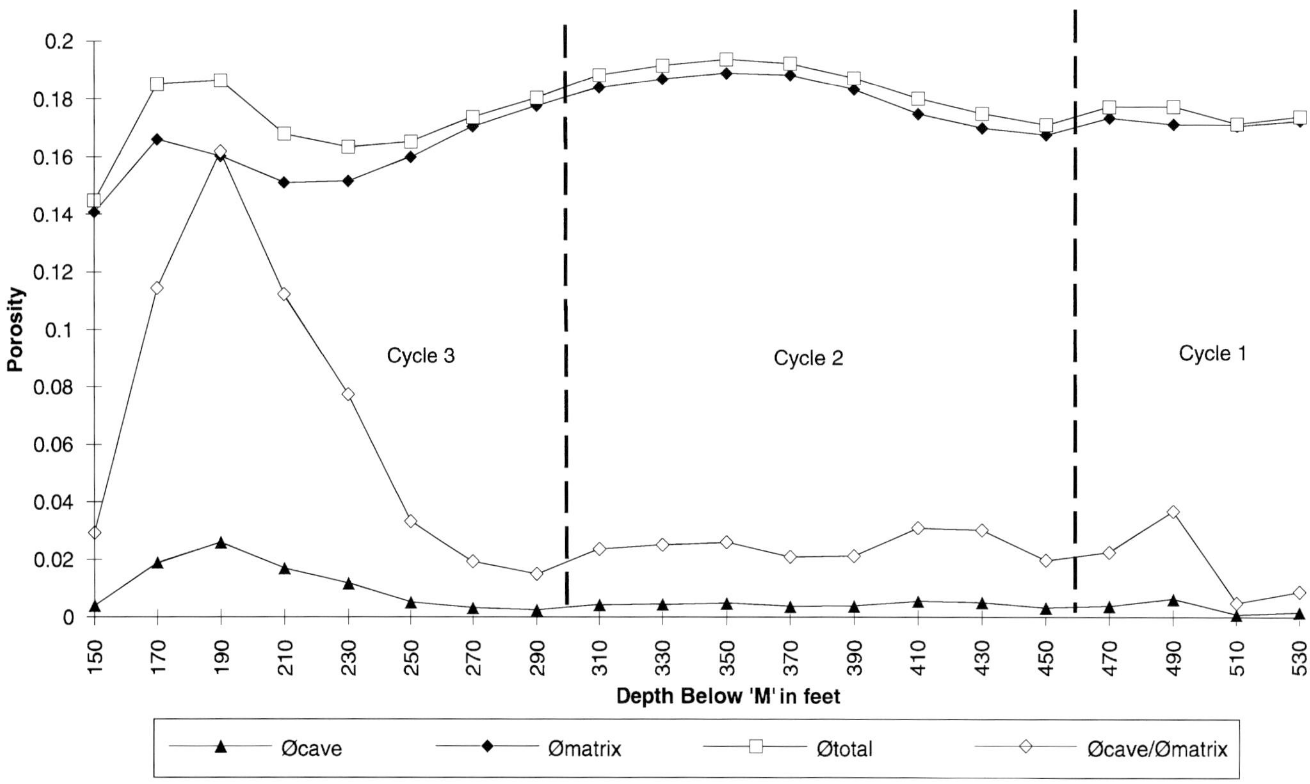

Figure 16. Cave porosity ($Ø_{cave}$), matrix porosity ($Ø_{matrix}$), total porosity ($Ø_{total}$), and $Ø_{cave}/Ø_{matrix}$ as a function of depth below Seven Rivers 'M.' Positions of major cycles are illustrated.

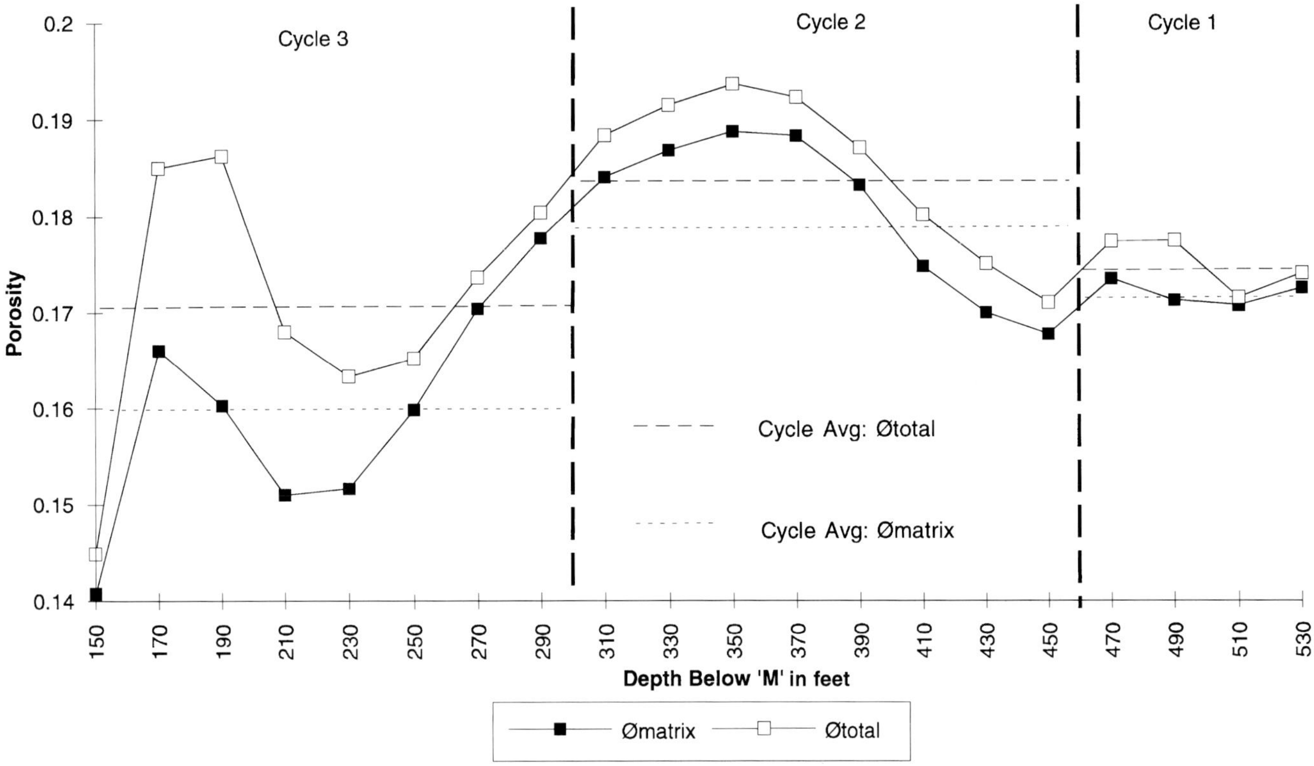

Figure 17. Matrix porosity ($Ø_{matrix}$) and total porosity ($Ø_{total}$) as a function of depth below Seven Rivers 'M.' Average values for $Ø_{matrix}$ and $Ø_{total}$ are shown for each major cycle.

Because Yates field is relatively shallow, most caves have not undergone mechanical collapse. Therefore, in contrast to the reduction in porosity observed in zones with many caves, the extensive network of open vertical caves and solution-enhanced fractures increases total vertical permeability of the system.

Cave Formation Model

Because open caves strongly influence fluid flow and porosity distribution in Yates field, applying an appropriate model for cave formation and understanding the temporal and spatial distribution of cave and karst features are important in order to characterize the reservoir. Several models could be used to explain the formation of the primary cave system in Yates field, including: (1) the island hydrologic model (Craig, 1988), (2) the "sulfuric acid oil-field karst" model (Hill, 1992), and (3) meteoric-marine mixing-zone dissolution resulting from the intersection of an active Permian regional meteoric aquifer with the coast at the edge of the platform (Moore, 1989).

The sulphuric acid model is not plausible for Yates field because it does not explain the paleodistribution of normalized cave feet and mean cave height, the surface karst features observed in core that are interpreted to be related to subaerial exposure, or the present-day normal distribution of cave feet. The single regional aquifer model (Moore, 1989) is attractive in terms of diagenetic drive, and may have enhanced the size of caves in Yates field. However, three factors indicate that this process did not play a large role in primary cave formation: (1) the relationship of cave distribution to the island trends, (2) the observed cyclicity of normalized cave feet, and (3) the limited number of cave feet to the west within each cycle, which was the source area for the freshwater regional aquifer.

Craig (1988) applied the well-known island hydrologic model (Thrailkill, 1968; Vacher, 1978; James and Choquette, 1984) to interpret the distribution of caves associated with the San Andres island complex. According to that model, varying patterns of water flow produce solution-enhanced features as a function of position relative to the meteoric water table and paleoshoreline. Consequently, caves tend to be elongated along directions of high matrix flow, and are concentrated at fresh water/sea water mixing-zone interfaces (Figure 18). In the Bahamas, "large dissolution voids, called flank margin caves, formed preferentially in the discharging margin of the fresh-water lens, primarily as a result of fresh water/salt water mixing corrosion at this site of high discharge. Inland from the lens margin, at the top of the lens where vadose and phreatic fresh waters mix, smaller dissolution voids may develop." (Mylroie and Carew, 1993). Like the Bahamas, the formation of caves and solution features within the San Andres islands was most aggressive at the discharging island margins and inland from the island margins at the vadose/phreatic mixing zone. In addition, cave formation in the vadose and phreatic zones was influenced by the movement of fresh water through vertical joint conduits with high permeability, forming "vadose shafts" (term from Mylroie and

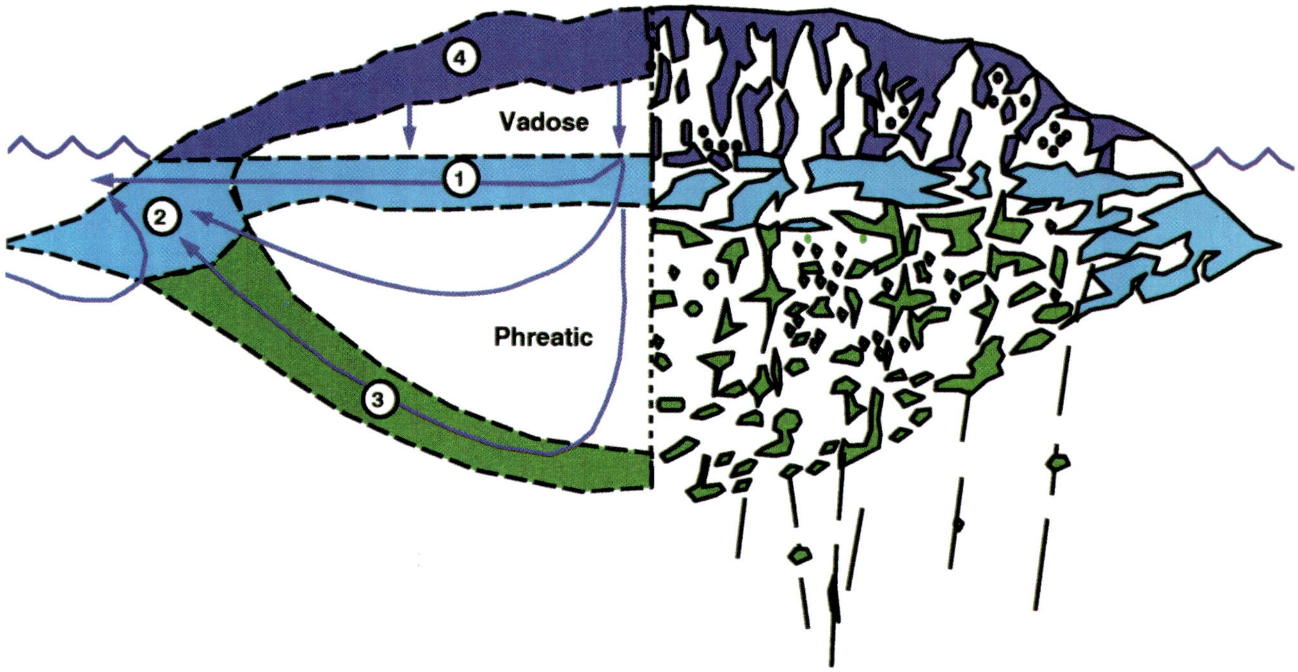

Figure 18. The island hydrologic model as applied to cave formation in Yates field. Cave formation is greatest at sites of mixed waters (1, 2, and 3) and in association with soil processes (4). The position of the vadose and phreatic zones and sea level for a static model are indicated, and probable water flow lines within a limestone island are drawn. Caves tend to be elongated along principal directions of flow (including joints and fractures), and concentrated along mixed-water interfaces. (Modified from Craig, 1988). Reprinted by permission of Springer-Verlag.

Carew, 1993). The resulting cave geometries are skewed toward vertical, and are oriented along preferred lineament trends.

With a relative fall in sea level at the end of each major cycle, the combined influence of mixed waters and joint control was imprinted on successively deeper rocks, resulting in a present-day network of interconnected caves that are narrow in one direction (1 to 3 ft; 0.3 to 1.0 m), elongated in a perpendicular direction (tens to hundreds of feet), and not particularly tall. A bar plot of cave heights (Figure 19) shows that nearly 50% of the caves have a mean height between 1 and 3 ft (0.3 to 1.0 m), and an additional 30% have a mean height between 3 and 5 ft (0.9 to 1.5 m), indicating that this is not a large-room cavern system. The tallest caves are found in Cycle 3. The distribution of caves and surface karst may, in part, reflect surface drainage patterns and the mechanisms by which fresh water was locally concentrated. Low-permeability sediments in the uppermost San Andres, particularly in the western portion of the field, may have played a significant role in the drainage of fresh water on exposed land surfaces. The exposed mudstone covering the western portion of the field would have had greater runoff potential, whereas surface water on the exposed packstone and grainstone shoal-island complex to the east would percolate downward into porous and permeable sediments, thus facilitating the development of karst.

By extracting only the Cycle 3 normalized cave feet and cave height data from the Figure 10 bar plot, quantification of the vertical aspect of the island hydrologic model is possible (Figure 20). Normalized cave feet increase from 22 to 66 cave ft/1000 ft and mean cave height increases from 2.5 ft to 5.3 ft (0.8 to 1.6 m) in the upper 60 ft (18 m) of the island. Normalized cave feet and cave heights decrease predictably from 210 to 430 ft (64 to 131 m) below 'M.' According to these data, the greatest number of cave feet and tallest caves are at the discharging margin of the freshwater lens and at the top of the lens where vadose and phreatic fresh waters mix. It is possible that caves in the meteoric zone are taller, but since they are very narrow (1 to 2 ft; 0.3 to 0.6 m), and follow early joints that are not always vertical (70–90°), they appear on logs as only 1 to 4 ft (0.3 to 1.2 m) tall. In other words, a vertical wellbore in the vadose zone may encounter only part of a dipping, solution-enhanced joint, whereas a wellbore in the vadose/phreatic mixing zone encounters the full extent of a 4 to 5 ft (1.2 to 1.5 m) room.

Importantly, the bar plots represent the dynamic, compound result of a falling sea level, whereas the island model is a static snapshot in time. Therefore, the high normalized cave feet count and tall caves observed above the 205 ft (63 m) sea level are a composite result of the shoreline and associated mixing zone moving through the island complex as relative sea level fell. The same process and similar results can be derived for the Cycle 2 island lens, although the number of normalized cave feet and cave heights are lower, indicating less-intense karst events.

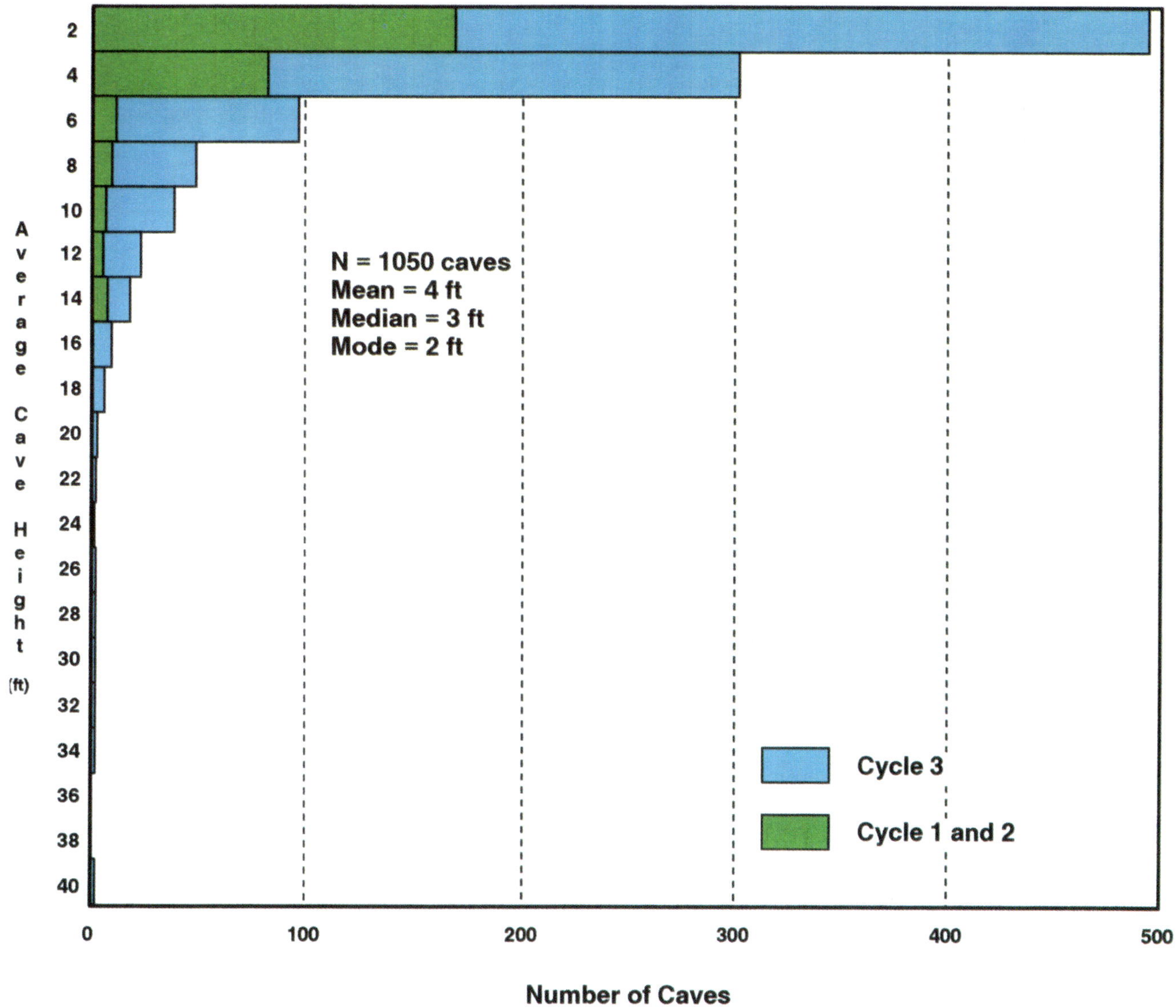

Figure 19. Bar plot of number of caves as a function of cave height. Cycle 1 and 2 are shown in green and Cycle 3 is shown in blue. Nearly 50% of the caves have a mean height between 1 and 3 ft (0.3 to 0.9 m), and an additional 30% have a mean height between 3 and 5 ft (0.9 to 1.5 m). This plot emphasizes that Yates caverns are probably not formed as tall "rooms," but more commonly as short, narrow caves with spatial geometries controlled by vertical joints. Vertical scale represents the midpoints of the bars.

SEQUENCE-STRATIGRAPHIC FRAMEWORK AND POSITION OF CAVES

Knowing the relationship between lithostratigraphy (lithofacies) and sequence stratigraphy (time) is important in order to understand the geometry of unconformity surfaces and related cave distribution in Yates field. An integrated analysis of logs, cores, facies cyclicity, diagenesis (caves and karst), and outcrop analogs was necessary in order to develop a field-wide, 3-D sequence-stratigraphic framework that could predict the distribution of unconformity-related karst features.

Cycle Stratigraphy

The major cycles, along with their component minor cycles, are illustrated on a log from the west side of Yates field (Figure 21). Minor cycles began with lagoonal shales that shoaled upward into subtidal dolomitic wackestone. These minor cycles can be bundled into major cycles (high-frequency sequences), with the thickest and most areally extensive shales and shale-carbonate minor cycles present at the base of a major cycle, and thinner, more laterally discontinuous shales and minor cycles present near the top. This overall thinning-upward stacking pattern indicates a decrease in accommodation toward the top of each major cycle, as all of the available space was filled with

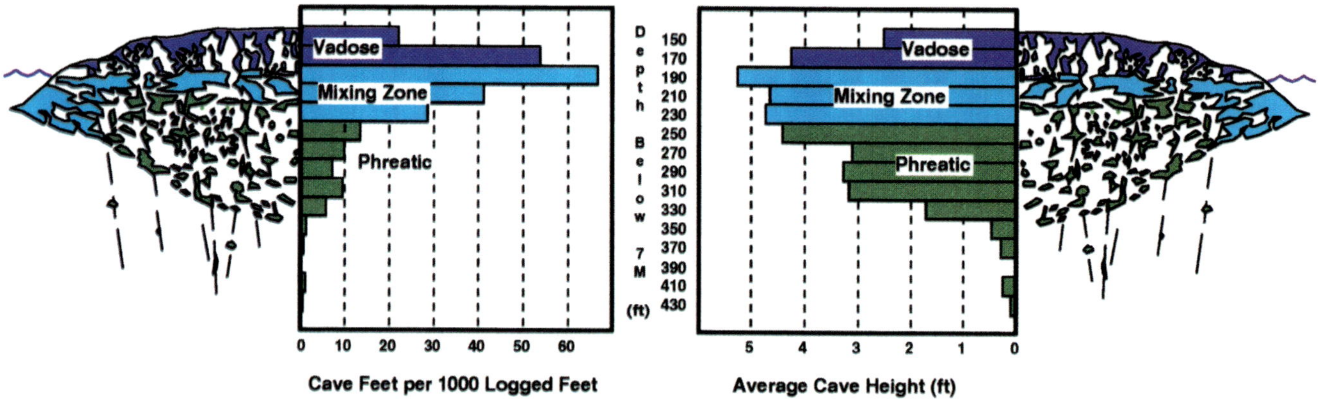

Figure 20. Bar plots of normalized cave feet and cave height, as a function of depth below Seven Rivers 'M' for Cycle 3 only. Vertical scale represents the midpoints of the bars. The plots are hung next to the island hydrologic model (modified from Craig, 1988) to illustrate the position of the vadose zone (dark blue), the vadose-phreatic mixing zone (light blue), the phreatic zone (green), and the resultant dynamic normalized cave feet and cave height distribution in Cycle 3 islands.

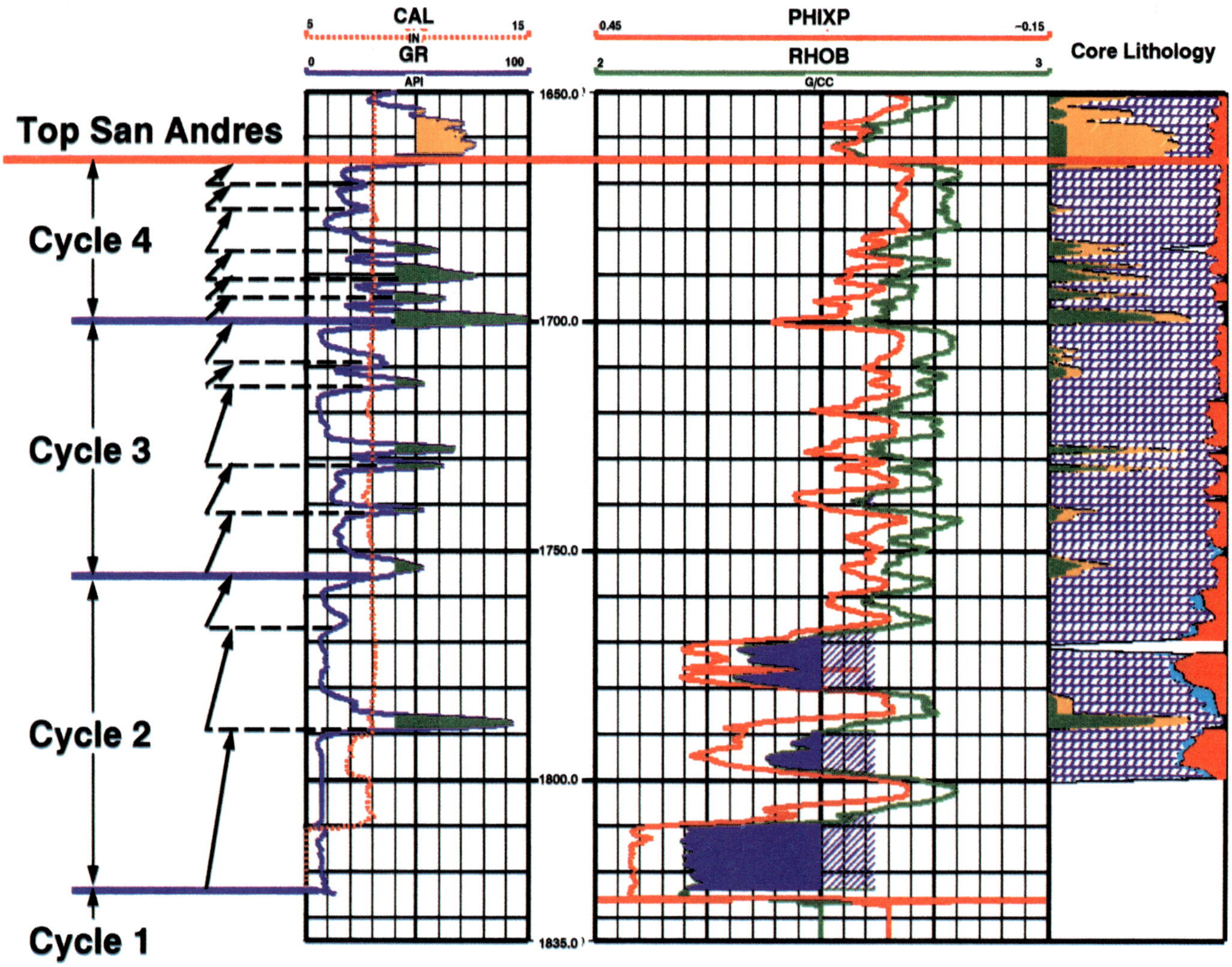

Figure 21. Cycle stratigraphy in Yates field illustrated on a west-side log. Major cycles (cycle sets) and component minor cycles are shown for Cycles 2 to 4. The GR log is shaded for shales and silts, and core lithology is shown on the far right.

sediment. In other words, the major subaerial-exposure events, both in terms of duration and relative sea level fall, should occur at the end, or top, of each major cycle. If karst and cavern systems were to form in the east-side islands, they would develop at these major cycle boundaries.

Sequence Stratigraphy

When the San Andres lithostratigraphy is superimposed on the sequence stratigraphy (Figures 4A and 22A), the relationship between depositional time lines and major lithofacies is apparent. Sequence 1 is interpreted as a ramp complex that prograded from west to east, with high porosity (20–25%) in the upper part of the sequence (Figure 4B). Sequence 2 began with a major transgression, followed by dominantly aggradational Cycles 1 and 2 and slightly more progradational Cycles 3 and 4. A major erosional unconformity, marked by significant formation of karst-related features, defines the top of the San Andres. Very little of Cycle 4 is preserved in the field area owing to erosion following deposition of the San Andres.

The position of caves along the same line of section (Figure 22B) shows the following top-to-bottom progression: (1) a lens of caves related to the unconformity

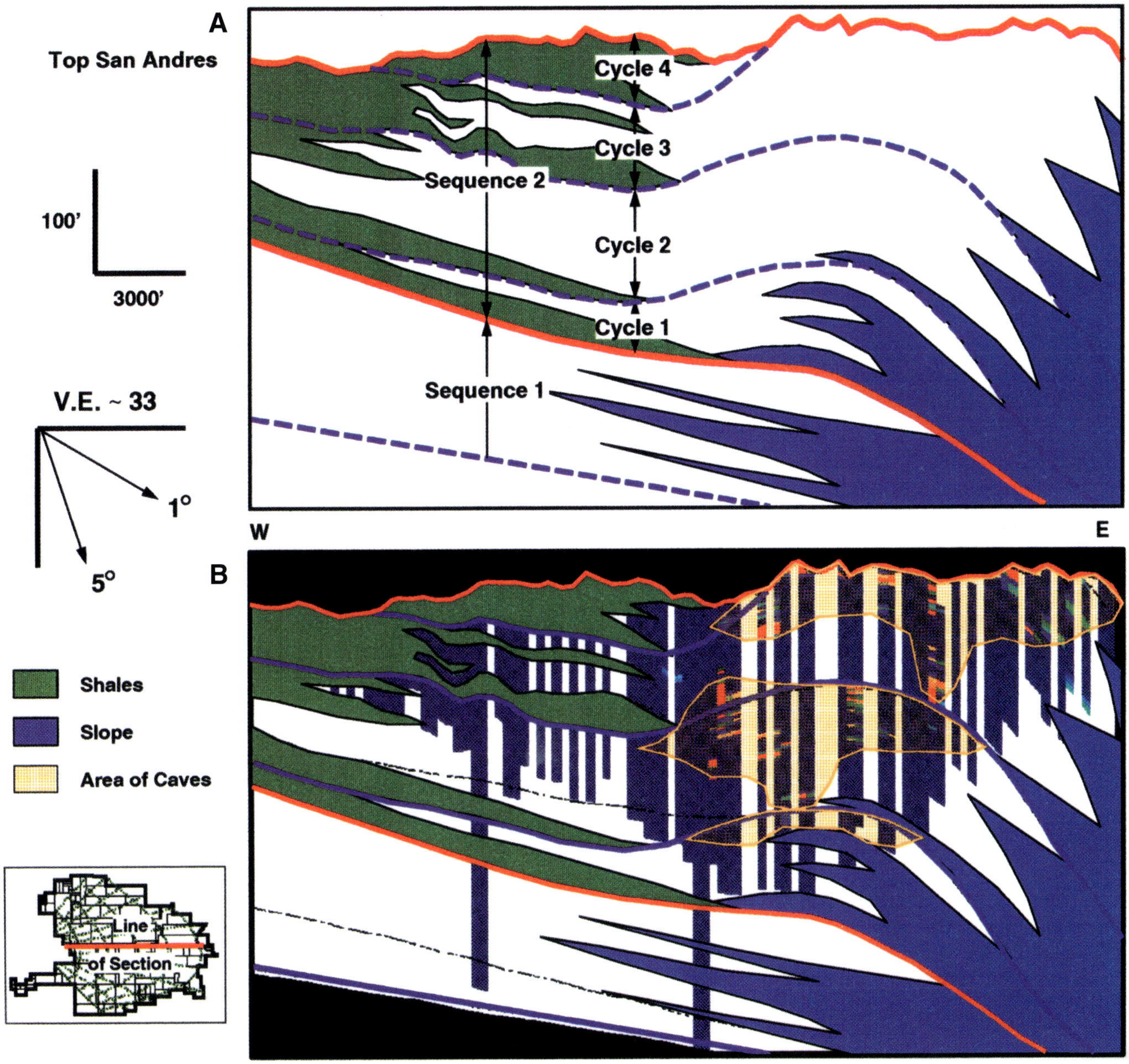

Figure 22. West-to-east stratigraphic sections, with a Seven Rivers 'M' datum, located in the same position as Figure 4. (A) Lithofacies superimposed on the sequence-stratigraphic interpretation, with sequence boundaries shown in red and major cycle boundaries in dashed blue. (B) Position of caves (hot colors with yellow overlay) in the sequence-stratigraphic framework.

at the top of the San Andres, (2) a relative paucity of caves, (3) a second lens of caves related to Cycle 2, (4) a relative paucity of caves, and (5) a few caves related to Cycle 1. Three-dimensional surfaces of Cycle 2 islands and islands following San Andres deposition (Figure 23A and 23B; compare to Figures 11 and 12) illustrate the shape and relief of the island complexes. The island hydrologic model is superimposed on the sequence-stratigraphic interpretation to illustrate the probable geometry and time-stratigraphic position of cave-forming lenses (Figure 23C). The section is reduced to a vertical exaggeration of 8× (Figure 23D) in order to provide a more accurate geometric visualization.

In essence, packstone and grainstone shoals aggraded and prograded from west to east. Caves, formed on these exposed island shoals, illustrate the same west-to-east distribution. The 3-D geographic distribution of caves in each cycle illustrates nicely the three major karst-forming events within the San Andres Formation (Figure 24) and how the position of each event aggraded and prograded to the east.

SUMMARY

An important economic issue in carbonate reservoirs is the effect of subaerial exposure on porosity and permeability. When the data in Yates field are examined by major stratigraphic cycle, the zones with abundant caves in Cycle 3 show lower average matrix porosity and lower total porosity than the zones with few caves in Cycles 1 and 2. This results from a combination of matrix porosity reduction due to the cave-forming process, and a lower pre-cave matrix porosity. Because Yates field is relatively shallow, most caves have not undergone mechanical collapse. Therefore, in contrast to the reduction in porosity observed in zones with many caves, the extensive network of open vertical caves and solution-enhanced fractures increases total vertical permeability of the system.

Understanding the 3-D distribution of caves and applying an accurate model for cave formation provides a significant tool for improved prediction of cave locations. The island hydrologic model describes the geometric and geographic distribution of caves and surface karst better than the "sulfuric acid oil-field karst" model or the meteoric-marine mixing-zone dissolution model. There are two possible explanations for the observed distribution of caves using the island hydrologic model: (1) multiple subaerial exposure events within the San Andres; or (2) a multistage fall in relative sea level following San Andres deposition whereby mean relative sea level fell to a position 210 ft (64 m) below 'M' and stood still, fell again to 350 ft (107 m) below 'M,' and again to 490 ft (149 m) below 'M.' Because caves are found up to 500 ft (152 m) below the top of the San Andres, a significant fall in relative sea level combined with substantial flow through the early joint system would be required if all of the caves observed were formed by exposure following San Andres deposition. If the model of multistage sea level fall were correct, the landmass of the island complex would grow as sea level fell. The resulting area of caves should get progressively larger with depth as the shoreline stepped out in all directions away from a central island core. This distribution is not observed in Yates field. In addition, a multistage, relative fall in sea level of 500 ft (152 m) in the shallow, intracratonic Midland basin is unlikely.

Although cave size could have been enhanced through time by both sulfuric acid oil-field karst processes and regional meteoric-marine mixing-zone dissolution, the most reasonable explanation for early cave formation is one of repeated subaerial exposure. A few critical observations and analyses were combined to arrive at this conclusion.

1. An understanding of outcrop geometries and cyclicity is necessary for accurate stratigraphic correlation. Without an outcrop analog, layer-cake correlation could easily be justified. East-side grainstone shoals had positive relief relative to coeval west-side lagoons, and could have formed islands while the San Andres was being deposited.
2. Stacking pattern analysis of shale-carbonate cycles shows an overall thinning-upward pattern within each of Cycles 2, 3, and 4, indicating decreased accommodation and a trend toward possible subaerial exposure at the end of each cycle, as all depositional space was filled. The most extensive and thickest shales in Yates field were preserved during the transgressions following Cycles 1, 2, and 3.
3. Cycles 1, 2, and 3 are capped by unconformities with associated longer-term subaerial exposure events, as evidenced by normalized cave feet counts and cave heights increasing toward the top of each major cycle. This supports meteoric diagenesis related to subaerial exposure as a cave-forming mechanism. The juxtaposition of caves and island trends in each major cycle provides additional evidence.
4. The 3-D distribution of cave feet, from deep in the west to shallow in the east, cannot be explained adequately by one exposure event following San Andres deposition, but can be explained by at least two, and probably three, separate island-exposure events during San Andres deposition.

In essence, separate cave lenses were formed as a result of meteoric processes acting on subaerially exposed island complexes, and the island complexes and associated cave lenses aggraded and prograded from west to east. Therefore, deeper drilling should encounter additional caves to the west of known shallower caves in the east.

Author's Note: Between the first writing and final edit of this manuscript, more than 40 wells have been deepened in Yates field. Of those, nearly 50% encountered a cave, many to the west of known shallower caves in the east.

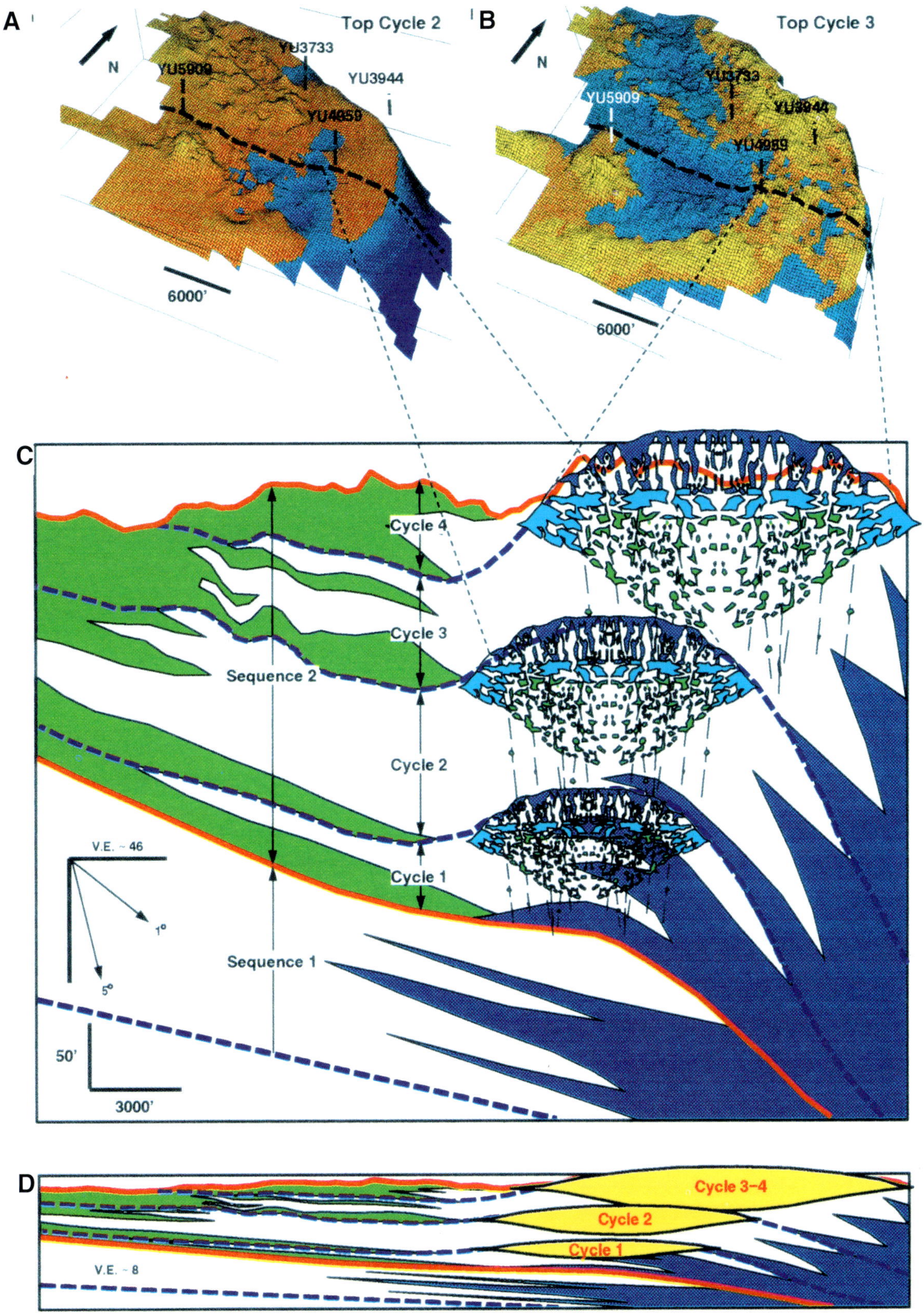

Figure 23. (A) 3-D surface of Cycle 2 islands with a shoreline 340 ft below 'M,' and (B) 3-D surface of post-San Andres islands with a shoreline 205 ft below 'M.' Line of section used in stratigraphic sections is illustrated in black dashed pattern. (C) Position of Cycle 1, Cycle 2, and Cycle 3 island lenses within the sequence-stratigraphic framework. (D) The same section condensed to a vertical exaggeration of 8×, with the cave lenses illustrated in yellow.

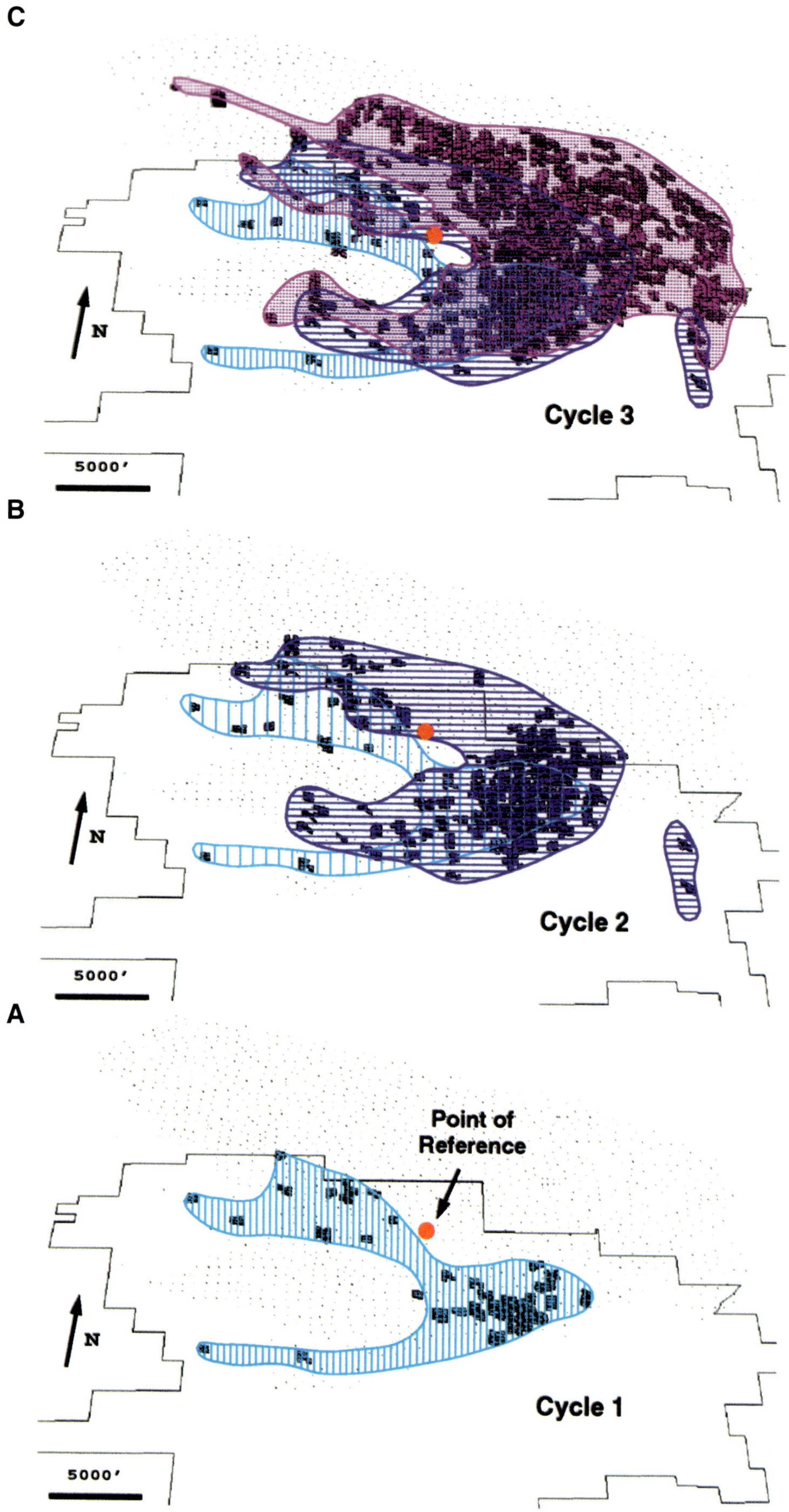

Figure 24. Position of caves in the 3-D model. Color overlays illustrate the minimal extent of islands and the red dot is a geographic reference point. Note the shift from west to east of islands and associated cave lenses.

ACKNOWLEDGMENTS

The authors thank Marathon Oil Company for permission to publish this paper. We thank G. M. Grammer, C. A. Hill, P. M. Harris, and L. Brinton for their careful technical reviews and constructive suggestions. Denise Mruk and Matt Parsley (Marathon Oil Company, Midland) contributed to the interpretation of the stratigraphic framework and cave/joint relationships. Don Caldwell and Mike Uland at Marathon's Petroleum Technology Center assisted with early statistical analyses, and in construction of the 3-D model, respectively. We extend our appreciation to Dexter Craig for numerous stimulating discussions during the course of our involvement in Yates field.

REFERENCES CITED

Adams, J. E., 1930, Origin of oil and its reservoir in Yates pool, Pecos County, Texas: AAPG Bulletin, v. 14, p. 705–717.

Choquette, P. W., and N.P. James, 1988, Introduction, *in* Choquette, P. W., and James, N. P., eds., Paleokarst: Springer-Verlag, New York, p. 1–21.

Craig, D. H., 1988, Caves and other features of Permian karst in San Andres dolomite, Yates field reservoir, west Texas, *in* Choquette, P. W., and N.P. James, eds., Paleokarst: New York, Springer-Verlag, p. 342–363.

Craig, D. H., 1990, Yates and other Guadalupian (Kazanian) oil fields, U. S. Permian Basin, *in* Brooks, J., ed., Classic Petroleum Provinces: Geological Society of London Special Publication No. 50, p. 249–263.

Galloway, W. E., T.W. Ewing, C.M. Garret, N. Tyler, and D.G. Bebout, 1983, Yates area, *in* Atlas of Major Texas Oil Reservoirs: University of Texas—Austin, Bureau of Economic Geology, p. 100–103.

Gester, G. C., and H.J. Hawley, 1929, Yates field, Pecos County, Texas, *in* Structure of typical American oil fields: AAPG Bulletin, v. 2, p. 480–499.

Hennen, R. V., and R.J. Metcalf, 1929, Yates oil pool, Pecos County, Texas: AAPG Bulletin, v. 13, p. 1509–1556.

Hill, C. A., 1992, Sulfuric acid oil-field karst, *in* Candelaria, M. R., and Read, C. L., eds., Paleokarst, karst related diagenesis and reservoir development: examples from Ordovician–Devonian age strata of west Texas and the mid-continent: SEPM Publication 92–93, p. 192–194.

James, N. P., and P.W. Choquette, 1984, Diagenesis 9. Limestones—the meteoric diagenetic environment: Geoscience Canada, v. 11, p. 161–194.

Kerans, C., and M.J. Parsley, 1986, Depositional facies and porosity evolution in a karst-modified San Andres reservoir—Taylor Link West San Andres, Pecos County, Texas, *in* Bebout, D. C., and P.M. Harris, eds., Hydrocarbon reservoir studies, San Andres/Grayburg Formations, Permian Basin: Permian Basin Section, SEPM Publication 86-26, p. 133–134.

Kerans, C., F. J. Lucia, and R. K. Senger, 1994, Integrated characterization of carbonate ramp reservoirs using Permian San Andres Formation outcrop analogs: AAPG Bulletin, v. 78, p. 181–216.

Moore, C. H., 1989, Carbonate diagenesis and porosity: Developments in Sedimentology 46, Amsterdam, Elsevier, 338 p.

Mylroie, J. E., and Carew, J. L., 1993, Karst development in carbonate islands, *in* Unconformities and Porosity Development in Carbonate Strata: Recognition, Controls and Predictive Strategies: AAPG Hedberg Research Conference Abstracts.

Shirley, K., 1987, Colorful history, odd geology—Yates field celebrates 60 years: AAPG Explorer, v. 8, no. 1, p. 4–5.

Sonnenfeld, M. D., 1991, Anatomy of offlap: upper San Andres sequence (Permian, Guadalupian), Last Chance Canyon, Guadalupe Mountains, New Mexico: Master's thesis, Colorado School of Mines, 375 p.

Thrailkill, J., 1968, Chemical and hydrologic factors in the excavation of limestone caves: Geological Society of America Bulletin, v. 79, p. 19–45.

Vacher, H. L., 1978, Hydrology of Bermuda—significance of an across-the-island variation in permeability: Journal of Hydrology, v. 39, p. 207–226.

Chapter 12

Identification of Subaerial Exposure Surfaces and Porosity Preservation in Pennsylvanian and Lower Permian Shelf Limestones, Eastern Central Basin Platform, Texas

J. A. D. Dickson
Cambridge University
Cambridge, U.K.

Arthur H. Saller
Unocal Energy Resources
Brea, California, U.S.A.

ABSTRACT

The southwest Andrews area on the eastern side of the Central Basin platform (west Texas) contains cyclic Pennsylvanian and Lower Permian shelfal limestones. Limestones were deposited in shallow marine environments during numerous highstands of sea level, but most cycles are bounded by subaerial exposure surfaces. Reservoir porosity is developed in only 10–45% of those depositional cycles in any given well. The purposes of this paper are to determine: (1) features useful for identifying subaerial exposure surfaces, (2) factors that affect stable-isotope profiles around subaerial exposure surfaces, and (3) circumstances critical to porosity preservation in subaerially exposed limestones.

1. Features commonly present at or below subaerial exposure surfaces include an abrupt change in depositional lithology, caliche crusts, micritic rhizoliths precipitated around roots, soil-related fractures, breccias, and mottling associated with plant roots. Rhizoliths, caliche crusts, and breccias have developed best in wackestones and packstones. Mottling associated with plant roots is distinct in grainstones and was caused by heterogeneous dissolution and cementation.

2. The stable isotope signature most characteristic of subaerial exposure is abrupt decreases in $\delta^{13}C$ of the carbonate immediately below subaerial exposure surfaces. This signature is displayed best in cycles with: (a) wackestones/packstones at the top, (b) moderate duration of subaerial exposure, (c) limited overprinting by later meteoric diagenesis, (d) little erosion during the subsequent transgression, and (e) negligible effects of late cements on the isotopic composition of the bulk rock.

3. Ultimate porosity of these subsurface limestones was largely determined by amounts of burial compaction and burial cement, rather than by the amount of porosity created during subaerial exposure. Most of the limestones were porous during shallow burial. Currently porous grainstones have significant early intergranular calcite cement, limited compaction-related features, and minor burial cements. In contrast, nonporous wackestones, packstones, and grainstones had much primary and secondary porosity occluded by compaction and/or cements derived from compaction-related pressure solution. Early marine and freshwater lithification apparently left currently porous strata with a rigid framework that resisted compaction during deeper burial. Early lithification and associated porosity are most common in the upper part of relatively thick depositional cycles.

INTRODUCTION

Subaerial exposure and freshwater diagenesis are partially to largely responsible for porosity in many carbonate reservoirs throughout the world (e.g., Arun field in the Miocene of Indonesia; Jordan and Abdullah, 1988, Bu Hasa field in the Cretaceous of Abu Dhabi; Harris et al., 1984, and Yates field in the middle Permian of west Texas; Craig, 1988). Lime sediments deposited in shallow marine waters require only a small fall in sea level for them to be subaerially exposed resulting in infiltration by meteoric waters. Meteoric waters, at least at recharge, are undersaturated with respect to carbonate and cause dissolution along the first parts of their flow paths producing new and commonly larger pores. However, pores in these sediments are also occluded by calcite cements during meteoric diagenesis. It is important, therefore, to know whether meteoric diagenesis of shallow marine sediments can create pores in areas large enough to become reservoirs and/or can create features that will prevent porosity loss during burial.

Sequence stratigraphy has provided better tools to predict which shallow marine sequences were subaerially exposed (Tucker, 1993), but prediction of porosity in subaerially exposed carbonates remains problematic. The purposes of this paper are to: (1) describe features used to identify subaerial exposure surfaces, (2) determine factors that affect stable carbon- and oxygen-isotope profiles near subaerial exposure surfaces, and (3) determine circumstances critical to porosity preservation in subaerially exposed limestones deposited during high-amplitude, high-frequency sea level fluctuations. The southwest Andrews area on the eastern side of the Central Basin platform in west Texas (Figure 1) is a good location to study porosity associated with subaerial exposure, because limestones deposited there were first repeatedly exposed and subjected to freshwater diagenesis, then were buried to depths of 2500–3000 m (8500–10,000 ft).

GENERAL SETTING

The southwest Andrews area (Parker, Andrews, and Deep Rock oil fields) is located on the eastern side of the Central Basin platform in Andrews County, Texas (Figure 1). The eastern part of the study area contains structural closures. Stratigraphic traps occur in the western part of the area where porous Pennsylvanian and lower Permian limestones pinch out updip (westward) into tight limestones and shales. Much of the existing structure in the upper Pennsylvanian and lower Permian interval is the result of compactional drape over tectonic structures formed during the Mississippian and early Pennsylvanian. The gross producing interval is approximately 275–365 m (900–1200 ft) thick at depths of 2500 to 3000 m (8500–10,000 ft). Oil is produced from numerous porosity intervals throughout the gross pay interval. Multiple oil/water levels indicate that many of those porosity zones are stratigraphically separate reservoirs. More than 95% of the wells in the field are oil producers. "Average" wells have ultimate primary recoveries of approximately 200,000 bbl of oil.

The eastern part of the Central Basin platform was subsiding fairly rapidly during late Pennsylvanian and early Permian time, resulting in deposition of approximately 360 m (1200 ft) of upper Pennsylvanian and lower Permian strata that thin onto a structural high to the west (Figure 2). The main stratigraphic intervals are (from top to bottom): (1) Wolfcamp "reef," (2) Wolfcamp detrital, (3) Cisco, (4) upper Canyon, (5) lower Canyon, and (6) Strawn (Figure 2). The Wolfcamp units are early Permian. Cisco, Canyon, and Strawn are Pennsylvanian, with Cisco approximately Virgilian in age, Canyon approximately Missourian, and Strawn approximately Des Moinesian (Boardman and Heckel, 1989). In the lower Canyon to Wolfcamp "reef" interval, 63 depositional cycles have been identified and most were subjected to subaerial exposure immediately after deposition. This paper focuses on two intervals, the Wolfcamp "reef" and

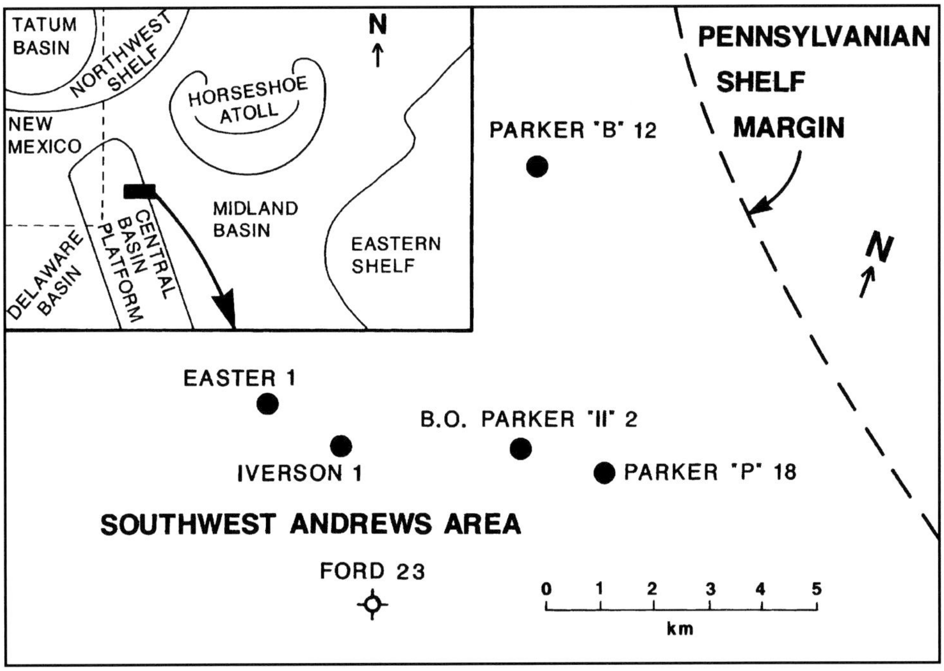

Figure 1. Maps showing location of the southwest Andrews area on the eastern side of the Central Basin platform and location of cored wells discussed in this study.

lower Canyon; however, limestones in the Strawn, upper Canyon, and Cisco have been studied, and information from those intervals is mentioned where appropriate.

Stratigraphic patterns in the southwest Andrews area were affected by moderate to rapid tectonic subsidence and eustatic sea-level fluctuations. The late Pennsylvanian and early Permian were times of high-amplitude (50–200 m) and high-frequency (100,000–400,000 yr) sea-level fluctuations, probably caused by repeated building and melting of continental glaciers (Heckel, 1986; Ross and Ross, 1987; Crowley and Baum, 1991). The Central Basin platform would have been between the equator and lat. 5°N during the Pennsylvanian (Ross and Ross, 1988; Walker et al., 1991), and hence would have had a warm tropical climate. The abundance of clay-rich shales and lack of depositional evaporites suggest that the Pennsylvanian and Wolfcamp in the Permian basin were dominated by a relatively humid climate. Algeo et al. (1992) also suggest a relatively warm, humid paleoclimate for the nearby Orogrande basin during the Pennsylvanian.

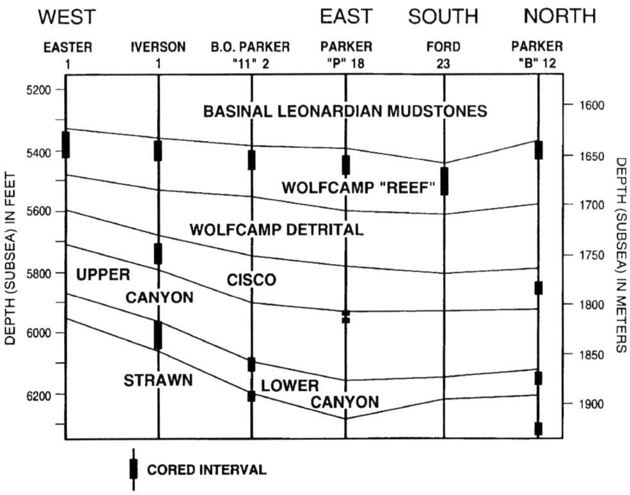

Figure 2. Stratigraphy across field showing major lithologic units and core control. Wolfcamp intervals are lower Permian. Strawn, Canyon, and Cisco are middle to upper Pennsylvanian.

METHODS

The southwest Andrews area was studied using wireline logs (mainly GR/neutron-density porosity logs), cores, and seismic data. Most porosity and permeability measurements were performed on whole (full diameter) pieces of core. Slabbed cores were lapped with 600 μ grit and/or etched with 10% hydrochloric acid, and described at 0.3 m (1 ft) intervals on a graphic logging form. Samples were then selected for thin sections and stable isotope analyses. Thin sections were cut from chips impregnated with blue plastic. Thin sections were first stained with a potassium ferricyanide/alizarin red-S solution (Dickson, 1965) and examined. The same thin sections were then polished and their cathodoluminescent properties noted using a Technosyn MK8200 cathodoluminescence stage chamber at conditions that included accelerating voltage of 14-18 kV, 0.5 mA beam current, and a vacuum of 60 mT. Stable isotope analyses were from: (1) bulk rock samples, generally collected at 1–2 m intervals, and (2) calcite cements scraped off slabs with an Exacto knife. Stable isotope analyses were performed at Brown University using methods described by Quinn (1991). Precision (2σ) of stable carbon and oxygen isotope analyses is estimated at ±0.2‰, and reported relative to a PDB standard.

LOWER CANYON

Stratigraphy

The lower Canyon section thickens from 24 m (80 ft) in the westernmost well to 33 m (110 ft) in the easternmost well. Five cycles were recognized in the lower Canyon in the southwest Andrews area (Figure 3). Shallowing-upward cycles commonly contain burrowed fossiliferous, abundantly spiculitic wackestones and packstones which pass upward to grainstones (bioclastic, peloidal, or oolitic). Lower parts of cycles are generally dark gray to brown, and the upper parts of cycles are generally light brown, light gray, or cream-colored. Root traces and other evidence of subaerial exposure are commonly present at the tops of shallowing-upward cycles. Burrowed fossiliferous wackestones and packstones were probably deposited at depths of 5–30 m (15–100 ft), whereas current-laminated grainstones were probably deposited in very shallow, subtidal to intertidal environments (0–5 m; 0–15 ft deep). Porosity occurs preferentially in grainstones in the upper parts of depositional cycles.

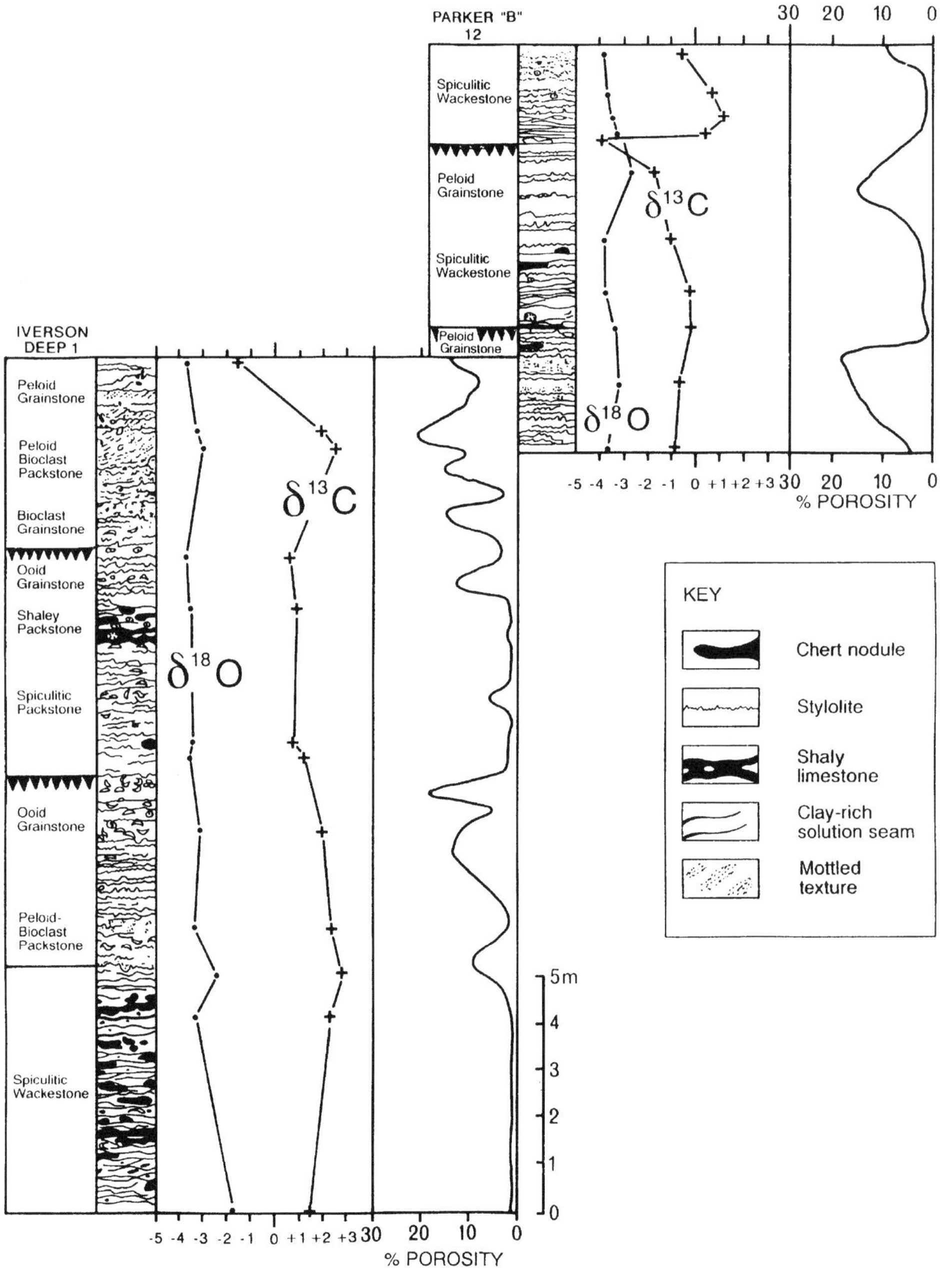

Figure 3. Descriptions of cores in lower Canyon strata along with core-measured porosity and isotope profiles for the Iverson 1 and Parker "B" 12 wells. Lithologies shown on left side of logs; solid line is cycle top, sawtooth lines are cycle tops with subaerial exposure. Logs of two wells correlate as shown with slight stratigraphic overlap.

Recognition of Subaerial Exposure Surfaces

Direct evidence for subaerial exposure at two prominent cycle tops in Parker "B" 12 core consists of laminated micrites capping the peloidal grainstones. These micrites are interpreted as caliche crusts. In addition, clasts containing dark brown rhizoliths occur at the base of the uppermost lower Canyon cycle (Figure 3). A color mottling caused by variable dissolution and cementation is also common below cycle tops. An example of this variable cementation below subaerial exposure surfaces is shown in Figure 4E and F. Densely cemented zones (Figure 4E, dark, 1–20 mm across) represent areas adjacent to roots where soil-zone water was being absorbed by the roots, enhancing calcite precipitation. Porous, light-colored parts of mottles have crusts of cements around grains that have been leached (Figure 4F).

Diagenetic Processes

The main diagenetic processes that affected porosity in lower Canyon limestones were calcite cementation, dissolution of aragonite, and compaction. Dolomitization and silicification affected these sediments, but had a minor influence on porosity. Late-burial dissolution is also apparently minor.

Calcite Cementation

Abundance, distribution, and timing of calcite cements are important factors determining porosity. Five different stages of calcite cement can be distinguished in the lower Canyon with petrography, staining, and cathodoluminescence (Figure 4). Those stages are described below in paragenetic order.

Stage 1—Nonferroan calcite cement with orange, patchy, flecked luminescence

Stage 1 cements occur as rims around many dissolved aragonitic grains in grainstones, indicating precipitation before aragonite dissolution (Figure 4B). Its patchy luminescence, similar to that of echinoderm fragments, suggests that stage 1 cement probably precipitated as a high-Mg calcite marine cement and subsequently recrystallized. Stage 1 cement is volumetrically insignificant in the lower Canyon.

Stage 2—Nonferroan calcite cement with dark luminescence interrupted by thin bands of bright yellow luminescence

Stage 2 cements occur mainly in primary pores (Figure 4), but also occur in some moldic pores, indicating that stage 2 cements precipitated before, during, and slightly after aragonite dissolution. This luminescent character is similar to the older banded cement of Meyers (1991) which is generally interpreted as precipitating in an active freshwater system (Walkden, 1987; Horbury and Adams, 1989), although its environmental significance is equivocal (Meyers, 1991). In the southwest Andrews area, aragonite dissolution probably occurred in freshwater environments during subaerial exposure, and probably so did stage 2 cementation.

Stage 3—Nonferroan calcite cement with red-orange luminescence and minor low-Fe zones (mauve color after staining)

Stage 3 cements (Figure 4A) occur in aragonite molds as well as intergranular pores, suggesting precipitation after most aragonite was dissolved. Red-orange luminescence suggests incorporation of significant amounts of Mn into the calcite, which requires reducing waters. Such reducing conditions generally require rocks that are not in an active (oxidized) freshwater aquifer. Therefore, stage 3 cements were probably precipitated at shallow to moderate burial depths from relatively stagnant freshwater and/or saline formation waters.

Stage 4—Low-Fe calcite cement (mauve color after staining) with yellow-orange luminescence

Stage 4 cement commonly fills moldic pores and the middle of some larger intergranular pores (Figure 4B). The presence of significant amounts of Fe (mauve staining) and Mn (causing yellow-orange luminescence) suggests precipitation in a reducing environment. Stage 4 cement fills collapsed fusulinid tests, and hence occurred after major physical compaction; however, stage 4 cements are cut by stylolites indicating chemical compaction occurred after stage 4 cements precipitated. Reducing conditions are common in burial environments. Therefore, stage 4 cements probably precipitated at moderate to deep burial (>300 m; >1000 ft?).

Stage 5—Ferroan calcite cement (mauve to purple color after staining) with red to orange luminescence and some broad zonation

Stage 5 cements fill intergranular, vuggy, and moldic pores (Figure 4B). Common Fe indicates precipitation of stage 5 cements in a reducing environment. Stage 5 cements grew over baroque dolomite cements. Baroque dolomites generally form at relatively high temperatures (>55°C) and are commonly a burial cement (Radke and Mathis, 1980). Stage 5 cements, like stage 4, probably precipitated at moderate to deep burial (>300 m; >1000 ft?).

Aragonite Dissolution

Aragonitic depositional grains (phylloid algae, gastropods, some bivalves, ooids, and some peloids) have been pervasively dissolved. Many of the pores remaining in these limestones are molds of original aragonitic grains (Figure 4C). As mentioned above, most aragonite dissolution occurred after stage 1 cementation, during stage 2 (meteoric) cementation, and before stages 3–5 cementation. A few aragonite molds have collapsed, especially in phylloid boundstones, but outlines of most original aragonite grains have been preserved by crusts of stage 1 and 2 cements (Figure 4B). Molds of aragonitic fossils are associated with subaerial exposure surfaces and meteoric cements, suggesting that they were produced by freshwater dissolution.

Compaction

Compaction includes early compression of micritic matrix and later pressure solution, resulting in

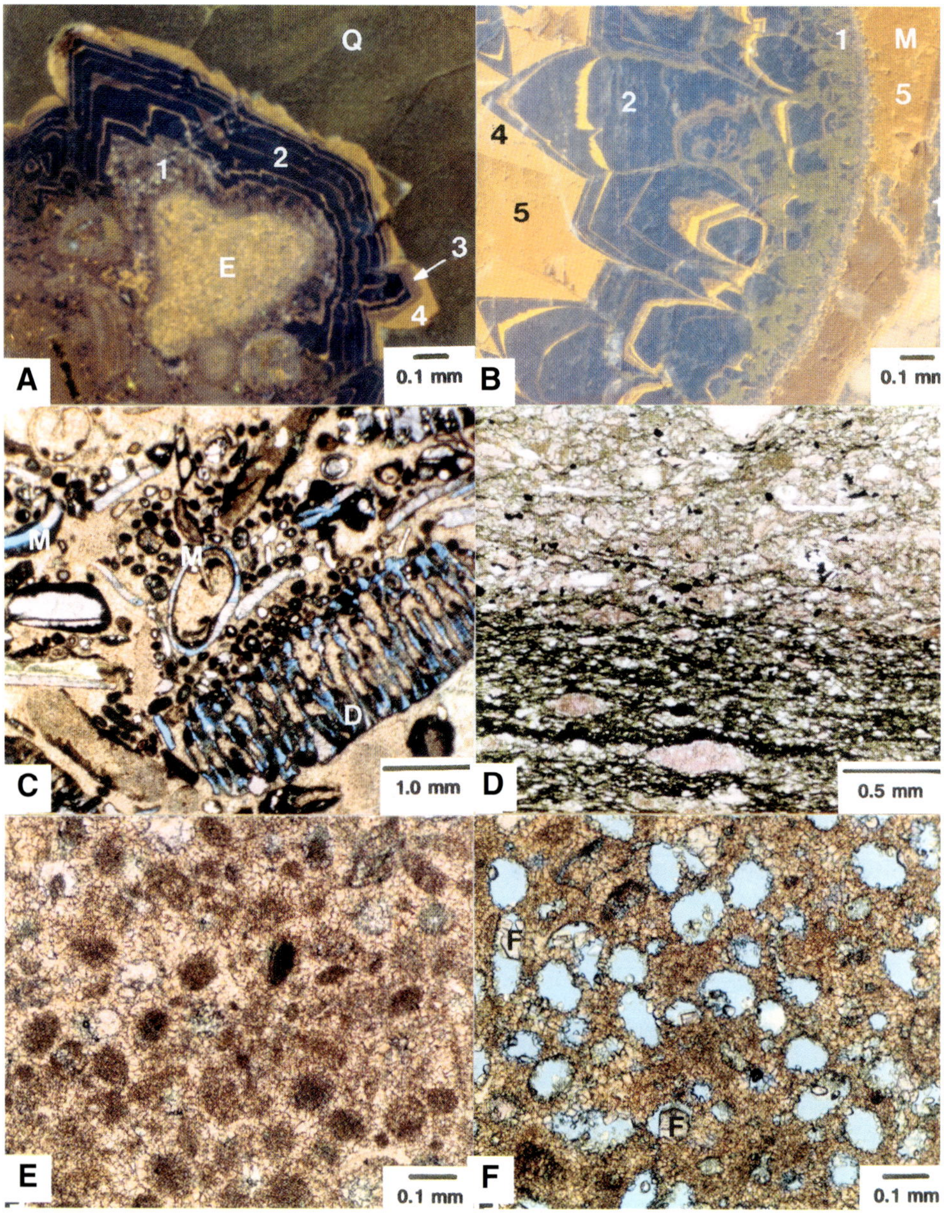

Figure 4. Photomicrographs of lower Canyon strata. (A) Cathodoluminescence (CL) photomicrograph of calcite overgrowth on an echinoderm grain (E) resting on floor of shelter cavity. Cement stages 1, 2, 3, and 4 are present. Stage 4 cement is overgrown by coarsely crystalline quartz (Q) (Iverson 1, 2811.1 m; 9223 ft). (B) CL photomicrograph of bivalve cast (M) with both surfaces preserved as light gray micrite. The mold is filled with stage 5 cement. Large shelter cavity to left of shell shows cement stages 1, 2, 4, and 5. Stages 1 and early 2 in the cavity are obscured in places by diagenetic chalcedony (Iverson 1, 2810.7 m; 9221 ft). (C) Photomicrograph of stained thin section. Skeletal-ooid grainstone. Intergranular pores filled by nonferroan early calcite cement which is absent and presumably predates formation of dasycladacean and mollusk molds. These moldic pores contain some late-stage ferroan calcite cement (mauve) (Iverson 1, 2811 m; 9222.6 ft). (D) Photomicrograph of stained thin section. Skeletal wackestone with clay-rich base. Rock intensely compacted. Wackestone shows draping of micrite around skeletal grains. Margins of skeletal grains removed by pressure solution where adjacent to clay (black), leaving calcite lenticles oriented parallel to bedding (Iverson 1, 2813.3 m; 9230 ft). (E) Plane polarized-light (PPL) photomicrograph of peloidal grainstone. Dark gray micritic peloids are largely intact. Intergranular pores are completely occluded by nonferroan calcite (Iverson 1, 2802.6 m; 9195 ft). (F) PPL photomicrograph 3 cm from the area shown in (A). Most peloids are dissolved to leave molds. Euhedral terminations of ferroan calcite cement (F, stage 5) occur in some molds (Iverson 1, 2802.6 m; 9195 ft).

stylolites (Figure 4D). Compression of micritic matrix is common in wackestones and is most obvious where compacted micrite occurs adjacent to uncompacted micrite. For example, micritic sediments within brachiopod shells are commonly peloidal and uncompacted, whereas micritic sediments outside the brachiopod are dense and draped around the shells. Stylolites occur in most of the limestone examined in the southwest Andrews area, but are most common in wackestones and many packstones. Stylolites are less common in grainstones at the top of depositional cycles. Stylolites reduce porosity: (1) directly by compacting the rock, and (2) indirectly by dissolving calcium carbonate which subsequently precipitated as calcite cement (stages 4 or 5).

Relation of Diagenesis and Porosity to Depositional Facies and Cycles

The intensity of diagenetic processes varies according to depositional facies and position in cycles (Figure 3). Wackestones, packstones, and transgressive grainstones had substantial porosity at their time of deposition and during shallow burial, but they did not retain porosity through deeper burial. Loss of porosity during deeper burial was caused largely by compaction (compression of micrite and pressure solution) and cementation (mainly brightly luminescing and/or ferroan calcite cements, stages 3–5).

Grainstones are the main reservoir rocks in the lower Canyon with 1–20% porosity. Grainstones have approximately 30–40% porosity when deposited (Enos and Sawatsky, 1981). Physical compaction was minor, and calcite cementation was the principal means of reducing porosity in grainstones in the upper parts of cycles. Porosity is commonly patchily developed, with high-porosity patches only millimeters across. Late ferroan calcite cements are not as abundant and, hence, do not completely occlude porosity in many grainstones, as they do in many nonporous rocks. Less intense compaction in these grainstones was apparently due to early intergranular cements forming a framework that resisted compaction during subsequent burial.

Stable Carbon and Oxygen Isotopes in the Lower Canyon

Bulk rock stable isotope data for the lower Canyon are plotted on Figure 3. Grossman et al. (1991) proposed that typical Pennsylvanian marine calcites had $\delta^{13}C$ values of +2.6 to +4.9‰ and $\delta^{18}O$ values of approximately –2.3‰ based on analyses of unaltered brachiopods. The $\delta^{13}C$ values for lower Canyon limestones (+2.8 to –3.9‰) vary from normal marine to substantially depleted in ^{13}C. Negative deflections in stable carbon isotopes are not present below several subaerial exposure surfaces. Lower $\delta^{13}C$ values occur at the top of the Iverson core close to a cycle top which was probably capped by a subaerial exposure surface, though no diagnostic soil structures are present at this surface (Figure 3). A second ^{13}C-depleted value occurs a few meters below the top of the Parker "B" 12 core. This sample is from a rhizolith-bearing micrite clast in a wackestone at the base of a lower Canyon cycle. The clast was probably eroded from a caliche at the top of the underlying cycle (Figure 3). The similarity of $\delta^{13}C$ values above and below exposure surfaces in the lower Canyon suggests that meteoric waters imported only a small amount of light organic/soil carbon due to brief exposure, and that this small amount was overwhelmed by carbon from the dissolving marine sediments. Light carbon isotopes imported during subaerial exposure may have also been partially offset by heavy carbon isotopes typical of ooids and peloids (+3 to +5‰; Lowenstam and Epstein, 1957; Budd and Land, 1990).

$\delta^{18}O$ values for the lower Canyon have a very small range (<1‰) with a mean value of –3.3‰. The proportion of original marine carbonate to late diagenetic calcite varies greatly in these bulk rock samples, suggesting a similar oxygen isotopic composition for both. The relatively heavy $\delta^{18}O$ values probably indicate that temperatures remained moderate (less than 70°C) in the southwest Andrews area while calcite was precipitated during burial.

WOLFCAMP "REEF"

Stratigraphy

The Wolfcamp "reef" interval contains eight depositional cycles (Figure 5). Subaerial exposure surfaces cap six of the eight cycles. The top Wolfcamp "reef" cycle ("A") was not subjected to subaerial exposure because the carbonate platform in the southwest Andrews area submerged, and the overlying shelf margin retreated approximately 16 km (10 mi) to the west. The Wolfcamp "reef" is overlain by black radioactive shales and limestones deposited in deep slope environments (Figure 5). The Wolfcamp "reef" succession thickens across the platform from west to east, from 41.4 m (138 ft) in Easter Deep 1 to 57.3 m (191 ft) in Parker "P" 18 (Figures 1 and 5).

The stratigraphy and diagenesis of the top five cycles in the Wolfcamp "reef" are discussed here (Figures 5 and 6). Cycles are designated "A," "B," "C," "D," and "E" from top to bottom. The top cycle, "A," which did not experience subaerial exposure, is an example where carbonate diagenesis was dominated by burial processes. Cycle "B" was flushed with freshwater during only one interval of subaerial exposure, in contrast to lower Wolfcamp "reef" and Canyon cycles, which were flushed with fresh water during several periods of subaerial exposure. Cycles "C" and "D" are thin and pinch out to the west (Figure 5).

Within the Wolfcamp "reef," prominent subaerial exposure surfaces occur at the top of cycles "B," "C," "D," and "E." The lower, more prominent emergence surface (top of "E") lies 0.3–1 m (1-3 ft) below the prominent GR peak (Figure 5). The rock below the emergent surface is, in places, extensively rooted and probably represents three amalgamated exposure events in the Easter 1 well (Figure 5). The upper exposure surface at the top of cycle "B" also shows rooting,

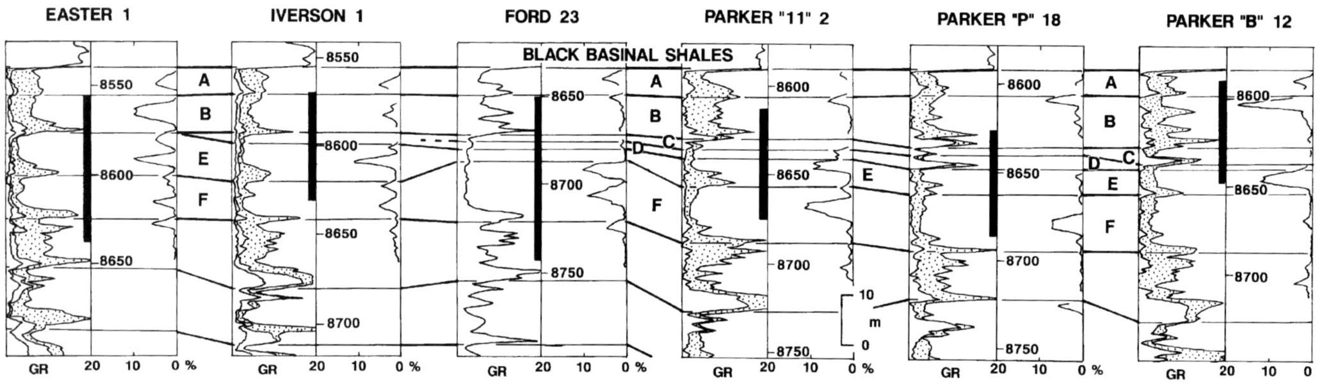

Figure 5. Cross section of wells cored in the Wolfcamp "reef." Datum is base of black basinal shales. Log depths are in feet. GR curves for each well are shown at the left, and neutron-porosity curves are shown at the right. Most wells have spectral GR logs. For wells with three GR curves (Easter 1 and Iverson 1), the left curve is potassium radiation, the middle curve is potassium plus thorium radiation, and the right curve is potassium + thorium + uranium radiation. For wells with two GR curves (Parker "11" 2, Parker "P" 18, Parker "B" 12), the left curve is potassium + thorium radiation, and the right curve is potassium + thorium + uranium radiation. In intervals with cores, porosity curves were calibrated with core-derived porosity values. No porosity values are shown for shaly intervals or intervals where the well bore was enlarged during drilling ("washed out"). The stratigraphic positions of cycles "A," "B," "C," "D," "E," and "F" are shown. Porosity zones in the lower part of cycle "F" are phylloid boundstones whose porosity is not directly related to subaerial exposure.

but rooting is limited to approximately 40 cm beneath the surface. The upper surface represents a single exposure event.

Wolfcamp "Reef" Cycle "A"

Cycle "A" does not show a distinct shallowing-upward succession of facies (Figure 6). Gray-green shale and dark-gray argillaceous packstones with abundant large fusulinids occur at the bottom of the cycle. Uppermost limestones cored in the top cycle are wackestones and packstones with compact oncoids, 1–3 cm across, which coat articulated brachiopods and other fossils. This unit is thoroughly burrowed. Clay-rich "horsetail" stylolites and dissolution seams are abundant. Cycle "A" does not have reservoir-quality porosity (Figure 5). The main diagenetic processes to affect the top Wolfcamp "reef" cycle were compaction and cementation. Compaction is represented mainly by stylolites, which formed along pressure-solution surfaces. Cements filled most primary and secondary (moldic) pores not occluded by compaction.

Primary and secondary pores, especially foraminiferan chambers, are filled with cement. The cements found in the top Wolfcamp "reef" cycle are dominantly calcite, although some small, intricately zoned dolomite crystals are also present. Calcite cements in the top Wolfcamp "reef" cycle can be subdivided by staining and cathodoluminescence into three stages. Stage 1 is the earliest nonferroan calcite cement with orange cathodoluminescence zones. Crystal faces are commonly rounded. This stage is unique to cycle "A." All other Wolfcamp "reef" cycles observed in this study have a different oldest calcite cement. Stage 2 is an intermediate ferroan calcite which stains mauve to purple, and has dull-orange, unzoned cathodoluminescence. In some places this stage is absent, and stage 1 is overlain by stage 3. Stage 3 is the latest highly ferroan calcite which stains blue and has dull-orange, unzoned cathodoluminescence. The nonferroan, nonluminescent calcite cements characteristic of freshwater diagenesis (lower Canyon, stage 2) are absent from the top cycle.

All of the cements present in the top cycle apparently formed during burial diagenesis, with calcium carbonate probably derived from pressure solution along stylolites. The top Wolfcamp "reef" cycle never experienced freshwater (meteoric) diagenesis. It lacks porosity because its argillaceous wackestones and packstones underwent substantial compaction. That compaction reduced porosity by simple compression, by pressure solution, and by generating calcium carbonate at stylolites which then precipitated as pore-filling calcite cement.

Wolfcamp "Reef" Cycle "B"

Cycle "B" can be divided into two or in some places three parts: (1) lower transgressive packstones and grainstones, (2) middle to upper fossiliferous wackestone to packstone, and sometimes (3) upper skeletal grainstone (cored only in the Parker "B" 12 well; Figure 6). The lower transgressive packstones and grainstones are commonly argillaceous and glauconitic. Wackestones and packstones in the middle to upper part of cycle "B" contain fusulinids, smaller foraminifera, solitary rugose corals, trilobite carapaces, bryozoa, crinoids, oncoids, glauconite pellets, phosphate clasts (black and yellow) and lime-mud matrix. Clay-rich dissolution seams, high-amplitude stylolites, and "horsetail" stylolites are common in the lower and middle parts of this cycle. The Parker "B" 12 well has skeletal grainstone in the uppermost part of cycle "B" (Figures 6 and 7C). Most of cycle "B" is burrowed.

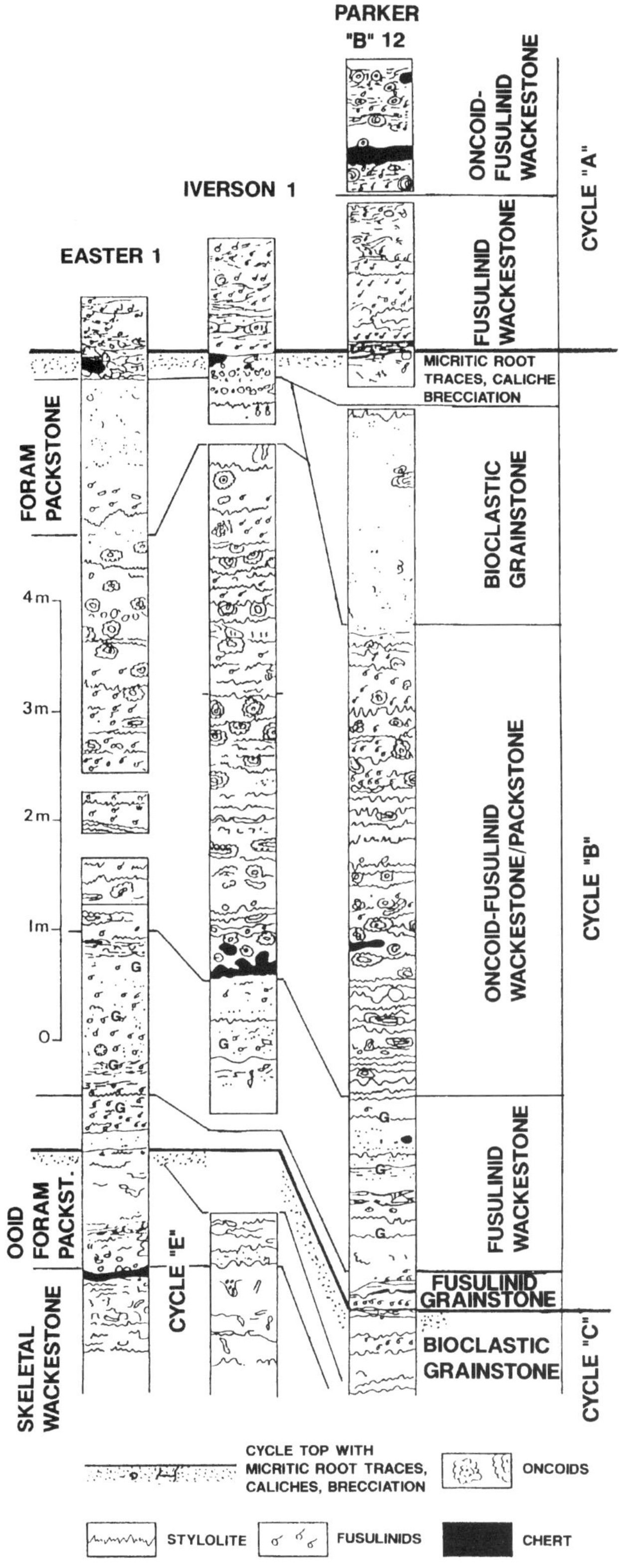

Figure 6. Lithologic logs from three cores which include all of Wolfcamp "reef" cycle "B."

Diagenesis is dominated by calcite cementation and compaction (stylolites). Stylolites are common in the lower part of the cycle, but their abundance diminishes near the top of the cycle. Three stages of sparry calcite cement were recognized in cycle "B": (1) nonferroan, nonluminescent, (2) nonferroan with bright yellow and orange luminescence, and (3) zoned ferroan with dull red and orange luminescence, and nonferroan with bright yellow and orange luminescence (Figure 7A–D). Stage 1 calcite cement does not occur above the subaerial exposure surface at the top of cycle "B." Stage 1 cements are composed of blocky, nonferroan calcite which has the luminescent characteristics of the older banded cement of Meyers (1991). It precipitated before most aragonite dissolution and probably precipitated in fresh water during subaerial exposure, shortly after deposition. Cement stages 2 and 3 probably precipitated during moderate to deep burial. Aragonite dissolution occurred during and after cement stages 1 and 2, which suggests that some aragonite in cycle "B" dissolved in a burial environment.

Grainstones at the top of cycle "B" in the Parker "B" 12 well have a rim of stage 1 cements on grains which partially filled many intergranular pores (Figure 7C, D). Dissolution of grains (probably aragonitic) followed precipitation of the stage 1 cements. Minor fracturing of these stage 1 cement crusts occurred and was followed by precipitation of stage 2 and 3 calcite cements. Stage 2 and 3 cements precipitated preferentially on echinoderm grains, causing very patchy cementation and allowing much porosity to remain.

Cycle "B" in the Wolfcamp "reef" contains variable porosity and diagenetic alteration. Porosity values >10% occur in some Wolfcamp "reef" wells, chiefly in the upper half of the succession (Figure 5). The highest values, up to 20%, occur in skeletal grainstones in the Parker "B" 12 well. Porosity in cycle "B" of the Wolfcamp "reef" is dependent on compaction and the abundance of the three stages of pore-filling cement. Compaction and associated porosity reduction are greatest in the lower part of cycle "B." Cement distribution in cycle "B" of the Wolfcamp "reef" is quite variable. The early (stage 1) cement is most abundant at the base of the cycle and diminishes upward. Only minor amounts of this cement were observed around grains in the porous skeletal grainstone. The second stage cement has the opposite distribution, being thinly developed at the base and common at the top. Precipitation of these two cement stages was followed by a final period of aragonite dissolution and minor fracturing of micrite envelopes. The late, stage 3 cements precipitated throughout the cycle. In porous parts of cycle "B," porosity was retained in intervals where compaction was minimal, and subsequent cementation was not sufficient to occlude all of the pores.

Wolfcamp "Reef" Cycles "C" and "D"

Cycles "C" and "D" are thin wedges which amalgamate into the top of cycle "E" to the west (Figure 5). Cycle "C" is separated from cycle "D" by a prominent shale up to 30 cm thick, whereas the base of cycle "D" is intermittently marked by a shale parting a few millimeters thick and is generally difficult to locate. Cycles "C" and "D" are crossed by brown tubular

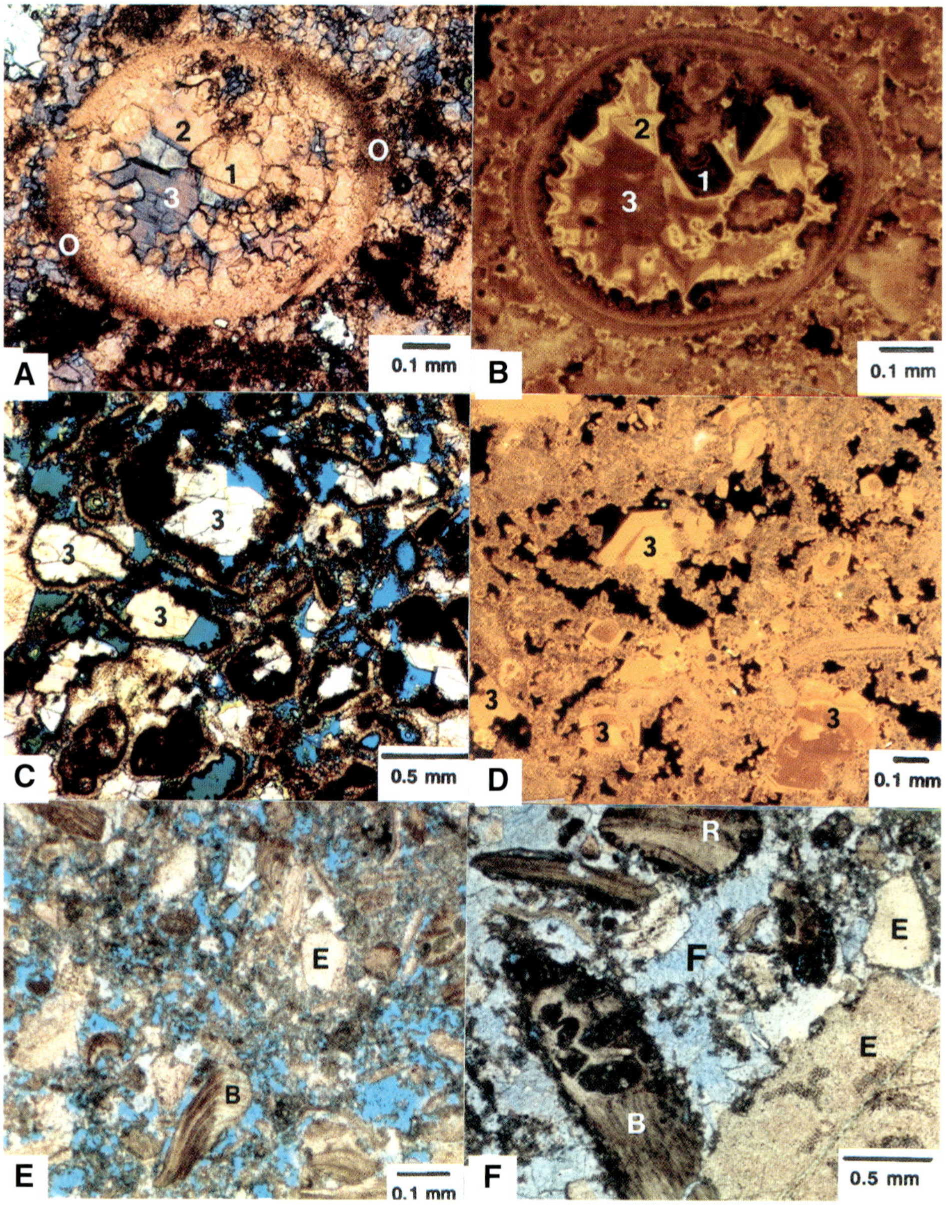

Figure 7. Wolfcamp "reef" strata. (A) Photomicrograph of a stained thin section showing three stages of cement filling the interior of an articulated ostracod (O) in a fossiliferous packstone in cycle "B." Cement stages 1 and 2 are nonferroan calcite. Cement stage 3 is a ferroan calcite cement (Iverson 1, 2619.1 m; 8593 ft). (B) Cathodoluminescence photomicrograph of thin section in (A) rotated 90°. Note the cathodoluminescence characteristics of cement stages 1–3 where they fill an articulated ostracod. (C) Photomicrograph of a stained thin section of a skeletal grainstone in cycle "B." A nonferroan rim of calcite cement (stages 1 and 2) occurs on grains. Many grains were removed by dissolution. Porosity appears blue. Late-stage, coarsely crystalline calcite cement with iron-rich zones (stage 3) partially fills primary and secondary porosity (Parker "B" 12, 2622.0 m; 8602.5 ft). (D) Cathodoluminescence photomicrograph of thin section in (C). Moldic porosity is common. Porosity appears black. Calcite cement stages 1 and 2 appear as thin rims of nonferroan calcite on grains. Stage 3 cements (3) are coarsely crystalline (Parker "B" 12, 2622.0 m; 8602.5 ft). (E) Photomicrograph of a stained thin section of a porous skeletal wackestone. Fibrous brachiopod (B) and echinoderm (E) grains preserved as nonferroan calcite. Micrite colored brown. Pores filled with blue plastic. A small amount of early nonferroan calcite cement occurs, but no late ferroan calcite cement is present (Iverson 1, 2624.8 m; 8611.6 ft). (F) Photomicrograph of a stained thin section showing a fossiliferous packstone with irregular, matrix porosity filled by late, coarsely crystalline, iron-rich calcite cement (stage 3; F). Fragments of echinoderms (E), bryozoa (B), and brachiopods (R) are common (cycle "F," Iverson 1, 2628.3 m; 8623 ft).

rhizoliths, and a thin caliche crust is locally present at the top of cycle "D." Cycles "C" and "D" are composed of tubular foram packstones and grainstones with minor wackestones. These cycles were substantially altered by precipitation of micrite during soil formation and the penetration of roots. Compaction is relatively minor in pure limestones but intense in shales and shaly limestone. Blocky, ferroan calcite fills many secondary matrix pores and, as a result, no significant porosity is present in cycles "C" and "D" (Figure 5).

Wolfcamp "Reef" Cycle "E"

Cycle "E" is composed mainly of burrowed, fossiliferous wackestones and packstones (Figure 7E) with grainstones near the top of the cycle in some wells. Fossils in cycle "E" include brachiopods, crinoids, tubular forams, and *Tubiphytes*. Diagenesis is highly variable. Porosity greater than 4% occurs in a 1–3 m interval near the top of cycle "E" in most wells (Figure 5). These porous wackestones (Figure 7E), packstones, and grainstones have microcrystalline to equant (up to 0.5 mm diameter), nonferroan calcite cement lining pores. The microcrystalline cement was precompaction and apparently related to freshwater infiltration. Pore types include molds, soil-related vugs, intercrystalline pores between microcrystalline calcites between grains, and intraparticle pores, mainly in fusulinids. These porous lithologies are surrounded by similar rocks with microcrystalline cements, but their pores are filled by coarsely crystalline, equant, ferroan calcite cements (Figure 7F). The source of this late ferroan calcite was probably adjacent strata which lack the early microcrystalline cement and have many stylolites and pressure-solution seams. The lower part of the cycle is composed of wackestones and packstones which (1) have no significant early cement, (2) are generally compacted, (3) lack porosity, and (4) have ferroan calcite cement. Stylolites are abundant in nonporous limestones, but are much rarer in porous rocks.

Identification of Subaerial Exposure Surfaces in the Wolfcamp "Reef"

The upper surfaces of Wolfcamp "reef" cycles "B," "C," "D," and "E," like the upper lower Canyon cycles, were subaerially exposed. However, the Wolfcamp "reef" cycles, in contrast to the lower Canyon, all show rhizolith zones (Figure 8C, D) and color mottling. Some rhizoliths show an alveolar septal fabric which is characteristic of beta calcretes (Wright, 1994). Thin laminar crusts of brown micrite (caliche crusts) are present at cycle tops in a few locations. Brecciation associated with roots and subaerial exposure is also present below subaerial exposure surfaces in several wells. Rhizoliths and brecciation are best developed in wackestones and packstones.

Stable Carbon and Oxygen Isotopes in the Wolfcamp "Reef"

Stable carbon and oxygen isotopic compositions of carbonates were analyzed in five cored wells in the Wolfcamp "reef" and are plotted against depth in Figure 9. A pronounced change in $\delta^{13}C$ exists at the top of cycle "B" where values change from +3‰ at the base of cycle "A" to –3‰ at the top of cycle "B," a distance of only a few millimeters (Figure 9). The $\delta^{13}C$ values change from –3‰ at the top to near 0‰ for most of cycle "B." Below cycle "B," $\delta^{13}C$ values again decrease to –2 to –4‰ at the tops of cycles "C," "D," and "E." Within cycle "E," $\delta^{13}C$ values gradually increase downward to approximately 0‰. $\delta^{18}O$ values change from –4‰ at the top of cycle "B" to –3‰ at the base of cycle "A" (Figure 9).

IMPORTANT FEATURES PRESENT IN UPPER CANYON AND CISCO LIMESTONES

While not studied in detail, upper Canyon and Cisco strata have features which are important to assessing factors critical to isotopic signatures and porosity development in carbonates associated with subaerial exposure. The upper Canyon and Cisco limestones have much less porosity than the lower Canyon and Wolfcamp "reef." Most cycles in the Cisco are relatively thin (Figure 10). Soils are prominently developed (Figure 8A, C), and in some places rhizolith zones penetrate entire cycles (Figure 10). Where present, porosity in the upper Canyon and Cisco occurs in grainstones more than 1.5 m (5 ft) thick. Thin grainstones, <1.5 m (5 ft) thick, are either compacted or, more commonly, their pores are filled with calcite cements whose luminescent characteristics match those of stage 2 from the lower Canyon or the older banded cement type of Meyers (1991), both of which are interpreted as early meteoric. Subaerial exposure surfaces are commonly developed on fossiliferous wackestones and packstones, but reservoir porosity is generally absent in those cycles because much primary and secondary porosity was occluded by soil-zone micrite and sparry meteoric cements (Figure 10). Stable carbon isotope compositions are often light throughout entire cycles in Cisco limestones (Figure 10). The thin and commonly nonporous Cisco cycles are thought to represent brief cycles of marine flooding and limestone deposition alternated with prolonged times of subaerial exposure leading to complete stabilization of the limestones.

PETROLOGIC FEATURES CHARACTERISTIC OF SUBAERIAL EXPOSURE SURFACES

Many petrologic features were used to identify paleosols and subaerial exposure in cores from the southwest Andrews area (Figures 8 and 11). One characteristic commonly associated with subaerial exposure surfaces is an abrupt change in depositional lithology, but this does not occur at all exposure surfaces, and may also occur at cycle tops that were not subaerially exposed. No single petrologic attribute is present beneath all subaerially exposed surfaces, but

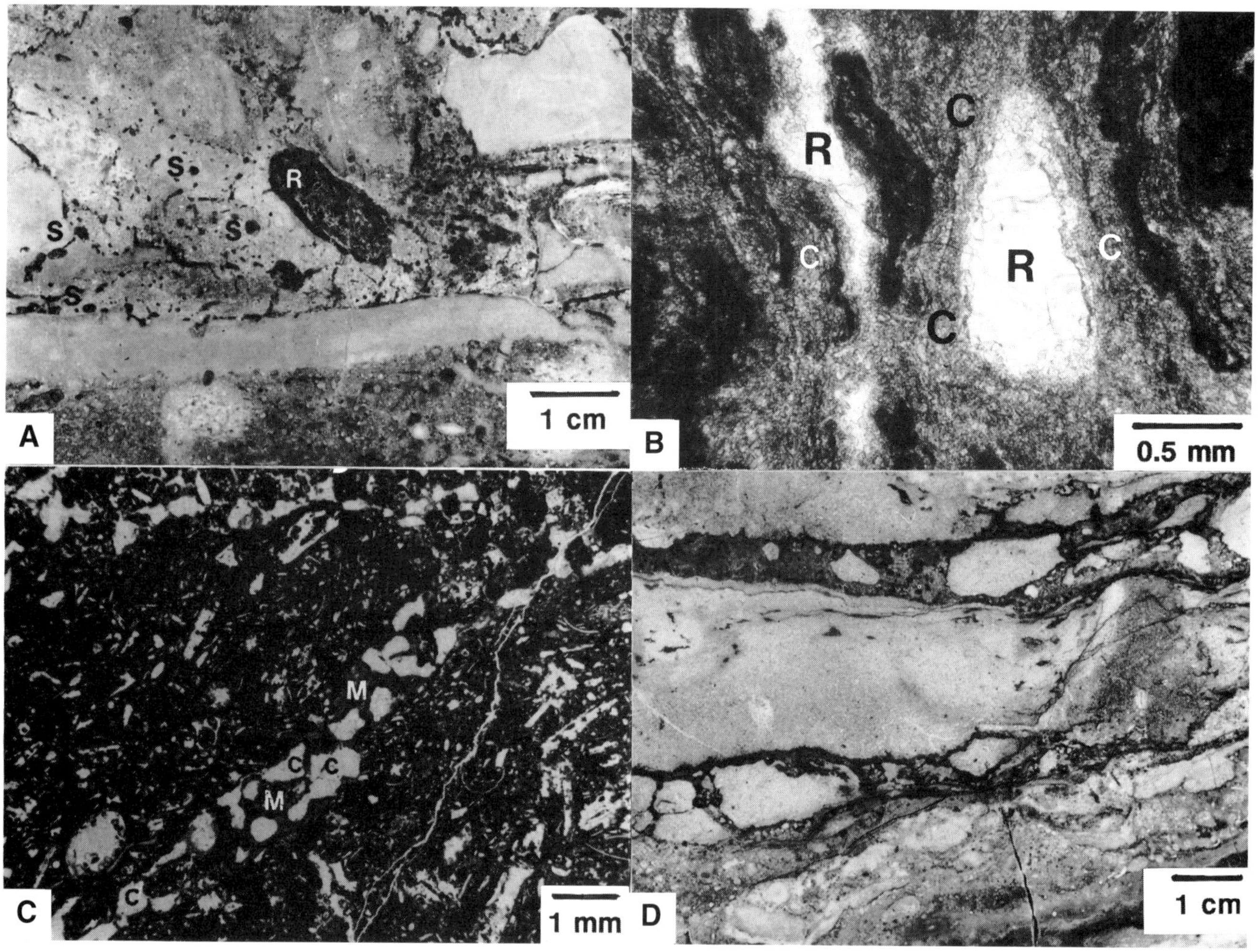

Figure 8. Soil-related features. (A) Core slab of mottled brecciated limestone ramified by large (R) and many small (S) rhizoliths. Cisco Group (Iverson 1, 2724.9 m; 8940 ft). (B) Thin-section photomicrograph of a small rhizolith. Layers of replacive microcrystalline calcite (C) surround elongate (tubular) pores filled with clear, equant, calcite cement (R). The elongate pores were probably rootlets around which replacive microcrystalline calcite (rhizoliths) precipitated. Photograph taken under plane polarized light. Cisco Group (Iverson 1, 2722 m, 8930 ft). (C) Photograph of thin section showing inclined rhizolith descending from peloidal caliche. Rhizolith composed of micritic screens (M), forming alveolar septal fabric, and late-stage calcite cement (C). Top of cycle "B," Wolfcamp "reef" (Iverson 1, 2614 m; 8575.9 ft). (D) Core slab of wackestone brecciated by root penetration. Dark matrix of two breccia layers composed of dark calcite characteristic of rhizoliths. Base of slab silicified. Top of cycle "B," Wolfcamp "reef" (Easter 1, 2616 m; 8584.7 ft).

most attributes of paleosols described by Wright (1994) are present at some location in the section. The exposure-related alteration is distributed according to host sediment composition. Mudstones, wackestones, and mud-rich packstones commonly developed different diagenetic features than mud-poor packstones and grainstones during subaerial exposure.

Rhizoliths (tubular root structures 1–40 mm in diameter) occur either individually with a vertical, inclined, or horizontal orientation, or as intertwined bundles forming banded caliches (Figure 8). Rhizoliths in the southwest Andrews area are generally dark brown calcite which surrounded the root and crossed it forming alveolar septal structures (Wright, 1994). Rhizoliths are especially distinctive in pale gray limestones. Large roots (40 mm diameter) commonly show only a dark micritic peripheral zone, with the root's center being filled by sediment or cement. The depth of rhizolith penetration varies from a few centimeters to 3 m below cycle tops. These are minimum amounts because the upper parts of many cycles may have been removed during the transgressive phase of the overlying cycle. The zone of individual rhizoliths is the most common indicator of subaerial exposure in the southwest Andrews area, occurring beneath 40% of the cycle tops in any given well. Rhizoliths are best preserved in micrite-rich lithologies.

Soil-related structures above the zone of separate rhizoliths are diverse but are rarely preserved, probably because of erosion. An exception to this is Wolf-

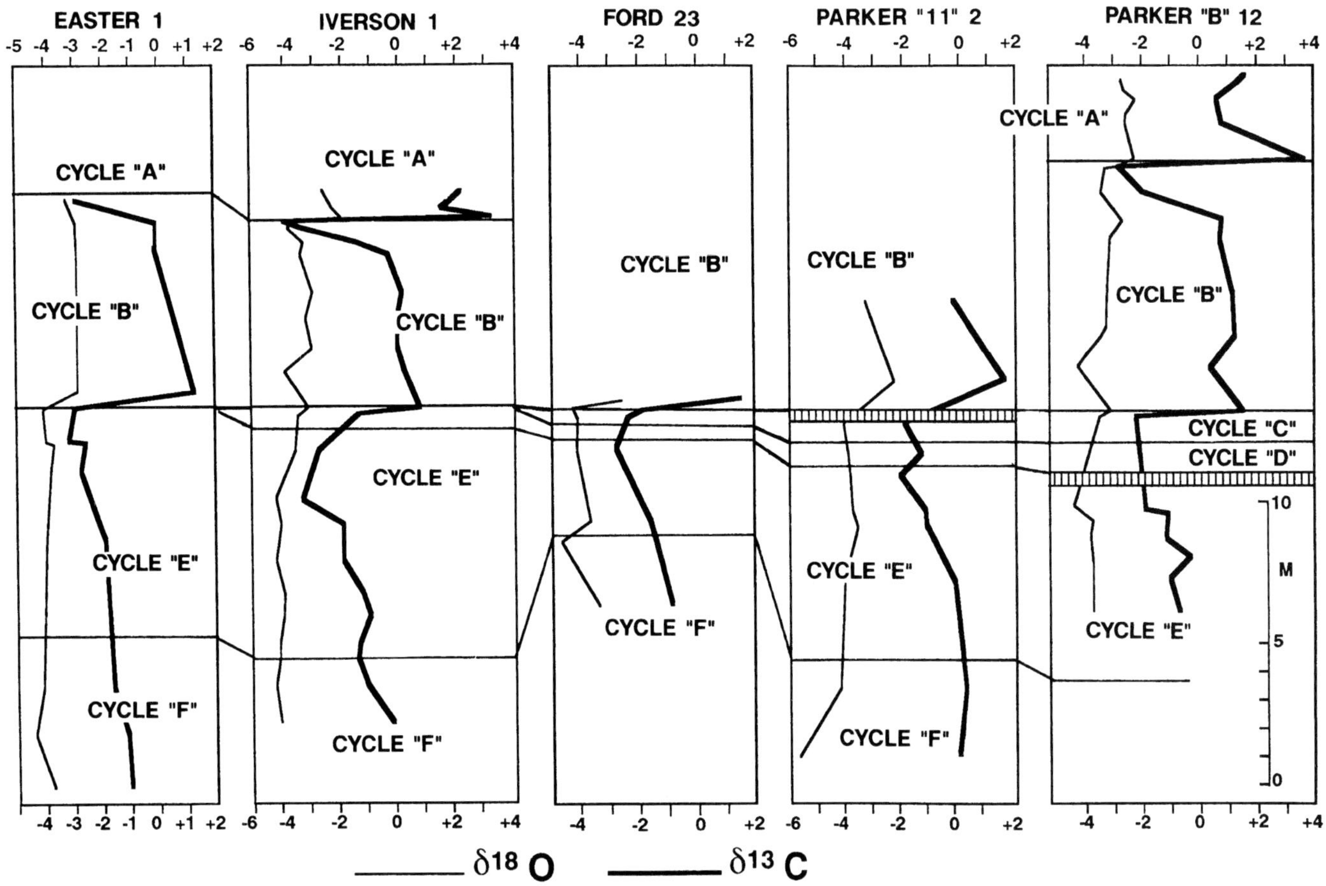

Figure 9. Plot of stable carbon and oxygen isotope compositions of bulk rock samples from five cores through the Wolfcamp "reef." Datum is the base of cycle "B."

camp "reef" cycle "B," which has a massive, dark brown caliche (30 cm thick) overlying a 40 cm thick, individual rhizolith zone. This caliche contains felted root masses, root-related brecciation and fracturing, and laminated micrite layers containing peloids and coated grains.

Color mottling is present in grainstones immediately below cycle tops in the lower Canyon and in many cycle tops in the Strawn and upper Canyon (Figure 4E, F; Figure 11). The mottled grainstones received patchy, early meteoric cement and patchy dissolution of grains. Some color mottling is related to burrowing and later diagenesis, rather than subaerial exposure. Locally (approximately 20% of occurrences), grainstones are capped by micrite with rhizoliths. Mottling due to patchy development of soil micrite as described by Riding and Wright (1981) has not been found, though cores have very limited lateral coverage.

DISCUSSION OF USING STABLE ISOTOPES TO DETECT SUBAERIAL EXPOSURE SURFACES

The use of stable carbon and oxygen isotope profiles to identify subaerial exposure surfaces from bulk rock samples was advocated by Allan and Matthews (1977, 1982). They observed several features in Pleistocene carbonates on Barbados including: (1) a negative $\delta^{13}C$ shift of –1 to –5‰ below subaerial exposure surfaces due to incorporation of soil-gas CO_2 into the carbonate, (2) a $\delta^{13}C$ shift of 2 to 4‰ at the water table with lighter (lower) values in the vadose zone, and (3) a shift toward heavier $\delta^{18}O$ values (1–2‰) immediately below the exposure surface due to evaporation. Similar profiles were presented by Allan and Matthews (1982) in some ancient examples, although all isotopic signatures were not present in every case. Other workers have had variable success in using stable carbon and oxygen isotopes to identify subaerial exposure surfaces. For example, Moshier et al. (1988) found no isotopic evidence of a subaerial exposure surface below a major subaerial unconformity in the Cretaceous of the Middle East.

Goldstein (1991) published stable isotope profiles for the cyclic Holder Formation from New Mexico which included two subaerial exposure surfaces indicated by soil-related features. One surface had a distinct negative shift in bulk rock $\delta^{13}C$ values, but the other did not. A wide range of $\delta^{13}C$ values was obtained from rhizoliths and other soil-generated calcites adjacent to these surfaces. Some components showed negative $\delta^{13}C$ values beneath both surfaces, but the ^{13}C-depleted materials were not present in suf-

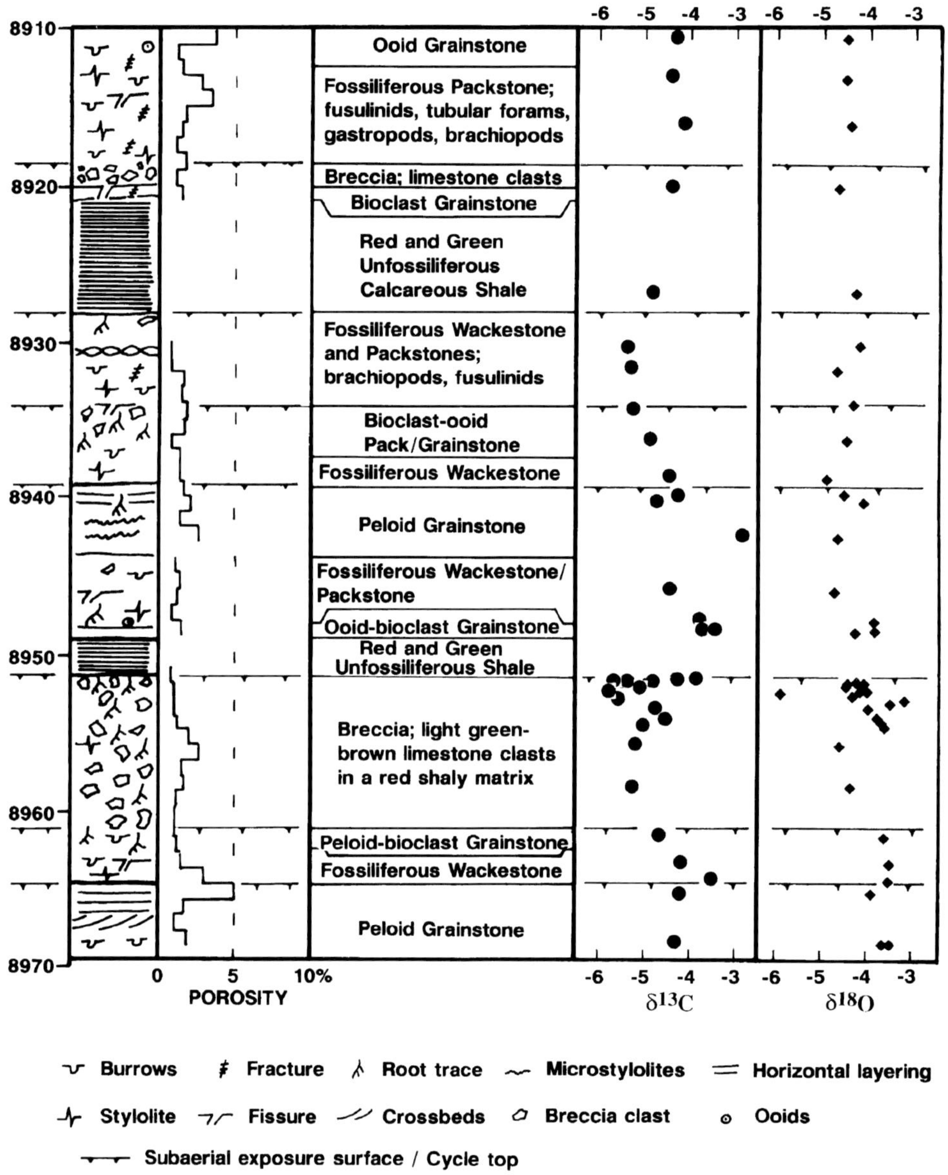

Figure 10. Description of Cisco strata cored in the Iverson 1 well showing associated cycle tops, core-derived porosity, stable carbon, and stable oxygen isotope values. Depths are in ft.

ficient amounts below some surfaces to affect bulk rock values. No significant shifts in $\delta^{18}O$ were noted in these profiles.

Stable isotopic profiles in lower Canyon and Cisco limestones lack distinctive carbon isotope deflections beneath many subaerial exposure surfaces (Figures 3 and 10). Lower Canyon limestones lack distinct carbon isotope deflections probably because of limited duration of subaerial exposure and/or depositional sediments with heavy carbon isotopes below subaerial exposure surfaces. Cisco limestones commonly lack carbon isotope deflections below subaerial exposure surfaces apparently due to light carbon isotopes occurring throughout depositional cycles (Figure 10).

The Wolfcamp "reef" profiles show an excellent match to part of the Allan and Matthews (1977, 1982) model. This is especially true for cycle "B," in which $\delta^{13}C$ values are approximately 0 to +1‰ throughout most of the cycle except within 1–2 m (3–6 ft) of the cycle top, where they shift to −3 to −4‰ before rebounding to +3‰ at the base of cycle "A" (Figure 9). This change across the top of cycle "B" is accompanied by a 1‰ shift in $\delta^{18}O$ to heavier values than in cycle "A." Beneath cycle "B" the zone of light $\delta^{13}C$ is not well defined; in Easter Deep 1 the $\delta^{13}C$ values are still negative 12 m below the top of cycle "E." It appears that soil-gas CO_2 was incorporated into calcite more deeply below the tops of cycle "C," "D," and "E" than below the top of cycle "B." Deeper influence of soil-related diagenesis is supported by the distribution of rhizoliths which penetrate less than 1 m below the top of cycle "B," but up to 3 m below the top of cycle "E."

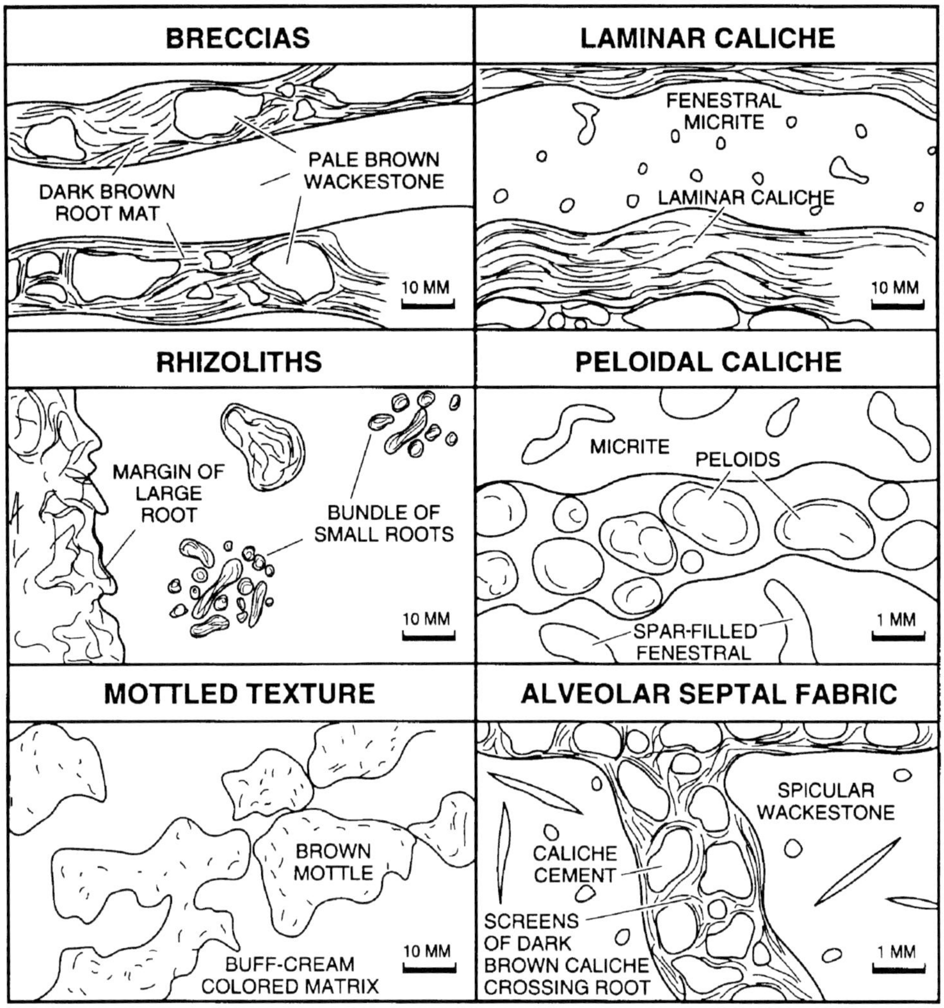

Figure 11. Features associated with subaerial exposure surfaces.

The excellent match between cycle "B" bulk rock stable isotope profiles in the Wolfcamp "reef" and the model of Allan and Matthews apparently resulted from a combination of factors. The southwest Andrews Wolfcamp "reef" and Barbados examples belong to glacial epochs when large-amplitude, high-frequency sea level fluctuations were/are ideal for repeatedly exposing shallow marine sediments long enough to develop mature soils. However, the Pennsylvanian profiles recorded by Goldstein (1991) belong to the same Gondwana glaciation as the Wolfcamp "reef" and are geographically close, yet they do not show the same good correlation with the Allan and Matthews model. Lack of consistent carbon isotope signatures in the Holder Formation could be due to abundant clastics in the Holder section or repeated subaerial exposure.

The general similarity of shape for the five Wolfcamp "reef" bulk rock profiles across 7 km of platform is remarkable and occurs despite thickness variations (Figure 9); cycle "B" thickens from 7.1 m in the west (Easter Deep 1) to 8.8 m in the east (Parker "B" 12). There is slight variation laterally in $\delta^{13}C$ values. Cycle "B" in the west (Easter Deep 1 and Iverson Deep 1) is 1 to 2‰ lighter than in the east (Parker "11" 2 and Parker "B" 12). This variation may be due to: (1) the eastern wells being closer to the seaward edge of the platform and farther downflow from the recharge area to the west, and hence bicarbonate in the water was influenced increasingly by dissolution of marine sediment, and/or (2) the western wells were farther updip and subaerially exposed longer.

The amalgamation of tops of cycles "C," "D," and "E" implies overprinting of diagenetic episodes in cycle "E." Overprinting can lead to complex cement sequences in cyclic carbonates as seen in the late Dinantian of Britain (Walkden, 1987; Horbury and Adams, 1989). Cycle "B," in contrast, has limited alteration. The soil features at the top of cycle "B," although thin, are constant across the area supporting a single exposure event. Unfortunately, no core was available from the top of the cycle "A." Lower diagenetic successions at the base of cycle "A" indicate that no meteoric cements are present, implying no meteoric overprinting of limestones in cycle "B." Therefore, the distinctive isotope profile from cycle "B" is due to its position in a series of cycles showing decreasing intensity of soil-related diagenesis. A distinctive stable-isotope profile can be produced when a single episode of subaerial exposure of sufficient duration to affect the upper quarter of a cycle exists with no overprinting from subsequent meteoric episodes.

Other cycles from glacial epochs may have developed distinctive stable-isotope profiles similar to the Pleistocene of Barbados, but the bulk rock signal was obscured by the addition of isotopically different late-stage diagenetic products. The Wolfcamp "reef" has a significant quantity of post-compaction calcite cement irregularly distributed in most cycles. Five of these late-stage cements from the Wolfcamp "reef" have a mean $\delta^{13}C$ of −0.3‰ and a mean $\delta^{18}O$ of −3.4‰. These calcites are close to the mean value for the entire Wolfcamp "reef" data set ($\delta^{13}C$ = 0.8‰, $\delta^{18}O$ = −3.6‰) and hence would only dampen variations. The similarity of $\delta^{13}C$ of late calcite cement to the mean host rock in the Wolfcamp "reef" is normal for most marine limestones where the later cement is formed by dissolution of the host rock. However, invariant $\delta^{18}O$ values are unusual because elevated temperatures during burial should cause late cements to have lighter $\delta^{18}O$ values. This effect was apparently counterbalanced by ^{18}O enrichment of the formation waters. Three analyses of formation waters from the study interval in the southwest Andrews area have a mean $\delta^{18}O$ value of +5.4‰ (SMOW), supporting this hypothesis.

CONTROLS ON POROSITY DEVELOPMENT AND RETENTION DURING BURIAL

The Pennsylvanian–Lower Permian succession shows that a considerable volume of calcite cement was precipitated and that aragonite was dissolved during meteoric diagenesis. Cycles in the lower Canyon and Wolfcamp "reef" probably had no significant change in pore volume due to meteoric diagenesis, but the modification of pore systems by dissolution of depositional material and cementation filling primary porosity was important at a reservoir scale within each cycle. Where present, porosity now occurs preferentially in the upper part of depositional cycles (Figures 3 and 5). Cycles that were subaerially exposed show substantial variations in porosity laterally (within the same cycle) and vertically (between different cycles).

In this study, compaction (physical and chemical) and calcite cementation were found to be the most important agents for pore destruction during burial. Of critical importance to mechanical compaction is the mechanical strength of the sediment on leaving the meteoric environment (Purser, 1978; Manley et al., 1993). Grainstones in the upper parts of cycles are the main reservoir rocks in the upper Canyon and Cisco of the southwest Andrews area as well as the lower Canyon. Porous grainstones in the southwest Andrews area generally have early fringing cements that do not fill the larger primary pores or the molds (Figure 4C, E, and F). Grainstones with sparse early cement (common in the lower transgressive parts of cycles) may have had high primary porosity during shallow burial, but that porosity was commonly eliminated during deeper burial because the weak rock could not withstand compaction (Figure 12). Extensive early cementation (common in the upper parts of cycles) protects the rock against compaction (Railsback, 1993), but too much early cement can eliminate porosity completely before burial begins.

Figure 12. Bioclastic grainstone which lacks early intergranular cements, but has undergone severe grain-to-grain compaction which eliminated porosity. Many grains are crinoid fragments (Easter 1, 2629.3 m; 8626.2 ft).

Porosity is relatively rare in micritic sediments, especially in the lower parts of cycles, in the southwest Andrews area. Significant porosity does occur in micritic rocks in the upper part of Wolfcamp "reef" cycles. Creating and retaining porosity in micritic rocks was apparently dependent on having enough carbonate precipitated during early diagenesis to lithify the rock, but not so much carbonate precipitated that all porosity was filled. Porous micritic rocks occur in the Wolfcamp "reef" where early meteoric diagenesis dissolved some grains and micrite, and precipitated microcrystalline calcite (Figure 7E). Precipitation of microcrystalline calcite lithified the micritic matrix without completely filling the matrix pore system (Figure 7E). In some areas, late burial cements eliminated porosity in micritic rocks that were porous after subaerial exposure (Figure 7F). In contrast, porosity in micritic limestones at the tops of cycles in Cisco strata was filled by microcrystalline calcite precipitated during freshwater diagenesis, resulting in nonporous micritic rocks at the top of cycles (Figure 10). Low $\delta^{13}C$ values throughout Cisco cycles (Figure 10) suggest more prolonged exposure than was experienced by Wolfcamp "reef" cycles which had low $\delta^{13}C$ values only in the upper parts of cycles (Figure 9). Unlithified micritic sediments compact and lose their porosity during burial. Burial compaction was especially severe where micritic carbonate was mixed with clay-rich material. Pressure solution associated with clay seams was common in the lower to middle parts of many cycles (Figures 3 and 6).

The abundance of late-stage burial cements is critical to the occurrence of reservoir porosity. Stylolites with 2 cm amplitude or more occur at a rate of 5–8 per meter in most lithologies, but occur at a rate of 1–3 per meter in porous grainstones and some porous wackestones and packstones. Stable-isotope data suggest a local source for the late-stage cement. Carbonate released by pressure dissolution would provide a source for cement if reprecipitated locally, and this would explain the scarcity of late cements in thick intervals that were lithified early and have few stylolites. Late burial calcite cement completely occludes porosity in thin intervals (<1.5 m; <5 ft) that were porous and lithified after subaerial exposure. In contrast, thick intervals (>1.5 m; > 5 ft) that were porous after exposure generally have substantial subsurface porosity because they have limited late-burial cements. Much late-burial cement in thin lithified intervals was probably derived from compaction of adjacent unlithified rocks.

CONCLUSIONS

Pennsylvanian and Lower Permian carbonates in the southwest Andrews area are characterized by cyclic sedimentation and repeated subaerial exposure. Freshwater diagenesis associated with repeated subaerial exposure had a direct influence on the diagenesis of the underlying sediment. Porosity was controlled by a combination of factors: (1) primary sediment texture, (2) primary sediment mineralogy, (3) subaerial emergence and penetration of meteoric water, (4) distance below the emergence surface, (5) duration of exposure, (6) extent of early cementation, (7) amount of compaction, and (8) amount of burial cement.

The lower parts of cycles have little porosity or permeability. These strata do have some freshwater cements indicating that fresh water did move through them. The lack of porosity in the lower parts of cycles can be attributed to (1) greater concentrations of lime mud and terrigenous clays, (2) greater compaction in the lower parts of cycles, and (3) occlusion of the remaining porosity by burial cements generated by pressure solution.

Grainstones in the upper part of depositional cycles are generally porous if they are >1.5 m thick. Three factors were critical for developing and retaining porosity in the grainstones. (1) Early fringing cements that precipitated in sea water and/or fresh water were thick enough to prevent major compaction. (2) Many aragonitic grains were leached by fresh water during subaerial exposure to create moldic porosity. (3) Late-burial cements had difficulty filling the thick porous grainstones, probably because: (a) the supply of calcium carbonate was limited within partially cemented grainstones, and (b) calcium carbonate from pressure solution in adjacent wackestones and packstones could not fill the large volume of porosity remaining in thick grainstones. Hence, porosity was preserved because relatively little late cement was precipitated in currently porous strata.

Porosity in the Wolfcamp "reef" interval also occurs in micritic limestones in the upper parts of cycles. Two factors apparently critical for development of porosity in micritic Wolfcamp strata were very brief duration of subaerial exposure and limited burial cementation. Brief subaerial exposure allowed dissolution and lithification in the upper part of cycles in the Wolfcamp "reef." In contrast, prolonged subaerial exposure apparently caused complete filling of pores in micritic strata below exposure surfaces in Canyon and Cisco strata. As in grainstones discussed above, lithification of these micritic rocks helped prevent compaction during burial. If the lithified zone were thick and porous, burial cements could not completely fill it.

ACKNOWLEDGMENTS

We thank Phil Choquette, Mark Longman, and Ray Mitchell for reviews which substantially improved this manuscript. Many people helped us start and complete this study, including Tim Anderson, Al Crawford, Tom Elliott, Stacie Boyd, George Moore, and Julie Saller. Jeff Brown prepared thin sections. Stable isotope analyses were performed at Brown University under the supervision of Robert Fifer. We thank Unocal Energy Resources for permission to publish this paper.

REFERENCES CITED

Algeo, T.J., B.H. Wilkinson, and K.C. Lohmann, 1992, Meteoric-burial diagenesis of middle Pennsylvanian limestones in the Orogrande basin, New Mexico: water/rock interactions and basin geotherm-ics: Journal of Sedimentary Petrology, v. 62, p. 652–670.

Allan, J.R., and R.K. Matthews, 1977, Carbon and oxygen isotopes as diagenetic and stratigraphic tools: surface and subsurface data, Barbados, West Indies: Geology, v. 5, p. 16–20.

Allan, J.R., and R.K. Matthews, 1982, Isotope signatures associated with meteoric diagenesis: Sedimentology, v. 29, p. 797–817.

Boardman, D.R., and P.H. Heckel, 1989, Glacial-eustatic sea-level curve for early Late Pennsylvanian sequence in north-central Texas and biostratigraphic correlation with curve for midcontinent North America: Geology, v. 17, p. 802–805.

Budd, D.A., and L.S. Land, 1990, Geochemical imprint of meteoric diagenesis in Holocene ooid sands, Schooner Cays, Bahamas: correlation of calcite cement geochemistry with extant groundwaters: Journal of Sedimentary Geology, v. 60, p. 361–378.

Craig, D. H., 1988, Caves and other features of the Permian karst in San Andres dolomite, Yates field reservoir, west Texas, *in* N. P., James, and P. W. Choquette, eds., Paleokarst: New York, Springer-Verlag, p. 342–363.

Crowley, T.J., and S.K. Baum, 1991, Estimating Carboniferous sea-level fluctuations from Gondwanan ice extent: Geology, v. 19, p. 975–977.

Dickson, J.A.D., 1965, A modified staining technique for carbonates in thin section: Nature, v. 205, p. 587.

Enos, P., and L. H. Sawatsky, 1981, Pore networks in Holocene carbonate sediments: Journal of Sedimentary Petrology, v. 51, p. 961–985.

Goldstein, R.H., 1991, Stable isotope signatures associated with paleosols, Pennsylvanian Holder Formation, New Mexico: Sedimentology, v. 38, p. 67–77.

Grossman, E.L., C. Zhang, and T.E. Yancey, 1991, Stable-isotope stratigraphy of brachiopods from Pennsylvanian shales in Texas: Geological Society of America Bulletin, v. 103, p. 953–965.

Harris, P.M., S.H. Frost, G.A. Seiglie, and N. Schneidermann, 1984, Regional unconformities and depostional cycles, Cretaceous of the Arabian peninsula, *in* J.S. Schlee, ed., Interregional Unconformities and Hydrocarbon Accumulations: AAPG Memoir 36, p. 67–80.

Heckel, P.H., 1986, Sea-level curve for Pennsylvanian eustatic marine transgressive-regressive depositional cycles along the midcontinent outcrop belt, North America: Geology, v. 14, p. 330–334.

Horbury, A.D., and A.E. Adams, 1989, Meteoric phreatic diagenesis in cyclic late Dinantian carbonates, northwest England: Sedimentary Geology, v. 65, p. 319–344.

Jordan, C.F., and M. Abdullah,, 1988, Lithofacies analysis of the Arun reservoir, north Sumatra, Indonesia, *in* A.J. Lomando and P.M. Harris, eds., Giant Oil and Gas Fields: A Core Workshop: SEPM Core Workshop 12, p. 89–118.

Lowenstam, H.A., and S. Epstein, 1957, On the origin of sedimentary aragonite needles of the Great Bahama bank: Journal of Geology, v. 65, p. 364–375.

Manley R.D., P.W. Choquette, and M.B. Rosa, 1993, Paleogeography and cementation in a Mississippian oolite shoal complex: Ste. Genevieve Formation Willow Hill field, southern Illinois basin, *in* B.D. Keith and C.N. Zuppman, eds., AAPG Studies in Geology 35, p. 91–113.

Meyers, W.J., 1991, Calcite cement stratigraphy: an overview in luminescence microscopy and spectroscopy: qualitative and quantitative applications, *in* C.E. Barker and O.C. Kopp, eds., SEPM Short Course 25, p. 133–148.

Moshier, S.O., C.R. Handford, R.W. Scott, and R.D. Boutell, 1988, Giant gas accumulation in a "chalky"-textured micritic limestone, Lower Cretaceous Shuaiba Fm., eastern United Arab Emirates, *in* A.J. Lomando and P.M. Harris, eds., Giant Oil and Gas Fields: A Core Workshop: Society of Economic Paleontologists and Mineralogists Core Workshop 12, p. 229–272.

Purser, B.M., 1978, Early diagenesis and the preservation of porosity in Jurassic limestones: Journal of Petroleum Geology, v. 1, p. 83–94.

Quinn, T.M., 1991, Meteoric diagenesis of Plio-Pleistocene limestones at Enewetak atoll: Journal of Sedimentary Petrology, v. 61, p. 681–703.

Radke, B.M., and Mathis, R.L., 1980, On the formation and occurrence of saddle dolomite: Journal of Sedimentary Petrology, v. 50, p. 1149–1168.

Railsback, L.B., 1993, Lithologic controls on morphology of pressure-solution surfaces (stylolites and dissolution seams) in Paleozoic carbonate rocks from the mideastern United States: Journal of Sedimentary Petrology, v. 63, p. 513–522.

Riding, R., and V.P. Wright, 1981, Paleosols and tidal flat/lagoon sequences on a Carboniferous carbonate shelf: sedimentary associations of triple disconformities: Journal of Sedimentary Petrology, v. 51, p. 1323–1339.

Ross, C.A., and J.R.P. Ross, 1987, Late Paleozoic sea levels and depositional sequences: Cushman Foundation for Foraminiferal Research Special Publication 24, p. 137–153.

Ross, C.A., and J.R.P. Ross, 1988, Late Paleozoic transgressive-regressive deposition, *in* C.K. Wilgus, B.S. Hastings, H. Posamentier, C.A. Ross, and C.G.S.C. Kendall, eds., Sea-Level Changes: An Integrated Approach: Society of Economic Paleontologists and Mineralogists Special Publication 42, p. 227–247.

Tucker, M.E., 1993, Carbonate diagenesis and sequence stratigraphy, *in* V.P. Wright, ed., Sedimentary Review: Oxford, Blackwell, p. 51–72.

Walkden, G.M., 1987, Sedimentary and diagenetic styles in Late Dinantian carbonates of Britain, *in* J. Miller, A.E. Adams, and V.P. Wright, eds., European Dinantian Environments: New York, John Wiley and Sons, p. 131–155.

Walker, D.A., J. Galonka, A.M. Reid, and S.A.T. Reid, 1991, The effects of Late Paleozoic paleolatitude and paleogeography on carbonate sedimentation in the Midland basin, Texas, *in* M.P. Candelaria, ed., Tomorrow's Technology Today, West Texas Geological Society Publication 91-89, p. 141–162.

Wright, V.P., 1994, Paleosols in shallow marine carbonate sequences: Earth-Science Reviews, v. 35, p. 367–395.

Chapter 13

Recognition and Significance of an Intraformational Unconformity in Late Devonian Swan Hills Reef Complexes, Alberta

Jack Wendte
Institute of Sedimentary and Petroleum Geology
Geological Survey of Canada
Calgary, Alberta, Canada

Iain Muir
Wascana Energy Inc.
Calgary, Alberta, Canada

ABSTRACT

Swan Hills reef complexes are isolated buildups up to 75 m thick that occur on an underlying drowned carbonate platform (approximately 60 m thick) in the subsurface of west-central Alberta and were studied in detail at the Judy Creek and Snipe Lake oil fields. Although these two reef complexes are 85 km apart, 8 to 10 m thick megacycles can be correlated between them. The top of the fourth reefal megacycle is a widespread subaerial unconformity (the intraformational Swan Hills unconformity [ISHU]) that separates an underlying rimmed-reef complex from an overlying ramp-bounded shoal complex.

Emergence at the ISHU was a result of a low-magnitude, relative sea level fall. This is substantiated by the following observations: (1) this surface exhibits a lithified nature continuously across both reefs; (2) shallower-water, mainly tidal-flat deposits overlie relatively deeper-water subtidal limestones at the contact; (3) solution vugs filled with marine sediments occur down to 2.3 m below the ISHU; and (4) oxidation of sediments occurs in some cores immediately beneath the unconformity.

Distinct and unique lithologic changes occur in lagoonal successions in the fourth megacycle below the ISHU. The middle and upper parts of this megacycle consist entirely of shallow lagoonal deposits and totally lack the "deep"-water lagoonal deposits that typify portions of the first three reefal megacycles. These distinct changes record the gradual and progressive loss of accommodation space prior to emergence and suggest that the withdrawal of the sea was not due to a Pleistocene-like, glacial eustatic lowering of sea level. This sea level fall and resulting emergence had little effect on reservoir quality of the limestones underlying the ISHU.

Our work shows that stacking patterns must be used with caution when predicting where subaerial unconformities *(sequence boundaries)* should occur. At both Judy Creek and Snipe Lake, the reefal megacycles below the ISHU exhibit a gradually retreating to a more pronounced backstepping of facies on their windward sides. This style contrasts with contemporary models which predict progradational stacking patterns of facies in the "highstand systems tract" prior to the relative sea level fall and a subaerial unconformity *(sequence boundary).*

INTRODUCTION

The development of seismic stratigraphy (Vail et al., 1977) and associated sequence-stratigraphic models (Jervey, 1988; Posamentier et al., 1988; Posamentier and Vail, 1988; Sarg, 1988) has renewed focus on the recognition and origin of unconformities. In carbonate successions, the significance of unconformities is especially important to understanding their diagenesis. Processes associated with subaerial exposure during times of unconformities may profoundly alter or modify the fabric of the rock. In particular, the creation of secondary porosity by freshwater dissolution may provide the storage capacity in hydrocarbon-producing pools. This mechanism is thought to be responsible for much of the secondary porosity in such oil and gas accumulations as the Arun field in Indonesia (Jordan and Abdullah, 1988), the Horseshoe atoll fields of west Texas (Vest, 1970; Schatzinger, 1983), Golden Lane fields, Mexico (Coogan et al., 1972), and numerous Lower Cretaceous fields of the Middle East (Wilson, 1975).

This paper discusses a subaerial unconformity that has widespread distribution within reefal and bank carbonates in the Swan Hills Formation of Late Devonian (earliest Frasnian) age in the subsurface of west-central Alberta. The conclusions in the paper are based on *independent* studies of two separate isolated reef complexes, Judy Creek and Snipe Lake (by J. Wendte and I. Muir, respectively) (Muir et al., 1990; Springate et al., 1992; Wendte, 1992c). These studies were undertaken by the authors in the mid-1980s (Judy Creek) and late 1980s (Snipe Lake) to provide reservoir geologists and engineers with three-dimensional models of reservoir continuity. Both studies incorporated the examination of thousands of meters of core (approximately 6000 m at Judy Creek and 3000 m at Snipe Lake) integrated with wireline logs. Reservoir units in both models correspond to genetic successions or cycles that were identified from core examination and correlated throughout each pool with core and log data.

This paper addresses three fundamental aspects of this unconformity. First, we identify criteria from core-based observations which permitted us to recognize the unconformity and to interpret exposure as the result of a relative fall in sea level, as opposed to the culmination of an upward-shoaling depositional phase. Second, we relate this unconformity to facies patterns in cycles both below and above the unconformity. We then compare these relationships to those predicted from published sequence-stratigraphic models. Third, we examine the effect of subaerial exposure and freshwater diagenesis on reservoir quality, especially the formation of secondary porosity by dissolution.

Most discussions of unconformities have focused on widespread surfaces of an interregional or, arguably, global extent. The unconformity that we describe in this paper is intraformational. It does not separate successions with the magnitudes of those described by Sloss (1963), nor does it have the spatial extent of those discussed in Schlee (1984).

GEOLOGICAL SETTING OF SWAN HILLS REEF COMPLEXES

Swan Hills reef complexes are isolated buildups that occur on an underlying, drowned carbonate platform in the subsurface of west-central Alberta. The distribution of these complexes is shown in Figure 1. These buildups are up to approximately 75 m thick, range from 10 to 30 km across and are time-equivalent to more areally widespread bank carbonates to the southwest. The underlying carbonate platform, from which these reefs evolved, is up to approximately 60 m thick. The margin of the lower platform at a 24 to 30 m thick stage is marked on this map. The accompanying cross section illustrates the overall stratigraphic relationships among the isolated reef complexes, the underlying carbonate platform, and the backstepped carbonate bank complex.

The Judy Creek and Snipe Lake complexes are approximately 85 km apart. The eastern margin of the Judy Creek complex is approximately 15 km back from the margin of the 24 to 30 m thick platform stage, whereas the margins of the Snipe Lake complex occur much closer to the edge of the equivalent carbonate platform (Figure 1). On its western side, Judy Creek is separated from the Judy Creek West complex by a narrow channel which was established during the deposition of the upper part of the underlying carbonate platform (Wendte, 1992c).

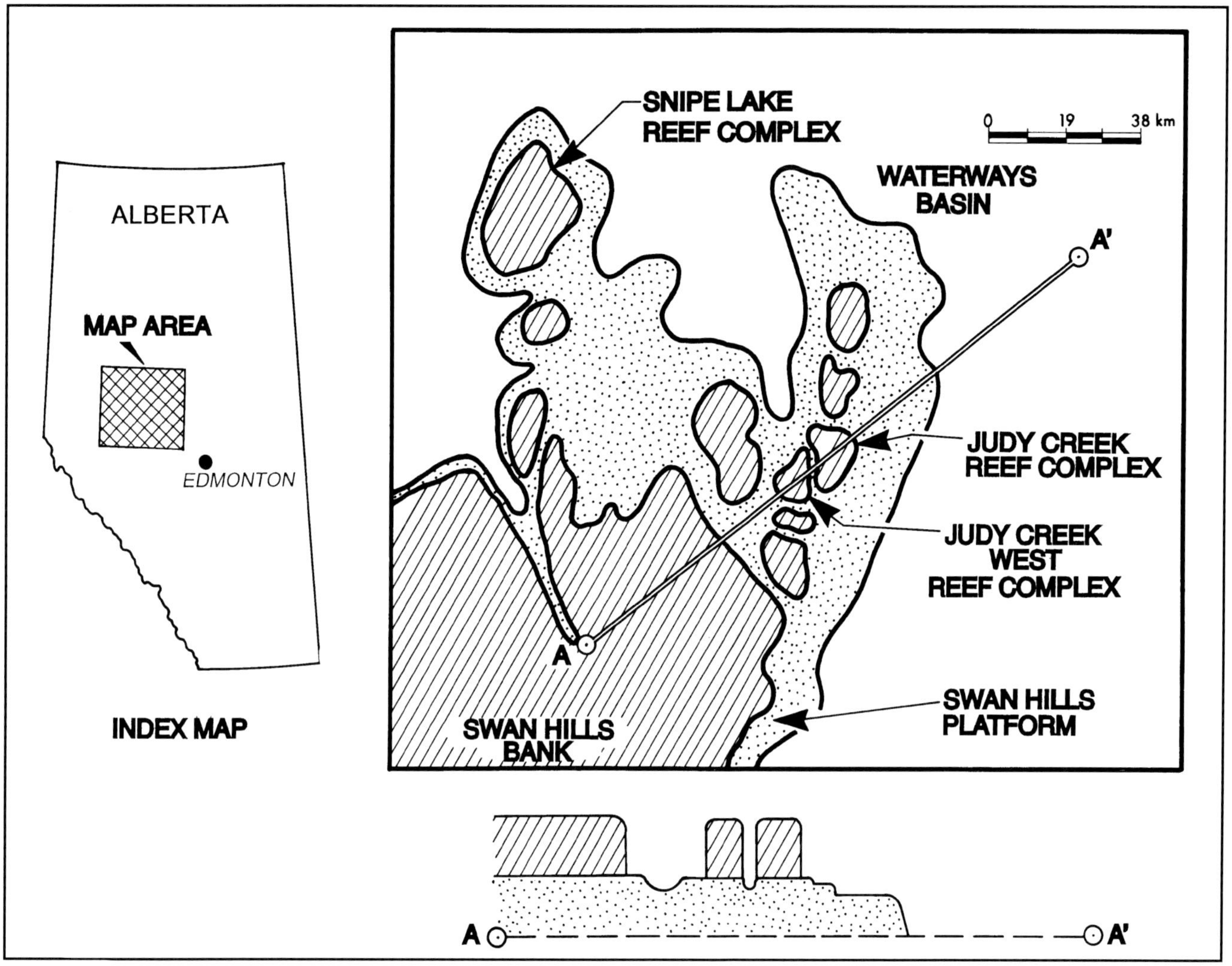

Figure 1. Location and paleogeography of the Swan Hills area and a diagrammatic profile of the general Swan Hills units. The Swan Hills Formation consists of a lower, areally widespread platform and overlying isolated reef complexes. The reef complexes are approximately time-equivalent to a backstepped, more areally widespread bank unit to the southwest. Modified from Wendte and Stoakes (1982).

Both Judy Creek and Snipe Lake contain large accumulations of oil (818 million bbl OOIP for Judy Creek and 195 million bbl OOIP for Snipe Lake) (Podruski et al., 1988). At Judy Creek about 75% of the reef complex was oil filled, with free formation water occurring only in the downdip, southwestern part of the complex. Well and core coverage extends throughout the complex and was used during the course of Wendte's study. In contrast, the accumulation of oil at Snipe Lake is limited to only the updip, northeast perimeter of the complex. Accordingly, the study by Muir was limited to this portion of the complex. Trapped oil accumulations at Snipe Lake occur in both the reef and platform successions.

Figure 2 shows a more detailed account of the Beaverhill Lake stratigraphy associated with isolated Swan Hills reef complexes. The Beaverhill Lake Group in west-central Alberta comprises, in ascending order, the Fort Vermilion, Swan Hills, and Waterways formations. Platform carbonates of the Swan Hills Formation overlie interbedded anhydrites and carbonates of the Fort Vermilion Formation, a succession about 10 m thick which, in turn, overlies shales of the Watt Mountain Formation of the Elk Point Group. The carbonate platform succession consists of a number of cycles, backstepping away from the platform edge shown on Figure 1. Thickness variations in the carbonate platform correspond not only to the level at which shallow-water carbonate growth was aborted, but also to basin position. Equivalent successions generally thicken to the east, away from the less rapidly subsiding western edge of the Devonian Alberta basin. As such, the maximum thickness of the platform at Judy Creek is 64 m, whereas the equivalent succession at Snipe Lake only attains a thickness of 52 m.

The position of the overlying isolated reef complexes at both Judy Creek and Snipe Lake correspond to areas with thick platformal successions. Results

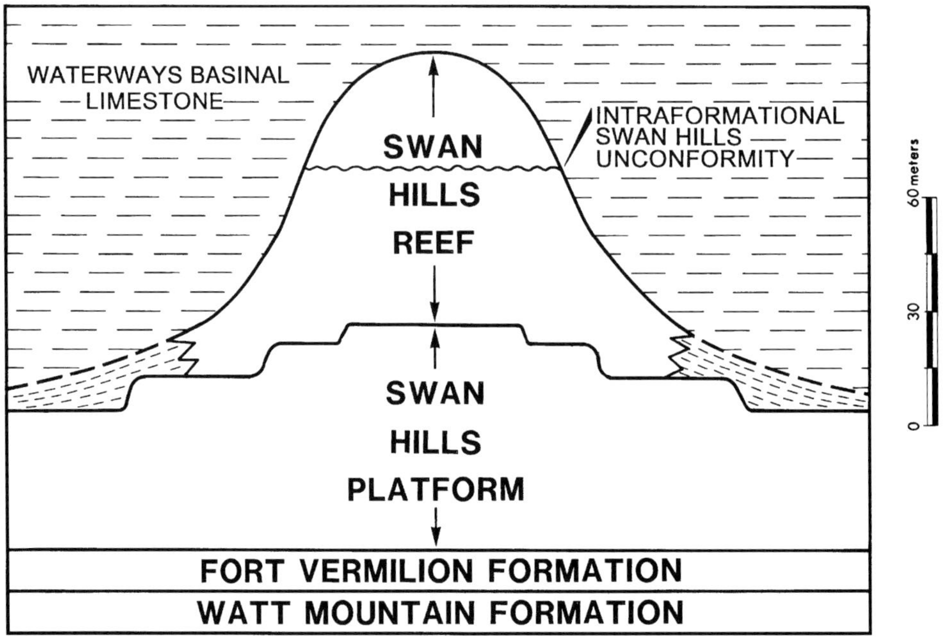

Figure 2. Beaverhill Lake stratigraphy associated with isolated Swan Hills reef complexes. The platform unit consists of a number of backstepping or retreating stages. The isolated reef complexes nucleated on the highest portion of the underlying carbonate platform and intertongue with and are overlain by basinal limestones of the Waterways Formation. The ISHU occurs at a level somewhat above the middle of the reefal succession. Modified from Wendte (1992c).

from both studies show that the reefs initially nucleated on the highest platform "step" and subsequently built laterally over lower and older platform levels. The reefal carbonates both intertongue with and are abruptly overlain by basinal limestones of the Waterways Formation.

Both Judy Creek and Snipe Lake consist of several growth stages or megacycles that can be correlated throughout the complexes. However, the only horizon that can be traced throughout each complex based on a consistent lithologic character is the ISHU. This unconformity occurs approximately 40 m above the base of the reef complex at Judy Creek and approximately 33 m above the base of the reef complex at Snipe Lake. At Judy Creek and Snipe Lake, up to 30 m of Swan Hills reefal limestones overlie the unconformable surface. The ISHU has also been identified in the backstepped Swan Hills bank complex to the southwest (Kaufman and Myers, 1988).

Regionally, the Swan Hills reef complexes occur along the western side of the Devonian Alberta basin. Figure 3 shows a composite schematic, east-to-west cross section across the Alberta basin. The section is divided into five major Devonian successions: Upper Elk Point, Beaverhill Lake, Woodbend, Winterburn, and Wabamun. The transect of the Beaverhill Lake succession crosses the Swan Hills area. Of particular significance to this paper is the contrasting style of sedimentation, or stacking patterns, of Beaverhill Lake strata on the east and west sides of the basin. On the eastern side of the basin, shallow-marine carbonates prograde over basinal limestones and shales. Conversely, Swan Hills carbonates on the western side of the basin show an overall backstepping and aggradational evolution. The difference in style has been attributed to a prevailing northeasterly windwave circulation system corresponding to a Devonian tradewind belt (Wendte, 1992a). The orientation and configuration of Swan Hills reef complexes in relation to the northeast prevailing water circulation has a significant impact on variation in stacking patterns within these complexes.

INTERNAL MAKEUP OF SWAN HILLS REEF COMPLEXES

Facies

Limestones that comprise both the Judy Creek and Snipe Lake complexes include a wide variety of facies. Eleven environmental facies displaying distinctive textures, sedimentary structures, fossils, and other constituents are recognized. These range from unfossiliferous, micritic laminites deposited on the floor of the surrounding basin to tidal-flat and beach limestones deposited on small islands within the interior reef lagoon. The disposition of these facies is summarized in terms of two general models with markedly different paleobathymetric profiles (Figure 4). A rimmed-reef complex (Figure 4A) characterizes the

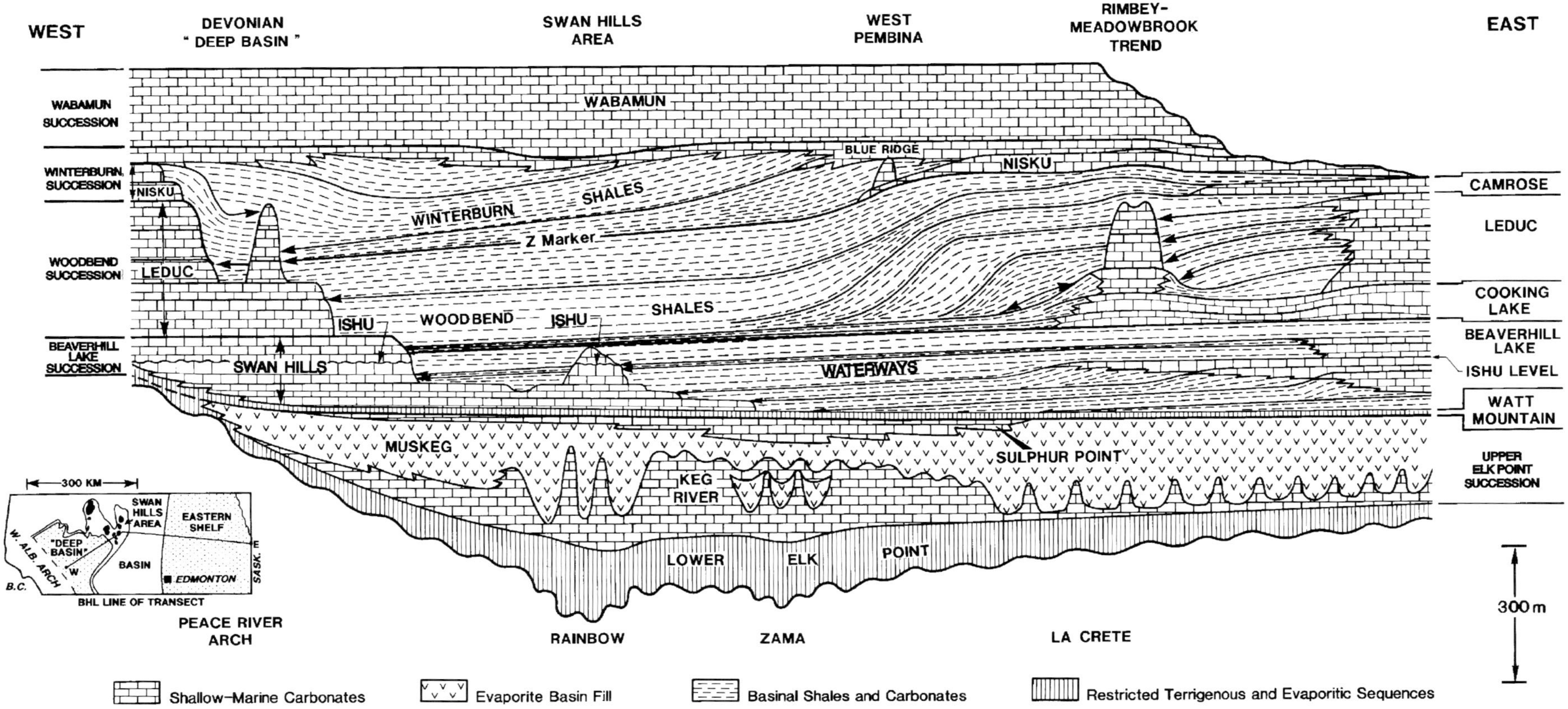

Figure 3. Composite schematic cross section across the Alberta basin illustrating the cyclicity of Devonian successions and the distribution of their major facies. The portion of the cross section above the base of the Watt Mountain Formation is from central Alberta and corresponds to the geographic locations listed above the section. The lower portion of the cross section, below the Watt Mountain Formation, is from northern Alberta and corresponds to the geographic locations listed below the section. The Beaverhill Lake transect extends from the structural Devonian "Deep basin" onto the eastern Beaverhill Lake shelf, whose margin is approximately coincidental with the overlying Leduc Rimbey-Meadowbrook reef trend. Note that the Beaverhill Lake strata on the east side of the Alberta basin display an overall forestepping pattern, whereas the time-equivalent Swan Hills strata on the west side of the basin have an overall backstepping or aggradational pattern. Modified from Wendte (1992b).

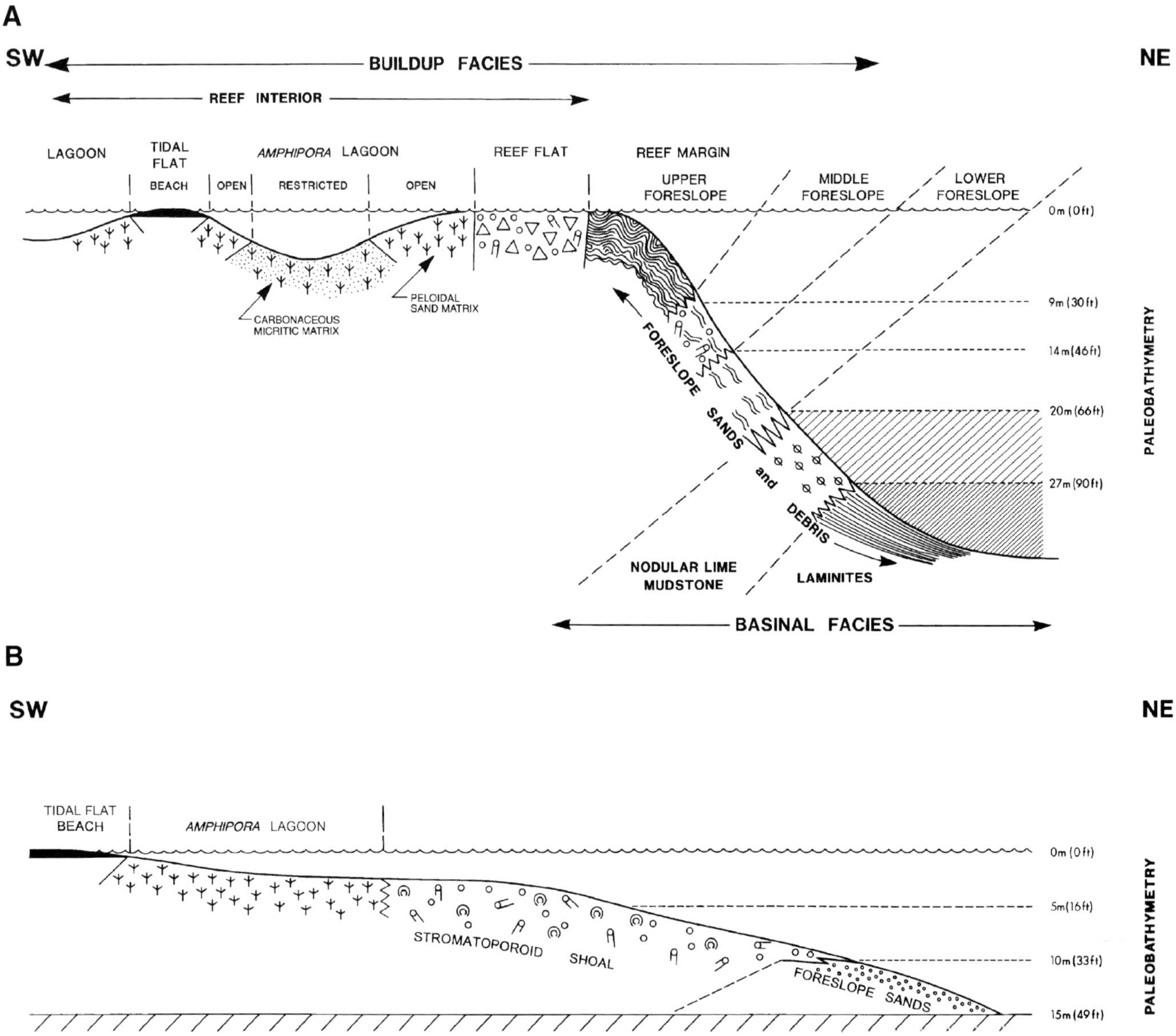

Figure 4. Paleobathymetric profiles summarizing the disposition of facies across the windward, northeasterly sides of the Judy Creek and Snipe Lake complexes. The disposition of facies below the IHSU corresponds to that of a rimmed-reef complex; the disposition of facies above to that of a ramp-bounded shoal complex. Figure 4A is modified from Wendte and Stoakes (1982).

portion of the reefal buildups below the ISHU. The disposition of facies along this profile corresponds to those of Holocene atoll reef complexes with well-developed reef margins and interior lagoons (Emery et al., 1954; Stoddart, 1962). Along this profile, wave energy is focused on the reef margin. The formation of encrusting thick-tabular stromatoporoid boundstones and interbedded debris reflects the wave-swept, turbulent conditions in upper foreslope and reef-margin settings. Basinward, the energy conditions become lower, and the facies are more micritic. This progression is also marked by a change in the general growth form of stromatoporoids. Lower-energy middle foreslope limestones are characterized by branching cylindrical stromatoporoids, most commonly *Stachyodes*. Further basinward, the thin-tabular or wafer morphology of the stromatoporoids in a lower foreslope setting appears to have been an adaptation to growing on a micritic substrate in a zone of slow sedimentation. The bioturbated nodular lime mudstones and relatively unburrowed laminites characterize the deeper-water, more oxygen-deficient environment in the surrounding basin. Foreslope debris and fine peloidal sands may occur anywhere along the reef foreslope because of their gravity-derived nature.

Lagoonward from the reef margin, a belt of debris generally narrower than 400 m characterizes a very shallow, almost emergent reef flat. Farther away from the margin, more micritic sediments were deposited in a protected lagoon. The stick-shaped stromatoporoid

Amphipora adapted to this environment. These limestones range from those with a darker carbonaceous matrix to a slightly higher-energy, shallower facies lacking significant carbonaceous matter and containing a peloidal sand matrix. Cryptalgal mats and beach gravels were deposited on or along small islands in the interior lagoon.

Variations in this model occur both laterally and vertically. For example, energy conditions along some "backstepped" margins and leeward sides of the complexes were not as high as on the windward, northeastern faces of these complexes. In these settings, upper foreslope reef-margin and reef-flat facies exhibit only minor textural and faunal variations and are collectively termed the shoal-margin facies.

A ramp-bounded shoal complex (Figure 4B) characterizes the portion of the reefal buildups above the ISHU. The major difference between depositional systems above and below the ISHU is not the composition but the disposition of facies. Instead of a rimmed-reef complex (Figure 4A), the distribution of facies along the flanks of the complexes above the ISHU corresponds to a ramp (Figure 4B). The facies occur in much wider belts and the transitions between facies are much more gradational. This is especially the case along the windward northeast faces of the shoal complexes where basinward slopes were very gentle. The highest-energy facies correspond to the stromatoporoid shoal limestones that typify relatively deeper and outer positions on the shoal. These limestones are mainly rudstones and floatstones containing a diverse stromatoporoid assemblage and generally a packstone to grainstone matrix. The more protected and shallower *Amphipora* lagoonal facies include rudstones and floatstones and interbedded tidal-flat deposits, analogous to those described beneath the ISHU. Foreslope sands occur in deeper-water flank positions in some locations.

Limestones that make up this shoal phase are time equivalent to deeper-water, more micritic limestones in the surrounding basin. However, because the edge of the shoal complex is backstepped from the margin of the reef phase immediately below the ISHU, *no* physical continuity could be ascertained between the shoal carbonates and the surrounding basinal limestones. Therefore, the basinal limestones are not depicted on the profile on Figure 4B.

Reef Megacycles

Major shifts and changes in facies belts occurred during the evolution of the Judy Creek and Snipe Lake complexes. A number of 8 to 10 m thick megacycles, whose bases correspond to major shifts in facies belts, can be traced across these complexes. In foreslope settings, each megacycle consists of an overall upward-shoaling succession of facies. The succession of facies in a well-developed megacycle in a foreslope to margin setting is illustrated in Figure 5A. The change from one megacycle to a succeeding megacycle, except across the ISHU, is marked by a shift to deeper-water facies.

The contacts between the reefal successions range from sharp to gradational. Below the reef tops, *only* the contact at the ISHU is continuously abrupt. All other contacts, including the base of the reefs, do not exhibit a consistently sharp surface and therefore lack evidence to support an interpretation of a relative sea level fall (see Wendte, 1992c). Therefore, shifts of facies belts at the base of the reefal megacycles, with the exception across the ISHU, are interpreted to be responses to increased rates of relative sea level *rise*. The top of both reefs is a *Trypanites*-bored submarine hardground.

Large changes in sea level, corresponding to a megacycle level, are made up of many small increments of sea level change most accurately recorded by deposition in the interior lagoon. Sediments in the lagoon are inferred to have accumulated at much lower rates than those in the more actively growing reef margin. Consequently, these sediments were outpaced by relative sea level rises that had little or no effect on the more rapidly growing margin. These minor relative rises of sea level produced lagoonal cycles 1 to 3 m thick. Where completely developed, each lagoonal cycle grades upward from dark carbonaceous and micritic *Amphipora* rudstones to light brown, peloidal *Amphipora* rudstones into tidal-flat or beach deposits (Figure 5B). These lagoonal cycles are difficult to correlate across the reefs because of variations in cycle frequency and facies due to the occurrence of small islands and variations in sediment accumulation rates. Therefore, they were not correlated throughout the reef interior. However, megacycle contacts from the reef margin were correlated to pronounced lagoonal cycle bases near the periphery of the reefs. The bases of these cycles were then correlated throughout each reef complex.

Figures 6 and 7 show the stacking pattern of the megacycles along the windward, northeastern margin of the Snipe Lake and Judy Creek complexes. Figure 6 is a cross section from the base of the underlying platform up to a level slightly above that of the ISHU at Snipe Lake. Figure 7 is a cross section along a comparable windward, northeastern face of the Judy Creek complex. The section encompasses the stratigraphic interval from the upper part of the platform up into the shoal complex above the ISHU.

The spatial configuration and the stacking patterns of the reefal megacycles up to the ISHU at both Snipe Lake and Judy Creek show a remarkable similarity (Figures 6 and 7). Four major successions occur in each complex. Each megacycle at Snipe Lake is approximately 8 m thick; those at Judy Creek are approximately 10 m thick. *No* thinning of megacycles approaching the ISHU was identified in either complex. The first megacycle in both complexes shows a progradational manner of sedimentation. The second megacycle in each complex displays an upbuilding style in relationship to the position of the pre-existing margin. The third megacycle shows a gradual retreat of the reef margins. The reef margins in the fourth megacycle backstep to the southwest. Above the

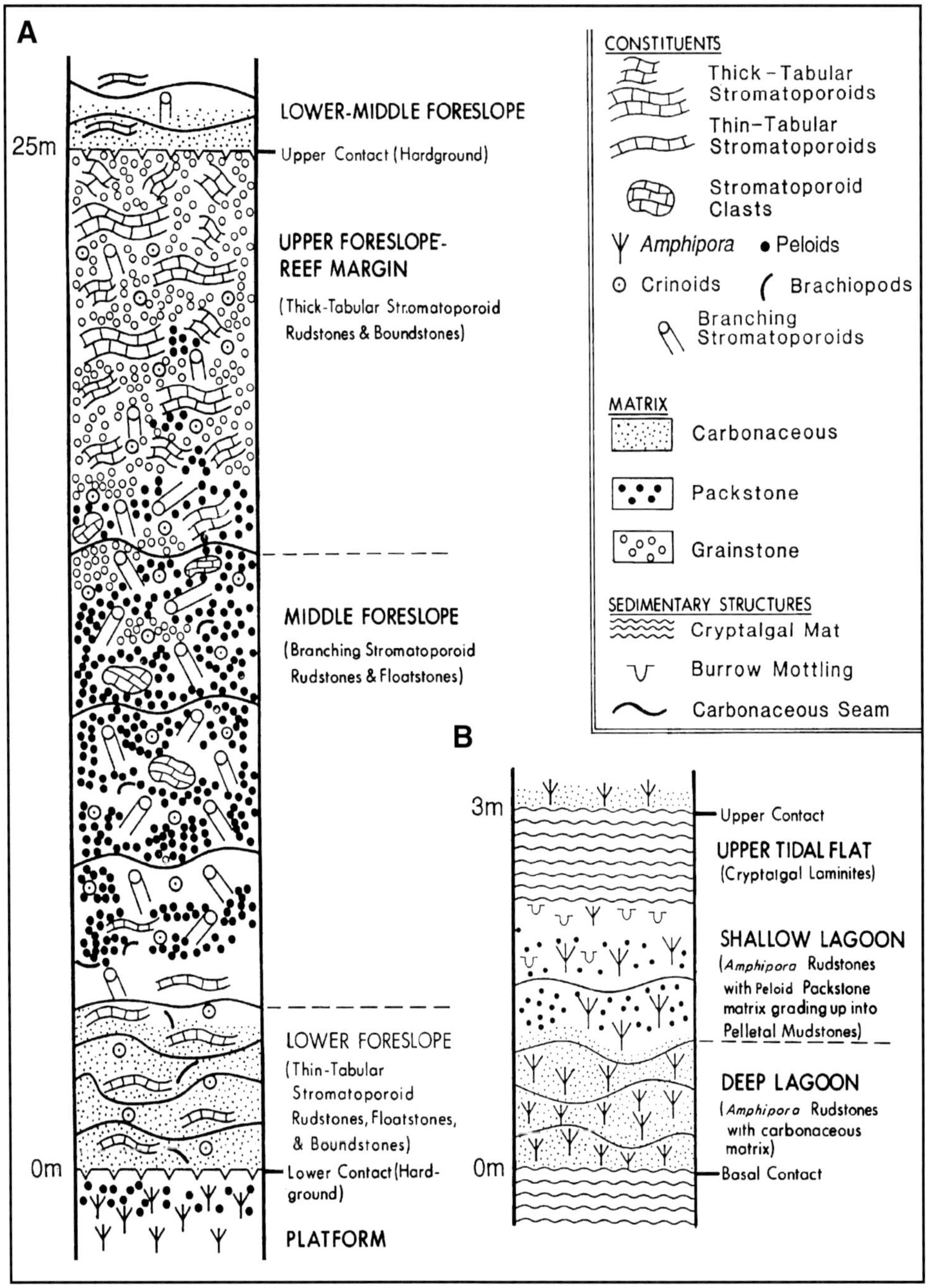

Figure 5. Succession of facies in a reefal megacycle along a foreslope to margin setting and that in a lagoonal cycle. Modified from Wendte (1992b).

ISHU, the Swan Hills succession shows a gradually retreating to a more pronounced backstepping style of sedimentation (see Wendte, 1992c, his figures 4 and 12). This style is reflected in the vertical succession in the 10-6 well at Judy Creek above the intraformational unconformity (Figure 7). In ascending order, tidal-flat deposits (too thin to show in Figure 7) are overlain by a beach-capped *Amphipora* lagoonal cycle which, in turn, is overlain by more open-marine stromatoporoid-shoal limestones of the succeeding cycle.

The stacking pattern along the windward portion of Judy Creek shows more variability than at Snipe Lake. This is due to the difference in the orientation of the eastern margin in each complex. At Snipe Lake, the entire front faces to the northeast, normal to the prevailing northeasterly wind-wave circulation. This results in a consistent stacking arrangement shown in Figure 6. At Judy Creek, the configuration of this side of the complex is more variable. Only a pronounced nose, from where the transect shown in Figure 7 was constructed, faces to the northeast. Elsewhere, this side faces due east and to the southeast. Cross sections through these lower-energy "sides" of the complex display a backstepping, rather than an upbuilding to

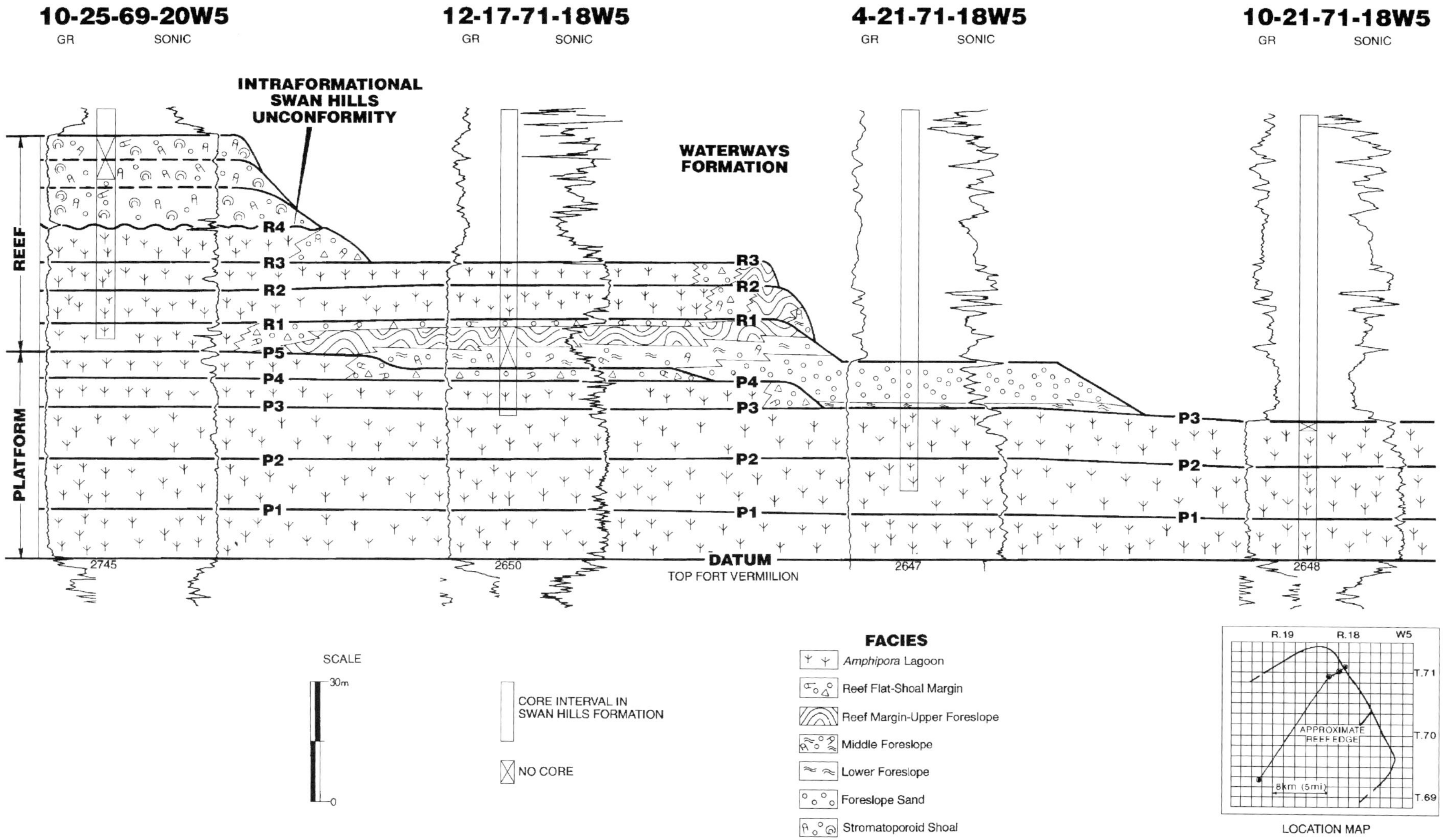

Figure 6. Megacycle-facies cross section across the windward side of the Snipe Lake platform and reef complex. The platform succession is divided into five megacycles, whose tops are numbered P1, P2, P3, P4, and P5. The overlying reef complex consists of a lower rimmed-reef succession consisting of four megacycles, whose tops are labeled R1, R2, R3, and R4, and an upper ramp-bounded shoal succession. The ISHU is at the R4 level and separates the two major reefal successions. Note that the unconformity is preceded by a major reefal backstep at the R3 level.

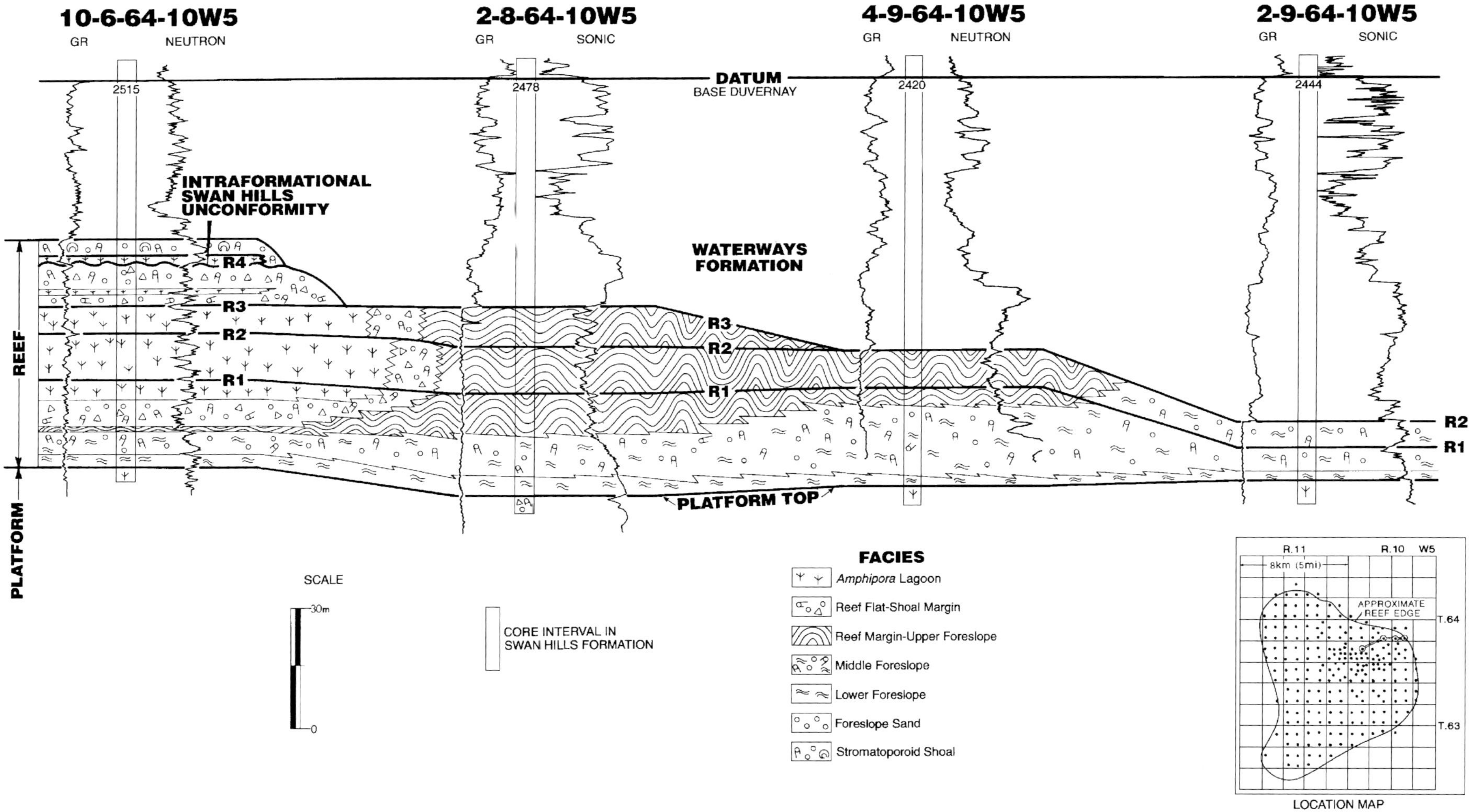

Figure 7. Megacycle-facies cross section across the windward northeasterly face of the Judy Creek reef complex. The reef complex, like that at Snipe Lake, nucleated on an underlying platform high (to the southwest of this section). The reef complex consists of a lower rimmed-reef succession consisting of four megacycles, whose tops are labeled R1, R2, R3, and R4, and an upper ramp-bounded shoal succession. The ISHU occurs at the R4 level and separates the two major reefal successions. Note that the unconformity is preceded by a major reefal backstep at the R3 level.

backstepping, pattern of stacking (see Wendte, 1992c, his figure 9).

The overall stacking of the reefal successions in relationship to the ISHU is summarized in Figure 8. This schematic section is based more on Judy Creek than Snipe Lake, because of the core control across the entire complex. The ISHU separates the rimmed-reef succession from the ramp-bounded shoal complex. Despite this difference, a backstepping to retreating deposition of depositional facies occurs both *beneath* and *above* the ISHU.

LITHOLOGIC CRITERIA FOR IDENTIFYING SUBAERIAL EXPOSURE DUE TO A RELATIVE SEA LEVEL FALL

The following four criteria from core examination led to the recognition of the emergent surface at the top of the fourth reef megacycle at both Judy Creek and Snipe Lake and to the interpretation of a relative sea level fall.

First, the contact at this level was *continuously* lithified across the Judy Creek and Snipe Lake reefs. The obviously cemented nature of the contact from a well at Snipe Lake is shown in Figure 9A. The contact is erosional, with a thin (1–6 cm) green shale bed overlying the contact. The near-vertical face of the contact and incorporation of angular lithoclasts in the overlying shale bed indicate early lithification, prior to the deposition of the shale bed. This green shale is interpreted as a storm-event deposit and is one of several thin-bedded green shale storm beds in the reef complexes.

Second, shallower-water facies, mainly tidal-flat deposits, overlie deeper-water subtidal limestones across the contact. This superposition of facies is opposite to all other megacycle and cycle contacts in both reef complexes. This unique superposition is illustrated in core photos in Figures 10 and 11 and on the close-up core photo in Figure 9B.

Third, solution vugs that are partially filled with geopetal green shale occur beneath the ISHU. The vugs are most abundant within 0.3 m of the unconformity and were not observed more than 2.3 m below the ISHU. Most of the vugs range from equidimensional voids up to 2 cm across (Figure 9B) to elongate oblique voids up to about 6 cm long (Figure 9C). A thin interval of poorly preserved core rubble containing mixed green shale and limestone occurs immediately beneath the ISHU in some locations (Figure 11). The green shales in these poorly preserved intervals are interpreted as the infill of vugs greater than the width of the core. The geopetal green clay infilling all these solution vugs was clearly derived from the storm event that deposited the thin green shale bed overlying the ISHU at other locations (see Figures 9A and 9E). These vugs, then, formed prior to the deposition of the storm bed. Leaching by meteoric water is the most likely explanation.

Fourth, oxidation of geopetal green clay occurs in some cores immediately beneath the unconformity. The close-up core photo in Figure 9D shows the oxidation of these deposits where they filled in honeycombed vugs in a geopetal manner beneath the contact. The lack of iron oxides in the same green clay illustrated in Figure 9E, on the reverse side of this core, emphasizes the patchy occurrence of the iron oxides.

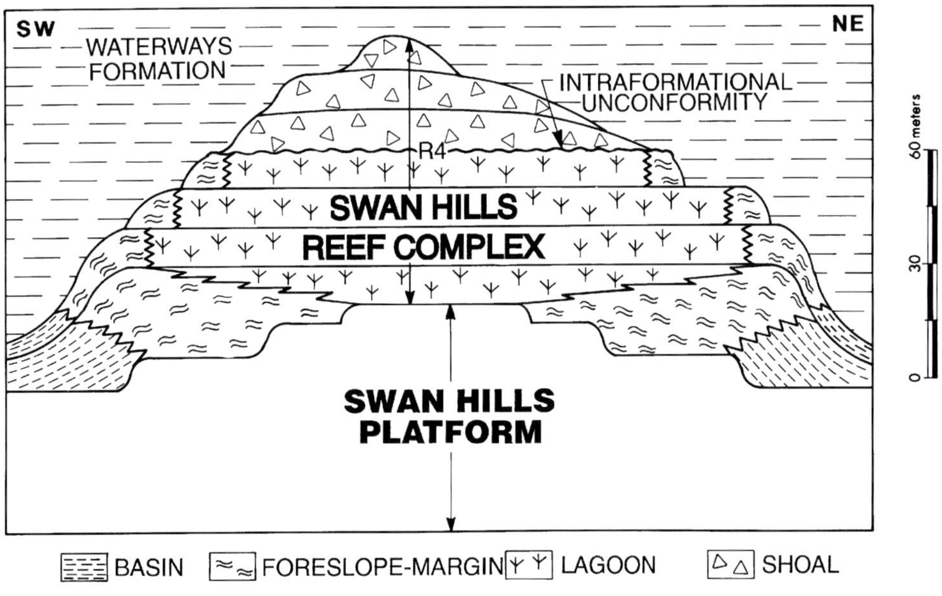

Figure 8. Schematic cross section summarizing the stacking relationships of the reefal megacycles both below and above the ISHU. The unconformity is both preceded and followed by an overall backstepping to retreating disposition of facies. The leeward, southwest side of this model is based solely on the study at Judy Creek.

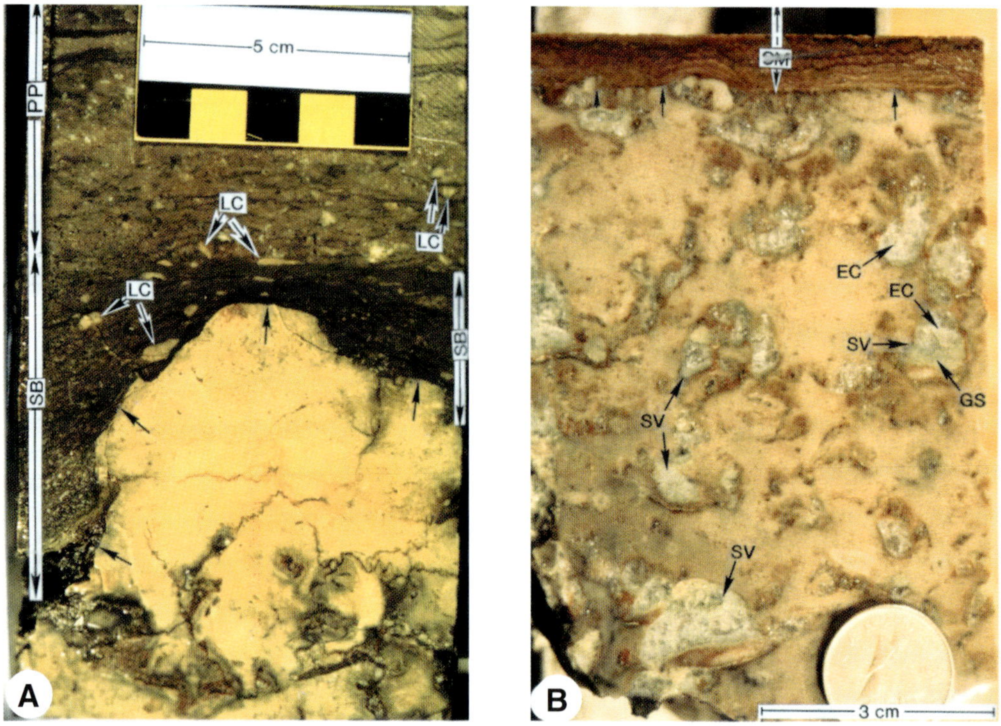

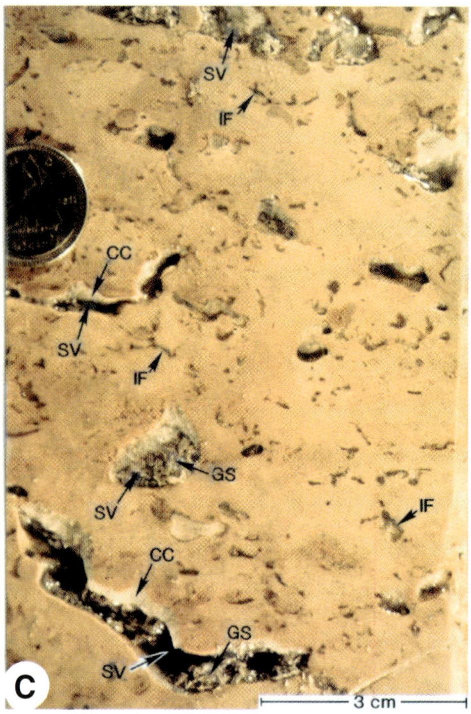

Figure 9. Core photographs of depositional and diagenetic features associated with the ISHU. (A) Unconformable surface with an erosional, near-vertical face (arrows). Limestone below the ISHU is a light brown, pelletal wackestone. The unconformable surface is overlain by a thin storm bed (SB) containing angular lithoclasts (LC) in an argillaceous micritic matrix. The lithoclasts are identical in composition to the underlying wackestone and are undoubtedly derived from it. The preservation of the near-vertical face and the angular shapes of the overlying lithoclasts indicate lithification of the underlying deposit prior to erosion and deposition of the storm bed. The interval above the storm bed is a peloidal packstone (PP) with some reworked lithoclasts from below and reflects deposition following the high-energy storm event. Cryptalgal limestones immediately overlie this sample (see Figure 11). Snipe Lake reef complex, 10-25-69-20W5, 2657.9 m (8720 ft). (B) Crinkly cryptalgal mat (CM) overlying the ISHU surface. The surface is sharp and scalloped (bored?) (arrows). The underlying deposit is a subtidal, bioturbated pelletal mudstone. A period of emergence is indicated by the lithified nature of the contact and by the occurrence of solution vugs (SV) below. The vugs are filled by geopetal green shale (GS) and by later coarse equant calcite cement (EC). This geopetal green shale records deposition from the same storm event that deposited the thin shale beds above the ISHU illustrated in Figures 9A and 9E. The superposition of shallow-water deposits (tidal-flat cryptalgal mats) over deeper-water (subtidal) deposits indicates emergence was due to a relative lowering of sea level. Judy Creek reef complex, 10-3-64-11W5, 2683.3 m (8803.5 ft). (C) Linear solution vugs in a fenestral pelletal wackestone 2.3 m below the ISHU. The linear solution vugs (SV) are completely filled by green argillaceous geopetal sediment (GS) and by overlying, possibly pendant calcite cements (CC). Most of the argillaceous geopetal sediment was removed by acid etching of the core slab. Small irregular fenestrae (IF) are calcite cemented. Snipe Lake reef complex, 10-25-69-20W5, 2660.1 m (8727.5 ft).

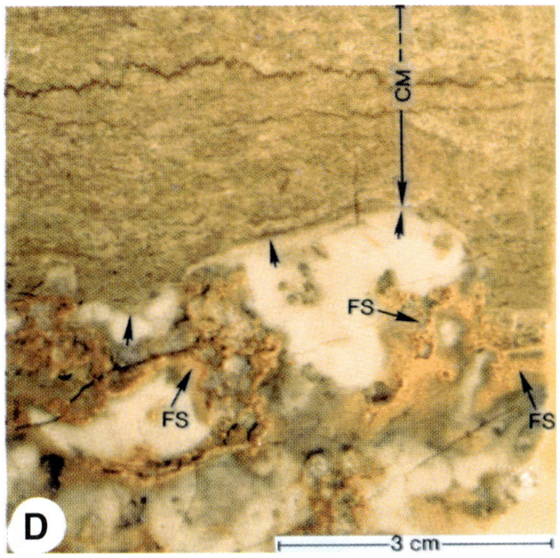

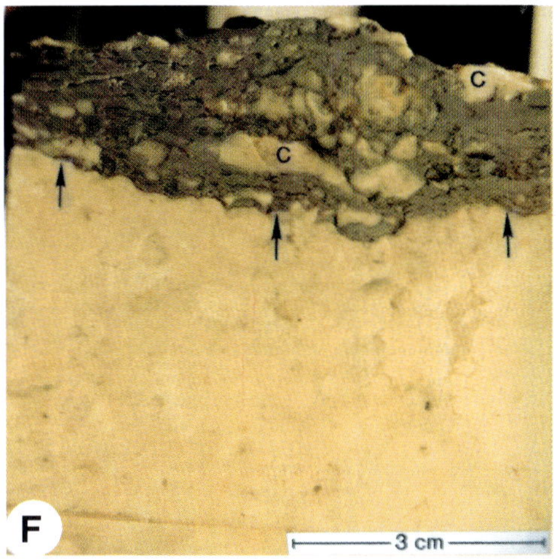

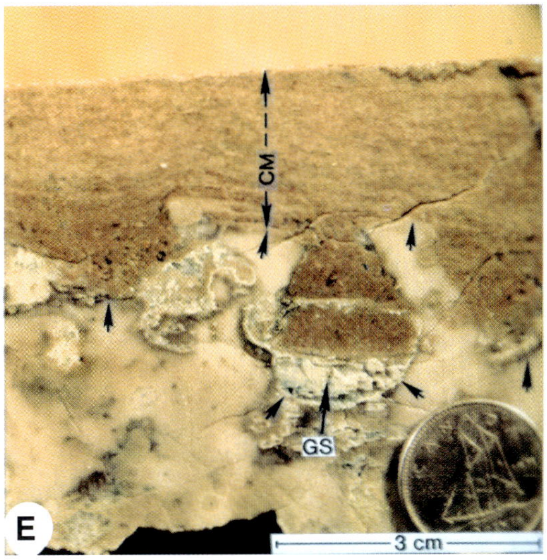

Figure 9 (*continued*). (D) Honeycombed vugs immediately beneath the ISHU surface (arrows). These vugs are filled by reddish, iron-oxidized argillaceous sediment (FS). On the other side of the core, this argillaceous deposit is unoxidized and retains its green color (see Figure 9E). Peloidal cryptalgal mats deposits (CM) overlie the ISHU surface. Judy Creek reef complex, 4-23-63-11W5, 2753.5 m (9034 ft). (E) Green shale storm-bed deposit (GS) filling a low above the ISHU surface (arrows). Peloidal cryptalgal mats (CM) overlie the ISHU surface; bioturbated pelletal wackestones occur below. Judy Creek reef complex, 4-23-63-11W5, 2753.5 m (9034 ft). (F) Thin green shale storm bed 1.5 m below the ISHU surface. The storm bed has an erosional base (arrows) and contains clasts (C) derived from the underlying, bioturbated pelletal wackestones. Judy Creek reef complex, 8-36-63-11W5, 2760.1 m (9055 ft).

At Snipe Lake the green shale storm bed is conformably overlain by a few centimeters of peloidal packstones and then cryptalgal mats (Figures 9A and 11). The top of the green shale storm bed was unlithified prior to deposition of the overlying sediments. Thus, we interpret that the green shale storm bed was deposited after emergence but just prior to reflooding of the reef complexes.

Jointly these observations led to the interpretation of a lowering of relative sea level (Wendte, 1987). The widespread superposition of tidal-flat facies unconformably overlying a subtidal limestone implies a period of emergence between two phases of submergence. Analogous evidence at the same level in the backstepped bank carbonate to the southwest led Kaufman and Myers (1988) to suggest a similar interpretation. The exact magnitude of the sea level drop is hard to determine and remains speculative. Two key observations, in this regard, are the lack of any "lowstand" reefal development on any prior reef terrace (or step) at either Judy Creek or Snipe Lake and that leached vugs have not been identified more than 2.3 m below the ISHU. Prominent drowned terraces approximately 8 m below the intraformational unconformity at Snipe Lake and approximately 10 m at Judy Creek would have provided ideal sites for carbonate rejuvenation. We therefore interpret a drop of only a few meters. We postulate the following succession:

1. Gradually diminishing rates of relative sea level rise during the fourth reefal megacycle.
2. A small increment of sea level drop, terminating growth along the reef tops.
3. A subsequent gradual rise in sea level allowing for onlap of tidal-flat deposits onto the previously emergent reef top.

The distribution of facies immediately overlying the ISHU supports a gradual rise in sea level following exposure. At Judy Creek, these facies occur in almost a concentric pattern as illustrated in Figure 12. Dark-colored cryptalgal mats (see Figure 10) occur in the center of the reef complex, and are surrounded by a belt of light-colored cryptalgal mats and an outer belt of subtidal deposits. The dark-colored tidal-flat sediments are interpreted to reflect the lower replenishment of seawater and, as a consequence, oxygen deficiency in the more restricted interior of the complex. These conditions permitted the preservation of carbonaceous matter. Improved circulation and exchange of waters

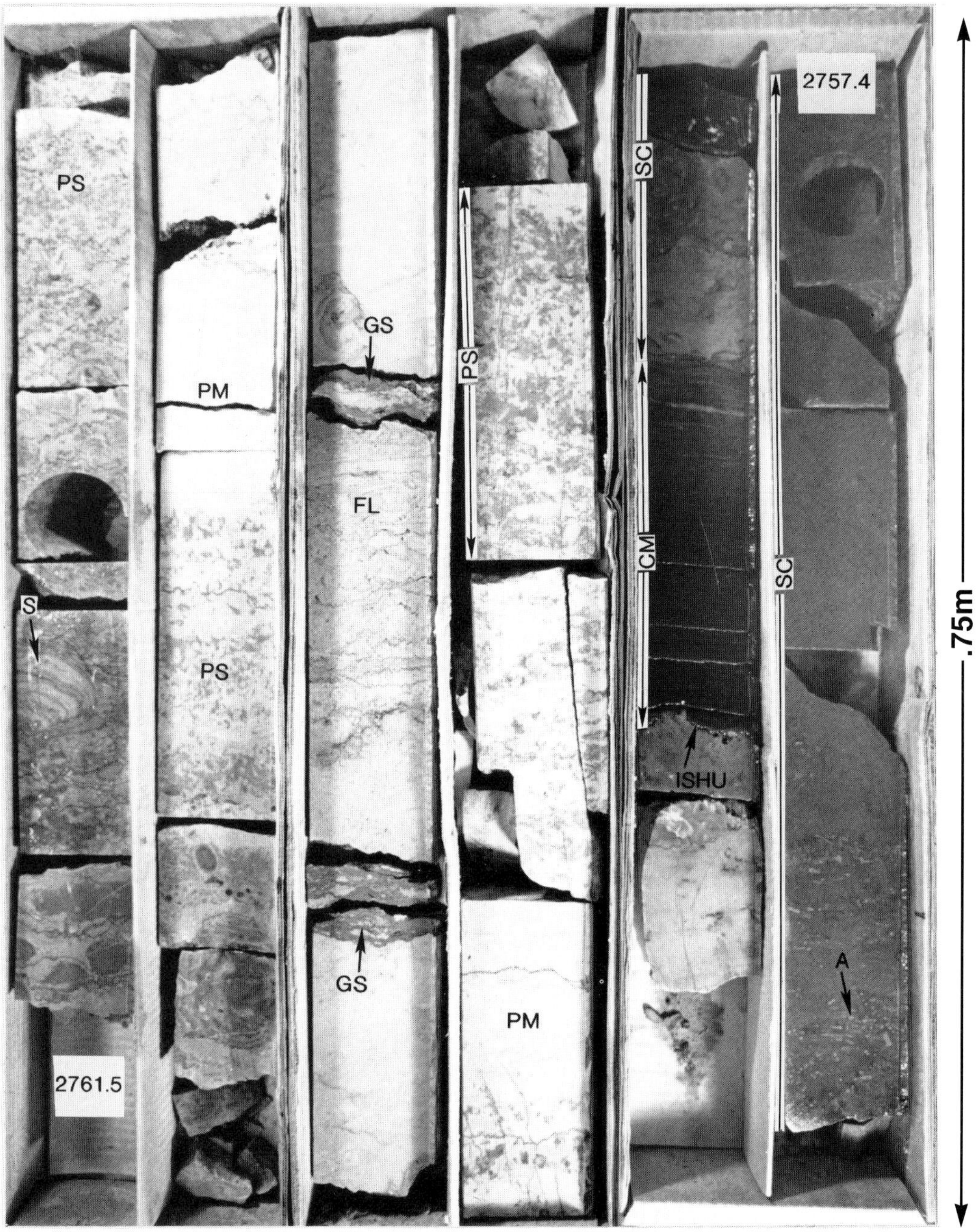

Figure 10. Succession of core below and above the ISHU in the Judy Creek reef complex. Stromatoporoid clasts (S) set in a peloidal sand matrix (PS) occur near the base of the core. These peloidal sands grade up into pelletal mudstones and wackestones (PM) and fenestral limestones (FL) with two thin green shale storm beds (GS). Figure 9F shows a close-up photograph of the upper green shale storm bed. The ISHU contact is marked by a stylolite. A dark cryptalgal limestone (CM) approximately 0.3 m thick overlies the unconformable surface and is, in turn, overlain by peloidal packstones with *Amphipora* (A) of the succeeding cycle (SC). Note the occurrence of patchy peloidal sands (PS) approximately 0.3 m below the ISHU surface. Dark oil-stained peloidal sands fill burrows in the lighter colored, low-porosity pelletal mudstone succession. Oil staining occurs in interparticle pores in the peloidal sands and along some vertical fractures. The presence of primary (interparticle) porosity in these peloidal sands and the lack of porosity in the associated pelletal mudstones only 0.3 m below the ISHU reflect facies control on porosity. Obviously, subaerial freshwater cements at most only partially filled the interparticle pores and did not create a dense, impermeable succession. Bottom of core is to lower left; top is to upper right. 8-36-63-11W5, 2757.4–2761.5 m (9037–9060 ft).

Figure 11. Succession of core below and above the ISHU in the Snipe Lake reef complex. Light-colored pelletal mudstones and wackestones occur beneath the ISHU contact. Green shale-filled solution vugs (SV) are present down to 2.3 m below the unconformity. Figure 9C shows a close-up photograph of some of these solution vugs. The unconformity is overlain by a 1–6 cm thick green shale storm bed (SB), illustrated in a close-up photo in Figure 9A. A light-colored, peloidal cryptalgal succession (CM) overlies the storm bed and, in turn, is overlain by a stromatoporoid-bearing limestone of the succeeding cycle (SC). The basal contact of this cycle is marked by a dashed line. Note the green shale zone (GS) less than 0.5 m below the unconformity. Limestone clasts (LC) separated by stringers of green shale occur at the base of this zone and are considered to be of solution-collapse origin. This green shale accumulation is interpreted to be the infill of a vug wider than the width of the core. Bottom of core is to lower left; top is to upper right. 10-25-69-20W5, 2556.5–2660.3 m (8387–8728 ft).

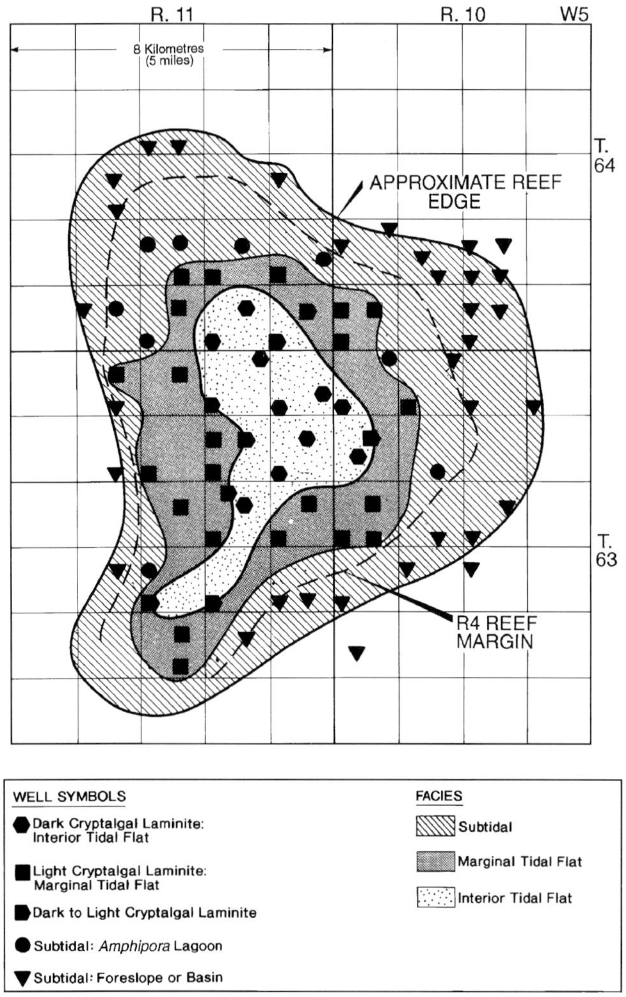

Figure 12. Map showing the nearly concentric distribution of tidal-flat and subtidal facies immediately above the ISHU at Judy Creek. The facies pattern records the initial disposition of facies after reflooding of the reef complex.

around the periphery of the complex resulted in oxidation of the carbonaceous matter. The outer belt merely records the permanently submerged setting around the entire complex. Overlying facies, above the tidal-flat deposits, exhibit the anatomy of a retreating and backstepping, ramp-bounded shoal complex (see Wendte, 1992c, his figure 4).

EVOLUTION OF LAGOONAL CYCLES BELOW THE ISHU

At both Judy Creek and Snipe Lake, the lagoonal cycles exhibit differences upon approaching the ISHU. Three changes in the fourth reef megacycle are practically identical in both reef complexes and are unique to this stratigraphic interval.

First, the dark carbonaceous *Amphipora* rudstones and floatstones that characterize the basal portions of many lagoonal cycles are not present. Below the ISHU, the highest occurrence of these limestones is in the basal part of the fourth megacycle. The upper 6 to 8 m of this succession lacks the carbonaceous facies. Second, the middle and upper portions of this megacycle consist entirely of light-colored limestones, including more tidal-flat cryptalgal mats and various fenestral limestones. These lighter colored limestones are mainly pelletal mudstones and wackestones, characterized by more abundant micrite and by a lower concentration of *Amphipora*. The dominantly micritic aspect of this interval results in the most widespread, low-porosity layer in these reef complexes. The core photographs in Figures 10 and 11 illustrate these differences in both the Judy Creek and Snipe Lake complexes. Third, storm beds containing green clay increase in abundance in the interval immediately beneath the ISHU. These beds are commonly 1 to 2 cm thick, have erosional bases, and consist of micrite and carbonate bioclasts, intraclasts, and lithoclasts as well as green clay (Figure 9F). These characteristics support our interpretation of episodic storm-event accumulations. The siliciclastic components in these thin-bedded shales are identical to those in the adjacent Waterways Formation (see Murray, 1965, 1966; Havard and Oldershaw, 1976) and are presumed to be derived from it.

Three green shale beds are common, but not ubiquitous. The lower two occur approximately 1 and 2 m below the ISHU (Figure 10). The upper green shale overlies the ISHU and is illustrated on the close-up core photographs in Figures 9A and 9E. The patchy distribution of the green shales is related to both their irregular depositional pattern and their preservation potential in various environments. Where large primary and secondary voids existed beneath the sediment-water interface, green shales partially to completely filled these voids in a geopetal manner. One particularly significant occurrence of geopetal green shale is beneath the ISHU (Figures 9B and 9C).

The abundant occurrence of green shale storm beds just below and above the ISHU corresponds to the maximum regression of shelfal carbonates and encroachment of basinal clinothem deposits from the east side of the Devonian Alberta basin (Figure 3). Wendte and Stoakes (1982) explained this relationship as follows. Occasional major storms rework sediment from the basinal clinothems. At times of stable sea level when the basinal clinothems encroach upon the reef complexes, the reworked clay and lime mud is transported onto the reef where it is incorporated with reef-derived debris. The abundance of the thin green shales on the reefs then records times of stable (still-stand or slightly falling) relative sea level.

These three changes foretell the "arrival" of the ISHU. The absence of the "deep"-water lagoonal facies, the corresponding increase in the abundance of shallower lagoonal facies, and the abundance of the thin green shale storm beds document the gradual and progressive loss of accommodation space prior to emergence. As such, the emergence records the culmination of a prolonged phase of either regional or eustatic lowering of sea level. No other widespread subaerial unconformities occur either immediately below or above the ISHU. This negates the possibility of other, more high-frequency relative sea level drops, corresponding to a level of the thin lagoonal cycles. This succession contrasts with Pleistocene carbonates affected by glacial eustatic sealevel fluctuations (Perkins, 1977; Beach, 1982; Wanless and Dravis, 1989). The Pleistocene subaerial surfaces repeat at a high frequency, and commonly no marked depositional lithological changes occur below the unconformable surfaces. We conclude, therefore, that the withdrawal of the sea that led to the formation of the ISHU was *not* due to a rapid, high-amplitude (Pleistocene-like) glacial eustatic lowering of sea level.

EFFECT ON RESERVOIR QUALITY

Our studies and those of other investigators (Wong and Oldershaw, 1981; Walls and Burrowes, 1985, 1989) show that subaerial exposure has had a comparatively minor effect on the reservoir quality of isolated Swan Hills reef complexes. Over 90% of the porosity within these reef complexes is of primary depositional origin. Secondary dissolution voids provide only minor storage capacity. Porosity is facies controlled (Jardine et al., 1976; Wendte and Stoakes, 1982), with higher-energy facies having more porosity than lower-energy facies which are commonly tight. This relationship is illustrated in the core photographs in Figure 13. Walls and Burrowes (1985) postulated that porosity in Swan Hills reef interior facies has been reduced, on average, from approximately 50 to 9% during diagenesis. They estimated subaerial cements represent about 20% of the total cement volume in the reef interiors and only resulted in a 5% porosity reduction overall. The one important consequence of subaerial diagenesis that Walls and Burrowes (1989) cited is the formation of thin discrete permeability barriers caused by a combination of depositional stratification and cementation.

Subaerial exposure at the ISHU had only a minor effect on the reservoir quality of underlying limestones at both Judy Creek and Snipe Lake. As previously

Figure 13. Core photographs illustrating facies control on porosity. High-energy facies deposited along the reef margin have high primary porosities, whereas lower-energy facies, commonly deposited in the reef lagoon, have lower porosities. (A) Highly porous reef margin-upper foreslope facies with stromatoporoid debris overlying a thick-tabular stromatoporoid (TS) which may be in growth position. Pores are dark areas and include abundant intraparticle and some fracture voids in the tabular stromatoporoid, and abundant interparticle, intraparticle, and shelter (S) voids in the overlying debris. Judy Creek reef complex, 4-9-64-10W5, 2491.4 m (8174 ft). (B) Lower-porosity lagoonal facies consisting of *Amphipora* rudstone with a dense carbonaceous micritic matrix. All original intraparticle pores within the *Amphipora* coenostea are completely filled by equant calcite cement. Judy Creek reef complex, 4-24-63-11W5, 2691.7 m (8831 ft).

noted, solution vugs are limited to the zone directly below the unconformity. Where present, these voids are filled by a combination of geopetal green shale and later coarse equant calcite cement (Figures 9B and 9C). In fact, the interval immediately beneath the unconformity is the most widespread, low-porosity zone in both reef complexes. The overall tight nature of this zone at both Judy Creek and Snipe Lake is illustrated on the core photographs in Figures 10 and 11.

The tight zone corresponds to the occurrence of pelletal lime mudstones and wackestones in both complexes. The lithified nature of the contact indicates that subaerial cementation resulted in some porosity loss (Figures 9A, 9B, 9D, and 9E). However, interstratified peloidal grainstones still retain significant primary porosity (Figure 10). Therefore, low porosity below the ISHU relates more to depositional facies control than subaerial lithification, and most porosity was lost during later burial.

Four possible explanations for the lack of enhanced porosity below the ISHU are possible. First, the apparent small magnitude of the sea level drop and presumably the brief interval of emergence provide only limited exposure to undersaturated fresh water. The amount of leaching should, therefore, be relatively minor and confined to the interval directly beneath the unconformity. Second, the prevailing arid climate during the Late Devonian would limit the degree of meteoric water recharge during the emergence. Time-equivalent anhydrite and salt deposits, equivalent to the Swan Hills Formation, occur in more restrictive settings in western Canada (see Meijer Drees, 1986). Third, the original mineralogy of Devonian limestones was dominantly calcitic (James and Choquette, 1990). As such, this system should be much less reactive than the mixed aragonite-calcite system in Holocene and Pleistocene tropical carbonates. As discussed by Matthews (1974), fresh water in calcitic systems tends to achieve equilibrium conditions relatively quickly with only minor diagenetic alteration. Porosity modification by either cementation or dissolution from diffuse flow should be minor. Fourth, limestones immediately below the ISHU tend to be micritic and, hence, less permeable to diffuse meteoric flow.

SUMMARY AND DISCUSSION

This paper focuses on three aspects of a regional intraformational unconformity. First, criteria to interpret emergence and subaerial exposure due to a relative drop of sea level were presented. Diagenetic features such as soil horizons, caliche crusts, vadose cements, and dissolutional fabrics have been documented from many subaerial surfaces (see reviews by Esteban and Klappa, 1983; James and Choquette, 1990). However, the occurrence of these features, alone, is insufficient to demonstrate a base-level drop in sea level, because deposition of shoaling-upward successions may also culminate in exposure. We emphasize integration of stratigraphic and facies relationships in addition to documenting the presence of diagenetic fabrics ascribed to meteoric diagenesis. The following observations at both Judy Creek and Snipe Lake were vital to our interpretation: (1) the occurrence of a continuously lithified contact; (2) evidence of meteoric diagenesis, such as the occurrence of dissolution vugs and iron oxidation in the underlying deposits; and (3) the superposition of shallower-water (tidal-flat) over deeper-water (subtidal) deposits across the contact. The presence of a widespread emergent surface imposed on subtidal deposits requires a lowering of sea level. The onlap of shoreline (tidal-flat) deposits onto this surface records the reflooding of both reef complexes.

Second, the occurrence of the ISHU is related to the stratigraphic successions both below and above the surface. On the windward sides of both the Judy Creek and Snipe Lake complexes, the correlatable and mappable 8 to 10 m thick megacycles beneath the ISHU exhibit a gradually retreating to more pronounced backstepping facies stacking pattern. This style contrasts with the stacking patterns postulated in the carbonate sequence-stratigraphic model of Sarg (1988). Sarg's model shows a seaward progradational phase of facies in the "highstand systems tract" prior to the major base-level drops forming the *sequence boundaries*. The retreating to backstepping facies stacking pattern above the ISHU conforms to the style of the "transgressive systems tract" in the Sarg model.

The stacking patterns at Judy Creek and Snipe Lake, especially beneath the ISHU, reflect the depositional setting of these two complexes. Time-equivalent Beaverhill Lake stages on the east side of the Alberta basin, unlike those on the windward, eastern sides of Judy Creek and Snipe Lake, comprise a forestepping succession into the basin (Figure 3). Also, "reefal overhangs" on the western side of some Swan Hills reef complexes, including Snipe Lake (Fischbuch, 1968) indicate markedly contrasting stacking styles between the windward and leeward sides of some of these buildups. We relate both the regional and local variations in stacking patterns to an additional control, the northeasterly wind-wave circulation driven by the Late Devonian trade winds. Thus, stacking patterns of depositional facies are controlled by factors other than changes in sea level and should not be used as the sole criterion for recognition of subaerial unconformities and other *sequence boundaries*.

Goldhammer et al. (1993) have identified *sequence boundaries* in carbonates based solely on accommodation changes expressed through "Fischer plots." This relationship is based largely on the published siliciclastic models presented in Jervey (1988) and Posamentier et al. (1988). These models postulate diminishing amounts of accommodation below the *sequence boundary* and increasing amounts of accommodation above the *sequence boundary*. At both Judy Creek and Snipe Lake, there is no progressive thinning at the megacycle level below the ISHU. However, the lagoonal cycles in both complexes show consistent and significant lithologic changes in the intermediate and upper portion of the fourth reefal megacycle. These changes show a total lack of "deep" lagoonal facies and a corresponding increase in shallow-water facies below the ISHU and are interpreted as representing diminished accommodation. These changes are consistent with those of the Jervey (1988) and Posamentier et al. (1988) models and that postulated by Goldhammer et al. (1993). We cannot, however, demonstrate the progressive thinning of lagoonal cycles because we could not correlate them in a three-dimensional manner throughout the entire buildups.

Third, our studies show that meteoric waters emanating from the ISHU have had little impact on reservoir quality at Judy Creek and Snipe Lake. In fact, the most widespread, low-porosity interval in both complexes corresponds to the pelletal lime mudstones and wackestones that underlie the ISHU. Low porosities in these deposits are largely due to depositional and facies controls. The minimal effect of meteoric diagenesis may be attributed to the small magnitude and brief duration of the unconformity, the prevailing arid climate, the dominant original calcitic mineralogy of the limestones, and the low permeability of the host sediments.

ACKNOWLEDGMENTS

Reviewers Ray Garber, Neil Hurley, and Art Saller made many constructive recommendations and suggestions. Their reviews helped to significantly improve the manuscript.

REFERENCES CITED

Beach, D.K., 1982, Depositional and diagenetic history of Pliocene-Pleistocene carbonates of northwestern Great Bahama Bank; evolution of a carbonate platform: Ph.D. dissertation, University of Miami, Florida, 447 p.

Coogan, A.H., D.G. Bebout, and C. Majjio, 1972, Depositional environments and geological history of Golden Lane and Poza Rica trends, Mexico, an alternative view: AAPG Bulletin, v. 56, p. 1419–1477.

Emery, K.O., J.I. Tracey, and H.S. Ladd, 1954, Geology of Bikini and nearby atolls: U.S. Geological Survey Professional Paper 260-A, 265 p.

Esteban, M., and C.F. Klappa, 1983, Subaerial exposure, *in* P.A. Scholle, D.G. Bebout, and C.H. Moore, eds., Carbonate Depositional Environments: AAPG Memoir 33, p. 1–54.

Fischbuch, N.R., 1968, Stratigraphy, Devonian Swan Hills reef complexes of central Alberta: Bulletin of Canadian Petroleum Geology, v. 16, p. 446–587.

Goldhammer, R.K., P.J. Lehmann, and P.A. Dunn, 1993, The origin of high-frequency platform carbonate cycles and third-order sequences (Lower Ordovician El Paso Group, west Texas): constraints from outcrop data and stratigraphic modeling: Journal of Sedimentary Petrology, v. 63, p. 318–359.

Havard, C., and A. Oldershaw, 1976, Early diagenesis in back-reef sedimentary cycles, Snipe Lake reef complex, Alberta: Bulletin of Canadian Petroleum Geology, v. 24, p. 27–69.

James, N.P., and P.W. Choquette, 1990, Limestones—the sea floor diagenetic environment, *in* I.A. McIlreath and D.W. Morrow, eds., Diagenesis: Geoscience Canada Reprint Series 4, p. 13–34.

Jardine, D., D.P. Andrews, J.W. Wishart, and J.W. Young, 1976, Distribution and continuity of carbonate reservoirs: Society of Petroleum Engineers, 6139, p. 1–7.

Jervey, M.T., 1988, Quantitative geological modeling of siliciclastic rock sequences and their seismic expression, *in* C.K. Wilgus, B.S. Hastings, C.G.St.C. Kendall, H.W. Posamentier, C.A. Ross, and J.C. Van Wagoner, eds., Sea Level Changes: An Integrated Approach: Society of Economic Paleontologists and Mineralogists Special Publication 42, p. 47–70.

Jordan, C.F., and M. Abdullah, 1988, Lithofacies analysis of the Arun reservoir, north Sumatra, Indonesia, *in* A.J. Lomando and P.M. Harris, eds., Giant Oil and Gas Fields: A Core Workshop: Society of Economic Paleontologists and Mineralogists, 12, p. 89–118.

Kaufman, J., and W.J. Myers, 1988, A backstepping platform reef, Swan Hills Formation, Rosevear Field, central Alberta, *in* H.H.J. Geldsetzer, N.P. James, and G.E. Tebbutt, eds., Reefs of Canada and Adjacent Areas: Canadian Society of Petroleum Geologists Memoir 13, p. 478–486.

Matthews, R.K., 1974, A process approach to diagenesis of reefs and reef associated limestones, *in* L.F. Laporte, ed., Reefs in Time and Space—Selected Examples from the Recent and Ancient: Society of Economic Paleontologists and Mineralogists Special Publication 18, p. 234–256.

Meijer Drees, N.C., 1986, Evaporitic deposits of Western Canada: Geological Survey of Canada Paper 85-20, 118 p.

Muir, I.D., G.L. Springate, and J.R. Mawdsley, 1990, Geological model of the Middle–Upper Devonian Snipe Lake 'A' Pool Platform-Reef Complex, Western Canada (abs.): AAPG Bulletin, v. 74, p. 726–727.

Murray, J.W., 1965, Stratigraphy and carbonate petrology of the Waterways Formation, Judy Creek, Alberta, Canada: Bulletin of Canadian Petroleum Geology, v. 13, p. 303–326.

Murray, J.W., 1966, An oil producing reef-fringed carbonate bank in the Upper Devonian Swan Hills Member, Judy Creek, Alberta: Bulletin of Canadian Petroleum Geology, v. 14, p. 1–103.

Perkins, R.D., 1977, Depositional framework of Pleistocene rocks in south Florida, *in* P. Enos and R.D. Perkins, eds., Quaternary Sedimentation in South Florida: Geological Society of America Memoir 147, p. 131–198.

Podruski, J.A., J.E. Barclay, A.D. Hamblin, P.J. Lee, K.G. Osadetz, P.M. Proctor, and W.C. Taylor, 1988, Part 1: Resource endowment of conventional oil resources of Western Canada (light and medium): Geological Survey of Canada Paper 87-26, p. 1–126.

Posamentier, H.W., and P.R. Vail, 1988, Eustatic controls on clastic deposition 2—sequence and system tract models, *in* C.K. Wilgus, B.S. Hastings, C.G.St.C. Kendall, H.W. Posamentier, C.A. Ross, and J.C. Van Wagoner, eds., Sea level changes: an integrated approach: Society of Economic Paleontologists and Mineralogists Special Publication 42, p. 125–154.

Posamentier, H.W., M.T. Jervey, and P.R. Vail, 1988, Eustatic controls on clastic deposition 1—conceptual framework, *in* C.K. Wilgus, B.S. Hastings, C.G.St.C. Kendall, H.W. Posamentier, C.A. Ross, and J.C. Van Wagoner, eds., Sea level changes: an integrated approach: Society of Economic Paleontologists and Mineralogists Special Publication 42, p. 109–124.

Sarg, J.F., 1988, Carbonate sequence stratigraphy, *in* C.K. Wilgus, B.S. Hastings, C.G. St. C. Kendall, H.W. Posamentier, C.A. Ross, and J.C. Van Wagoner, eds., Sea level changes: an integrated approach: Society of Economic Paleontologists and Mineralogists Special Publication 42, p. 155–182.

Schatzinger, R.A., 1983, Phylloid algal and sponge-bryozoa mound-to-basin transition: a Late Paleozoic facies tract from the Kelly-Snyder field, west Texas; *in* P.M. Harris, ed., Carbonate Buildups—A Core Workshop: Society of Economic Paleontologists and Mineralogists Core Workshop 4, p. 244–303.

Schlee, J.S., 1984, Interregional unconformities and hydrocarbon accumulation: AAPG Memoir 36, 184 p.

Sloss, L.L., 1963, Sequences in cratonic interior of North America: Geological Society of America Bulletin, v. 74, p. 93–114.

Springate, G.L., I.D. Muir, and T.R. Caldwell, 1992, A performance evaluation of Canada's Snipe Lake/Beaverhill Lake 'A' Pool: SPE Reservoir Engineering Journal, p. 390–396.

Stoddart, D.R., 1962, Three Caribbean atolls: Turneffe Islands, Lighthouse Reef and Glover's Reef, British Honduras: Atoll Research Bulletin, v. 87, 151 p.

Vail, P.R., R.M. Mitchum, R.G. Todd, J.M. Widmier, S. Thompson, J.B. Sangree, J.N. Bubb, and W.G. Hatelid, 1977, Seismic stratigraphy and global changes of sea level, *in* C.E. Clayton, ed., Seismic stratigraphy—applications to hydrocarbon exploration: AAPG Memoir 26, p. 49–212.

Vest, E.L., 1970, Oil fields of Pennsylvanian–Permian, Horseshoe Atoll, west Texas, *in* M.T. Halbouty, ed., Geology of Giant Petroleum Fields: AAPG Memoir 14, p. 185–203.

Walls, R.A., and G. Burrowes, 1985, The role of cementation in the diagenetic history of Devonian reefs, *in* N. Schneidermann and P.M. Harris, eds., Carbonate Cements: Society of Economic Paleontologists and Mineralogists Special Publication 36, p. 185–220.

Walls, R.A., and O.G. Burrowes, 1989, Diagenesis and reservoir development in Devonian limestone and dolostone reefs of western Canada, *in* G.R. Bloy, M.G. Hadley, and B.V. Curtis, compilers and organizers, The Development of Porosity in Carbonate Reservoirs—Short Course Notes: Canadian Society of Petroleum Geologists, 40 p.

Wanless, H.R., and J.J. Dravis, 1989, Carbonate environments and sequences of Caicos platform: American Geophysical Union Field Trip Guidebook T374 in conjunction with 28th International Geological Congress, 75 p.

Wendte, J.C., 1987, Inception, growth and termination of the Judy Creek reef complex, Middle to Upper Devonian, central Alberta (abs.), *in* J.C. Packard, ed., Program and Abstracts Book—Reef Research Symposium: Canadian Society of Petroleum Geologists, p. 64.

Wendte, J.C., 1992a, Overview of the Devonian of the Western Canada Sedimentary basin, *in* J.C. Wendte, F.A. Stoakes, and C.V. Campbell, Devonian–Early Mississippian Carbonates of the Western Canada Sedimentary Basin: A Sequence Stratigraphic Framework: Society for Sedimentary Geology Short Course Book 28, p. 1–24.

Wendte, J.C., 1992b, Cyclicity of Devonian strata in the Western Canada Sedimentary basin, *in* J.C. Wendte, F.A. Stoakes, and C.V. Campbell, Devonian–Early Mississippian Carbonates of the Western Canada Sedimentary Basin: A Sequence Stratigraphic Framework: Society for Sedimentary Geology Short Course Book 28, p. 25–39.

Wendte, J.C., 1992c, Evolution of the Judy Creek complex, a Late Middle Devonian isolate platform-reef complex in west-central Alberta, *in* J.C. Wendte, F.A. Stoakes, and C.V. Campbell, Devonian–Early Mississippian Carbonates of the Western Canada Sedimentary Basin: A Sequence Stratigraphic Framework: Society for Sedimentary Geology Short Course Book 28, p. 89–125.

Wendte, J.C., and F.A. Stoakes, 1982, Evolution and corresponding porosity distribution of the Judy Creek reef complex, Upper Devonian, central Alberta, *in* W.G. Cutler, ed., Core Conference Manual: Canada's Giant Hydrocarbon Reservoirs: Canadian Society of Petroleum Geologists, p. 63–81.

Wilson, J.L., 1975, Carbonate facies in geologic history: New York, Springer-Verlag, 471 p.

Chapter 14

Lower Paleozoic Cavern Development, Collapse, and Dolomitization, Franklin Mountains, El Paso, Texas

F. Jerry Lucia
Bureau of Economic Geology
The University of Texas at Austin
Austin, Texas, U.S.A.

ABSTRACT

The superb exposures of the Lower Ordovician El Paso Group in the Franklin Mountains, El Paso, Texas, provide an excellent opportunity to investigate the effects of unconformities on porosity and permeability of carbonate rocks. Unconformities at cycle, sequence, and supersequence boundaries represent time gaps ranging from thousands to millions of years. Unconformities at cycle and sequence boundaries are marked by tidal-flat facies and reflux dolomitization. No significant karsting is found at these boundaries.

A large cavern system was developed in the upper 300 m (1000 ft) of the El Paso Group during the 33 m.y. time gap marked by the supersequence boundary between the Lower Ordovician El Paso Group and the Upper Ordovician Montoya Group. In the upper 75 m (250 ft), the El Paso caverns were tabular and horizontal and were formed near the phreatic–vadose interface. In the lower 225 m (750 ft), the caverns were linear and vertical and were formed in the deep phreatic zone along vertical fractures oriented N20°W and spaced 900 m (3000 ft) apart. A stratiform dolomite unit separated the two cave systems. Collapse was initiated during cave development and continued through Silurian time. Collapse of the El Paso caverns formed large fracture systems and megacollapse breccias 300 m (1000 ft) thick, 450 m (1500 ft) wide, and several kilometers long. Collapse of the cavern roof produced brecciation and fracturing in the overlying Montoya strata. Much of the breccia and adjacent country rock was dolomitized by fluid migrating through the collapsed caverns after Silurian time.

Cavern development, collapse, and dolomitization of the El Paso and Montoya groups has completely altered the original porosity and permeability distribution from one controlled by depositional patterns to one controlled by diagenetic processes. Karst-related dissolution resulted in cavernous porosity comprising up to 30% of some intervals. However, infilling sediment and collapse during burial destroyed most of the cavernous porosity by the end of Silurian time; by the end of Pennsylvanian time, much of the fracture and interbreccia-block pore space had been occluded by saddle dolomite. The Ellenburger Group of the Permian basin, the subsurface equivalent of the El Paso Group, produces from fractures and interbreccia-block pore space similar to that found associated with the collapse breccias of the El Paso Group, although the total porosity is only 1 to 3%.

INTRODUCTION

Dissolution of carbonate rocks at unconformities has been considered an important mechanism proposed for porosity development for a hundred years. More recently it has been discovered that primary porosity in carbonate sediments is high (Enos and Sawatsky, 1981), and that many carbonate reservoirs produce from secondary porosity formed by diagenetic modification of primary pore space (Murray, 1960). Porosity distribution is controlled by depositional processes overprinted by diagenetic processes. The purpose of this paper is to explore the degree to which diagenetic processes related to unconformities affect the nature and distribution of pore space in carbonate rocks.

The almost complete exposures of the 450 m (1500 ft) thick El Paso Group of Early Ordovician age in the Franklin Mountains, El Paso, Texas (Figure 1), provide an excellent location to investigate the effects of unconformities on carbonate rocks. Unconformities, with time gaps ranging from thousands to millions of years, can be examined for over 8 km (5 mi) laterally on the faulted eastern face of the southern Franklin Mountains with few significant breaks in exposures. Extensive cavern development and subsequent cavern collapse have been described in detail. Three-dimensional relationships have been developed by mapping in canyons cut into the dip slope. The results of the detailed mapping have been summarized in field guidebooks and a local symposium (Lucia, 1968, 1970). The complete results, however, are presented here.

The Lower Ordovician Ellenburger Group is equivalent to the El Paso Group and is a major hydrocarbon producer in the Permian basin located some 320 km (200 mi) east of the Franklin Mountains. The Ellenburger reservoirs are dolomite with several different times of dolomitization from early tidal-flat dolomite to late saddle dolomite (Kupecz and Land, 1991). There is little intercrystalline pore space in the dolomite, suggesting that dolomitization destroyed the matrix porosity. The average porosity of these reservoirs is reported to be between 1 and 3% and production is reported to be from fractures (Galloway et al., 1983). The fractures are considered tectonic in origin, but recent studies attribute much of the fracturing and vuggy pore space to large karsted and collapse features related to the Middle Ordovician unconformity (Kerans, 1989a, b). Kerans (1989a) suggests these karsted features are related to a single unconformity at the top of the Lower Ordovician while others (Loucks and Anderson, 1985; Mazzullo, 1989) suggest multiple events at internal unconformities within the Lower Ordovician.

Karst features are common in the Lower Ordovician carbonates of North America and are interpreted to have been formed during a major relative sea level drop during the Middle Ordovician. Walters (1946) described karst features in the Kraft-Prusa (Arbuckle) field in Kansas, and Kay (1951) illustrated extensive karst development during this relative sea level fall. More recently, lead-zinc deposits of the Mississippi Valley type in the south-central United States have been found to be associated with karsting of Lower Ordovician carbonates (Kyle, 1976; Ohle, 1985; Mussman and Read, 1986).

STRATIGRAPHY

The complete geologic section exposed in the Franklin Mountains is about 4000 m (13,000 ft) thick and ranges from Precambrian to Cretaceous (Figure 2). In the southern Franklin Mountains, Precambrian through Upper Ordovician strata are exposed. The Lower Ordovician succession from near-shore, shallow-water Bliss sandstones through shallow-water carbonates of the El Paso Group forms a second-

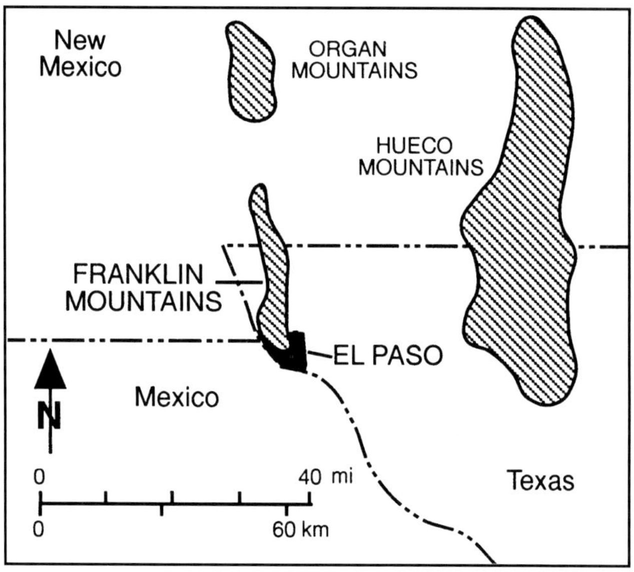

Figure 1. Location map of the Franklin Mountains, El Paso, Texas.

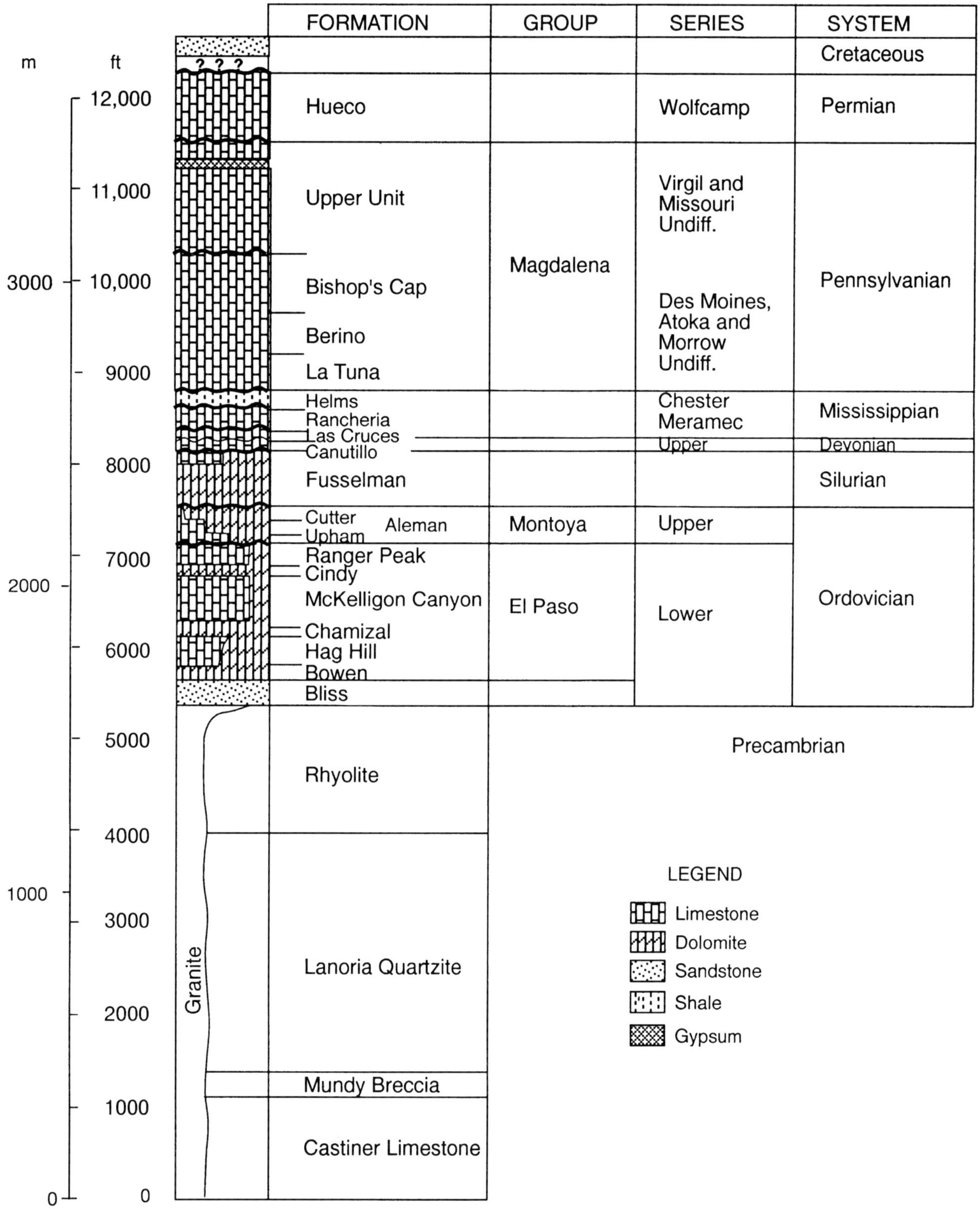

Figure 2. Franklin Mountains stratigraphic section.

order supersequence capped by an unconformity that represents a time interval of 33 m.y. (Goldhammer et al., 1993). The Upper Ordovician Montoya Group and the Silurian Fusselman Formation overlie this major unconformity, and a second major unconformity is located at the top of the Fusselman Formation representing a time interval of about 30 m.y. (LeMone, 1987).

The El Paso Group has been divided into both genetic depositional units (Lucia, 1968) and biostratigraphic

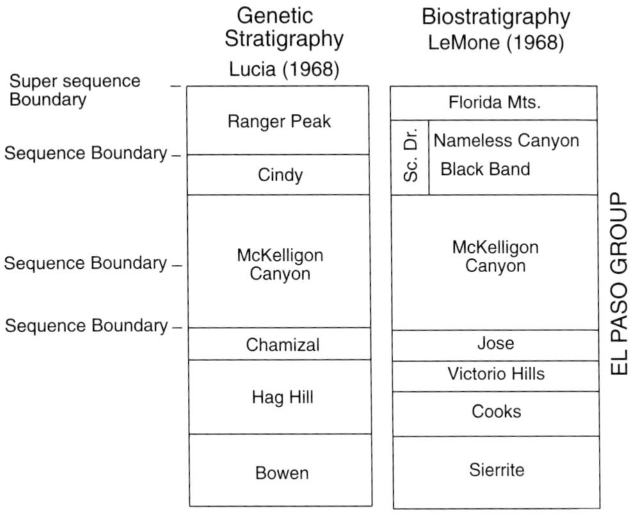

Figure 3. Stratigraphy of the El Paso Group comparing genetic stratigraphic units based on depositional concepts with biostratigraphic units.

units (LeMone, 1968). These two approaches are correlated in Figure 3. The biostratigraphic units are based on Flower's (1964) biostratigraphic correlations whereas the genetic units are based on depositional sequence considerations. Genetic depositional units are used in this paper.

The lowest formation overlying the transgressive Bliss Sandstone is the 50 m (170 ft) thick Bowen Formation (Figure 2). The Bowen Formation is a deepening-upward succession of cross-bedded siliciclastic dolomite and peritidal sediments to marine dolowackestones. Overlying the Bowen Formation is the 80 m (275 ft) thick Hag Hill Formation. The Hag Hill Formation is a shallowing-upward succession of fossiliferous wackestones with thrombolitic bioherms to interbedded subtidal and peritidal sediments. Overlying the Hag Hill Formation is the 27 m (90 ft) thick Chamizal Formation composed of interbedded siliciclastic cross-bedded dolomite and algal dolowackestones and capped by a supratidal interval. Next in the succssion is the McKelligon Canyon Formation, which is composed of 220 m (700 ft) of subtidal limestone and is overlain by 33 m (110 ft) of dolomitized siliciclastic peritidal cycles named the Cindy Formation. The upper formation, the Ranger Peak Formation, is composed of 75 m (250 ft) of dolomitic shallow-water limestone.

The Montoya Group is divided into four formations (in ascending order): Cable Creek, Upham, Aleman, and Cutter (Howe, 1959) (Figure 2). The Cable Creek is not present in the southern Franklin Mountains. The Upham Formation is transgressive and is composed of 30 m (100 ft) of subtidal, open-marine carbonate containing large corals, cephalopods, and gastropods. The Aleman Formation is comprised predominantly of 30 m (100 ft) of dark cherty subtidal fossiliferous wackestone and represents a maximum flood-back. The Cutter Formation is composed of 45 m (150 ft) of shallow-water to tidal-flat fine-crystalline dolostone (Lucia, 1970).

UNCONFORMITIES, TIME SCALE, AND DIAGENETIC EFFECTS

Unconformities of varying time intervals have been described and mapped in the southern Franklin Mountains. While the time intervals cannot be determined with accuracy, the possible exposure time is likely to be the least at cycle boundaries, intermediate at third-order sequence boundaries, and maximum at supersequence boundaries. Exposure time at any one unconformity will vary from a maximum value in a landward direction to zero in a seaward direction, and the resulting diagenetic effects will vary depending upon the duration of the exposure time and the physical and chemical setting.

Depositional Cycle Boundaries

Depositional cycles in the El Paso Group have recently been described by Goldhammer et al. (1993) and Kerans and Lucia (1989). Both describe shallowing-upward cycles capped by tidal-flat sediments. These cycles contain coarse siliciclastics and are dolomitized. Dolomitization is interpreted to be the result of penecontemporaneous hypersaline reflux dolomitization. Examination of tidal-flat–capped cycles by walking 8 km (5 mi) laterally reveals no significant karsting effect associated with these cycle boundaries.

Subtidal cycles are composed of beds of thrombolitic sponge-algal bioherms, ribbon rock, coarse-grained packstone/grainstone, and fossiliferous wackestone. Goldhammer et al. (1993) argued that the ribbon rock indicated the top of the shallowing-upward cycles whereas Kerans and Lucia (1989) interpreted the packstone/grainstone and thrombolitic beds as cycle tops. However, no significant karsting has been observed associated with either cycle top in 8 km (5 mi) of near-continuous outcrops. The thrombolitic bioherms commonly exhibit truncated upper surfaces that could be mistaken for karsting, but detailed examination shows they result from biohermal growth, submarine hardgrounds, and submarine erosion.

The boundaries between subtidal cycles represent no sedimentation break whereas the boundaries between tidal-flat–capped cycles have unconformities of variable duration. While the cycle period cannot be established with any high degree of reliability (see Goldhammer et al., 1993), an estimate of 20 to 200 k.y. probably encompasses the expected range, with the exposure duration of the tidal-flat–capped cycles being considerably less, perhaps on the order of tens of thousands of years. The Bonaire reflux model as described by Deffeyes et al. (1965) suggests that these time periods are sufficient to produce the volume of reflux dolomite associated with the tidal-flat–capped cycles. However, the length of exposure time is not sufficient to produce significant karsting effects.

Third-Order Sequence Boundaries

Three third-order sequence boundaries are defined within the El Paso Group by Goldhammer et al. (1993), and two by Kerans and Lucia (1989). The concentration of siliciclastic-rich, tidal-flat–capped cycles in the Chamizal and Cindy formations is interpreted to represent eustatic lowstand deposits and to clearly define sequence boundaries. A third sequence boundary is defined by Goldhammer et al. (1993) within the subtidal McKelligon Canyon Formation based on cycle stacking patterns. This sequence boundary correlates with a package of tidal-flat–capped cycles described from the middle of the El Paso Group in the Beach Mountains located 190 km (120 mi) to the east (Lucia, 1968, 1970; Goldhammer et al., 1993).

No significant karsting has been found at the three third-order sequence boundaries. No karsting would be expected at the subtidal sequence boundary located within the subtidal McKelligon Canyon Formation. The sequence boundary in the Chamizal and Cindy formations is located at the tops of the formations by Kerans and Lucia (1989) and at the top of internal cycles by Goldhammer et al. (1993). However, no significant karsting has been observed at any of these boundaries.

Supersequence Boundary

The boundary between the El Paso Group and the Montoya Group is a major time gap of some 33 m.y., during which the El Paso Group underwent extensive subaerial exposure (Lucia, 1968, 1970) due to tectonic uplift (Goldhammer et al., 1993). Cavern development associated with this exposure and subsequent cavern collapse had a major effect on the porosity and permeability distribution of the El Paso Group and on the overlying Montoya and Silurian strata and controlled the distribution of late-stage dolomitization. A description of this major unconformity-related diagenetic event is the principal subject of this paper and is presented next.

CAVERN FORMATION AND COLLAPSE BRECCIA

Distribution of Collapse Breccia

The Franklin Mountains consist of a faulted horst block dipping to the west at about 30° with the Hueco bolson on the east and the Mesilla bolson on the west. The mountains can be divided into a structurally high central area composed primarily of Precambrian formations and northern and southern areas where Paleozoic strata are exposed. This report describes the southern area. Selective dolomitization of collapse brecciated units by tan dolomite simplifies distinguishing the collapse breccia from the blue-gray unbrecciated limestone.

The distribution of collapse brecciation in the lower Paleozoic strata in the southern Franklin Mountains is presented in three sections. The first section describes the distribution of collapse breccia on the eastern face of the southern Franklins, the second section describes brecciation on the western dip slope, and the third section describes brecciation in McKelligon Canyon located at the north boundary of the southern Franklin Mountains.

Collapse Brecciation on the East Face

The distribution of collapse brecciation can best be seen along the eastern face of the southern Franklin Mountains (Figure 4). Within the El Paso Group, brecciation occurs in two modes (Figure 5). Breccia in the Ranger Peak Formation tends to be laterally continuous and the overlying Upham Formation is fractured and brecciated. Within the McKelligon Canyon Formation the collapse brecciation is laterally confined. The Cindy Formation is brecciated only where the underlying McKelligon Canyon Formation is brecciated. In one location, a breccia pipe extends from the McKelligon Canyon Formation through the Cindy, Ranger Peak, and Montoya units into the Fusselman Formation.

The most spectacular exposures on the eastern face are breccia bodies that extend down into the McKelligon Canyon Formation, referred to as Lechuguilla Breccia, Quarry Breccia, and the Great McKelligon Sag (Figure 4). These breccias have limited lateral dimensions, the largest being about 450 m (1500 ft) wide and extending down 300 m (1000 ft) beneath the El Paso top.

The most complete exposure is the Great McKelligon Sag (Figures 4 and 6). Unfortunately, it is also the most inaccessible breccia and has not been mapped in great detail. The major portion of the collapse breccia is in the McKelligon Canyon, Cindy, and Ranger Peak formations. Below the McKelligon Canyon Formation no significant brecciation has been found. At the level of the McKelligon Canyon Formation, there is one large area of brecciation about 60 m (200 ft) thick that occurs immediately below the Cindy Formation and is composed primarily of blocks of the Cindy Formation. Brecciation below this level is less distinct and less intense. The Cindy Formation protrudes over and is folded and faulted down into the main breccia. Fragments of the Cindy Formation are easily recognized because of their distinctive tidal-flat sedimentary structures and have been found about 120 m (400 ft) below their stratigraphic position.

Breccia is continuous from the McKelligon Canyon through the Cindy and Ranger Peak formations. Blocks of the Upham Formation are present in the breccia at the level of the Ranger Peak Formation. Brecciation in the Ranger Peak Formation spreads laterally over the unbrecciated Cindy Formation.

A spectacular collapse of the Montoya Group into the El Paso Group occurs over the main area of El Paso brecciation (Figure 6). The collapsed Upham Formation extends downward about 75 m (250 ft) to the stratigraphic level of the Cindy Formation. On the north side, the Montoya is brecciated and faulted and the breccia includes blocks of the overlying Fusselman Formation.

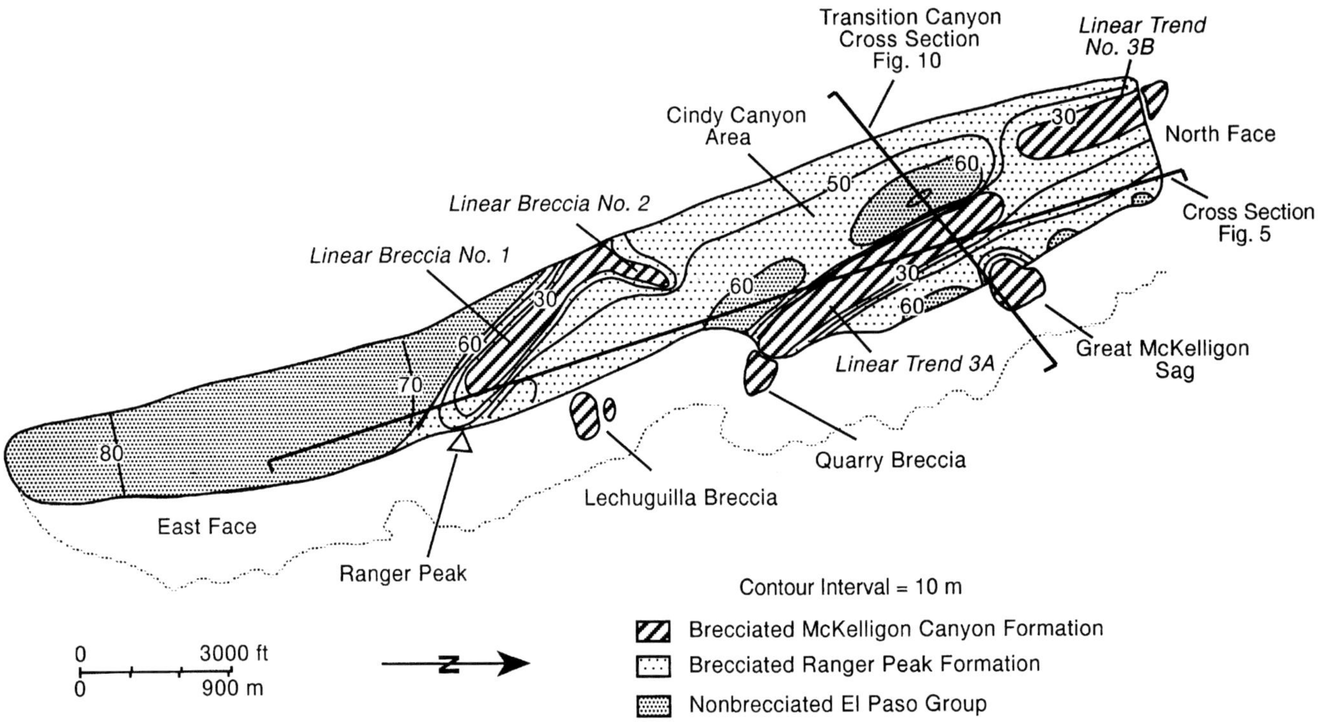

Figure 4. Map of the southern Franklin Mountains showing the locations of breccia outcrops, an isopach of the Ranger Peak Formation, and an interpretation of the distribution of collapse breccia.

These relationships suggest that a 60 m (200 ft) high cavern existed in the Ranger Peak Formation during deposition of the Montoya Group, and that collapse of this large cavern was still occurring after deposition of the Fusselman Formation. The concentration of Cindy blocks below the stratigraphic position of the Cindy Formation suggests that a large cavern was also located immediately below the Cindy Formation.

The most accessible and most completely studied laterally restricted breccia is Lechuguilla Breccia (Figures 4 and 7). The geometry is similar to the Great McKelligon Sag except that folding of the Montoya Group into the brecciated area cannot be seen. This may be due to the breccia having the geometry of a pipe extending vertically from the east-facing slope and the exposed Montoya beds being adjacent to the breccia pipe. Brecciation is developed in the McKelligon Canyon, Cindy, and Ranger Peak formations and not in lower formations. In the McKelligon Canyon Formation, the breccia is about 210 m (700 ft) wide. A cave 2 m (6 ft) in diameter filled with red to tan cave sediment is located below a narrow, vertical,

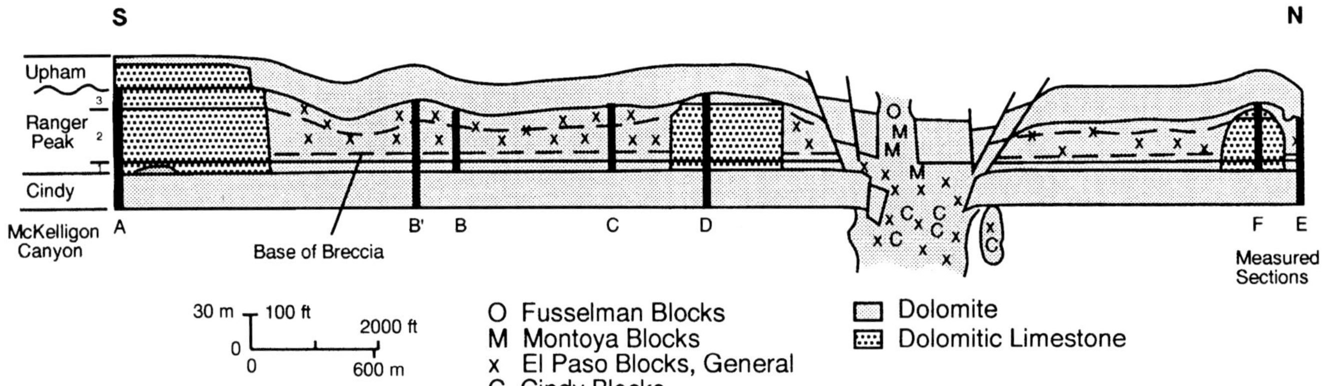

Figure 5. Cross section of the eastern face of the southern Franklin Mountains showing laterally continuous dolomitized breccia in the Ranger Peak Formation and laterally discontinuous dolomitized breccia beneath the Cindy Formation. Note that the Upham Formation is dolomitized above the dolomitized Ranger Peak Formation.

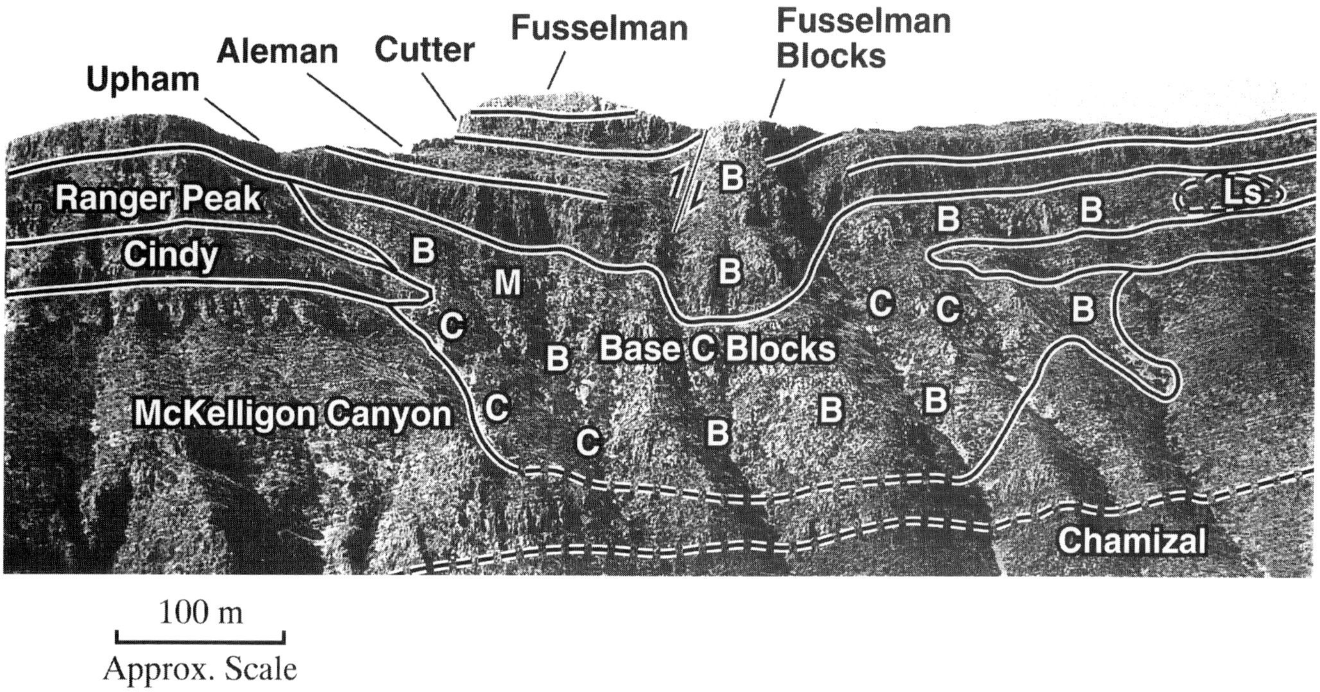

Figure 6. Photograph of the Great McKelligon Sag in McKelligon Canyon along the eastern face of the southern Franklin Mountains, showing the distribution of collapse breccia and collapse of the Montoya Group into the Ranger Peak Formation. B = breccia, C = blocks of Cindy Formation, M = blocks of Montoya Group.

Figure 7. Photograph of Lechuguilla Breccia north of Ranger Peak along the eastern face of the southern Franklin Mountains, showing the distribution of collapse breccia. B = breccia, C = blocks of Cindy Formation, M = blocks of Montoya Group.

breccia dike north of the main breccia body (Figure 7). A small fracture filled with sediment can be traced from the base of the breccia dike to this cave. The Cindy Formation extends over most of the brecciated McKelligon Canyon Formation, and blocks of Cindy Formation are found as far as 120 m (400 ft) below their stratigraphic position. The Ranger Peak Formation is more or less brecciated for several thousands of meters north and south of Lechuguilla Breccia. Blocks of the Upham Formation are present at the Ranger Peak level and the Upham Formation is fractured and brecciated. A pillar of unbrecciated limestone about 60 m (200 ft) wide is present in the Ranger Peak Formation (Figure 7).

Similar to the Great McKelligon Sag, the most chaotic breccia is found at stratigraphic levels of the upper McKelligon Canyon, Cindy, and Ranger Peak formations. The Cindy Formation extends into the brecciated area from both sides, and the underlying chaotic breccia is composed primarily of Cindy Formation blocks. Some of the Cindy blocks are automobile size and are found 45 to 60 m (150 to 200 ft) below their stratigraphic position, indicating a very large cavern. In the middle and lower stratigraphic equivalent of the McKelligon Canyon Formation, chaotic breccia is found in vertically continuous dikes, whereas the breccia between these dikes appears to be composed of blocks that have rotated in place. Similar to the Great McKelligon Sag, it appears that the main caverns at Lechuguilla Breccia were developed immediately below the Cindy Formation in the Ranger Peak Formation. Lower down, the caverns were mainly vertical openings with some lateral passages.

The third well-exposed area of collapse brecciation is the Quarry Breccia (Figure 4). This breccia is complicated by the presence of a normal fault with 60 to 90 m (200 to 300 ft) of stratigraphic displacement. Two other large faults cut the eastern face, but this is the only fault that intersects collapse breccia. The shear zone and brecciation formed by the faulting are about 3 m (10 ft) in width and, therefore, are a small part of the Quarry Breccia. The collapse breccia is highly fractured next to the fault, showing that the fault occurred after the formation of the collapse breccia. Although this collapse breccia has not been mapped in detail, a reconnaissance suggests that it has a geometry very similar to that of the other two collapse breccias. The Montoya drapes into the area of brecciation. Most of the brecciation occurs in the upper part of the McKelligon Canyon Formation and in the Cindy and Ranger Peak formations. The Cindy Formation is less brecciated than either the McKelligon Canyon or the Ranger Peak formations. Brecciation within the Ranger Peak Formation again extends laterally for several thousands of feet whereas brecciation in the McKelligon Canyon Formation is laterally restricted.

The Ranger Peak Formation contains more widespread areas of brecciation than the McKelligon Canyon Formation. These breccias are continuous with the vertical breccias. The brecciation is not as easy to observe as in the vertical breccias, so the distribution of brecciation was not mapped in detail. However, it is clear that much of the dolomite in the Ranger Peak Formation is not brecciated.

Seven sections of the Ranger Peak Formation on the eastern face were measured, four in the brecciated areas and three in the unbrecciated areas (Table 1). The Ranger Peak Formation was divided into three members (Figure 5). The lower member (1) is extensively brecciated only in areas where the McKelligon Canyon Formation is brecciated. The upper member (3) is very cherty and is brecciated over the entire area of Ranger Peak breccia, with blocks of this member found below their stratigraphic position. The middle unit (2) is partially brecciated over the entire area of Ranger Peak breccia. The Ranger Peak/Upham contact is very sharp in areas where the Ranger Peak is not brecciated, but, in brecciated areas, the contact commonly is a breccia of Ranger Peak and Upham blocks, demonstrating that collapse occurred after deposition of the Montoya Group.

The brecciated sections average 13 m (43 ft) thinner than the unbrecciated sections (Table 1). The base of the Ranger Peak Formation is the top of a supratidal surface, which is assumed to represent a level surface. The thinning of the Ranger Peak, therefore, is a structural sag in the Ranger Peak/Upham contact.

The Ranger Peak thinning could be due to either erosion or collapse of caverns in the Ranger Peak. Studies by Howe (1959) have shown regional thinning of the El Paso Group to the north. This explains the 27 m (90 ft) of thinning seen in the unbrecciated Ranger Peak sections from south to north. If local relief existed before the deposition of the Montoya Group, it should be reflected in the thickness of the Upham Formation. The Upham Formation, however, is about 30 m (100 ft) thick and,

Table 1. Thickness and general lithology of seven sections in the Ranger Peak Formation.

Section	Thickness	Lithology
A	89 m (294 ft)	Unbrecciated dolomitic limestone
B	54 m (177 ft)	Brecciated dolomite
B′	66 m (217 ft)	Slightly brecciated dolomite
C	58 m (192 ft)	Brecciated dolomite
D	67 m (223 ft)	Unbrecciated dolomitic limestone
E	43 m (143 ft)	Brecciated dolomite
F	62 m (204 ft)	Unbrecciated dolomitic limestone

from photo and reconnaissance mapping, does not appear to vary significantly in thickness, demonstrating that Ranger Peak thinning occurred after Upham deposition. A few small faults appear in the Upham where it overlies the Ranger Peak collapse breccia, but the faults do not carry throughout the Ranger Peak Formation, suggesting that they result from collapse of caves in the Ranger Peak. Therefore, the thinning of the Ranger Peak Formation in the brecciated areas relative to the unbrecciated areas is due to the collapse of caves after the deposition of the Montoya Group.

These thickness relationships indicate that the upper member of the Ranger Peak Formation was not removed by erosion but instead formed a roof over caverns developed in the middle member of the Ranger Peak Formation. Because 13 m (43 ft) of section is missing, the total vertical thickness of the caverns must have been at least 13 m (43 ft), or about 30% of the upper and middle members.

Several small areas of less well preserved collapse brecciation, which are believed to be related in time and origin to the major collapse brecciation, are located in the lower member of the Ranger Peak Formation south of Ranger Peak (Figure 5). One location is easily accessible from Scenic Drive. The Ranger Peak Formation is limestone in this area except for several local areas of dolomitized breccia in the lower member. A small breccia body is located near the base of the Ranger Peak Formation just above the letter "J" painted on the mountain, and others can be found at the same stratigraphic horizon northward along the mountain front. The breccias are small, and the associated bodies of dolomite are triangular-shaped bodies 6 m (20 ft) high and 30 m (100 ft) wide extending vertically from the top of the underlying Cindy Formation.

Collapse Brecciation on the West Dip Slope

Collapse brecciation in the Montoya and El Paso groups is exposed in four areas on the west dip slope of the southern Franklin Mountains. From south to north these areas are: (1) west of Ranger Peak, (2) Cindy Canyon, (3) Transition Canyon, and (4) the North Face (Figure 4). With the exception of the North Face, only the Ranger Peak Formation and locally the upper part of the Cindy Formation are exposed. Brecciation in the Cindy and McKelligon Canyon formations is inferred from the presence of structural sags and collapse brecciation in the Montoya Formation similar to that observed on the eastern face associated with brecciation in the McKelligon Canyon Formation.

On the dip slope west of Ranger Peak, the Ranger Peak Formation and the upper part of the Cindy Formation are exposed over a large area in a scar left by a rock slide. The trend of the breccia is NNW-SSE (Figure 4). At the north edge of this scar the Upham/Ranger Peak contact is well exposed (Figure 8). The contact sags into the Ranger Peak Formation in a manner similar to that observed in the Great McKelligon Sag. Blocks of the Aleman Formation are present in the brecciated Ranger Peak Formation. The Montoya is brecciated and highly fractured overlying the brecciated Ranger Peak. By analogy to the eastern face, these relationships suggest that the brecciation extends down into the McKelligon Canyon Formation beneath this area. This sag can be traced northwestward across the west slope to the base of the mountain; however, it ends in a northeasterly direction before it reaches the eastern face (Figure 4). This linear trend of collapse breccia is labeled "Linear Breccia No. 1" on the isopach map presented in Figure 4. In the

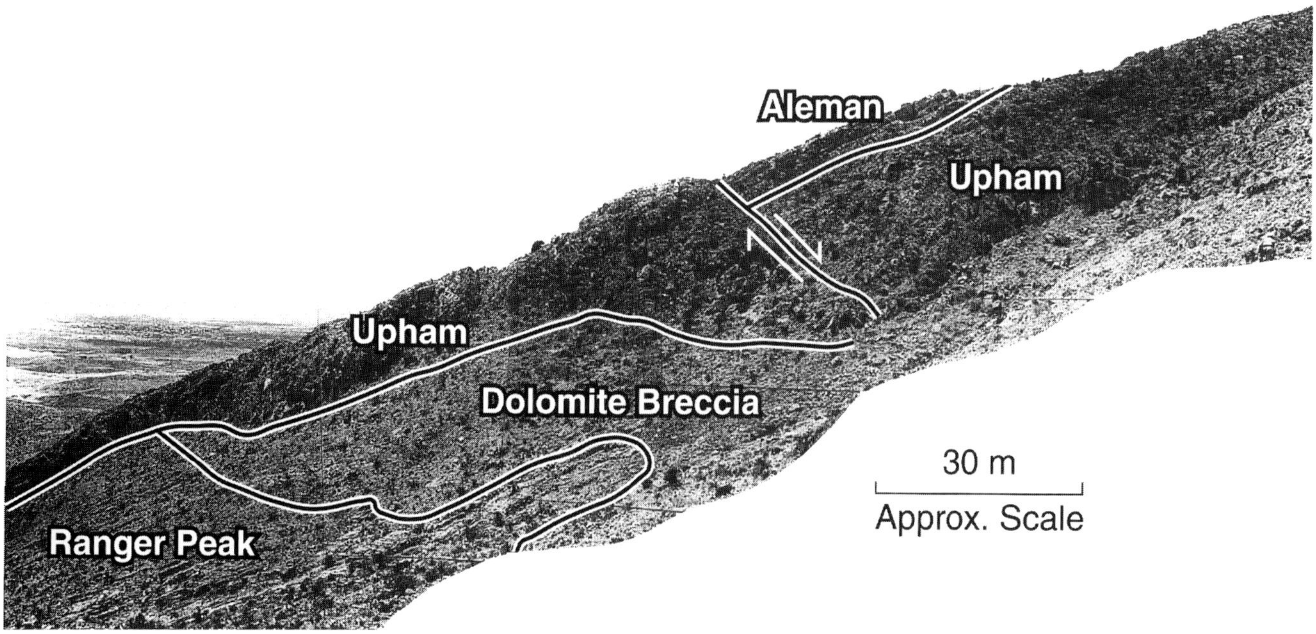

Figure 8. Photograph showing the distribution of collapse breccia in the Montoya Group and Ranger Peak Formation on the dip slope west of Ranger Peak. View is north.

next canyon north, a vertical collapse breccia is exposed in the Montoya Group. Blocks of Fusselman Formation are mixed with blocks of the Montoya Group, suggesting that this breccia also extends down into the McKelligon Canyon Formation. The breccia trends north-northeasterly and can be traced into linear breccia No. 1 and a short distance to the north-northeasterly direction. This breccia is labeled "Linear Breccia No. 2" on Figure 4.

In Cindy Canyon, the Ranger Peak Formation and a small outcrop of uppermost Cindy Formation are exposed. The Ranger Peak Formation contains collapse breccia except for an unbrecciated limestone in the southeastern corner of the canyon. This unbrecciated limestone is probably continuous with unbrecciated limestone in the Ranger Peak Formation mapped on the eastern face south of the Quarry Breccia (Figure 4). The Montoya sags into the Ranger Peak Formation at the very head of the canyon, and this sag can be traced to the Quarry Breccia on the eastern face of the mountain. This breccia is labeled "Linear Trend 3A" on Figure 4. On the north side of Cindy Canyon, there is a narrow discontinuous vertical breccia that extends up into the Fusselman Formation.

The middle and upper units of the Ranger Peak Formation are exposed in Transition Canyon (Figure 4). The Ranger Peak is brecciated at the mouth of the canyon, for a short distance in the middle of the canyon, and at the head of the canyon (Figure 9). At the mouth of Transition Canyon, Upham sediments were deposited around blocks of Ranger Peak, suggesting the presence of a Middle Ordovician sinkhole and karsting before Upham deposition. Most of the Ranger Peak Formation is unbrecciated limestone except for two areas: a vertical breccia dike with a 30 m (100 ft) halo of dolomite that can be traced across the canyon floor in a NNW-SSE trend, and a major collapse feature at the head of the canyon where the Montoya sags at least 45 m (150 ft) into the Ranger Peak Formation in a manner similar to that seen in the Great McKelligon Sag (Figures 9 and 10). Extensive Montoya brecciation is associated with this deep sag. These relationships suggest brecciation down into the McKelligon Canyon Formation. This sag is continuous with the sag in the head of Cindy Canyon and with the Quarry Breccia on the eastern face. This linear trend is labeled "Linear Trend 3A" on the isopach (Figure 4). There is a fault with about 60 m of stratigraphic

Figure 9. Photograph of the north wall of Transition Canyon showing the collapse of the Montoya Group into the Ranger Peak Formation and a small, vertical breccia dike and associated dolomite in the Ranger Peak limestone.

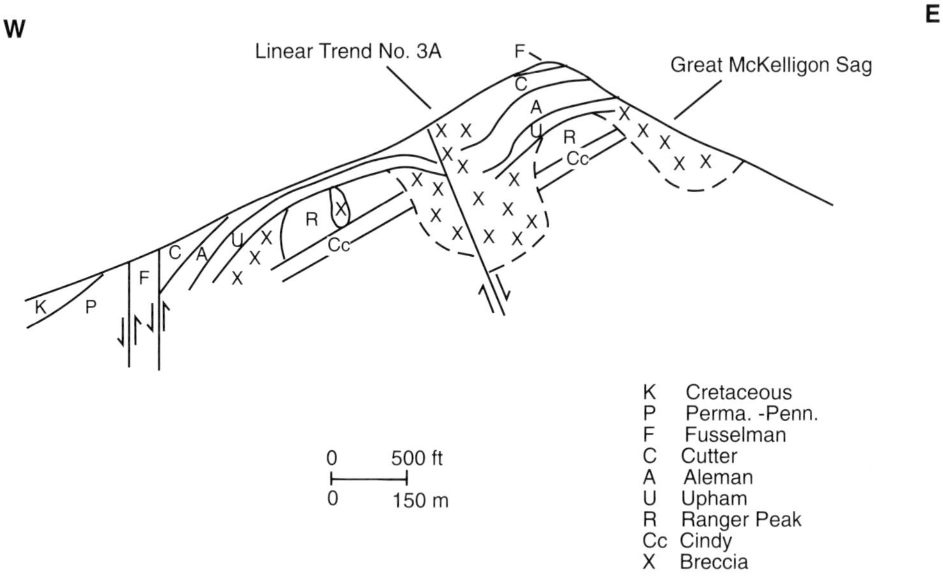

Figure 10. East-west cross section of Transition Canyon illustrating dolomitized collapse breccia at the mouth of the canyon, collapse breccia extending down below the Cindy Formation, and separation between the Great McKelligon Sag and Linear Trend No. 3A.

displacement located within this sag that can be traced to the fault exposed in the Quarry Breccia and is clearly younger than the collapse brecciation. The Great McKelligon Sag is located on the eastern face opposite this canyon. Structural mapping in the Montoya Group indicates that the Great McKelligon Sag is not continuous with linear breccia No. 3A (Figure 10).

The North Face is an excellent exposure of the upper half of the McKelligon Canyon Formation, and the Cindy and the Ranger Peak formations (Figure 4). The Ranger Peak Formation contains collapse breccia throughout the entire exposure. A collapse breccia in the Cindy and McKelligon formations is found at the western, downdip limit of the exposure. The El Paso/Montoya contact is fractured and brecciated overlying this laterally restricted breccia. The laterally restricted breccia can be traced SSE across the west dip slip by structural mapping in the Montoya. Whether or not this breccia trend actually connects to linear breccia No. 3A is not clear. Therefore, it is labeled "Linear Trend No. 3B" on the isopach map presented in Figure 4.

Collapse Brecciation in McKelligon Canyon

A large area of collapse brecciation in the McKelligon Canyon and Cindy formations is found at the head of McKelligon Canyon in a downfaulted block (Figure 11). The breccia is mostly dolomitized with small undolomitized areas within the breccia and at the breccia edge. Fracturing associated with faulting cuts the breccia fabric, indicating that faulting is postcollapse brecciation. East of this location along the northeastern ridge of McKelligon Canyon, the Ranger Peak, Cindy, and McKelligon Canyon formations are unbrecciated limestone.

Distribution of Collapse Breccia

In order to show the distribution of the collapse brecciation and to illustrate its effect on structure, an isopach map of the Ranger Peak Formation was constructed (Figure 4). In the linear breccias the Ranger Peak Formation is shown as less than 30 m (100 ft) thick because of the difficulty of picking the El Paso/Montoya and the Ranger Peak/Cindy contacts. The measured sections are used as control points for the rest of the map. Where no measured sections are available, the areas of Ranger Peak brecciation are shown as 9 to 15 m (30–50 ft) thinner than the areas of nonbrecciated Ranger Peak based on the measured sections.

The cross section in Figure 5 is based on the measured sections on the eastern face and illustrates the effect of the collapse on the local structure and dolomitization. The Ranger Peak Formation is thinner where it is brecciated dolomite than where it is unbrecciated limestone. This relationship holds true for the several small unbrecciated limestone pillars exposed on the eastern face as well as for large areas of limestone. Where the Ranger Peak Formation is partially brecciated and dolomitized, the overlying Upham Formation is also partially brecciated and dolomitized. Ranger Peak strata are replaced by brecciated Ranger Peak Formation and Montoya Group where brecciation extends down into the McKelligon Canyon Formation. Breccia pipes containing blocks of Montoya and Fusselman carbonate extend up into the Fusselman. The Cindy Formation resists dissolution and extends over the brecciated McKelligon Canyon Formation, where it disappears into a chaotic breccia.

From the mapping (Figure 4) it is clear that there are two linear breccias that can be traced for some distance—breccias Nos. 1 and 3A–3B. These breccias are oriented N20°W. Linear breccia No. 2 shows a minor NNE trend. Mapping also shows that the Ranger Peak Formation contains breccia in all but the southern part of the southern Franklin Mountains, the two large elongated areas in the northern part, the small limestone pillars, and on the ridge northeast of McKelligon Canyon.

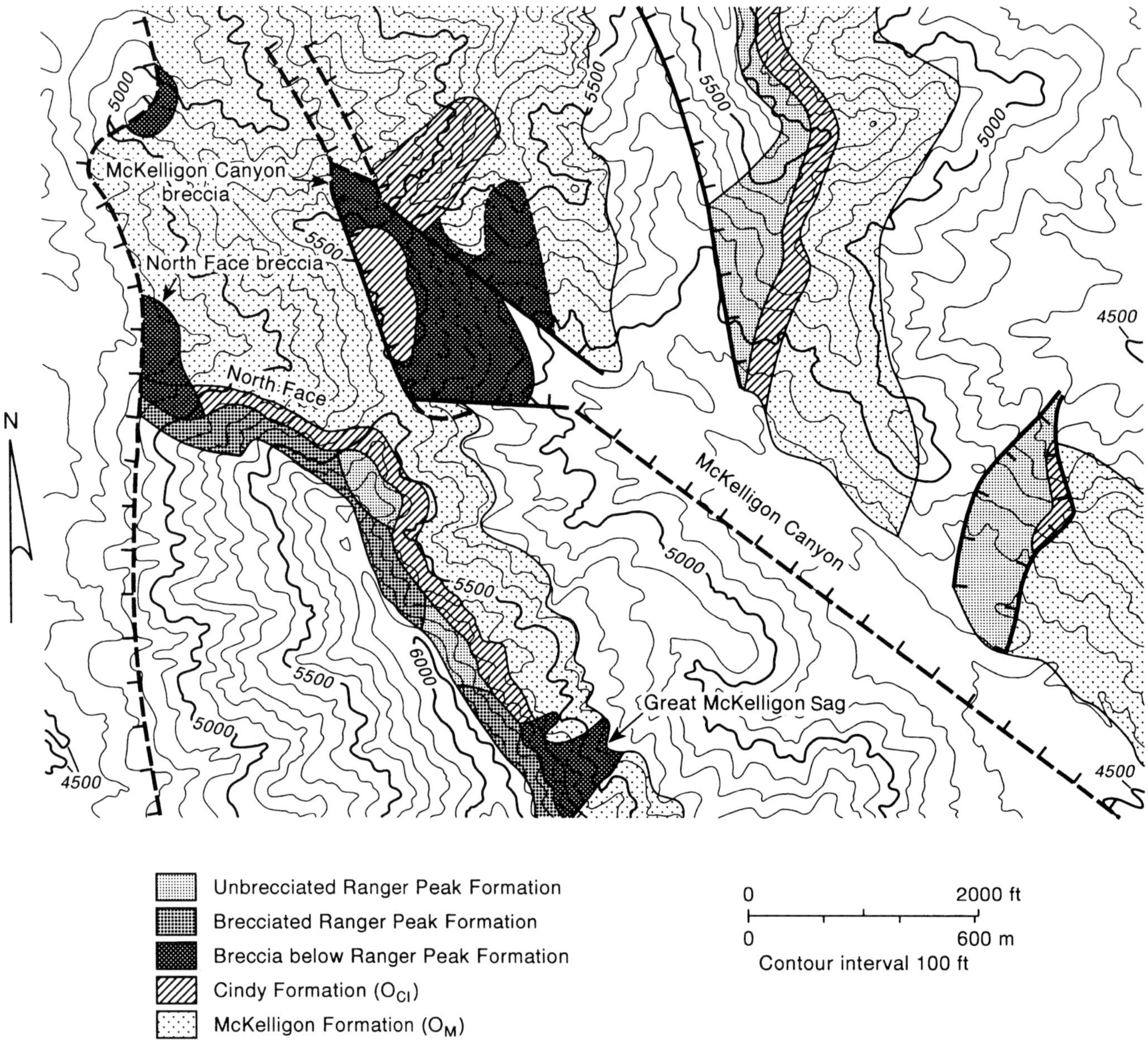

Figure 11. Map showing location of breccia located in a downfaulted block at the head of McKelligon Canyon and east of the North Face location.

The fact that collapse breccias have N20°W trends suggests that the two laterally restricted collapse breccias on the eastern face which do not continue into the mountain (Lechuguilla and the Great McKelligon Sag breccias) are remnants of eroded linear trends. When the stratigraphic position of the McKelligon Canyon collapse breccia is restored by palinspastic reconstruction of the faults assuming simple dip-slip movement, the Great McKelligon Sag can be extended in a NNW direction to connect to the McKelligon Canyon breccia (Figure 12). The Lechuguilla Breccia cannot be connected to any known linear trend.

Texture of the Collapse Breccia

The breccia is composed of angular to subrounded blocks of dolomite, dolomitic limestone, and chert in a matrix of dolomite, chert, calcite, and feldspar (Figure 13). The rock fragments range in size from microscopic to meters on a side. Where present, dolomitic limestone blocks are found toward the base of collapse breccia in the McKelligon Canyon Formation.

A crude stratigraphy displaced downward can be seen in the succession of blocks in the vertical breccia. Most of the blocks at the base of the breccias and all of the dolomitic limestone blocks are from the McKelligon Canyon Formation. Overlying these blocks is a zone composed primarily of blocks from the Cindy Formation (Figure 14). This zone grades upward into blocks that are mainly from the Ranger Peak Formation. Some Montoya blocks are mixed in toward the top of the Ranger Peak Formation (Figure 15). In the brecciated Montoya Formation, Fusselman blocks appear approximately halfway down through the Montoya Formation. All of the blocks have moved down from their relative stratigraphic position.

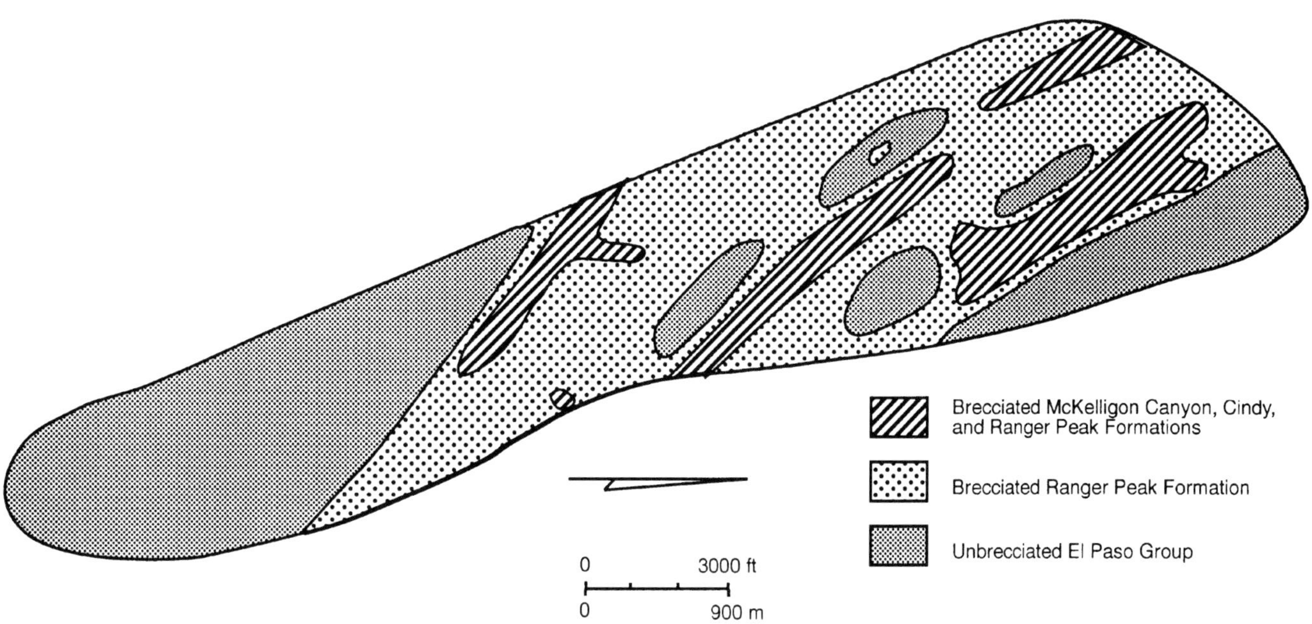

Figure 12. Reconstruction of the collapse breccia in the southern Franklin Mountains, El Paso, Texas. Map view.

The matrix between the breccia blocks is a jumbled mixture of dolomite, chert, calcite, and small amounts of clay minerals. In association with the limestone blocks the matrix is commonly, but not everywhere, composed of calcite, dolomite, and chert. Between all dolomite blocks and some places where limestone blocks are present, the matrix is composed entirely of dolomite and chert (Figure 16). Much of the matrix shows no sedimentary structures. In some areas, however, the matrix is evenly bedded or laminated (Figures 17 and 18). Thin-bedded, laminated internal sediment is found primarily in the McKelligon Canyon Formation but is also present in the Ranger Peak and Upham formations. These areas of laminated internal sediment are up to 2 m (6 ft) in diameter. The dip of the laminae is often conformable with the structural dip, indicating that they were deposited before structural tilting. In some locations, the laminated sed-

Figure 13. Photograph of dolomitized collapse breccia in the McKelligon Canyon Formation, Lechuguilla Breccia.

Figure 14. Photograph of dolomitized collapse breccia composed of blocks of the Cindy Formation in the middle of Lechuguilla Breccia.

Figure 15. Photograph of collapse breccia near the top of Lechuguilla Breccia. Dark blocks are Montoya lithology and are mixed with El Paso blocks.

Figure 16. Photograph of a slab of dolomitized collapse breccia showing poorly sorted texture and massive internal sediment. Slab is 12 cm wide.

Figure 17. Photograph of rust/tan-colored bedded internal sediment near base of Lechuguilla Breccia. The internal bedding has a structural attitude compatible with the dip of the mountains.

imentary structures suggest periodic transport of sediment into water-filled openings. The laminations are commonly graded, have microflame structures, and in some places contain clasts of dolomite and chert (Figure 19).

The color of the laminated internal sediment ranges from rust and tan to almost white. The rust/tan-colored internal sediment is exclusively found in the El Paso breccia and the white sediment is found primarily in the brecciated Montoya and Ranger Peak. At the El Paso–Montoya contact in two locations, a few decimeters of paleosol are present that have a mineralogy, texture, and rust color similar to the laminated internal sediment found exclusively in the El Paso breccia. Therefore, it seems likely that the rust/tan-colored laminated internal sediment was transported into the breccia by meteoric water running off the exposed El Paso surface. The white internal sediment must

Figure 18. Photograph of white-colored bedded internal sediment in the Upham (Montoya) Formation showing bedding conformance to structural dip.

A

Figure 19. Laminated internal sediment in the McKelligon Canyon Formation. (A) Photograph of a slab of laminated internal sediment showing millimeter-scale laminations. Slab is 12 cm wide. (B) Photomicrograph of (A) showing graded bedding and microflame structures. Photomicrograph is 2.2 cm wide.

B

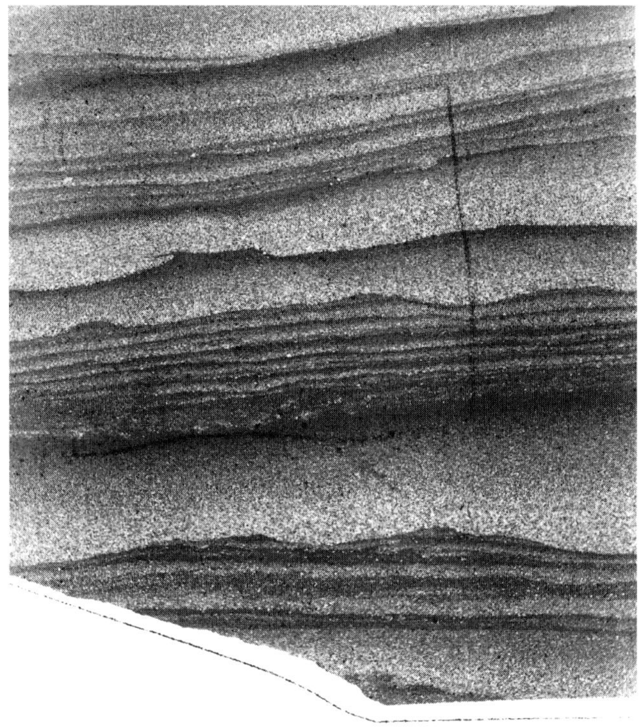

have been deposited in the Ranger Peak Formation after the Montoya was deposited. No source has been found for the white internal sediment, but it must have had a source from above the Montoya Group.

Reconstruction of Cavern Development and Collapse

The geometry of the El Paso caverns is reconstructed schematically in Figure 20. Folding and faulting of the Montoya down into the areas of brecciation in the El Paso Group and the downward displacement of blocks from their stratigraphic position show that El Paso brecciation was formed by roof collapse of a system of caves developed mainly in the upper 300 m (1000 ft) of the El Paso Group. Rust/tan-colored internal sediment similar to a paleosol at the El Paso–Montoya contact and Montoya sediment enclosing El Paso breccia demonstrate that the caverns were formed between deposition of the El Paso and deposition of the Montoya.

Caves occupied about 30% of the Ranger Peak Formation based on the difference in thickness between the brecciated and unbrecciated Ranger Peak measured sections. The Ranger Peak caves were tabular in character (Figure 20A). Dissolution in the McKelligon Canyon Formation was principally along vertical fractures with minor dissolution along bedding plains forming caverns with a vertical character (Figure 20A). Where dissolution extended deep into the El Paso Group, caverns as large as 60 m (200 ft) high and 200 m (700 ft) wide were formed in the Ranger Peak Formation and in the McKelligon Canyon Formation immediately beneath the Cindy Formation. The dolomitized Cindy Formation resisted dissolution and formed a ceiling (Figure 20A). Few sinkholes were present, and they were probably located in the areas of greatest vertical dissolution. The presence of rust/tan internal sediment from the El Paso surface filling small caves and between breccia blocks in the McKelligon Canyon Formation indicates that some collapse occurred during cave formation (Figure 20A). Collapse of the El Paso Caverns was probably continuous until after the Fusselman was deposited (Figure 20b) as shown by the presence of Fusselman blocks in the upper parts of the vertical breccias.

In the southern Franklin Mountains the top of the Fusselman is not exposed. In the northern Franklin Mountains, however, the Fusselman is well exposed and is overlain by a major unconformity with a time gap of about 30 m.y. (LeMone, 1987). Collapse brecciation similar to the El Paso brecciation has been observed in the Fusselman Formation in the northern Franklin Mountains (personal observation). Therefore, two periods of cave development occurred, one between deposition of the El Paso and the Montoya and a second between deposition of the Fusselman and the Canutillo. The laterally restricted breccias in the El Paso Group are connected with the Fusselman breccias by vertical breccia pipes, suggesting that the collapse of the El Paso caverns affected cavern development in the Fusselman Formation (Figure 20B). No collapse brecciation has been reported above the Fusselman Formation.

Collapse of the cave roof produced a structural sag with 13 m (43 ft) of relief in the top of the El Paso, and

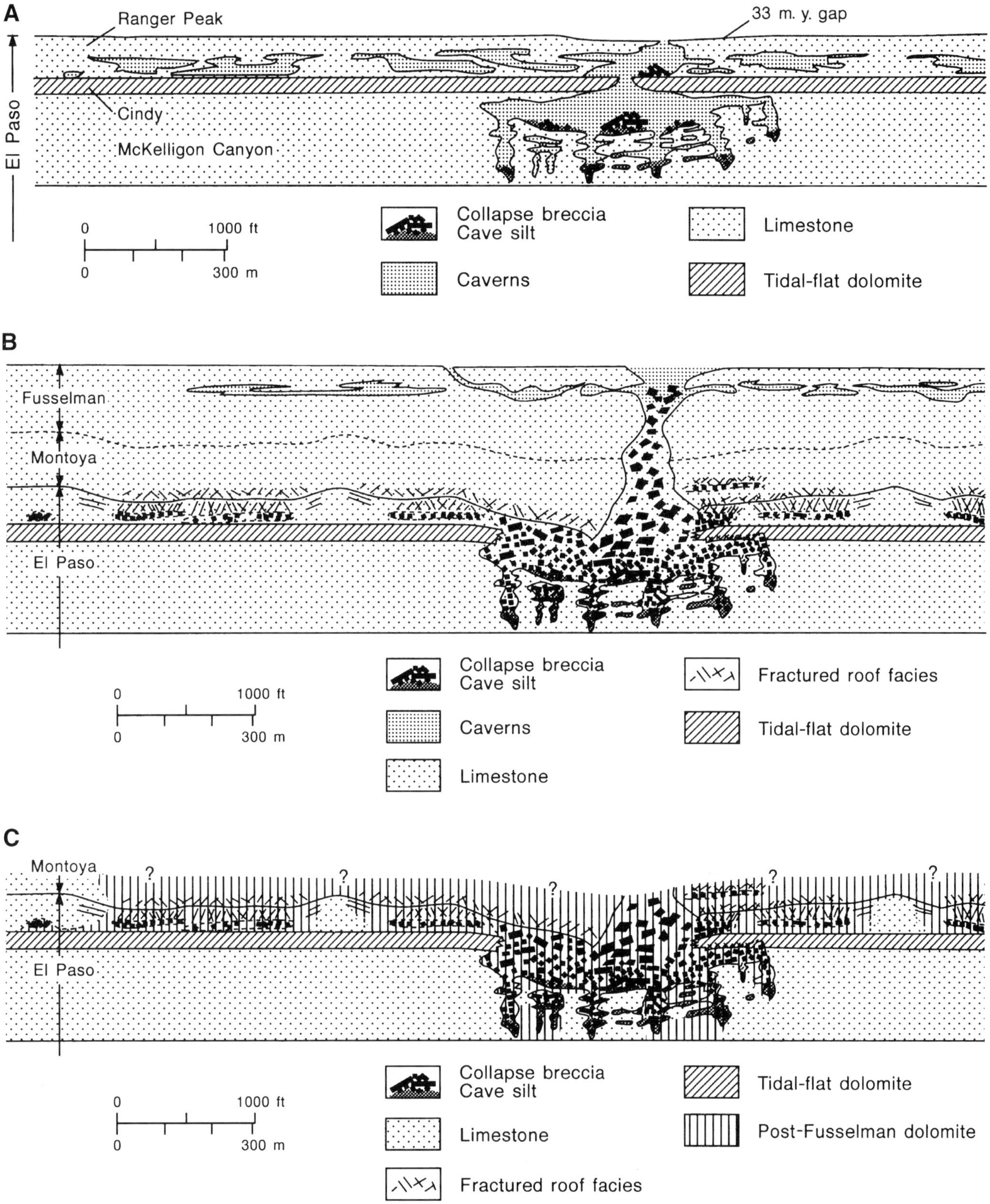

Figure 20. Reconstruction of El Paso caverns. (A) Penecontemporaneous dolomitization of the Cindy Formation and development of tabular, laterally continuous caverns in the Ranger Peak Formation and vertical, laterally discontinuous caverns in the McKelligon Canyon Formation. (B) Collapse of the El Paso caverns showing collapse of the Montoya, development of breccia pipes up into the Fusselman Formation, and development of caverns in the Fusselman Formation. (C) Late-stage dolomitization of the El Paso and Montoya groups controlled by fluid flow through collapse breccia, fractures, and into adjacent carbonates.

fracturing and minor faulting in the overlying Upham Formation (Figure 20B). The greatest amount of collapse was in areas of greatest vertical cave development. Collapse in this area produced long, linear, structural sags with 60 m (200 ft) of relief that extend into the Montoya section.

Origin of the El Paso Caverns

The presence of laminated internal sediment suggests that the caves were formed in the vadose or high phreatic groundwater zones. Only in these environments can groundwater move with sufficient velocity to transport sediment (Dunham, 1969). The texture of the laminated internal sediment suggests that it was deposited in the phreatic zone rather than the vadose zone. The laminae are commonly graded, suggesting periodic influx of sediment-laden water. Flame or pull-over structures are present at the contacts of some of these graded laminae. These structures are commonly interpreted to be formed by drag on a watery clay sediment or by squeezing of a highly mobile foundation due to rapid loading. Therefore, the laminae did not dry out between periods of deposition as might be expected if deposited in the vadose zone. The mechanism of transport may have been by density currents set up by the introduction of flood waters containing large amounts of sediment into the usually clear, less dense groundwater of the cave system.

The distribution of the two colors of internal sediment suggests two stages of infill, one from the El Paso surface and one after the Montoya was deposited. Although the presence of laminated internal sediment does not necessarily mean that the caves were formed in the phreatic zone, the fact that most of the rust/tan internal sediment thought to be derived from the El Paso surface is found in the brecciated McKelligon Canyon Formation suggests that the McKelligon Canyon Formation was consistently within the phreatic zone. The presence of the later-deposited white internal sediment in the brecciated Ranger Peak Formation suggests that the Ranger Peak Formation was only consistently in the phreatic zone after flooding of the El Paso surface at the beginning of Montoya deposition.

The widespread collapse brecciation in the Ranger Peak Formation suggests dissolution near the phreatic–vadose interface. The best areas for extensive cave development are generally thought to be in the vicinity of the water table where groundwater and meteoric recharge mix. Data on groundwater levels in the southeastern and south-central parts of the United States (USGS, 1955, 1965) show that water tables are seldom more than 60 m (200 ft) below the surface and generally around 15 to 30 m (50–100 ft) deep regardless of the amount of rainfall. In these areas, topography is gently rolling, which is similar to the topography on the El Paso surface. Assuming that a regional water table occurred around 15 to 30 m (50–100 ft) below the El Paso surface would explain the tabular character of cave development in the Ranger Peak Formation.

Assuming a water table around 15 to 30 m (50–100 ft) below the surface requires major cave formation in the phreatic zone to account for the presence of collapse breccia 300 m (1000 ft) below the El Paso surface. Back's work (1963) on the groundwater of central Florida has shown that the groundwater is undersaturated with respect to calcite over 100 m below the water table. Therefore, dissolution of limestone and thus cave formation could occur some 200 to 250 m (600–800 ft) below the water table. This dissolution would occur preferentially along high-velocity flow pathways such as those formed by vertical fractures. High-velocity flow along vertical fractures oriented N20°W would account for the linearity and the vertical character of the brecciation in the McKelligon Canyon Formation.

DOLOMITIZATION

The dolomite in the southern Franklin Mountains is found in four general modes based on field relationships: (1) stratiform dolostones; (2) medium-crystalline, laterally discontinuous dolostones; (3) coarsely crystalline saddle dolomite; and (4) medium-crystalline dolomite as dolomite mottles in limestone. From detailed mapping it is calculated that 30% of the McKelligon Canyon, Cindy, and Ranger Peak formations is dolostone.

Three stratiform dolostones are found in the southern Franklin Mountains, at the base, in the middle, and toward the top of the El Paso Group. All three are associated with tidal-flat facies and are presumed to be formed by reflux dolomitization. The upper stratiform dolostone is within and below the Cindy Formation (Figure 20A), which is composed of a series of tidal-flat–capped cycles. No evaporite minerals are found in this part, but irregular vugs suggest dissolution of sulfate nodules. The $\delta^{18}O$ signature is –3.3 to –4‰ (PDB). The volume of upper stratiform dolostone is calculated to be 2×10^8 m^3 or 11% of the volume represented by the McKelligon Canyon, Cindy, and Ranger Peak formations.

The laterally discontinuous, medium-crystalline dolostones are everywhere associated with collapse breccia and formed after collapse brecciation (Figure 20C). Since brecciation was in progress during the Silurian (when the Fusselman was deposited), this dolomitization event is post-Silurian in age. Dolomitization of the Upham Formation of the Montoya Group belongs to this event. The collapse breccia is nearly all dolomite, and the dolomitization extends into the nonbrecciated country rock for variable distances. The widest extent is in the Ranger Peak and Upham formations (Figure 20C). The limestone/dolomite boundary in the Ranger Peak and Upham formations is irregular, vertical, and very abrupt. In the McKelligon Canyon formation dolomitization does not extend far into the unbrecciated country rock (Figure 20C). Indeed, some of the collapse breccia in the McKelligon Canyon Formation is still limestone. Locally, however, beds of dolomite extend up to 400 m (1320 ft) into the unbrecciated areas.

The $\delta^{18}O$ signature of dolomitized country rock ranges from –5 to –8‰ (PDB), whereas the signature from the breccias ranges from –3 to –5‰ (PDB). The values in the breccia are probably contaminated by dolomite from the Cindy Formation, which has a heavier isotopic value than the dolomitized country rock. The volume of breccia dolomite is calculated to be 3.4×10^8 m^3 or 20% of the McKelligon Canyon, Cindy, and Ranger Peak formations.

Saddle dolomite occurs in vugs and fractures, between breccia blocks, and as replacement saddle dolomite in the vicinity of limestone/dolomite boundaries. In Transition Canyon, pore-filling saddle dolomite rests on a geopetal reddish infilling sediment and is overlain by another reddish infilling sediment. The infilling sediment is most likely early Paleozoic in age, fixing the minimum age of dolomite formation. Stepanek (1987) has shown that the saddle dolomites have an inner zone with a $\delta^{18}O$ value of –8.5‰ (PDB) and an outer zone with a $\delta^{18}O$ value of –14 to –16‰ (PDB) separated by a reddish infilling sediment, suggesting two periods of saddle dolomite formation. The similarity of the $\delta^{18}O$ values of the inner zone with the breccia dolomite and the association with red infilling sediment suggest that the inner zone dolomite is part of the breccia dolomitization. The very light isotopic values for the outer zone suggest a later, hotter, and deeper dolomitization event. Because saddle dolomite is found principally in fractures in the breccia dolomite, it is estimated to be <1% of the late dolomite volume, which is only a trace amount of the total carbonate represented by the McKelligon Canyon, Cindy, and Ranger Peak formations.

Brecciation apparently produced permeable pathways for water flow. Fluids must have moved laterally through the breccia, and dolomitization of the unbrecciated strata adjacent to the breccias suggests lateral flow from the breccia into the country rock. There are few field observations, however, that suggest the direction from which dolomitizing fluids came: above, below, or laterally. However, the presence of limestone breccias at the base of some of the vertical breccias and the widespread dolomitization in the Ranger Peak and Upham formations compared with the laterally restrictive dolomitization in the underlying McKelligon Canyon Formation points to a source from above. No dolostones are found above the Devonian in this area, suggesting that no overlying source for dolomitizing water was available after Devonian time. Therefore, the late dolomitization event occurred after mid-Silurian and before Mississippian time; between 75 and 100 m.y. after deposition of the El Paso Group.

Kupecz and Land (1991) have concluded that late-stage saddle dolomite composes 10% of the Ellenburger Group of the Permian basin and central Texas, is Pennsylvanian in age, and formed from deep-basin fluids introduced laterally from the Ouachita thrust belt. The late-stage saddle dolomite in the El Paso Group may also be Pennsylvanian in age and have flowed laterally.

IMPLICATIONS FOR POROSITY AND PERMEABILITY HISTORY

The distribution of karst features and diagenetic dolomite is directly related to fluid flow. Therefore, the history of their development can be used to infer the fluid transmissibility of the containing formations. Transmissibility can be expressed in terms of permeability, which is a rock property, and can be related to porosity through rock-fabric considerations (Lucia, 1983). By relating karsting and dolomitization events to time and space, the history of porosity and permeability modification in the El Paso Group can be inferred.

The Cindy Formation is a stratiform dolostone formed by hypersaline reflux, a mechanism that requires fluid flow through young sediments. The porosity and permeability of the sediment were probably similar to modern carbonate sediments with mud-dominated sediments estimated at 50% porosity and 100 md permeability and grain-dominated sediments at 50% porosity and darcys of permeability (Enos and Sawatsky, 1981). Dolomitization probably reduced porosity and permeability by overdolomitization, a process recently described by Lucia and Major (1994). However, the Cindy dolostone remained permeable through Silurian time (some 75 m.y. after El Paso deposition) because late dolomitization of small breccia tubes in the lower Ranger Peak Formation is continuous with the Cindy Formation, suggesting that the Cindy dolostone was a conduit for late dolomitizing fluids.

Most of the El Paso Group is composed of mud-dominated sediments suggesting original porosity on the order of 50% and permeability values around 100 md. During the Middle Ordovician, open vertical fractures developed that extended at least 300 m (1000 ft) into the El Paso Group. Groundwater flow was concentrated in these fractures, particularly in the McKelligon Canyon Formation.

Extensive late dolomitization of the limestone surrounding the breccia in the Ranger Peak and Upham formations suggests moderate porosity and permeability values (perhaps 30% and 10 md) at the end of the Silurian. These values are consistent with the values for Eocene carbonates presented by Halley and Schmoker (1983). However, only a few unbrecciated beds in the McKelligon Canyon Formation are dolomitized, suggesting <1 md of permeability at the end of the Silurian.

Major karsting occurred during the Middle Ordovician creating large voids and greatly increasing the permeability of the El Paso Group. The dissolution was concentrated in the limestone whereas the dolomitized Cindy Formation was left intact. Locally, the porosity was significantly increased by the development of caverns, but there is no indication that the overall porosity of the El Paso Group increased. It is likely that the calcite dissolved during cave formation precipitated as calcite cement in the surrounding El Paso carbonates. Pore space was simply redistributed whereas permeability was greatly enhanced. Assuming that the porosity of the El Paso Group was 40% at the end of El Paso deposition, and that 30% of the late dolomite was cavernous at the end of Middle Ordovician time, a porosity value of 32% for the unbrecciated El Paso limestone would be required to retain an average porosity value of 40% for the total El Paso Group. It is likely that the porosity of the El Paso limestone was <32% because of the addition of carbonate from dissolution at the unconformity, further suggesting that the average porosity of the El Paso Group was reduced, rather than increased, by the karsting event.

Cave infill and collapse occurred during the Middle Ordovician and Silurian. Cave infill and collapse occurred during the Middle Ordovician in response to infiltration of surface waters and cavern enlargement. The main collapse occurred during the Silurian. During this time, the roof collapsed, producing extensive fracturing and brecciation in the Upham Formation. Infilling also occurred during this time. Porosity and permeability of the El Paso caverns were reduced and permeability in the Upham Formation was increased as a result of fracturing and brecciation. By the end of the Devonian (100 m.y. after deposition of the El Paso Group and 300–600 m [1000–2000 ft] of overburden), the El Paso caverns had collapsed, forming a brecciated and fractured roof facies in the overlying Montoya Group. Late dolomitization fluids were channeled through the collapse breccia, indicating that the breccia, fracture, and cavern porosity in the breccia contained the highest permeability in the El Paso Group at the end of the Silurian.

The last dolomitization event is the formation of saddle dolomite which precipitated in fractures and interbreccia-block porosity. Only local replacement of limestone has been observed. Assuming this to be a Pennsylvanian event suggests that the unkarsted El Paso Group had little permeability after about 200 m.y. had passed and 1500 m (5000 ft) of overburden were deposited. The only significant permeability left after 200 m.y. of diagenesis was associated with fracture and breccia-block porosity. Porosity and permeability were further reduced by pore-filling saddle dolomite.

SUMMARY AND DISCUSSION

Unconformities of varying duration are present in the Lower Ordovician El Paso Group exposed in the southern Franklin Mountains at El Paso, Texas. The diagenetic effects that can be directly related to these unconformities are dolomitization and karsting. Tidal-flat–capped cycles and associated third-order sequence boundaries are marked by penecontemporaneous dolomitization but do not show significant karsting.

The unconformity at the top of the El Paso Group, which represents a time gap of 33 m.y., is marked by significant karsting that profoundly affects the diagenetic and reservoir history of the lower Paleozoic strata. A major cavern system was formed in the upper 300 m (1000 ft) of the El Paso Group. Collapse of these caverns produced brecciation in the overlying Montoya Group and eventually connected the El Paso caverns with karsting in the Fusselman Formation 270 m (900 ft) above the El Paso Group. A late dolomitization pattern mimics collapse brecciation in the El Paso and Montoya groups and probably in the Fusselman Formation.

The caverns were formed during the time of the unconformity at the top of the El Paso Group. Vertical fractures with a N20°W trend were formed in the El Paso carbonates, and meteoric groundwater preferentially flowed down and laterally along fractures that were spaced about 900 m (3000 ft) apart, dissolving carbonate as far as 300 m (1000 ft) below the surface. The water table was about 15 to 30 m (50–100 ft) below the surface, and laterally extensive tabular caves developed near this phreatic–vadose interface. At greater depths, cave geometry was controlled by deep dissolution along vertical fractures, producing long, linear cave systems with minor dissolution along bedding planes. This combination of water table and vertical fracture control on cave geometry produced a "T" bar shape to the El Paso caverns.

At least 13 m (43 ft) or 30% of the Ranger Peak Formation was dissolved, producing tabular caverns. Where dissolution extended deep into the El Paso Group, caverns as large as 60 m (200 ft) high and 200 m (700 ft) wide were formed in the Ranger Peak Formation and in the McKelligon Canyon Formation and are separated by the dolomitized Cindy Formation which resisted dissolution. Collapse of the Ranger Peak and Cindy formations produced the majority of the collapse breccia. Collapse of strata within the McKelligon Canyon Formation was not as extensive and took the form of dislodged large blocks. Much of the breccia found within the McKelligon Canyon stratigraphic interval was derived from the overlying Cindy and Ranger Peak formations.

Some collapse and sediment infilling occurred during cave formation. Sediment-laden flood water periodically flowed into the caves, depositing laminated rust/tan-colored internal sediment below the water table between early-formed breccia blocks and in small caves located in the McKelligon Canyon Formation. Most of the internal sediment is massive, poorly sorted carbonate and chert that was probably produced internally during collapse. Collapse of the El Paso caverns was probably continuous during deposition of the Montoya Group. The main collapse of the cavern roof, however, appears to have occurred during and after the Silurian. A laminated white-colored internal sediment was deposited in the pore space between breccia blocks in the Ranger Peak and Upham formations, probably during the Late Silurian–Early Devonian. Thus, the major collapse occurred some 50 m.y. after cave formation and under 300 to 600 m (1000–2000 ft) of overburden.

Collapse of the cavern roof produced a structural sag with 13 m (43 ft) of relief at the top of the El Paso and fracturing, brecciation, and minor faulting in the overlying Upham Formation. The greatest amount of collapse is in areas of deepest cave development. Collapse in this area produced long, linear, structural sags with 60 m (200 ft) of relief that extend into the Montoya Group. Vertical breccia pipes extend from the El Paso Group up into the Fusselman Formation connecting Fusselman and El Paso collapse breccias. The breccias formed permeable pathways for the migration of dolomitizing fluids. It is postulated that these fluids originated from a modified seawater source during late Silurian or Devonian (75–100 m.y. after El Paso deposition) and migrated downward and laterally through the fractured and brecciated Fusselman, Montoya, and El Paso.

Deposition, cavern development, collapse, and dolomitization of the El Paso and Montoya groups probably took place over a 100 m.y. duration and with a maximum overburden of about 600 m (2000 ft), spanning Middle Ordovician to middle Devonian time. During this time the porosity and permeability distribution was drastically modified. The original 50% porosity in the limestone was reduced to about 30%; however, sufficient permeability remained to allow limited passage of the late dolomitizing fluids. Early reflux dolomitization reduced the original porosity and permeability of the Cindy Formation, but the Cindy dolostone retained sufficient permeability to act as a conduit for late dolomitizing waters. The Cindy dolostone appears to have been more permeable than the El Paso limestone after 75 m.y. because dolomitizing fluids flowed through the Cindy Formation, whereas there is no indication that fluid flowed through the El Paso limestone. This is consistent with observations that dolomite tends to be more porous and permeable than adjacent limestones in Paleozoic formations. Both the El Paso limestone and Cindy dolostone had become practically impermeable by Pennsylvanian time (200 m.y. after El Paso deposition).

The pattern of late dolomitization suggests that by the Devonian (75–100 m.y. after El Paso deposition) collapse breccias and associated fracturing had become the main conduits for fluid flow. The Middle Ordovician karsting event created a large system of caverns with as much as 30% of the section as large caverns. This pore space was gradually destroyed by collapse and sediment infill until, by Devonian time, the porosity of the El Paso caverns was probably less than 10%. However, the permeability was significantly higher than the unkarsted El Paso because of breccia and fracture porosity. By Pennsylvanian time, only the breccia and fracture pore space was open to fluid flow and it was further occluded by saddle dolomite.

As a result of compaction, cementation, dolomitization, and extensive karst-related diagenesis, very little porosity or permeability remain in the El Paso carbonates. The similarity to the Ellenburger reservoirs in the Permian basin suggests that the El Paso breccias could contain producible hydrocarbons. Ellenburger reservoirs produce from fractures, breccias, and vugs related to extensive karsting and collapse and have little porosity or permeability (Kerans, 1989a, b; Kupecz and Land, 1991). Similarly, the remaining pore space found in the El Paso is in collapse-breccia fractures and between breccia blocks. The breccia dolomite in the southern Franklin Mountains covers 835 acres and could contain 35 bcf of gas, assuming the breccia has 1% porosity and a gas volume similar to volumes found in Ellenburger gas fields of the Permian basin. The highest porosity and permeability are found between breccia blocks and in fractures in the collapsed Upham Formation that overlies the El Paso. The collapse in the Upham Formation results from roof collapse of the El Paso caverns, similar to the origin of the roof reservoir facies in Ellenburger reservoirs as described by Kerans (1989a). Therefore, even after the extensive diagenesis observed in the Lower Ordovician El Paso carbonates, reservoir conditions are maintained because of the major karsting event associated with a supersequence unconformity.

ACKNOWLEDGMENTS

The El Paso Cavern story is the result of research initiated while the author was with Shell Oil Company and revised while with the Bureau of Economic Geology, The University of Texas at Austin. The author is grateful for discussions with Charles Kerans and for reviews by Art Saller, Dave Osleger, and Scott Tinker. Published with permission of the Director, Bureau of Economic Geology, The University of Texas at Austin.

REFERENCES CITED

Back, W., 1963, Preliminary results of a study of calcium carbonate saturation of a ground water in Central Florida: International Association Scientific Hydrology VIIIe, Annee no. 3, p. 43–51.

Deffeyes, K. S., Lucia, F. J., and Weyl, P. K., 1965, Dolomitization of Recent and Plio-Pleistocene sediments by marine evaporate waters on Bonaire, Netherlands Antilles, *in* Pray, L. C., and Murray, R. C., eds., Dolomitization and limestone diagenesis—a symposium: SEPM Special Publication 13, p. 71–88.

Dunham, R. J., 1969, Early vadose silt in Townsend mound (reef), New Mexico, *in* Friedman, G. M., ed., Depositional environments in carbonate rocks—a symposium: SEPM Special Publication 14, p. 139–181.

Enos, P., and Sawatsky, L. H., 1981, Pore networks in Holocene carbonate sediments: Journal of Sedimentary Petrology, v. 51, p. 961–985.

Flower, R. H., 1964, The Nautiloid order Ellesmeroceratida (cephalopoda): New Mexico Bureau of Mines and Mineral Resources, Memoir 12, 234 p.

Galloway, W. E., Ewing, T. E., Garrett, C. M., Tyler, N., and Bebout, D. G., 1983, Atlas of major Texas oil reservoirs: The University of Texas at Austin, Bureau of Economic Geology Special Publication, 139 p.

Goldhammer, R. K., Lehmann, P. J., and Dunn, P. A., 1993, The origin of high-frequency platform carbonate cycles and third-order sequences (Lower Ordovician El Paso Gp., west Texas): constraints from outcrop data and stratigraphic modeling: Journal of Sedimentary Petrology, v. 63, p. 318–359.

Halley, R. B., and Schmoker, J. W., 1983, High-porosity Cenozoic carbonate rocks of South Florida: Progressive loss of porosity with depth: AAPG Bulletin, v. 67, p. 191–200.

Howe, H. J., 1959, Montoya Group stratigraphy (Ordovician) of Trans-Pecos Texas: AAPG Bulletin, v. 43, p. 2285–2332.

Kay, M., 1951, North American geosynclines: Geological Society of America Memoir 48, 143 p.

Kerans, C., 1989a, Karst-controlled reservoir heterogeneity in Ellenburger Group carbonates of west Texas: AAPG Bulletin, v. 72, p. 1160–1183.

Kerans, C., 1989b, Karst-controlled reservoir heterogeneity and an example from the Ellenburger Group (Lower Ordovician) of west Texas: The University of Texas at Austin, Bureau of Economic Geology Report of Investigations 186, 40 p.

Kerans, C., and Lucia, F. J., 1989, Recognition of second, third, and fourth/fifth order scales of cyclicity in the El Paso Group and their relation to genesis and architecture of Ellenburger reservoirs, *in* Cunningham, B. K., and Cromwell, E. W., eds., The Lower Paleozoic of West Texas and Southern New Mexico—Modern Exploration Concepts: SEPM Permian Basin Section Publication 89-31, p. 113–122.

Kupecz, J. A., and Land, L. S., 1991, Late-stage dolomitization of the Lower Ordovician Ellenburger Group, west Texas: Journal of Sedimentary Petrology, v. 61, p. 551–574.

Kyle, J. R., 1976, Brecciation, alteration, and mineralization in the central Tennessee zinc district: Economic Geology, v. 71, p. 892–903.

LeMone, D. V., 1968, The Canadian (Lower Ordovician) El Paso Group of the southern Franklin Mountains, El Paso County, Texas, *in* Delaware Basin exploration: West Texas Geological Society Guidebook 68-55, p. 76–81.

LeMone, D. V., 1987, Sequence stratigraphy of the Tobosa basin—related Paleozoic sediments of the Franklin Mountains, El Paso County, Texas and Dona Ana County, New Mexico, *in* Cunningham, B. K., and Cromwell, E. W., eds., The Lower Paleozoic of West Texas and Southern New Mexico—Modern Exploration Concepts: SEPM Permian Basin Section Publication 89-31, p. 71–84.

Loucks, R. G., and Anderson, J. H., 1985, Depositional facies, diagenetic terranes, and porosity development in Lower Ordovician Ellenburger Dolomite, Puckett field, west Texas, *in* Roehl, P. O., and Choquette, P. W., eds., Carbonate petroleum reservoirs: New York, Springer-Verlag, p. 10–38.

Lucia, F. J., 1968, Sedimentation and paleogeography of the El Paso Group, *in* Delaware Basin Exploration: West Texas Geological Society Guidebook 68-55, p. 61–75.

Lucia, F. J., 1970, Lower Paleozoic history of the western Diablo platform of west Texas and south-central New Mexico, *in* Seewald, K., and Sundeen, D., eds., Symposium in honor of Professor Ronald K. DeFord: West Texas Geological Society Publication, p. 39–56.

Lucia, F. J., 1983, Petrophysical parameters estimated from visual description of carbonate rocks: a field classification of carbonate pore space: Journal of Petroleum Technology, p. 626–637.

Lucia, F. J., and Major, R. P., 1994, Porosity evolution through hypersaline reflux dolomitization, *in* Purser, B., Tucker, M., and Zenger, D., eds., Dolomites: International Association of Sedimentologists Special Publication 21, p. 325–341.

Mazzullo, S. J., 1989, Formational and zonal subdivisions of the Ellenburger Group (Lower Ordovician), southern Midland basin, Texas, *in* Cunningham, B. K., and Cromwell, E. W., eds., The Lower Paleozoic of West Texas and Southern New Mexico—modern exploration concepts: SEPM Permian Basin Section Publication 89-31, p. 113–122.

Murray, R. C., 1960, Origin of porosity in carbonate rocks: Journal of Sedimentary Petrology, v. 30, p. 59–84.

Mussman, W. J., and Read, J. F., 1986, Sedimentology and development of a passive-to-convergent margin unconformity–Middle Ordovician Knox unconformity, Virginia Appalachians: Geological Society of America Bulletin, v. 97, p. 282–295.

Ohle, E. L., 1985, Breccias in Mississippi Valley-type deposits: Economic Geology, v. 80, p. 1736–1752.

Stepanek, B. E., 1987, Dolomitization of paleokarst collapse breccias—the Ordovician El Paso Group, Franklin Mountains, west Texas, *in* Mega-Collapse Breccia and Associated Late Stage Dolomitization of Ordovician Carbonates, Franklin Mountains: SEPM Mid-Year Meeting Guidebook, unpaginated.

U.S. Geological Survey (USGS), 1955, Ground-water levels in the United States, southcentral states, 1955: Water-Supply Paper 1407.

U.S. Geological Survey (USGS), 1965, Ground-water levels in the United States, 1959–63, southeastern states, 1965: Water-Supply Paper 1803, 263 p.

Walters, R. F., 1946, Buried Precambrian hills in northeastern Barton County, central Kansas: AAPG Bulletin, v. 30, p. 660–710.

Chapter 15

H_2S-Related Porosity and Sulfuric Acid Oil-Field Karst

Carol A. Hill
Consulting Geologist
Albuquerque, New Mexico, U.S.A.

ABSTRACT

"H_2S-related porosity" refers to porosity created in a H_2S system where dissolution can be produced by the mixing of waters of different H_2S content or by the oxidation of H_2S. "Sulfuric acid oil-field karst" refers to a specific kind of H_2S-related porosity where carbonate reservoirs of cavernous size have been dissolved by a sulfuric acid mechanism. In a H_2S system, porosity can be produced entirely in the deep subsurface and does not have to represent a paleokarst surface or dissolution in the shallow-phreatic or vadose zones.

H_2S-related porosity is characterized by the large volume of hydrocarbons it can host, by extensive fracture permeability interconnected with "spongework" cavities or caves of tens to hundreds of meters in extent, by porosity related to structural and/or stratigraphic traps, and by the presence of high uranium and/or iron. Possible examples of H_2S-generated porosity systems are the Lisburne field, Prudhoe Bay, Alaska, and some of the extremely productive fields of the Middle East.

INTRODUCTION

While the importance of karst and karst-related processes to reservoir porosity has become increasingly recognized in recent years, the importance of hydrogen sulfide in the creation of that porosity has not. Traditionally, "karst" has denoted a type of topography characterized by sinkholes, caves, and underground drainage, and karst reservoir systems have denoted porosity formed by fresh water and carbonic acid in surface to shallow-subsurface environments and in marine–freshwater mixing zones (e.g., Craig, 1988; Wigley et al., 1988). However, more recently the concept of karst has been expanded to encompass topics like deep-seated dissolution, cavernous porosity, or the processes by which limestones are dissolved (Sweeting, 1973; Ford and Williams, 1989). Karst-forming mechanisms have also been expanded to include dissolution by sulfuric acid (e.g., Hill, 1987, 1990). It is in this broader, more general sense that "H_2S-related porosity" and "sulfuric acid oil-field karst" are discussed. By understanding the different characteristics of freshwater karst-related porosity and H_2S-related porosity, we can better predict their occurrence in the subsurface.

POROSITY DEVELOPMENT IN A H_2S SYSTEM

Porosity in a H_2S system may be created by the oxidation of hydrogen sulfide or by the mixing of waters of different H_2S content. Mixing of two saturated solutions of H_2S can produce an undersaturated solution by the method of "mixture corrosion" (Figure 1). Mixture corrosion in a H_2S system operates on the same

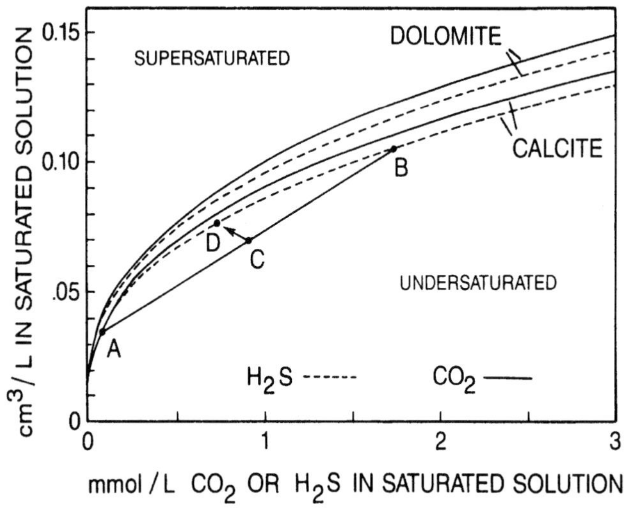

Figure 1. Saturation concentration of calcite and dolomite as a function of CO_2 or H_2S concentration in volume of dissolved solid per liter. Mixing of two saturated solutions (e.g., A and B) produces an undersaturated solution (C) and renewed aggressiveness. Subsequent dissolution follows line C–D. After Palmer (1991).

principle as does CO_2 mixture corrosion: that is, the mixing of two saturated solutions (for example A and B, Figure 1) produces an undersaturated solution (C) and renewed aggressiveness (Palmer, 1991). Aggressiveness caused by the mixing of waters is probably more important than oxidation in the evolution of deep-seated carbonate petroleum reservoirs, and in such cases porosity development can take place entirely in the deep subsurface.

Porosity development in a H_2S system can also take place by the oxidation of hydrogen sulfide to sulfuric acid in an evaporite-rich hydrocarbon environment. When natural gas reacts with sulfate ions derived from evaporite rocks, hydrogen sulfide is formed:

$$Ca^{2+} + 2SO_4^{2-} + 2CH_4 + 2H^+ = 2H_2S + CaCO_3 + 3H_2O + CO_2 \quad (1)$$

In this reaction the evaporite ($CaSO_4$) is directly replaced by bio-epigenetic carbonate rock ($CaCO_3$) and H_2S is produced. When this H_2S reacts with oxygenated groundwater it can be converted to sulfuric acid:

$$H_2S + 2O_2 = HSO_4^- + H^+ \quad (2)$$

The sulfuric acid attacks carbonate rock and creates pores/caves suitable for hosting hydrocarbons. Hill (1992) called this type of H_2S-related porosity "sulfuric acid oil-field karst." Sulfuric acid porosity development commonly takes place in the shallower subsurface where oxidized water has access to the H_2S produced in equation 1; however, this oxidation type of porosity may become superimposed over mixture-corrosion porosity if an area becomes uplifted or if groundwater patterns change over time.

CHARACTERISTICS OF H_2S-RELATED POROSITY

In order to distinguish between H_2S and non-H_2S porosity/karst systems it is necessary to describe the features characteristic of a H_2S system:

1. *Pore size.* Porosity generated by a H_2S mechanism can vary from the microscopic to macroscopic. Incipient, slow, deep-phreatic mixture corrosion can create an interconnected system of microscopic porosity, but shallow-phreatic, sulfuric acid corrosion is capable of quickly dissolving immense amounts of carbonate bedrock and of creating karstic porosity on a large scale. An example of a large karst chamber dissolved by sulfuric acid is the Big Room of Carlsbad Cavern, New Mexico, which is over 600 m long, 330 m wide, and 87 m high (Hill, 1987).

2. *Spatial distribution.* Typically, H_2S-generated porosity consists of a matrix of "spongework" surrounding fractures of open or karstic porosity. "Spongework" consists of interconnected solution cavities of varied size in a seemingly random three-dimensional pattern like the pores of a sponge (Palmer, 1991). Extensive fracture permeability may be encountered, but outward from these open channels the rock exhibits interconnected porosity on any scale of examination from less than millimeters to tens, and sometimes even hundreds, of meters.

Also characteristic of a H_2S-related system is the discontinuity of sulfuric acid karst. Typical carbonic acid–dissolved caves display extensive linear passageways which can be traced from recharge areas to specific discharge points (e.g., Flint–Mammoth Cave, Kentucky). Palmer (1991) termed such caves "epigenic" because they are formed in the near-surface where carbonic acid is derived from CO_2 in the atmospheric and soil zones. In contrast, caves related to a H_2S-mechanism are "hypogenic" and have a deep-seated origin. Hypogenic karst is seemingly unrelated to meteoric recharge and discharge points and is instead related to H_2S gas-injection points along joints, fractures, faults, or other avenues available for gas movement (Hill, 1990). Thus, hypogenic cave passages begin and end abruptly and none display extensive linear passageways (e.g., the Big Room of Carlsbad Cavern terminates without a single passage extending from it).

3. *Structural and stratigraphic traps.* H_2S will rise and accumulate in structural and/or stratigraphic traps and therefore the porosity created by this mechanism is commonly directly related to such traps. Hill (1990) related sulfuric acid cave development in the Guadalupe Mountains, New Mexico, to structural and stratigraphic traps. H_2S derived from hydrocarbons in the Delaware basin ascended from the basin into the reef to accumulate in anticlinal traps and below the impermeable Yates siltstone (Figure 2). This H_2S then oxidized to H_2SO_4 and dissolved the rock to form the caves of the Guadalupe Mountains. Hill (1993) also attributed the sulfide (pyrite, sphalerite, galena) mineralization in these same structural/stratigraphic traps to a H_2S mechanism (Figure 2).

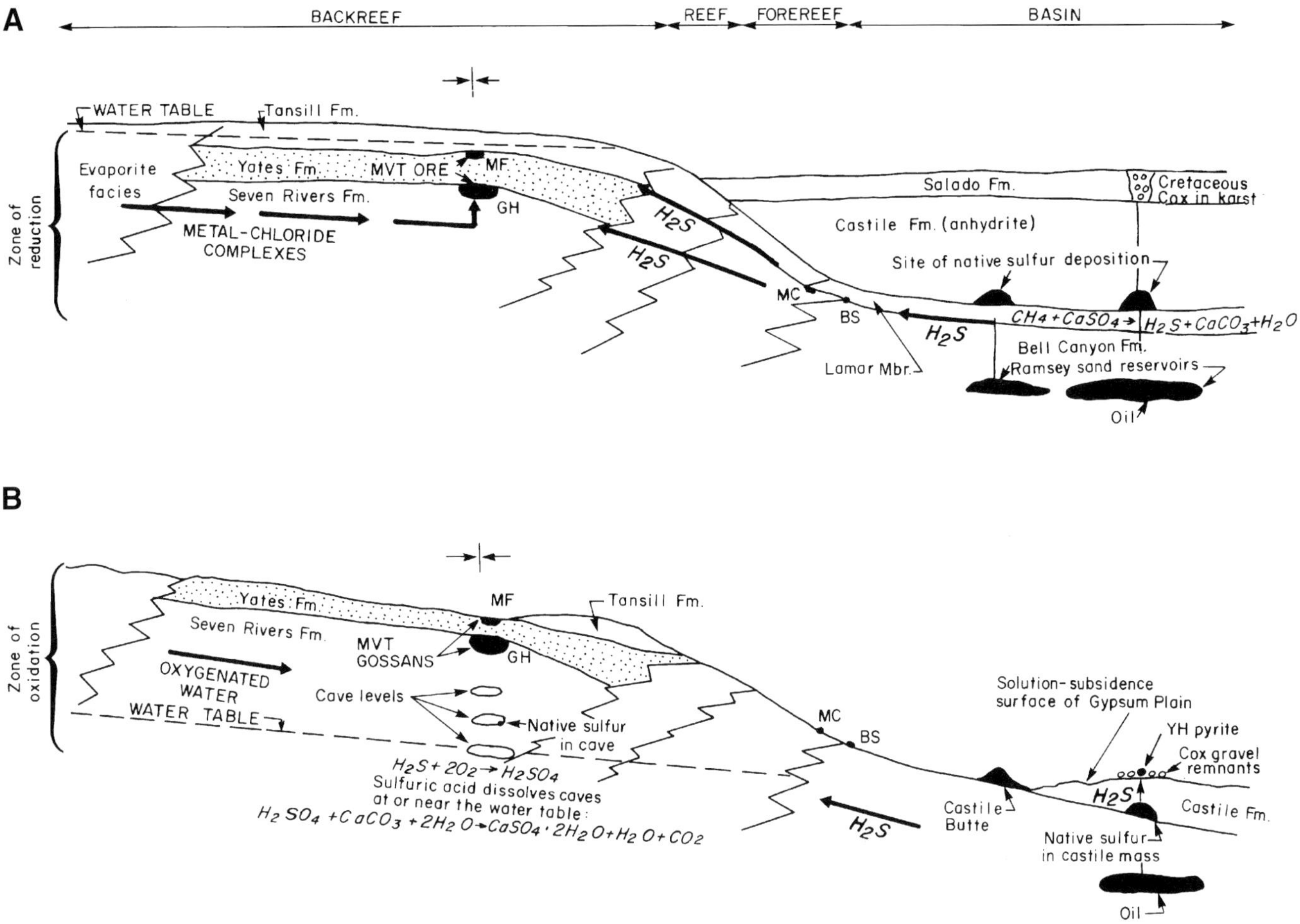

Figure 2. Diagram showing the ascension of H_2S in the Delaware basin to produce the sulfuric acid caves in the Capitan reef. (A) Oligocene–Miocene (40–20 Ma). (B) Pliocene–Pleistocene (5–0 Ma). MVT = Mississippi Valley type; MF = Middle Fork waterhole marcasite; GH = Grisham Hunter prospect; MC = McKittrick Canyon pyrite; BS = Bell Springs prospect; YH = Yeso Hills. After Hill (1993).

4. *Uranium enrichment*. Uranium (and vanadium) readily precipitate in a H_2S environment. Uranium is soluble in groundwater as long as this groundwater remains oxidizing, but when it encounters and percolates through a reducing (H_2S) environment the uranium (and vanadium) precipitate, commonly as "roll-front"–type deposits along the redox boundary interface. Hill (1987) reported high levels of uranium in montmorillonite clay (320 ppm) and speleothems (238 ppm) of Carlsbad Cavern. High radon levels were also found associated with the montmorillonite because of the uranium "fixed" by this clay under low pH conditions. Recently, metatyuyamunite ($Ca(UO_2)_2(VO_4)_2{\cdot}3\text{–}5H_2O$) has been reported from Carlsbad Cavern and other caves in the Guadalupe Mountains (Cunningham et al., 1994).

5. *Iron enrichment*. Not only does a H_2S system precipitate uranium, it can also precipitate iron as pyrite:

$$2H_2S_{aq} + Fe^{2+}_{aq} = FeS_{2s} + 4H^{+}_{aq} \tag{3}$$

Roll-front deposits of uranium are often associated with pyrite, geothite, and hematite, where the iron is concentrated by H_2S present in the host sedimentary rock.

DISCUSSION

This is one of the first attempts to classify porosity of specific oil fields with regard to their H_2S genesis, and as such should be viewed as preliminary in nature. Some oil-field karst/porosity systems seem to display the characteristics of a H_2S genesis, while others do not. It is hoped that other fields will be evaluated with respect to this model in the future.

1. *Permian basin, southwestern USA*. Three good examples of oil-field karst exist in the Permian basin: the Ordovician Ellenburger (Arbuckle) of New Mexico, Texas, and Oklahoma, the Silurian Dollarhide field of southeastern New Mexico and west Texas, and the Permian Yates field, Pecos County, west Texas.

a. *Ellenburger*. The Ellenburger is an example of oil-field karst that is probably *not* related to a H_2S or sulfuric acid mechanism. Kerans (1989, 1990) demonstrated that porosity development in Ellenburger Group carbonates was directly related to a prolonged period of subaerial exposure that coincided with a

Middle Ordovician eustatic lowstand. This latter period occurred before transgression of Simpson Group siliciclastics, during which a widespread system of caves, sinkholes, joint-controlled solution features, and collapse breccias developed. During this time a regionally extensive (New Mexico, Texas, Oklahoma) cave system formed 30–90 m beneath an exposed Ellenburger surface.

The Ellenburger has none of the features that would classify it as a H_2S-related karst system. It is regionally extensive and seems obviously related to a past epigenic paleokarst surface where meteoric water was entering the system from above. This system does not contain high concentrations of H_2S, nor does it have significant enrichments of uranium or iron.

b. *Dollarhide field.* Stormont (1949), Mygdal (1949), and Keener (1957) reported an extensive cavern system in the Dollarhide field, on the western edge of the Central Basin platform. Here, oil is trapped in huge caverns comparable in size to Carlsbad Cavern (Stormont, 1949). Completion maps in the Dollarhide (Mendenhall, 1967) show cave passages from 1 to 15 m in height and covering an area of >15 km^2. Cave development is predominantly along the more gently dipping, western side of a faulted anticline, but seems to predate rather than postdate this structure and be genetically related to it. The dominant cave horizon is located directly above a shale layer which separates the Fusselman Formation from the underlying Montoya Formation (Bruce McPherson, 1994, personal communication). The Dollarhide is the only field in the vicinity known to display cavernous porosity.

The origin of the porosity/karst in the Dollarhide field is not so easily determined as that in the Ellenburger Group carbonates. The large size of the passages and the fact that the caves seem isolated from a regional karst system favor a H_2S/sulfuric acid origin; however, other important criteria are missing. The caves do not appear related to structural or stratigraphic traps. In fact, major cave development is *above* an impermeable shale horizon, not below it, suggesting that it is epigenic rather than hypogenic in origin. In addition, H_2S in the field is low (<2%) and there is no appreciable amount of uranium or iron enrichment in the rock. Therefore, the Dollarhide field is probably not an example of a H_2S-related karst system.

c. *Yates field.* Another karst system in the Permian basin is the Yates field, located along the southeastern corner of the Central Basin platform. Here, 285 caves, ranging in height from 0.3–6.4 m and averaging 0.6 m, have been encountered in 142 out of 898 wells drilled in the middle Permian San Andres Formation (Craig, 1988). Craig favored an island hydrologic model for karst development in the middle of San Andres deposition for the Yates field, where freshwater lenses formed beneath a cluster of low-relief limestone islands.

The size of the caves and their spatial distribution definitely favor Craig's (1988) island model; however, the Yates field is known to contain abundant H_2S and "hot" (high-uranium) dolomite (S. Tinker and N. Hurley, 1993, personal communication), two factors that favor a H_2S-genesis. It is possible that the Yates field has a composite genesis: that relatively late H_2S-generated porosity is superimposed over earlier karst development.

2. *Lisburne field, Prudhoe Bay, Alaska.* Jameson (1994) described the Lisburne field of Prudhoe Bay, Alaska, as a combination structural/truncation trap in which porosity decreases away from two major unconformities. The lower, a sub-upper Permian unconformity, is marked by late dolomitization and uranium enrichment. The higher unconformity is lower Early Cretaceous in age, and is marked by extensive microporosity formation in limestone. From Jameson's (1994) description of the Lisburne field, at least some porosity in this field may have a H_2S genesis:

a. "Hot" (uranium-rich) dolomite occurs directly beneath one of the two major unconformities at Lisburne. This uranium is concentrated along clay-rich stylolitic seams in dolomite. Major unconformities at Lisburne could be acting as stratigraphic traps for the H_2S, which could be "fixing" the uranium in clay minerals directly beneath the unconformity.

b. High iron enrichment (>1000 ppm) has been found in most Lisburne dolomite, and this dolomite also occurs directly beneath a major unconformity, as was the case with the "hot" dolomite. This iron enrichment takes the form of ferroan dolomite, ankerite, and Fe-sulfides. Again, these unconformities could have acted as stratigraphic traps where the H_2S reacted with iron wherever both constituents were available.

c. On the eastern side of the field, porosity enhancement in limestones is enormous (up to 40–50% and totaling 1.01×10^9 m^3; Jameson, 1994). Porosity in this part of the field consists of microporosity and macroporosity, with the macroporosity due to the coalescence of microporosity. Megafractures and open faults contain large envelopes of porosity. This distribution of voids is similar to the sulfuric acid caves of the Guadalupe Mountains, where spongework surrounds fractures which are avenues for H_2S and major groundwater movement.

d. The latest stage of porosity in the Lisburne field (mid-early Tertiary) is deep seated and related to tilting and a time when the rock was just beginning to enter the oil window. Faults and fractures were opened during the same late-stage dissolution event that created the bulk of the porosity. Cave development in the Guadalupe Mountains was also late and related to hydrocarbons and tectonic tilting (Hill, 1990). Hydrocarbons are needed for the generation of H_2S (equation 1) and tilting provides the tectonic drive for ascension of H_2S into structural and stratigraphic traps.

e. Reportedly, minor amounts of H_2S are produced from the eastern side of the field, and SF_6 tracer studies have demonstrated rapid communication along Lisburne faults (Missman and Jameson, 1991).

All of the above characteristics suggest that the Lisburne field may be an example of a H_2S-generated porosity system. Tilting and faulting in the mid-early Tertiary could have caused oil to migrate into evaporite rock below the Lisburne Group carbonates, producing H_2S as a by-product. The H_2S could have moved up along faults to mix with oxygenated groundwater to form (entirely in the subsurface) the extensive, interconnected, macroscopic porosity and an open fracture system. The H_2S could have also been responsible for concentrating U and Fe in "hot," iron-rich dolomites on the western side of the field where the east and west sides were connected.

3. *Middle East*. Some Middle East fields are well known for their ready fluid communication over many kilometers and for the fact that they are H_2S-rich and associated with evaporite-carbonate sequences. The most productive oil wells on record are from Agha Jari in Iran: 10^5 m^3 daily from 36 wells (Ford and Williams, 1989). Inter-reservoir communication up to 30 km has been proven for some Iranian fields and up to 95 km for the Kirkuk field, Iraq. The Kirkuk field is extremely porous and permeable, and every well drilled has encountered a network of super-capillary channels cutting the Main Limestone, seemingly in all directions and in all areas (Daniel, 1954). Both vertical and horizontal caves exist within this field; caverns of mainly lateral extent are developed chiefly where backreef-reef limestone beds overlie forereef beds. Lost circulation while drilling wells is frequent. To re-establish circulation "huge quantities of camel thorn, straw, cottonseed hulls, bundles of reeds, old gunny sacks, etc. have been pumped down into these cavities without success" (Daniel, 1954, p. 796). All of these factors make the Kirkuk field a possible candidate for porosity development in a H_2S, sulfuric acid system.

SUMMARY

A H_2S-related mechanism of creating porosity has not been well studied, but such a mechanism may be important to understanding the origin of some hydrocarbon reservoirs, especially those which contain substantial quantities of sour gas. This mechanism can perhaps help answer perplexing questions such as how porosity enhancement can occur within the deep subsurface and how it can coincide with hydrocarbon emplacement (problems mentioned by Mazzullo and Harris, 1991, 1992). In such situations where no evidence of meteoric diagenesis is observed, H_2S reactions may be responsible for mesogenetic, deep-burial dissolution. The petroleum geologist should keep this mechanism in mind when trying to determine the origin of porosity in carbonate systems.

ACKNOWLEDGMENTS

I thank Art Saller and Bruce McPherson of UNOCAL for supplying completion maps of the Dollarhide field, and Jeremy Jameson for his discussions of the Lisburne field.

REFERENCES CITED

Craig, D. H., 1988, Caves and other features of Permian karst in San Andres dolomite, Yates field reservoir, west Texas, *in* N. P. James and P. W. Choquette, eds., Paleokarst: New York, Springer-Verlag, p. 342–363.

Cunningham, K. I., H. R. DuChene, C. S. Spirakis, and J. S. McLean, 1994, Elemental sulfur in caves of the Guadalupe Mountains, New Mexico (abs.), *in* I. D. Sasowsky and M. V. Palmer, eds., Breakthroughs in Karst Geomicrobiology and Redox Geochemistry: Abstracts and Field-trip Guide, Symposium, Colorado Springs, Colorado, February 16–19, p. 11–12.

Daniel, E. J., 1954, Fractured reservoirs of the Middle East: AAPG Bulletin, v. 38, p. 74–85.

Ford, D. C., and P. W. Williams, 1989, Karst geomorphology and hydrology: London, Unwin Hyman, 601 p.

Hill, C. A., 1987, Geology of Carlsbad Cavern and other caves in the Guadalupe Mountains, New Mexico and Texas: New Mexico Bureau of Mines and Mineral Resources Bulletin 117, 150 p.

Hill, C. A., 1990, Sulfuric acid speleogenesis of Carlsbad Cavern and its relationship to hydrocarbons: AAPG Bulletin, v. 74, p. 1685–1694.

Hill, C. A., 1992, Sulfuric acid oil-field karst, *in* M. P. Candelaria and C. L. Reed, Paleokarst, Karst-Related Diagenesis, and Reservoir Development: Permian Basin Section, Society of Economic Paleontologists and Mineralogists, 1992 Field Trip Guidebook, Publication 92-33, p. 192–194.

Hill, C. A., 1993, Sulfide/barite/fluorite mineral deposits, Guadalupe Mountains, New Mexico and west Texas: New Mexico Geology, v. 15, p. 56–65.

Jameson, J., 1994, Models of porosity formation and their impact on reservoir description in the Lisburne field, Prudhoe Bay, Alaska: AAPG Bulletin, v. 78, p. 1651–1678.

Keener, M. H., 1957, Dollarhide field; *in* Occurrence of oil and gas in west Texas: University of Texas Publication 5716, p. 87–96.

Kerans, C., 1989, Karst-controlled reservoir heterogeneity and an example from the Ellenburger Group (Lower Ordovician) of west Texas: University of Texas Bureau of Economic Geology, Report of Investigations 186, 40 p.

Kerans, C., 1990, Depositional systems and karst geology of the Ellenburger Group (Lower Ordovician), subsurface west Texas: University of Texas Bureau of Economic Geology, Report of Investigations 193, 63 p.

Mazzullo, S. J., and P. M. Harris, 1991, An overview of dissolution porosity development in the deep-burial environment, with examples from carbonate reservoirs in the Permian basin, *in* M. P. Candelaria, ed., Permian Basin Plays—Tomorrow's Technology Today: West Texas Geological Society Symposium Publication 91-89, p. 125–138.

Mazzullo, S. J., and P. M. Harris, 1992, Mesogenetic dissolution: its role in porosity development in carbonate reservoirs: AAPG Bulletin, v. 70, p. 607–620.

Mendenhall, G. V., 1967, Dollarhide Silurian unit, Dollarhide Silurian field: UNOCAL Internal Report.

Missman, R. A., and J. Jameson, 1991, An evolving description of a fractured carbonate reservoir: the Lisburne field, Prudhoe Bay, Alaska, *in* R. Sneider,

W. Massell, R. Mathis, D. Loren, and P. Wichmann, eds., The integration of geology, geophysics, petrophysics, and petroleum engineering in reservoir delineation, description, and management: AAPG-SPE-SPWLA Arche Conference, Houston, Texas, p. 204–224.

Mygdal, K. A., 1949, The geology at Dollarhide field: Pure Oil News, January, p. 8–11.

Palmer, A. N., 1991, Origin and morphology of limestone caves: Geological Society of America Bulletin, v. 103, p. 1–21.

Stormont, D. H., 1949, Huge caverns encountered in Dollarhide field make for unusual drilling conditions: Oil and Gas Journal, April 7, p. 66–68, 94.

Sweeting, M. M., 1973, Karst landforms: New York, Columbia University Press, 362 p.

Wigley, P. L., J. D. Bouvier, and J. M. Dawans, 1988, Karst and mixing-zone porosity in the Amposta Marino field, offshore Spain (abs.): AAPG Bulletin, v. 72, p. 1031.

Index